AF602075

Crop Improvement for Sustainability

The Editors

Dr. Pawan Kumar Yadav is serving as Lecturer Biology at Education Department, Govt. of Haryana, India. He did his M.Sc. & Ph.D. in Genetics from Maharishi Dayanand University, Rohtak, India and CCS Haryana Agricultural University, Hisar, Haryana, India respectively. He also qualified the CSIR-UGC-NET (JRF), UGC-SRF, ICAR-NET. He is an ex-faculty at Department of Biotechnology, Inderprastha Engineering College & Dr. K.N. Modi Institute of Pharmaceutical Education & Research, Ghaziabad, Uttar Pradesh, India. He has published several research articles in peer-reviewed journals of repute and also is an editor of three books.

Dr. Sunil Kumar is working as Assistant Professor and Head (Botany), at Chhaju Ram Memorial Jat PG College, Hisar. He did his M.Sc., M.Phil. and Ph.D. from Kurukshetra University. He has published several national and international research and review articles in peer-reviewed journals; also is an editor of three international books.

Dr. Sandeep Kumar is working as a Research Fellow at Haryana Kisan Ayog, Panchkula, Haryana. He did his Ph.D and M.Sc in Genetics from CCSHAU, Hisar and also qualified NET in Genetics and Plant Breeding. At present he is the Nodal officer of three Working Groups in Haryana. He has published several research article and review papers in peer-reviewed journal of repute and also is a co-editor of an international book.

Dr. Ram C. Yadav is currently working as Director (Technical) at Centre for Plant Biotechnology, Hisar, India. Earlier, he served as Professor and Head, Department of Molecular Biology, Biotechnology & Bioinformatics at CCS Haryana Agricultural University, Hisar, India. Dr. Yadav has many national and international awards to his credit. He has published more than 100 research papers in National & International journals.

Crop Improvement for Sustainability

Editors:
Dr. Pawan Kumar Yadav
Dr. Sunil Kumar
Dr. Sandeep Kumar
Dr. Ram C. Yadav

2018
Daya Publishing House®
A Division of
Astral International Pvt. Ltd.
New Delhi – 110 002

ISBN 9789388027984 (Int. Edition

Publisher's Note:

Every possible effort has been made to ensure that the information contained in this book is accurate at the time of going to press, and the publisher and author cannot accept responsibility for any errors or omissions, however caused. No responsibility for loss or damage occasioned to any person acting, or refraining from action, as a result of the material in this publication can be accepted by the editor, the publisher or the author. The Publisher is not associated with any product or vendor mentioned in the book. The contents of this work are intended to further general scientific research, understanding and discussion only. Readers should consult with a specialist where appropriate.

Every effort has been made to trace the owners of copyright material used in this book, if any. The author and the publisher will be grateful for any omission brought to their notice for acknowledgement in the future editions of the book.

Published by : **Daya Publishing House®**
A Division of
Astral International Pvt. Ltd.
– ISO 9001:2015 Certified Company –
4736/23, Ansari Road, Darya Ganj
New Delhi-110 002
Ph. 011-43549197, 23278134
E-mail: info@astralint.com
Website: www.astralint.com

Digitally Printed at : **Replika Press Pvt. Ltd.**

Prof. Ramesh Kumar Yadava
Chairman, Haryana Farmers Commission

Foreword

In ancient time Mahrishi Kanad and some other great Yogis believed that the universe is formed by the Panch Maha Bhut viz. Agni, Vayu, Prithvi, Jal and Aakash, it signifies the importance of the protection of nature and to sustain the life. As the agricultural practices are the basic to the civilization of humans and cultivation of food crops is the first scientific practice performed by the human. Farmer is an institution of inspiration, dedication and humanity.

Genetic improvement is the area in which to look for the major breakthroughs in last century for securing the food requirement of human population. The innovations in science and technology lead to change in the lifestyle of every person. From last two-three centuries, we are increasingly exploiting the natural resources at a large scale without caring the mother Earth. Now, the world is facing life threatening problems like food insecurity, health hazards, global warming etc. To solve these, we should adopt the traditional ways of agriculture with new approaches in sustainable manner. Sustainable agriculture to be adopted and promoted by farmer so that it would help in meeting the demand for vegetables, fruits, flowers, milk and milk products, other food products and poultry products for humans and reducing the rural/urban poverty. It plays important role in increasing the farmer's income as it is based on zero budget and ensuring environmental sustainability.

The authors have felt the urge to meet this challenge and present the topics, which believed as, most important in broader understanding and having implications in agricultural science. They seem to have strong conviction that cutting edge technologies will be beneficial for human welfare in present or future era. Hence, presented the same along a logical sequence from simple to complex form of the understanding of the concepts and technologies.

The contents of various topics covered in the edited book entitled "Crop Improvement for Sustainability" permit stimulating, selective, and detailed compilation of each of the various aspects of the broad field of agricultural science. Thus, the book is designed to cover the contemporary scientific thought, serving as a thorough, up-to-date technologies support in the field of crop improvement. I hope pointing, the reader toward advance technologies even have more perceptive than those described here.

The authors namely Dr. Pawan Kumar Yadav, Dr. Sunil Kumar, Dr. Sandeep Kumar and Dr. Ram C. Yadav have fittingly justified the contents for this edited book. I congratulate them in bringing out this anthology for the benefit of young researchers and academicians in particular.

(Prof. Ramesh Kumar Yadava)

Preface

Since, the civilization of human being needed the domestication and cultivation of wild plants as food crop by simple selection for desired traits. Under the changing climate conditions, it is focused and vision for developing a sustainable biobased economy to enhance farmer's income. It is urged that the agricultural research and development programs to look beyond food, feed, and fiber production and to address sustainable biobased industries.

The new advancement in tools and techniques like genomics, proteomics, transcriptomics, and Metabolomics in crop improvement for enhancing the yield, quality, taste, and nutritional composition of food crops; increasing agricultural production for food, feed, and energy; playing a significant role in crop protection; and significantly affecting agricultural economics in sustainable way means to create rural and urban job opportunities; improve the quality of air, water, and soil; improve the healthfulness of food; and produce human health-related products in plants, microbes, and animals for sustainability.

The young researchers and academicians along with society including the active participation of farmers to discuss exchange the knowledge, experiences and collaborate for achieving the overall sustainable development and fulfilling the food and nutrient requirements. All these aspects are in focus when preparing this book entitled as "Crop Improvement for Sustainability". This edited book covers different aspects of Crop Improvement which provides updated recent information on modern approaches/methods as possible with in the limitation of the space.

The editors are greatly thankful to Prof. Dr. Ramesh Kumar Yadava, Chairman, Haryana Kisan Aayog, Panchkula, Haryana, India for his kind support and patronage. We indebted to Prof. Dr. Rajender Kumar, Director General, Uttar Pradesh Council of Agricultural Research, Lukhnow, India, Prof. Narender Singh, Chairman, Botany Department, Kurukshetra University, Kurukshetra, Haryana,

India and Prof. Dr J.P. Yadav, Professor & Head, Department of Genetics, Maharishi Dayanand University, Rohtak, Haryana, India for their critical reviewing and immense help for preparation of this manuscript.

We also express gratitude to young innovative minds/thinkers for their scientific efforts and valuable suggestions which are incorporated in this edited book. The editors are confident that this book will be a stepping stone for developing the new ideas and innovations among young researchers, academicians, policy planners and corporate persons.

Dr. Pawan Kumar Yadav
Dr. Sunil Kumar
Dr. Sandeep Kumar
Dr. Ram C. Yadav

Contents

1

Agricultural Distress and Coping Strategies in Haryana

Sandeep Kumar[1*], Gajender Singh[1], Pawan Kumar[2], Deepak Kumar Yadav[3] and Ramesh Kumar Yadava[1]

[1]Haryana Kisan Ayog, Anaj Mandi, Sector – 20, Panchkula, Haryana – 134116

[2]Department of Secondary Education, GSSS Firozpur Zirka, Nuh, Haryana

[3]Department of Agriculture, Cooperation and Farmers' Welfare, New Delhi

*Corresponding author: sjangra.07@gmail.com

Abstract

The State has made great strides in food production during last decades. However, this success resulted in several second generation problems such as declining natural resources, reduction in soil organic carbon content, multiple nutrient deficiencies, poor soil health, hydrological imbalance, decline in water table climate change and pests & diseases are posing serious challenge in agriculture. The rice-wheat cropping system in the State, which has covered approximate 58% cultivated area, is now seen as major cause of soil health deterioration. Alternate options for sustainable farming should be the top priority for agricultural research and development in the State. Diversification of agriculture through horticulture, vegetables, mushroom cultivation, organic farming, dairying, bee-keeping and fisheries should be adopted at large scale. Also, need to create a balance by introducing climate smart technologies, which may help in conserving natural resources, reducing costs of cultivation, inputs requirements and risk management to make farming more resilient and profitable.

Keywords: Natural resource degradation, Agricultural diversification, Peri-urban farming, organic farming, Post Harvest Management

Introduction

Haryana, as a State, emerged on the political map of India on 1 November, 1966. The region has been playing a vital role in the economic growth and agricultural development of the country. Geographically, Haryana is bounded by the Shiwalik Hills in the North, the Aravali Hills in the South, Yamuna River in the East and the Thar-desert in the West. It has 44.2 lakh hectares of land, which is 1.34% of the total geographical area of the country. The average height ranges from 700 to 950 feet above sea-level. The climate of the State varies from arid, semi-arid and humid with annual average rainfall of 617 mm (http:www.rainwaterharvesting.org/urban/rainfall.htm). Major rainfall is received during July to September. A comprehensive review of history of agriculture showed that agriculture and animal husbandry have always been the mainstay of its economy through the ages, infact since the Aryans inhabited this region in the second millennium B.C. Archaeological evidences reveal that agriculture was being practised in this region even earlier than the Harappan civilisation, but the earliest literary reference is found in the Vaman-Purana.

Fig. 1: Map showing ecology and cropping pattern zones

The State surrounds National Capital from three sides. It is located near to international air port, which is an additional advantage, as it enhances the reach to domestic and global markets. Based on ecology and cropping pattern, the state is delineated (Fig 1) into the following three zones. These zones have their own strength and weaknesses. Accordingly, the farming systems and cropping systems have emerged.

Zone	Districts	Area %	Agricultural options
I	Panchkula, Ambala, Kurukshetra, Yamunanagar, Karnal, Kaithal, Panipat and Sonepat	32	Wheat, rice, sugarcane, maize, cows, buffaloes, and poultry
II	Sirsa, Fatehabad, Hisar, Jind, Rohtak, Faridabad and Palwal	39	Wheat, cotton, rice, sugarcane, *bajra*, buffaloes, cows and poultry
III	Bhiwani, Mahendergarh, Rewari, Jhajjar, Gurgaon and Mewat	29	Pearl millet, rapeseed & mustard. Mewat area is also suitable for agro-forestry, sheep and goat rearing

Note: Zone I and II have better irrigation facilities and overall infrastructure

The process of growth has been technologically highly dynamic in the State. Adoption of improved varieties/hybrids of wheat, rice, *bajra* and oilseeds and associated production technologies in late 1960s & 1970s onwards and Bt cotton after 2002-03 have changed the landscape of Haryana agriculture. Further, it is quite visible that transformation in State agriculture has also boosted the growth of agro-industries. The significance of agriculture sector is not restricted to its contribution to GDP but it has far reaching impact on poverty alleviation and rural development.

Despite very small acreage under farming, Haryana is a major contributor to the national food reserves. The major crops of the State are wheat, rice, maize, sugarcane, cotton, pearl millet, rape seed & mustard, guar, gram etc. and the cropping intensity is about 185%. The State has received the prestigious **"KRISHI KARMAN AWARD"** from Prime Minister of India for the outstanding performance in wheat (in 2010-11and 2011-12) and rice (2014-15) productivity. Increasing productivity and profitability of farmers has always been the focus of Haryana agriculture. The Government is also keen to provide the required technical and policy support. These initiatives have helped in achieving almost seven fold increase in food grains production since the Green Revolution period.

Despite above achievements, several second generation problems particularly depletion of soil and water resources have cropped-up. The problems like low organic carbon in soils, deficiency of various major and micro-nutrients, declining water table in rice-wheat cropping areas, decreasing availability of good quality water for irrigation, disposal of raw sewer water into agricultural fields, climate change and pests & diseases are posing serious challenge in agriculture. Decline in TFP, size of holdings and capacity of employment are also important problems of this sector. Also there is need a to further improve the productivity of all crops by improved efficiency, timely operations, resilience and much needed crop diversification, including promotion of secondary agriculture. As total cultivable land is limited, only option left is to go vertical by out scaling farm innovation and by adopting well planned strategy in a "Mission Mode" approach.

Emerging Challenges for Sustainable Agriculture

Climate Change

Climatic variability directly impacts yields of crops besides affecting soil quality, water availability and increasing vulnerability to pests and diseases. For instance, a short period of exposure of wheat crop to high temperatures results in sharp decrease in productivity, ranging from 10-20%. As per report published by CRIDA (http://www.crida.in/Climate%20change/network.htm), high thermal stress during post-flowering duration manifested 18, 60 and 12 percent reduction in economic yield of wheat, mustard and potato, respectively.

Unpredictable moisture stress during critical crop growth stages is major factor in reducing productivity of crops. According to Haryana State Action Plan on Climate Change (2011) the mean maximum temperature is likely to increase by 1.3^{0}C and mean minimum temperature by 2.1^{0}C by 2050. The increase in mean maximum temperature is projected to be 4.2^{0}C and mean minimum temperature 4.7^{0}C by 2100. Mean annual rainfall is projected to decrease marginally by about 63 mm (3%) by mid century and increase by about 347 mm (17%) by the end of century. The parts of Bhiwani, Faridabad, Fatehabad, Gurgaon, Jhajjar, Jind, Karnal, Kurukshetra, Mahendragarh, Rohtak, Sirsa and Sonipat showed decreasing trend in the monsoon rainfall. Evapo-transpiration and Green House Gases (GHG) are projected to increase, whereas, negligible changes, in ground water recharge have been anticipated.

The scientists and planners are concerned about the erratic rainfall pattern in the State. For example, of the last four seasons, three received deficit rainfall. This adversely affected the production and productivity of various crops in the State.

Natural Resource Degradation

The total factor productivity (TFP) is declining due to resource degradation, high cost of inputs & labour, natural calamities and gaps in technology. The continuous degradation of natural resources and intensive cultivation is affecting the sustainability of agriculture. The soils of the State are low in various major and micro-nutrients. About 70% soils have low Organic Carbon Matter (OCM) and large scale deficiency of nutrients. The organic carbon and nutrient deficient soils cannot afford pressure of intensive agriculture for longer time. About 15 percent of the net sown area of the State is now salt affected. The rice-wheat rotation is disturbing the balance of available nutrients in the soil and also causing micronutrients deficiency. Water is becoming a scarce commodity in Haryana. Agriculture's share of water is declining at a faster rate due to increasing competition for good quality water from urban and industrial sector.

Effect of Sewage Water and Industrial Effluents

Soil contamination by sewage and industrial effluents has affected adversely both soil health. The toxic metals like Cd, Cr and Ni were found to be accumulated in the soil and plant due to long-term use of sewage irrigation. The elements Cd, Cr and Ni are more likely to become health hazards for consumers of the crops grown in sewage irrigated soils. With the rapid urbanization and industrialization, the amount of sewer water is increasing substantially. The disposal of raw sewer water in to agricultural fields is becoming a serious problem. It is adversely affecting soil, plants and animal/human health. It has been found that about 60% sewer water in Haryana is unsafe for irrigation and requires treatment before use. To avoid such problems, continuous monitoring of quality of sewage and industrial effluents available in the Sate and their impact on soil plant health is required in order to make use of sewage waters as a cheap potential alternative source of plant nutrients in agriculture.

Declining Size of Holdings

In Haryana, the size of holdings is reducing continuously. The average of operational holdings is 2.25 ha. About 65% of the totals farming families are small and marginal, owning 21% area of operational holdings. The farmers of rainfed area also own about 19% of the total cultivated area and they have harnessed comparatively fewer benefits of advanced technologies. The small size of holdings hinders in farm mechanization and adoption of new and potential technologies due to lack of economic and technological viability. Hence, increasing the productivity and profitability of these farms is a challenge and needs to be addressed appropriately.

Imbalanced Input and Energy Use

The energy use is fast increasing in agriculture sector. The increase is mainly due to intensification of agriculture. The growth in farm power is due to fast increase in number of tractors and tube wells. The increase in tractors and tubewells has enhanced the consumption of diesel and electricity. Agriculture consumes the highest KWH electricity of total consumption followed by industry, domestic sector, commercial establishment etc. To meet the demand of energy and also to reduce the energy cost, new options of energy like bio-fuel and efficient machines and tools will have to be explored and exploited. Further, the growth in farm machinery needs to be rationalised. Therefore, there is a need to develop energy efficient technology and cost-effective machines and tools to effectively use energy in order to reduce the cost of cultivation.

It has been observed that there is over use or injudicious use of agro-chemicals and fertilizers in the State which adds to the cost of cultivation. The use of chemicals needs to be minimized with the adoption of appropriate technology keeping in view the trade and health concerns. Haryana is well placed for targeting global market. Hence, it has to ensure that its products meet the International quality standards.

Crop Losses due to Insects/Pests

In monetary terms, the Indian agriculture currently suffers heavy losses due to insect pests. The proportionate losses, which could be too high in Haryana, may be extrapolated from the trend of losses in the country. The composition and competition by weeds are dynamic and dependent on soil, climate, cropping and management factors. The changing scenario of insect pest problems in agriculture as a consequence of green revolution technology has also been well documented by the scientists. New problems like yellow rust in wheat, *orobanche* weed in mustard and nematodes in wheat & other crops are emerging. These problems have to be managed with the technological interventions.

Profitability Issues

The rice-wheat rotation has emerged as the most preferred cropping system across Haryana eliminating many of the cropping patterns due to its comparative economic advantages, assured marketing and stable productivity levels. As a result, rice-wheat cropping system continues to occupy more than 58% of gross area sown in 2011-12 in the State. These crops are labour intensive and require more water and other inputs. Keeping in view the sustainability aspects, diversification in cropping system needs emphasis. The farmers are hesitant to adopt diversification in agriculture because of marketing problems and therefore, the options are limited to enhance profitability. The sharp decline in net farm profitability is due to increase in the cost of cultivation particularly due to increased cost of inputs, machines and shortage of labor.

Supply of Inputs

Farmers have complaints about non-availability of various inputs in time and sometimes supply of spurious inputs i.e. seed; fertilizers, pesticides etc. These factors adversely affect the productivity of crops.

Crop Residue Management

A huge quantity of residue of crops in various cropping systems is left unutilized, particularly in rice-wheat system. The farmers are not aware of benefits of its recycling. Unutilized residue causes several problems including insect, disease etc. The crop residues are good sources of plant nutrients and are important components for the stability of agricultural ecosystems. In areas where mechanical harvesting is practised, a large quantity of crop residues is left in the field, which can be recycled for nutrient supply. About 25% of nitrogen (N) and phosphorus (P), 50% of sulphur (S), and 75% of potassium (K) uptake by cereal crops are retained in crop residues, making them valuable nutrient sources. Traditionally, wheat straw is removed from the fields for use as cattle feed and for other purposes. The rice straw is left un-utilized. With the advent of mechanized harvesting, farmers have been burning large quantities of crop residues left in the field. As crop residues interfere with tillage and seeding operations for the next crop, farmers often prefer to burn the residue in situ, causing loss of nutrients and organic matter in the soil. The Government of Haryana has banned the burning of crop residues, which is a welcome initiative.

Post Harvest/Value Additions

Post harvest and storage facilities have not been established in the commodity production areas and therefore, most of the produce of the farmers is sold as raw material. This directly reduces the return on investment and also does not generate employment for rural youth. Lack of value addition, processing, post harvest technology and marketing support.

Mitigating the Agricultural Distress

The remarkable progress in enhancing agricultural production and productivity in Haryana has been achieved mainly due to research and development, infrastructure development and policy interventions. Now, the Prime Minister of

India has recently called upon planners, scientists, farmers and other stakeholders to take necessary initiatives to double the farmers' income by the year 2022. This call carries clear message to strengthen agriculture sector in order to alleviate rural poverty and enhance purchasing power of farmers and youth which, ultimately, would work as game changer to uplift the overall economic growth of the country. Although doubling the farmers' income in the State in a stipulated period is a challenge, yet the State government has taken this challenge as mission to achieve the target. Therefore, it is mandatory to shift focus of research, extension and human resource development to address the emerging problems. For mitigating the agricultural distress attention may be given on following practices.

Natural Resources Management

Scaling-up the research to develop the efficient technologies for soil and water management and enhancing nutrient use efficiency is an urgent need. Feasible, efficient and cost-effective technologies for micro-management of natural resources and INM in agriculture at block level are to be made available. Water scarcity has to be addressed through improved efficiency of irrigation systems and promotion of low water requiring varieties in different cropping systems. Benefits of conservation agriculture have been acknowledged world over for resource conservation and enhancing agricultural production as well as quality of produce. However, more research efforts are required to develop these technologies for various agro-climatic conditions of Haryana. Development of technology for crop residue management also needs immediate attention.

Scaling-up of IPM

Keeping in view the health hazards, cost of production and environmental pollution issues, IPM is being recommended. It is estimated that insect-pests and diseases cause about 15-20 per cent losses in production of field and horticultural crops. The frequent use of pesticides to control pests and diseases escalate the cost of production. Cotton, rice, vegetable and fruit crop growers face recurring problem of huge losses due to high cost of pest control. Besides, high use of pesticides increases the pesticide residue in products and thereby such products are not accepted in global agri-trade. Therefore, intensive efforts will have to be done to develop suitable low cost IPM technology. IPM is knowledge intensive technology and therefore, matching human resource needs to be developed.

Out-scaling Farm-led Innovations

The farmers have been silently contributing to the agricultural development and evolved innovative methods of farming and also the crop varieties and adopted them to get better production. It is always argued that farmers' innovations are inexpensive, easily accessible, and locally appropriate and tested in real farm situation. The technologies developed by them have rarely been verified, refined, demonstrated and documented. There is urgent need to assess farm innovations decisively and make them available for larger benefits of agriculture.

Diversification in Agriculture

Diversification has become the general recommendation to solve many problems in agriculture. Intensive studies are needed on diversification to replace or intensify cropping systems. The rice-wheat cropping system in the State, which has covered approximate 58% cultivated area, is now seen as major cause of soil health deterioration. Alternate options for sustainable farming should be the top priority for agricultural research and development in the State. Therefore, we need to create a balance by introducing climate smart technologies, which may help in conserving natural resources, reducing costs of cultivation, inputs requirements and risk management to make farming more resilient and profitable. There are still several untapped opportunities that will help to increase profit in agriculture and to create employment. Diversification of agriculture by adoption of peri-urban farming including horticulture, vegetables, mushroom cultivation, organic farming, dairying, bee-keeping, fisheries, post-harvest management, processing, value addition and agriculture based industry have enormous scope to increase the farm income.

Development of Horticulture and Protected Cultivation

Haryana is well suited for the promotion of horticulture, especially in view of it's vicinity to the National Capital Region and other big cities around, besides having easy access to both domestic and external markets. Increased production of horticultural crops (fruits and vegetables) can help in improving both food and nutrition security, enhancing rural employment, alleviating poverty and promoting agricultural exports. Also in view of increased urbanization and living standards, and considering the convenience needs of dual income families, much greater thrust is needed to exploit the market potential of horticultural crops, both as fresh and processed products.

Protected cultivation of vegetables and flowers also offers much needed option for agricultural diversification needed to increase farm income and to generate employment. High quality nursery raising of vegetables is a new option in meeting the increasing demand for kitchen gardening by the affluent urbanites interested in growing their own healthy foods. Similarly, growing of good planting materials of fruits and flowers offers yet other options of increasing farmer's income. The technology is also scale neutral as it benefits both the large-scale and small scale farmers and ensures higher productivity as well as income.

Periurban Farming

It is a practice of growing, processing and marketing of vegetables and ornamental plants in and around peripheries of the city and towns for household consumption as well as for the rapidly growing urban population largely in response to daily demand of consumers within a town, city or metropolis. It provides livelihood support by making fresh fruits and vegetables available in urban areas and helps to manage solid wastes. Haryana has huge potential for organic/speciality and *peri*-urban farming due to its geographical location which provide easy availability of national and international markets. *Peri-urban* farming with organic practices of vegetables like potato, onion, tomato; pea, cucurbits, etc. are paying rich dividends to the farmers with higher benefit-cost ratio than rice-wheat sequence. Floriculture is also a good cash crop for small and marginal farmers of peri-urban areas but a suitable policy support for infrastructure, processing, value addition and marketing is required. Seed production of flowers need to be focused. Hence, a suitable policy support for promotion and adoption of peri-urban farming and research for development of varieties suitable for organic farming is the need of the hour.

Beekeeping

Beekeeping provides excellent source of employment for the rural unemployed, enhances income of farmers, and the landless beekeepers. It enhances the productivity levels of agricultural, horticultural and fodder crops through pollination services. Honey and bee products find use in several industries which are under; pharmaceuticals, bees wax industries, bee venom, royal jelly, bee nurseries, bee equipment and hives etc. Promotion and development of Beekeeping in the State should be encouraged.

Organic Farming

Organic farming is another area for small farmers which decrease the input cost provides high benefit to the farmers and has enormous options rural employment. Organic farming production system totally avoids the use of synthetic fertilizers, pesticides, growth regulators, etc. and relies heavily on animal manures and on farm production of biomass by cultivation of legumes or green manure crops. Organic foods are becoming popular both in domestic and overseas market. India today produces range of organic products from fruits and vegetables, spices to food grains, pulses, milk and organic cotton. Organic Farming is mainstreaming in the country faster than expected. The factors attracting public and private attention include increasing prospects of organic agribusiness trade because of increasing demand for safe food and an approach to sustainable development of farming based rural livelihoods in marginal areas and for small farmers. Even though India is a late starter but during the past few years organic as farming and agribusiness option has spread across the country. Furthermore, since organic fertilizers and pesticides can be produced locally, the yearly costs incurred by the farmer are also low.

Vermi-compost Technology

The vermin-compost or vermin-culture technology should be adopted by the farming community. It has a visible impact on the economic upliftment of them and provided with self employment opportunities to the youth and farmwomen. The market for the compost be ensured with the intervention organizations promoting organic farming. Many farmers have realized reasonable income in a short period through the sale of worms and compost. This technology can easily meet the requirement of good quality organic manures of each farmer on a permanent basis and additional income with the effective utilization of family labour alone.

Post Harvest/Value Addition Technologies, Packaging and Branding

Post harvest management and branding of products are crucial for profitability. Besides capacity building of farmers through education and training, technological back-up should be provided to the farmers. In Indian food industry, the primary processing accounts for almost 80 % of the value addition. To enter the global market, the State food industry will have to move to secondary and tertiary processing. Obviously, processing industry and entrepreneurs would need technological support. Changing market trends are influencing branding and packaging with eco-friendly, recyclable and bio-degradable material and there is a growing sophistication in it. The research institutions also need to come out with technologies for value addition/processing along with packaging in harmony with international standards for promoting export.

Mechanization of Small Farms

Increasing agricultural production and value addition to the farm produce are two important factors for enhancing the rural prosperity. Farm mechanization can be decisive to realize this goal. Efficient and low cost machines/equipments for crop residue management of field crops, water harvesting, resource conservation also need to be appropriately promoted to make agriculture profitable. The "Mission on Mechanization of Small Farms" needs to be launched to make farming efficient and remunerative. The custom hiring services for farm machinery is to be promoted in order to take forward the mission of small farm mechanization.

Entrepreneurship Development

In this era of change the State government has focused upon achieving higher returns on per unit of investment in different sectors of agriculture. Entrepreneurship development and self-employment orientation in agricultural sector should receive high priority. There are several areas like seed production, hybrid seed production, dairying, food processing, bee keeping, peri-urban farming, organic farming, protected cultivation, nursery raising, machines & tools manufacturing, value added agriculture etc where enormous opportunities are emerging for income generation and therefore, the capacity building of farmers and rural youth is urgently required. Haryana has a good potential for the export of value added agricultural commodities and to realize this, Processing, value addition, branding and marketing of the produce is required to ensure better realization to farmers, minimization of postharvest losses, employment generation and more investment in creation of infrastructures for production and postproduction agriculture.

Linking Farmers with Industry

Linking farmers with processing industry through appropriate mechanism such as contract and/or cooperative farming and automation of operations in agricultural marketing such as primary processing (cleaning, grading, drying, storage, etc.) should be done so that farmers get better return on investment. Participation of private sector in handling, storage and supply chain in collaboration with farmers needs to encourage. There is a need for creation of specialty agricultural hubs with production, processing, storage and marketing facilities for exports.

Conclusion

Doubling the farmers' income up to 2022 is a Mission for the Scientists, Planners and the State Govt. This target could be achieved by mitigating the emerging challenges to the agriculture such as depletion in the soil heath, declining water table, quality of ground water, climate change and crop residue management. Adoption of diversification of Agriculture, organic farming, peri-urban farming, integrated pest management, post harvest management, value addition, branding and marketing will ultimately help in achieving this target. The State of Govt of Haryana and Haryana Kisan Ayog also keen to provide required technical and policy support for farmers' welfare.

References

Working Group Report on Conservation Agriculture for Sustainable Crop Production in

Haryana, published by HKA (http://www.haryanakisanayog.org/Reports / Working_ Group_ Report_CA.pdf)

Working Group Report on Productivity Enhancement of Crops in Haryana, published by Haryana Kisan Ayog. (http://www.haryanakisanayog.org/Reports/WG%20Report% 20on% 20Productivity%20Enhancement%20 of%20Crops%20in%20Haryana.pdf)

Working Group Report on Natural Resource Management in Haryana, published by HKA (http://www. haryanakisanayog.org/Reports/Report_on_ Horticulture.pdf)

Working Group Report on Horticulture Development in Haryana, published by HKA (http://www. haryanakisanayog.org/Reports/Report_on_Horticulture. pdf)

Working Group Report on Post Harvest Technolgy and value addition in Haryana

2

Role of Crop Wild Relatives in Crop Improvement Under Changing Climatic Conditions

Vikender Kaur[1*], Kumari Shubha[1], Pardeep Kumar[2], Manju[3]

[1]Division of Germplasm Evaluation, [2]Division of Plant Quarantine,
ICAR- National Bureau of Plant Genetic Resources, New Delhi-110012, India
[3]Department of Botany, Baba Mastnath University, Asthal Bohar, Rohtak-124021, India
**Corresponding author: vikender.kaur@icar.gov.in*

Abstract

Future crops will need to thrive in a drier, warmer, and more variable climatic conditions followed by emergence of new pathogens and pest. These diverse environmental stresses will pose serious threat to food production for exponentially increasing human population. The development of novel crop varieties with increased resistance to biotic and abiotic stresses as well as maintenance of natural diversity is very important to feed future population. Crop wild relatives (CWRs) have great potential to combat future challenges. CWRs are an important source of the genes for breeding stress tolerant high yielding varieties as they have been evolving for thousands of years in adverse environmental conditions and possess a much higher degree of adaptability. The threats posed by climate change and increasing human populations have led to increased momentum worldwide regarding important role of CWRs in crop improvement. In this chapter, the significance of CWRs for crop improvement has been highlighted by providing examples of wild and weedy relatives that have been used to increase biotic and abiotic stress resistance/tolerance as well as for traits of various agronomic importance including yield in various crops.

Keywords: Biotic and abiotic stress, Climate change, Crop wild relatives, Plant genetic resources

Introduction

Crop wild relative (CWR) is a "wild plant taxon that has an indirect use derived from its relatively close genetic relationship to a crop" (Maxted *et al.*, 2008). The importance of CWRs as a critical resource for future food security has not been fully recognized despite their potential as gene donors for crop improvement was clearly recognized much early in the 1920s and 1930s by the renowned Russian plant geneticist Nicolai Vavilov. Climate change is predicted to bring about increased global temperature, along with associated carbon dioxide (CO_2) increase, altered pattern of rainfall and salinity, emergence of new pest strains and diseases (Tester and Langridge, 2010). The timing and pattern of rainfall is also predicted to change although uncertainty exists about the expected degree of changes. These predicted climatic changes are expected to have fairly widespread impacts on agriculture for example, rice flowers show increased sterility at high temperatures, maize is very sensitive to drought at the time of flowering, wheat senescence starts earlier and faster under warmer conditions etc. (Lobell *et al.*, 2012). Thus, future crops species will need to be able to thrive in a drier, warmer, and more variable climatic conditions. "Adapting" agriculture by shifting planting dates or switching to different crop varieties can combat this problem to a lesser extent only. To meet these challenges, plant breeders need all the genetic diversity available in the form of germplasm, landraces and wild or weedy forms. Some of the genetic diversity may be found in landraces, traditional/farmer's varieties that are still being cultivated by farmers around the world. However, a much wider spectrum of biodiversity is found in wild plant species that are closely related to domesticated crops. They are of key importance to breeding efforts that help in adopting to climate change. Therefore, CWRs need to be acknowledged as an important future source of the genes for breeding high yielding/stress tolerant varieties.

Importance of CWRs

The process of domestication has resulted in reduced diversity in modern crops. For example, more than half of the genetic variation has been lost in cultivated soybean (Hyten *et al.*, 2006; Zhou *et al.*, 2015), significant reduction in maize (Wright *et al.*, 2005), and cultivated rice (Xu *et al.*, 2012). Positive selection on target locus controlling a trait of interest can also result in a reduction in the diversity of closely linked loci, known as *selective sweep*, resulting in loss of specific alleles. The increasing genetic uniformity of crop varieties combined with climate change effects makes crops more vulnerable to various biotic and abiotic stresses. There had been examples of large scale devastations of crops due to genetic uniformity. The Irish famine in 1840's caused by late blight of potato led to the death of around one million people due to starvation and epidemic disease across Ireland, Europe and North America. Similarly, rice blast in the Philippines, Indonesia, and India leading to the Great Bengal famine resulted in rice crop failures. CWRs retain high levels of genetic diversity compared to their domesticated descendants. The importance of CWR in crop improvement was reviewed by Prescott-Allen and Prescott-Allen (1987, 1988) and Hajjar and Hodgkin (2007), who conducted a comprehensive survey of the use of CWRs genes in domesticated plants and identified more than 100 beneficial traits from 60 wild species of 13 major crops. Vincent *et al.* (2013) prioritized the CWRs of food crops, identifying 1667 wild species that are significant resources for 173 globally important crop species.

Utilization of CWRs for crop improvement

Utilization of CWRs has enjoyed a great success only in few crops despite having valuable genes with immense value for crops improvement and adaptation to changing environmental conditions. Many genes still lie untapped in these genetic resources, presumably due to the lack of useful genetic information and genetic bottlenecks. The plant breeders have not fully exploited the potential of CWRs as they rely on searching genes for beneficial traits associated with certain CWRs rather than searching more generally for beneficial genes. Also, plant breeders prefer to use pre-breeding lines containing the desirable wild traits in domesticated genetic backgrounds. Major efforts have been concentrated primarily on use of CWRs in certain food crops such as wheat, rice, barley, tomato, soybean and mustard. The use of CWRs lags far behind its potential due to certain hindrances such as cross incompatibilities, infertility in the F_1 and subsequent progeny, non-availability or poor conservation of CWRs; under-utilized and under-explored wild germplasm, lack of reliable evaluation data, and the expression of desirable traits in cultivated genetic background. Accessibility of CWR resources to researchers and breeders is another important issue for crop improvement utilizing CWRs. Many international collections such as International Center for Tropical Agriculture (CIAT), Crop Trust and Kew as well as National Gene Banks of respective countries, conserve a number of CWRs (Table 1) and provide information on pre-breeding data, distribution and potentially useful traits. Many global online portals serve for the purpose of CWR inventories such as database Crop Wild Relative Global Portal (http://www.cropwildrelatives.org/), Crop Wild Relatives and Climate Change (http://www.cwrdiversity.org/), Gateway to Genetic Resources (http://www.genesys-pgr.org/), Global Crop Diversity Trust (http://www.biodiversityinternational.org/cwr/).

Table 1: Collections of crop wild relatives of important food crops

Crop	Wild relatives	Storage location
Rice	*Oryza rufipogon*, *O. officinalis*, *O. granulata*	Chinese Academy of agriculture Sciences; International Rice Research Institute
Wheat	*Triticum, Aegilops*, *Dasypyrum villosum*	The Wheat Genetics Resource Center (14,000)*
Barley	*Hordeum spontaneum* and other *Hordeum* wild species	International Barley Core Collection (300)*; USDA-ARS National Small Grains Collection
Soybean	*Glycine soja*	USDA Soybean Germplasm Collection (1,100)*; Chinese National Crop Genebank (6,172)*

Crop	Wild relatives	Storage location
Sorghum	23 wild sorghum species	International Crops Research Institute for the Semi-Arid Tropics (449)*
Tomato	Wild *Lycopersicon* and *Solanum* species	Tomato Genetics Resource Center (1,196)*
Potato	187 wild *Solanum* species	International Potato Center

**Brackets give the number of conserved wild relatives (accessions or species) for each crop. (Zhang et al., 2017)*

There are a number of examples of successful gene discovery and transfer of superior alleles from CWRs to domesticated crops despite a variety of difficulties in using CWRs. The CWRs along with their utilization traits has been summarized crop-group wise under following subheads:

Cereals: Cereals are staple foods to billions of peoples and their production is increasingly threatened by the recent changes in weather patterns due to global warming. The impact is more severe in less-developed countries where the consequences of changing climate have devastating socio-economic impact. A summary of the potential genes identified in CWRs of some important cereals are given in Table 2.

Wheat: *Triticum dicoccoides* is one of the best known member in wild emmer wheat family, has been found to be a rich source for *Fusarium* head blight resistance (Oliver *et al.*, 2007). Krugman *et al.* (2011) underlined the importance of *Triticum turgidum* ssp. *dicocooides* for drought tolerance genes, metabolites and high protein content. Among diploid wheat *Triticum monococcum* (A genome progenitor of common hexaploid wheat) has been used to mark the traits related to resistance genes against powdery mildew (Yao *et al.*, 2007) and leaf rust (Sodkiewicz *et al.*, 2008). Using marker assisted selection, one *adult plant resistance* (*APR*) gene for leaf rust and stripe rust has been transferred from *T. monococcum* to bread wheat WL711 and one gene for leaf rust has been transferred to PBW343 background (Singh *et al.*, 2007). Sohail *et al.* (2011) have reported diploid wheat progenitors *Aegilops tauschii* (D genome progenitor) for drought and salt tolerance in wheat and *Aegilops crassa* for enhanced physiological performance under water scarcity. *Aegilops tauschii* for Hessian fly resistance genes, Russian leaf rust resistance gene (*Lr21*), cereal cyst nematode disease controlling locus Cre3, *Aegilops geniculata* for powdery mildew resistance and barley yellow dwarf virus resistance, *Aegilops variabilis* for nematode resistance have contributed towards wheat improvement (Yumurtaci, 2015).

Rice: Bacterial leaf blight caused by *Xanthomonas oryzae* pv. *oryzae* has been one of the most widely distributed and devastating rice disease worldwide. In 1990, the transfer of *Xa21* gene from wild rice (*Oryza longistaminata*) for resistance to bacterial blight in IR72 started the systematic use of wild rice gene pool. Wild rice, *Oryza rufipogon* has been used for blast resistant gene, *Pi33* introgression into rice variety IR64, which in the most used rice blast resistant variety (Ballini *et al.*, 2007). The resistance to grassy stunt virus utilizing *Oryza nivara* has been incorporated in most rice crop germplasm emanating from International Rice Research Institute (IRRI), Philippines through utilization of wild relatives (Leung *et al.*, 2002). Many unique traits such as weed competitiveness, drought tolerance and ability to grow under low input conditions has been transferred from *O. glaberrima* to *O. sativa* and combined with high yield to develop NERICA (NEw RIce for AfriCA), which is high yielding, drought, pest resistant and adapted to the growing conditions of West Africa (Sarla and Mallikarjuna Swamy, 2005). *Oryza officinalis* has been used to change the time of flowering of *O. sativa* rice cultivar to mitigate high temperature induced spikelet sterility at anthesis. Similarly, *O. longistaminata* has potential for drought tolerance (Brar, 2005). Another wild relative of rice *Porteresia coarctata* (*O. coarctata*), is a halophyte known for its adaptation to high salinity (20-40 dS/m) and its complete tolerance to long term submergence in saline water has been utilized in producing intergeneric sexual hybrid with *O. sativa* (Jena, 1994).

Table 2: CWRs of some important cereals utilized in breeding for biotic and abiotic stress tolerance

CWR of cereals	Traits	Utilization	Reference
Triticum monococcum	Heat tolerance Salt tolerance Powdery mildew resistance	Controlling thermal tolerance through heat shock protein (HSP) gene Sodium extrusion mechanism Mapping of Pm resistance markers	Vierling and Nguyen, 1992 James *et al.*, 2011 Yao *et al.*, 2007
Aegilops uniarisfata	Aluminum tolerance	Wheat substitution lines production	Miller *et al.*, 1997
Oryza rufipogon	Blast resistance	Blast resistant gene *Pi33* introgressed in rice var. IR64	Ballini *et al.*, 2007
O. longistaminata	Bacterial blight resistance, drought tolerance	*Xa21* gene transfer from wild to variety IR72	Brar, 2005
O. glaberrima	Weed competitiveness, drought tolerance and high yield	NERICA (NEw RIce for AfriCA)	Sarla and Mallikarjuna Swamy, 2005
Hordeum spontaneum	Severe salt and dehydration stress; aluminum tolerance; *Fusarium* resistance	*Hordeum spontaneum* x two-rowed malting barley population	James *et al.*, 2008; Cai *et al.*, 2013; Huang *et al.*, 2013

CWR of cereals	Traits	Utilization	Reference
H. marinum	Salt tolerance	Salt tolerant amphiploid production	Islam *et al.*, 2007; Seckin *et al.*, 2010; Alamri *et al.*, 2013
H. bulbosum	Resistance to powdery mildew and leaf rust; leaf scald	Advanced back cross population of wild x cultivated	Schmalenbach *et al.*, 2008; Friedt *et al.*, 2011
Avena barbarata	Crown rust resistance	Seedling resistance	Cabral and Park, 2014
Tripsacum dactyloides	Corn root-worm resistance	Maize x *Tripsacum* introgression line production	Prischmann *et al.*, 2009

Maize: Hajjar and Hodgkin (2007) reported some introgression initiatives in maize breeding performed by introducing *Tripsacum* resistance genes, such as *Helminthosporium, Puccinia,* rootworm, drought and aluminum stress into cultivated maize. Blight resistant alleles transferred from wild relative of Mexican maize (*Tripsacum dactyloides* L.) into commercial corn lines is a well-known success story which resolved corn blight of U.S.A. Prischmann *et al.* (2009) introduced genes from *Tripsacum dactyloides* into cultivated corn for resistance to rootworm. Genetically modified insect resistant corn varieties have been produced for resistance against two main corn borer infestation diseases known as European corn borer (*Ostrinia nubilalis*) and Mediterranean corn borer (*Sesamia nonagrioides*). *Bacillus thuringiensis* (*Bt*) transgenic corn, expressing different Cry1 proteins, has been generated against to Asian corn borer pathogen (He *et al.*, 2003). Recently, Du *et al.* (2014) developed transgenic insect resistant corn varieties carrying *cry1C* gene.

Barley: *Hordeum spontaneum,* the wild progenitor of cultivated barley, has been utilized for acquiring resistance against many biotic stresses such as *Fusarium* (Chen *et al.*, 2013), leaf stripe (Biselli *et al.*, 2010), powdery mildew and leaf rust (Schmalenbach *et al.*, 2008), leaf scald (Friedt *et al.*, 2010) as well as abiotic stresses such as drought (Kalladan *et al.*, 2013), and salt tolerance (Pakniyat and Namayandeh, 2007; Shavukav *et al.*, 2010). In *H. bulbosum,* important sources of resistance against powdery mildew, leaf rust and leaf scald have been identified (Friedt *et al.*, 2011; Morrell and Clegg, 2011). Halophytic relatives of barley, like sea barley grass (*H. marinum*) is known as a provider line for oxidative stress defense with its reactive oxygen species (ROS) and superoxide dismutase (SOD) scavenging enzymes which has been used for the production of *Hordeum marinum-Triticum aestivum* salt tolerant amphiploid production (Islam *et al.*, 2007; Seckin *et al.*, 2010; Alamri *et al.*, 2013). Wild species have already proven to be very fruitful source of genes for widening the genetic base of Indian barley (Singh *et al.*, 2016) and improvement for resistance to spot blotch (Verma *et al.*, 2013); barley yellow rust (Yadav and Kumar, 1999; Selvakumar *et al.*, 2013) through utilization of exotics and constructed wild populations in barley germplasm international nurseries. Wild barley accessions have been used routinely in the International Center for Agricultural Research in the Dry Areas (ICARDA) crossing programme for stress environments (Lakew *et al.*, 2013; Ceccarelli, 2014).

Oat: *Avena* has three different ploidy levels in the form of diploid, tetraploid and hexaploid like wheat. Diploid *Avena strigosa* carry important genes for multiple herbicide resistance, leaf rust resistance (Lehnhoff *et al.*, 2013). *Avena barbata* is a tetraploid wild relative of cultivated oat (*Avena sativa*) is a useful gene reservoir for powdery mildew pathogen resistance and survival under mesic and xeric conditions (Swarbreck *et al.*, 2011). *A. fatua* has been studied extensively for seed dormancy regulation genes and seed vigor.

Oilseeds: Oilseeds are major food crops, known for their unique protein and oil rich characteristics. Major biotic and abiotic stresses are the most serious constraint for global oilseed production, and are predicted to worsen with anticipated climate change. Utilization of some important CWRs of oilseed Brassicas and other minor oilseed crops are mentioned below.

Brassicas: Sexual incompatibility barriers between different *Brassica* species make gene introgression from CWRs into cultivars difficult. However, these barriers have been overcome by using somatic hybridization via protoplast fusion. Wild black mustard (*Brassica nigra*) represents wild gene reservoir to improve cultivated *Brassicas* against several pathogens *e.g.* interspecific allopolyploids between *B. fruticulosa, B. nigra* and *B. rapa* facilitate the use of CWRs in *Brassicas* as bridge species (Chen *et al.*, 2011). Monosomic addition lines derived from backcrosses of somatic hybrids between *B. oleracea* var. *botrytis* and *B. nigra* serve for the purpose of resistance breeding in *B. oleracea* (Wang *et al.*, 2012). *B. fructiculosa* (twiggy turnip) is a potential source for resistance to cabbage aphid (*Brevicoryna brassicae*) and cabbage root fly (*Delia radicum*) (Pink *et al.*, 2003). *Brassica oxyrhina, Moricandia arvensis, Trachystoma balli* and *Diplotaxis catholica* have been identified as potential sources of cytoplasmic male sterility (CMS) while *B. tournefortii, Diplotaxis acris, D. harra and Eruca sativa* for drought tolerance in *Brassica.*

Linseed: *Linum tenuifolium,* a CWR of linseed has potential as genetic resource for lowering the linolenic acid content of linseed oil to make it edible. *Linum perenne* has potential for drought and cold hardiness, *L. grandiflorum* for linseed bud fly and *Alternaria* blight. Molecular study have established that the genetic diversity of the stearoyl-ACP desaturase II (*sad2*) locus in cultivated flax is low compared to wild pale flax (*L. augustifolium*) (Allaby *et al.*, 2005).

Sunflower and Sesame: The CWR of sunflower, *Helianthus argophyllus* is used for tolerance to drought stress in sunflower and *Helianthus paradoxus* for tolerance to salinity (Miller and Seiler, 2003). Gene sources have been identified in *Sesamum laciniatum* to leaf phyllode, S. *malabaricum, S. mulyanum* and *S. alatum* for powdery mildew.

Pulses: Among CWRs of pulses, *Vigna tribolata, V. mungo* var. *sylvestris, V. radiata* var. *sublobata* have provided resistance to yellow mosaic virus. *V. vexillata* has high protein and resistance to cowpea pod sucking bug and bruchids and it is crossable with *V. unguiculata* and *V. radaita*. Cytoplasmic male sterile systems were developed for pigeon pea exploiting the cross-pollination mechanism and utilizing wild *Cajanus* species (Mallikarjuna *et al.,* 2012). High protein and seed size breeding lines such as HPL 2, HPL 7, HPL 40 and HPL 51 were developed from *C. sericeus, C. albicans* and *C. scarabaeoides* (Saxena *et al.* 1987, Jadhav *et al.,* 2012). *Phaseolus coccineus* is a source of resistance to anthracnose as well as root rots, white mold, and bean yellow mosaic virus (BYMV) in common bean (Sharma and Rana, 2012). In chickpea, productivity enhancement related traits have been introgressed from *C. reticulatum, C. echinospermum* (Sandhu *et al.,* 2006), and for resistance to *Ascochyta* blight, pod number and short internode from *C. reticulatum* and *C. echino* (Singh *et al.,* 2014). *Cicer arietinum* has been reported for improved heat and drought tolerance utilizing *Cicer reticulatum* and *Cicer echinospermum* genepool (Canci and Toker, 2009). *Cicer microphyllum* have been identified to carry genes for cold hardiness, drought tolerance and seeds/pod. Stable recombinant inbred lines (RILs) were developed for resistance to rust, powdery mildew and pod number from *Lens orientalis, L. odomensis* and *L. ervoides* (Singh *et al.,* 2013).

Table 3. Improved traits in grain legumes characterized under various abiotic and biotic stresses using wild relatives

Grain legume	Utilization Traits	Population strategy	Reference
Soybean	Salt tolerance	RILs (cultivated x wild) *Glycine soja*	Qi *et al.,* 2014
Soybean	*Sclerotinia* stem rot resistance	Germplasm lines	Iquira *et al.,* 2015
Soybean	Soybean cyst nematode resistance	Wild soybean (*Glycine soja*)	Winter *et al.,* 2007; Zhang *et al.,* 2016
Common bean	White mold resistance	Wild / Landrace	Mkwaila *et al.,* 2011
Peanut	Root-knot ematode resistance, drought tolerance	Wild peanut (*Arachis stenosperma*)	Burow *et al.,*1996; Leal-Bertioli *et al.,* 2016
Peanut	Late leaf spot resistance	F2 (*A. duranensis* x *A. stenosperma*)	Leal-Bertioli *et al.,* 2009
Pigeon Pea	Drought tolerance and Pod borer insect resistance	Wild x F2 cultivated (*Cajanus scarabaeoides* x C. *cajan*)	Saxena *et al.,* 2011
Lentil	Anthracnose resistance	RILs (wild x cultivated) (*Lens ervoides* x L. *culinaris*)	Tullu *et al.,* 2013

Vegetables and fruits: The wild relatives of vegetable crops constitute an increasingly important resource for improving vegetable production and critical resource in ensuring food security. Among different vegetable crops, potato is more genetically uniform which make it more vulnerable to biotic and abiotic stresses. Late blight caused by the *Phytophthora infestans* (Mont.) de Bary is the most important disease of potato production worldwide as exemplified by the Irish potato famine in the mid-nineteenth century. Late blight resistance genes were introgressed from the wild species *S. demissum, S. stoloniferum* and the cultivated *S. tuberosum* ssp. *andigena* and *S. phureja* into common potato in different parts of the world (Bradshaw *et al.*, 2006). Many novel genes for late blight resistance, virus resistance, high dry matter content and other useful traits are available in diploid wild potato species like *Solanum pinnatisectum, S. etuberosum, S. cardiphyllum, S. acaule, S. brachistotrichum, S. jamesii, S. polyadenium, S. stoloniferum* etc. Similarly in tomato, more than 40 resistance genes have been derived from *S. peruviannum, Solanum pennellii* Correll var. *pennellii, S. cheesmanii, and S. pimpinellifolium* for traits such as increased soluble solid content, fruit color, and adaptation to harvesting (Rick and Chetelat, 1995). In case of eggplant, *Solanum viarum,* a close relative of *Solanum melongena* was found highly resistance to shoot and fruit borer (Pugalendhi *et al.*, 2010) while wild species from Andaman *S. torvum* showed the recessive gene action for resistance to bacterial wilt (Bainsla *et al.*, 2016).

In okra, yellow vein mosaic disease (YVMD) is the most serious disease among the biotic stresses (Samarajeewa and Rathnayaka, 2004; Kumar and Reddy, 2015) whereas shoot and fruit borer, and leaf hopper are the major insect pests (Dhankar and Mishra, 2004). Resistance genes for YVMD, shoot and fruit borer, and leaf hopper are available in okra wild species *A. manihot, A. tuberculatus* and *A. moschatus*, respectively (Rana *et al.*, 1991; Singh *et al.*, 2006). Gangopadhyay *et al.*, (2016) also reported *A. caillei, A. manihot* and *A. moschatus* resistant to YVMD while *A. caillei, A. manihot, A. moschatus, A. tuberculatus* to shoot and fruit borer and leaf hopper

Among cucurbits, wild *Cucumis figarei* exhibited absolute resistance to cucumber green mottle mosaic virus (CGMMV), *Fusarium* wilt and high level resistance to downy mildew (Pan *et al.*, 1996). *C. figarei, C. myriocarpus, C. myriocarpus, C. africanus, C. africanus, C. meeusii, C. ficifolius* and *C. zeyheri* were also reported resistance to CGMMV virus (Rajamony *et al.*, 1990). However wild species *Cucumis hardiwickii* has been reported with high level of resistance to powdery and downy mildew diseases (Pitchaimuthu *et al.*, 2013) and it is potential source for increased yield in pickling cucumbers (Horst *et al.*, 1978). Some important crop wild relatives along with their important traits in different vegetable crops are listed in Table 4.

Table 4. Crop wild relatives in different vegetable crops utilized for various biotic/ abiotic stress tolerance

Vegetable crop	Wild species	Important traits	References
Tomato	*Solanum pimpinellifolium* *S. peruvianum* *S. pimpinellifolium* *S. chilense* *S. habrochaites* *S. pimpinellifolium* *S. cheesmanii*	Bacterial wilt Bacterial spot Tomato leaf curl virus High lycopene content Salt tolerance	Hanson *et al.*,1998 Yang and Francis., 2007 Kasrawi, 1989; Pilowsky and Cohen, 1990; Zamir *et al.*, 1994 Hanson *et al.*, 2000 Fernández-Ruiz *et al.*, 2002 Rick, 1986
Brinjal	*Solanum linnaeanum* *S. torvum,* *S. Sisymbriifolium* *S. indicum, S. violaceum, S. aethiopicum, S. incanum* *S. violaceum*	Verticillium wilt Powdery mildew Bacterial wilt Fruit and Shoot Borer Male sterility	Liu *et al.*, 2015 Bubici *et al.*, 2008 Mian *et al.*,1995 Behera and Singh, 2002 Isshiki and Kawajiri, 2002
Cucumber	*Cucumis hystrix* *C. metuliferous*	Gummy stem blight (*Didymella* bryoniae), Downy mildew (*Pseudoperonospora cubensis*), Cucumber mosaic virus (CMV-C), Zucchini yellow mosaic virus (ZYMV), and Papaya ringspot virus watermelon strain (PRSV-W) *Meloidogyne* sp.	Chen *et al.*, 2009 Norton *et al.*,1980
Crucifer	*Brassica tournefortii*	Cytoplasmic male sterility	Pradhan *et al.*, 1991

Among CWRs of fruits, *Malus baccata* in apple; *Pyrus pashia* and *P. pyrifolia* in pear; *Prunus cerasoides* in cherry and *P. mira* in peach are used as rootstocks with multiple disease, insect resistance and drought tolerance. In citrus, *Citrus jambhiri, C. limonia* and *C. karna* are used as rootstocks for cultivated species and are tolerant to citrus tristeza virus (CTV) and most promising rootstocks for mandarin, orange and kinnow in the lower hills. *Musa acuminata* ssp. *burmannica* found resistant to leaf spot and *Fusarium oxysporum* f. sp. *cubense* (Foc) wilt and Rhodochlamys such as *M. laterita, M. velutina, M. ornate* and *M. aurantiaca* have resistance to leaf spot, wilt and tolerance to nematodes. *Phylanthus acidus* has tolerance for rust and frost. *Vitis parviflora* showed multiple disease resistance while *V. himalayana* is cold hardy, drought tolerant and late ripener, hence escape fruit cracking in rainy season in grapes.

Useful Gene Sources from Alternative CWRs

The genome assisted breeding coupled with next generation sequencing approaches has opened an important corridor for evaluating alternative crop genomes for different stress tolerance mechanisms. The invention of model plant, *Arabidopsis thaliana* helped in evaluation of alternative crop genome for serving new gene resources, for example; diploid wild grass *Brachypodium distachyon* with fully sequenced small genome size (International Brachypodium Initiative, 2010) is a valuable model plant for wheat and barley for drought tolerance owing to high phylogenetic similarity reported by Mochida *et al.* (2013). Moreover, there was no report on negative effects of drought stress on growth and development of *Brachypodium distachyon* (Verelst *et al.*, 2013). Some important alternative CWRs utilized in different crops are listed in Table 5. *Haynaldia villosa* L. is another example that possesses many beneficial genes for improving resistance to powdery mildew, leaf and stem rusts, wheat streak mosaic virus (Chen *et al.*, 2002). *Elytrigia elongata* is a potential source of salt tolerance gene from tribe Triticeae. Similarly, Sheepgrass (*Leymus chinensis*), desiccation model plant (*Sporobolus stapfianus*) and *Agropyron cristatum* are beneficial reservoir of stress tolerance genes for introduction into cultivated crops such as wheat and barley.

Table 5. CWRs of Alternative Crop Species Utilized in Breeding for Climate Resilient Agriculture

CWR	Traits	Utilization	Reference
Agropyron elongatum	Rust resistance	Marker validation in wheat for leaf rust resistance	Gupta *et al.*, 2006
Agropyron cristatum	Drought and cold tolerance	Antioxidant mechanism activation for drought tolerance; Fructan biosynthesis for cold tolerance	Chatterton and Hardson, 2003; Shan and Liang, 2010
Agrostis stolonifera	Drought tolerance	QTL detection for drought tolerance	Merewitz *et al.*, 2014
Brachypodium distachyon	Cold and drought tolerance	Fructan accumulation under low temperature (CBF3 genes); osmoprotectan sugar biosynthesis	Li *et al.*, 2012; Verelst *et al.*, 2013
Haynaldia villosa	Powdery mildew resistance	Pm resistance line development in wheat	Cao *et al.*, 2011

CWR	Traits	Utilization	Reference
Elymus repens	*Fusarium* head blight resistance	Wheat introgression line production	Zeng *et al.*, 2013
Leymus chinensis	*Fusarium* head blight resistance, salt tolerance	Wheat-*Leymus* introgression lines, induction of salt stress tolerance genes	Qi *et al.*, 2008; Xianjun *et al.*, 2011
Sporobalus stapfianus	Drought tolerance	Leaf specific desiccation gene	Le *et al.*, 2007

Enhanced Use of CWRs through Advanced Biotechnology and Molecular Breeding Techniques

Omics approaches: Omics-scale technologies including genomics, transcriptomics, proteomics, and metabolomics have provided alternative opportunities for global analysis of regulatory genes, expressed proteins and metabolites regulating important traits in CWRs. High throughout next generation sequencing technology has led to increased discovery of single nucleotide polymorphisms (SNPs) and their association with important traits. Combining the phenotypic data with genotypic data, researchers have been able to validate and fine-map previously identified genes and to discover novel genomic regions underlying valuable agronomic traits in CWRs by association mapping (Li *et al.*, 2014; Qi *et al.*, 2014; Zhou *et al.*, 2015). Thus, it is evident that availability of genomic data along with efficient phenotypic evaluation is the key factor in discovery of genes controlling superior traits in CWRs. These omics approaches are particularly suitable for dissection of complex traits such as drought tolerance and pest resistance by characterizing CWRs under diverse treatments. For example, dehydrin genes in both wild barley (*H. spontaneum*) and wild tomato (*S. chilense* and *S. peruvianum*) as well as ABA/water stress/ripening induced (*Asr*) gene family members (*Asr2* and *Asr4*) from wild *Solanum* species are known to be involved in drought tolerance (Fischer *et al.*, 2013). An omics pathway after its identification can be traced by searching available annotations in an omics database. It is also feasible to transfer an appropriate metabolic pathway from CWRs into cultivated plant species to increase resistance against biotic stress (*e.g.* terpenoid biosynthetic pathway in wild tomato to cultivated one) (Bleeker *et al.*, 2012). Although the use of proteins and metabolites provides a deeper understanding of mechanism of gene action, the high quantification cost and relatively low levels of heritability of metabolites limit the application of metabolomics-assisted breeding.

Genetic modification: Genetic modification (GM) technology to transfer target genes to crop cultivars has been considered a revolutionary technique to produce transgenic crops with desired traits however, the safety of foods developed from transgenic crops and risk of environmental contamination continues a main concern. Another approach is induced mutagenesis in existing genes rather than introduction of new genes. This includes cisgenesis, intragenesis, genome editing, and RNA-dependent DNA methylation. Cisgenesis have been successfully used to confer late blight resistance in potatoes (Haverkort *et al.*, 2009) and scab resistance in apples (Vanblaere *et al.*, 2011) through transfer of single desired gene from native or cross compatible species avoiding linkage drag and achievement in less time compared with traditional breeding.

Conclusion

Crop wild relatives support the genetic improvement of crops in multiple ways. To cope with problems arising from intensive modern agriculture and climate change, it is essential to utilize genetic diversity within crop wild gene pools. Intense pre-breeding to widen the genetic base of the crops coupled with priorities evaluation of major genes with marker system for identification of desirable chromosomal segments should be developed to further enhance the utilization of CWRs in breeding programme. In the past, crop breeders had been struggling with the problem of linkage drag while dealing with CWRs. Nevertheless, now the advances in DNA sequencing technology particularly combination of *de novo* sequencing and resequencing are being used efficiently to explore useful genetic variation in CWR. Improved technologies especially genomics assisted breeding are facilitating the introgression of favorable traits from wild species into cultigens despite of incompatibility barriers. Therefore, plant breeders have to work closely with genetic engineers to develop new cultivars. Under the assumption that a vast reservoir of beneficial alleles for crop improvement exist in the wild gene pool, the continued development and implementation of strategies for the efficient utilization of these resources are underway.

References

Alamri, S.A., Barrett-Lennard, E.G., Teakle, L.N. and Colmer, T.D. 2013. Improvement of salt and water logging tolerance in wheat: comparative physiology of *Hordeum marinum-Triticum aestivum* amphiploids with their *H. marinum* and wheat parents. *Funct. Plant Biol.*, 40: 1168–1178.

Allaby, R.G., G.W. Peterson, D.A. Merriwether, and Y.B. Fu. 2005. Evidence of the domestication history of flax (*Linum usitatissimum* L.) from genetic diversity of the *sad2* locus. *Theor. Appl. Genet.*, 112:58–65.

Bainsla, Singh, N.K., Singh, S., Kumar, P.K., Singh, K., Gautam, A.K. and R.K. 2016. Genetic Behaviour of Bacterial Wilt Resistance in Brinjal (*Solanum melongena* L.) in Tropics of Andaman and Nicobar Islands of India. *American J. Plant Sci.*, 7: 333–338.

Ballini, E., Berruyer, R., Morel, J.B., Lebrun, M.H., Notteghem, J.L. and Tharreau, D. 2007. Modern elite rice varieties of the 'Green Revolution' have retained a large introgression from wild rice around the *Pi33* rice blast resistance locus. *New Phytologist*, 175: 340–350.

Behera, T.K. and Singh, G. 2002. Studies on resistance to shoot and fruit borer (*Leucinodes orbonalis*) and interspecific hybridization in eggplant. *Indian J. Hort.* 59 (1): 62–66.

Biselli, C., Urso, S., Bernardo, L., Tondelli, A., Tacconi, G., Martino, V., Grando, S. and Val Valè G., 2010. Identification and mapping of the leaf stripe resistance gene Rdg1a in *Hordeum spontaneum*. *Theor. Appl. Genet.*, 120: 1207–1218.

Bleeker, P.M., Mirabella, R., Diergaarde, P.J., VanDoorn, A., Tissier, A., Kant, M.R. and Schuurink, R.C. 2012. Improved herbivore resistance in cultivated tomato with the sesquiterpene biosynthetic pathway from a wild relative. *PNAS USA*, 109: 20124–20129.

Bradshaw, J.E., Bryan, G.J. and Ramsay, G. 2006. Genetic resources (including wild and cultivated *Solanum* species) and progress in their utilisation in potato breeding. *Potato Res.*, 49: 49–65.

Brar, D. 2005. Broadening the genepool and exploiting heterosis in cultivated rice. In *Rice is life: Scientific perspectives for the 21st century*, Toriyama, K., Heong, K.L. and Hardy, B. (eds.) Proceedings of the World Rice Research Conference November, 2004. Tokyo and Tsukuba, Japan. pp 4–7.

Bubici, G. and Cirulli, M. 2008. Screening and selection of eggplant and wild related species for resistance to Leveillula taurica. *Euphytica*, 164 (2): 339–45.

Burow, M.D., Simpson, C.E., Paterson, A.H. and Starr, J.L. 1996. Identification of peanut (*Arachis hypogaea* L.) RAPD markers diagnostic of root-knot nematode (*Meloidogyne arenaria* (Neal) Chitwood) resistance. *Mol. Breed.*, (2): 369–379.

Cabral, A.L. and Park R.F. 2014. Seedling resistance to *Puccinia coronata* f. sp. *avenae* in *Avena strigosa, A. barbata* and *A. sativa*. *Eupyhtica,* 196: 385–395

Cai, S., Wu, D., Jabeen, Z., Huang, Y., Huang, Y. and Zhang, G. 2013. Genome-wide association analysis of Aluminum tolerance in cultivated and Tibetan wild barley. *PLoS ONE*, 8(7): e69776.

Canci, H. and Toker, C. 2009. Evaluation of annual wild *Cicer* species for drought and heat resistance under field conditions. *Genet. Resour Crop Evol.*, 56: 1–6.

Cao, A., L. Xing, X. Wang, X. Yang, W. Wang, Y. Sun, C. Qian, J. Ni, Y. Chen, D. Liu, X. Wang and P. Chen. 2011. Serine/threonine kinase gene Stpk-V, a key member of powdery mildew resistance gene Pm21, confers powdery mildew resistance in wheat. *PNAS USA*. 108: 7727–7732.

Ceccarelli, S. 2014. Drought. In: Jackson, M., Ford-Lloyd, B., Parry, M., (eds.) Plant genetic resources and climate change. Wallingford, UK: CAB International, pp 221–235.

Chatterton, N. J. and Hardson, P.A. 2003. Fructans in crested wheatgrass leaves. *J. Plant Physiol.*, 160: 843–849.

Chen, G., Liu, Y., Ma, J., Zheng, Z., Wei, Y., McIntyre, C.L., Zheng, Y.L. and Liu, C. 2013. A novel and major quantitative trait locus for *Fusarium* crown rot resistance in a genotype of wild barley (*Hordeum spontaneum* L.). *PLoS ONE*, 8: e58040.

Chen, J.F., Chen, L.Z., Zhuang, Y., Chen, Y.G. and Zhou, X.H. 2009. Cucumber breeding and genomics: Potential from research with *Cucumis hystrix*, In:

Pitrat, M., (ed.) Proceedings of the IX EUCARPIA meeting on genetics and breeding of Cucurbitaceae. INRA, Avignon, France. pp 95–100.

Chen, Q., Conner, R.L., Li, H., Laroche, A., Graf, R.J. and Kuzyk, A.D. 2002. Expression of resistance to stripe rust, powdery mildew and the wheat curl mite in *Triticum aestivum- Haynaldia villosa* lines. *Can. J. Plant Sci.,* 82: 451–456.

Chen, S., Nelson, M.N., Chèvre, A., Jenczewski, E., Li, Z., Mason, A.S. *et al.* 2011. Trigenomic bridges for *Brassica* improvement. Crit. Rev. *Plant Sci.,* 30: 524–547.

Dhankar, B.S. and Mishra, J.P. 2004. Objectives of okra breeding. In: Singh, P.K. (ed.) *Hybrid vegetable development.* Haworth Press, Binghamton, NY, pp 195–209.

Du, D., Geng, C., Zhang, X., Zhang, Z., Zheng, Y., Zhang, F., Lin, Y. and Qiu, F. 2014. Transgenic maize lines expressing a *cry1C** gene are resistant to insect pests. *Plant Mol. Biol. Rep.,* 32: 549–557.

Fernández-Ruiz, V., Sánchez-Mata, M.C., Cámara, M., Torija, M.E., Roselló, S. and Nuez, F. 2002. Lycopene as a bioactive compound in tomato fruits. Symposium on "Dietary Phytochemicals and Human Health". The Phytochemical Soc. of Europe. Salamanca, pp 193–194.

Fischer, I., Steige, K. A., Stephan, W. and Mboup, M. 2013. Sequence evolution and expression regulation of stress- responsive genes in natural populations of wild tomato. *PLoS ONE,* 8: e78182.

Friedt, W., Horsley, R.D., Harvey, B.L., Poulsen, D.M.E., Lance, R.C.M., Ceccarelli, S., Grando, S. and Capettini, F. 2011. Barley breeding history, progress, objectives, and technology. In: Ullrich SE, (ed.) *Barley: production, improvement, and uses.* Oxford, UK: Wiley-Blackwell, pp 160–220.

Friedt, W., Horsley, R.D., Harvey, B.L., Poulsen, D.M.E., Lance, R.C.M., Ceccarelli, S., Grando, S. and Capettini, F. 2010. Barley breeding history, progress, objectives, and technology. In Ullrich, S.E., (ed.), *Barley.* Hoboken, NJ, USA: Wiley-Blackwell pp. 160–220.

Gangopadhyay, K.K., Singh, A., Bag, M.K., Ranjan, P., Prasad, T.V., Roy, A. and Dutta, M. 2016. Diversity analysis and evaluation of wild *Abelmoschus* species for agro-morphological traits and major biotic stresses under the north western agro-climatic condition of India. *Genet. Resour. Crop Evol.,* 1–16.

Gupta, S.K., Charpe, A., Koul, S., Haque, Q.M.R. and Prabhu, K.V. 2006. Development and validation of SCAR markers co-segregating with an *Agropyron elongatum* derived leaf rust resistance gene *Lr24* in wheat. *Euphytica,* 150: 233–240.

Hajjar, R. and Hodgkin, T. 2007. The use of wild relatives in crop improvement: A survey of developments over the last 20 years. *Euphytica,* 156: 1–13.

Hanson, P. M., Bernacchi, D., Green, S., Tanksley, S. D., Muniyappa, V., Padmaja, A. S., Chen, H. M., Kuo, G., Fang, D. and Chen, J.T. 2000. Mapping a wild tomato introgression associated with Tomato yellow leaf curl virus resistance in a cultivated tomato line. *J. Am. Soc. Hort. Sci.*, 125: 15–20.

Hanson, P.M., Licardo, O., Hanudin, Wang, J.F. and Chen, J.T. 1998. Diallel analysis of bacterial wilt resistance in tomato derived from different sources. *Plant Dis.*, 82: 74–78.

Haverkort, A.J., Struik, P.C., Visser, R.G.F. and Jacobsen, E. 2009. Applied biotechnology to combat late blight in potato caused by *Phytophthora infestans*. *Potato Res.*, 52: 249–264.

He, K., Wang, Z., Zhou, D., Wen, L., Song, Y. and Yao, Z. 2003. Evaluation of transgenic *Bt* corn for resistance to the Asian corn borer (Lepidoptera: pyralidae). *J. Econ. Entomol.*, 96: 935–940.

Horst, E.K. and Lower, R.L. 1978. *Cucumis hardwickii*: A source of germplasm for the cucumber breeder. *Cucurbit Genet. Coop. Rpt.*, 1: 5.

Huang, Y., Millett, B.P., Beaubien, K.A., Dahl, S.K., Steffenson, B.J., Smith, K.P. and Muehlbauer, G.J. 2013. Haplotype diversity and population structure in cultivated and wild barley evaluated for *Fusarium* head blight responses. *Theor. Appl. Genet.*, 126: 619–636.

Hyten, D. L., Song, Q. J., Zhu, Y.L., Choi, I.Y., Nelson, R.L., Costa, J.M. and Cregan, P.B. 2006. Impacts of genetic bottlenecks on soybean genome diversity. *PNAS USA*, 103: 16666–16671.

Iquira, E., Humira, S. and François, B. 2015. Association mapping of QTLs for *sclerotinia* stem rot resistance in a collection of soybean plant introductions using a genotyping by sequencing (GBS) approach. BMC *Plant Biol.*, 15: 5.

Islam. S., Malik, A.I., Islam, A.K. and Colmer, T.D. 2007. Salt tolerance in a *Hordeum marinum-Triticum aestivum* amphiploid, and its parents. *J. Exp. Bot.*, 58: 1219–1229.

Isshiki, S. and Kawajiri, N. 2002. Effect of cytoplasm of *Solanum violaceum* Ort. on fertility of eggplants (*Solanum melongena* L.). *Sci. Hort.*, 93: 9–18.

Jadhav, D.R., Mallikarjuna, N., Rathore, A. and Pokle, D. 2012. Effect of Some Flavonoids on Survival and Development of *Helicoverpa armigera* (Hübner) and *Spodoptera litura* (Fab) (Lepidoptera: Noctuidae). *Asian J. Agri. Sci.*, 4: 298–307.

James, R.A., Blake, C., Byrt, C.S. and Munns, R. 2011. Major genes for Na+ exclusion, *Nax1* and *Nax2* (wheat HKT1;4 and HKT1;5), decrease Na^+ accumulation in bread wheat leaves under saline and waterlogged conditions. *J. Exp. Bot.*, 62 (8): 2939–47.

James, V.A., Neibaur, I. and Altpeter, F. 2008. Stress inducible expression of the DREB1A transcription factor from xeric, *Hordeum spontaneum* L. in turf and forage grass (*Paspalum notatum* Flugge) enhances abiotic stress tolerance. *Transgenic Res.*, 17: 93–104.

Jena, K.K. 1994. Production of intergeneric hybrid between *Oryza sativa* L. and *Porteresia coarctata Curr. Sci.*, 67: 9–10.

Kalladan, R., Worch, S., Rolletschek, H., Harshavardhan, V., Kuntze, L., Seiler, C., Sreenivasulu, N. and Röder, M. 2013. Identification of quantitative trait loci contributing to yield and seed quality parameters under terminal drought in barley advanced backcross lines. *Mol. Breed.*, 32: 71 –90.

Kasrawi, M.A. 1989. Inheritance of resistance to tomato yellow leaf curl virus (TYLCV) in *Lycopersicon pimpinellifolium*. *Plant Dis.*, 73: 435–437.

Krugman, T., Peleg, Z., Quansah, L., Chagué, V., Korol, A.B., Nevo, E., Saranga, Y., Fait, A., Chalhoub, B. and Fahima, T. 2011. Alteration in expression of hormone-related genes in wild emmer wheat roots associated with drought adaptation mechanisms. *Funct. Integr. Genom.*, 11: 565–583.

Kumar, S. and Reddy, M.T. 2015. Morphological characterization and agronomic evaluation of yellow vein mosaic virus resistant single cross hybrids for yield and quality traits in Okra (*Abelmoschus esculentus* (L.) *Moench.*). *Open Access Libr. J.*, 2: e1720.

Lakew, B., Henry, R.J., Ceccarelli, S., Grando, S., Eglinton, J. and Baum, M. 2013. Genetic analysis and phenotypic associations for drought tolerance in *Hordeum spontaneum* introgression lines using SSR and SNP markers. *Euphytica*, 189: 9–29.

Le, T.N., Blomstedt, C.K., Kuang, J., Tenlen, J., Gaff, D.F., Hamill, J.D. and Neale, A.D. 2007. Desiccation-tolerance specific gene expression in leaf tissue of the resurrection plant *Sporobolus stapfianus*. *Funct. Plant Biol.*, 34: 589–600.

Leal-Bertioli, S.C., José, A.C., Alves-Freitas, D.M., Moretzsohn, M.C., Guimarães, P.M., Nielen, S., Vidigal, B.S., Pereira, R.W., Pike, J. and Fávero, A.P. 2009. Identification of candidate genome regions controlling disease resistance in *Arachis*. *BMC Plant Biol.*, 9: 112.

Leal-Bertioli, S.C., Moretzsohn, M.C., Roberts, P.A., Ballén-Taborda, C., Borba, T.C., Valdisser, P.A., Vianello, R.P., Araújo, A.C.G., Guimarães, P.M. and Bertioli, D.J. 2016. Genetic mapping of resistance to *Meloidogyne arenaria* in *Arachis stenosperma*: A new source of nematode resistance for peanut. *G3: Genes.Genomes.Genetics*, 6 (2): 377–390.

Lehnhoff, E.A., Keith, B.K., Dyer, W.E. and Menalled, F.D. 2013. Impact of biotic and abiotic stresses on the competitive ability of multiple herbicide resistant wild oat (*Avena fatua*). *PLoS ONE*, 8: 64478.

Leung, H., Hettel, G.P. and Cantrell, R.P. 2002. International Rice Research Institute: Roles and challenges as we enter the genomics era. *Trends Plant Sci.*, 7: 139–142.

Li, C., Rudi, H., Stockinger, E.J., Cheng, H., Cao, M., Fox, S.E., Mockler, T.C., Westereng, B., Fjellheim, S., Rognli O.A. and Sandve, S.R. 2012. Comparative analyses reveal potential uses of *Brachypodium distachyon* as a model for cold stress responses in temperate grasses. *BMC Plant Biol.*, 12: 65.

Li, Y.H., Zhou, G.Y., Ma, J.X., Jiang, W.K., Jin, L.G., Zhang, Z.H. and Qiu, L.J. 2014. *De novo* assembly of soybean wild relatives for pan-genome analysis of diversity and agronomic traits. *Nature Biotechnol.*, 32: 1045–1052.

Liu, J., Zheng, Z.S., Zhou, X.H., Feng, C. and Zhuang, Y. 2015. Improving the resistance of eggplant (*Solanum melongena*) to *Verticillium* wilt using wild species *Solanum linnaeanum*. *Euphytica*, 201: 463–469.

Lobell, D.B., Sibley, A. and Ivan Ortiz-Monasterio, J. 2012. Extreme heat effects on wheat senescence in India. *Nature Climate Change*, 2: 186–189.

Mallikarjuna, N., Jadhav, D.R., Saxena, K.B. and Srivastava, R.K. 2012. Cytoplasmic male sterile systems in pigeonpea with special reference to A_7CMS. Elect. *J. Plant Breed.*, 3: 983–986.

Maxted, N., Ford, B.V., Kell, S.P. and Turok, J. 2008. Crop wild relatives: conservation and use. CABI Publishing, Wallingford, UK. pp 23–28.

Merewitz, E., Belanger, F., Warnke, S., Huang, B. and Bonos, S. 2014. Quantitative trait loci associated with drought tolerance in creeping bentgrass (*Agrostis stolonifera* L.).

Mian, I.H., Ali, M. and Akhter, R. 1995. Grafting on *Solanum* rootstocks to control root-knot of tomato and bacterial wilt of eggplant. *Bull. Inst. Trop. Agric., Kyushu. Univ.*, 18: 41–47.

Miller, J.F. and Seiler, G.J. 2003. Registration of five oilseed maintainer (HA 429–HA 433) sunflower germplasm lines. *Crop Sci.*, 43: 2313–2314.

Miller, T.E., Iqbal, N., Reader, S.M., Mahmood, A., Cant K.A. and King, I.P. 1997. A cytogenetic approach to the improvement of aluminum tolerance in wheat. *New Phytol.*, 137: 93–98.

Mkwaila, W., Terpstra, K.A., Ender, M. and Kelly, J.D. 2011. Identification of QTL for agronomic traits and resistance to white mold in wild and landrace germplasm of common bean. *Plant Breed.*, 130: 665–672.

Mochida, K., Uehara-Yamaguchi, Y., Takahashi, F., Yoshida, T., Sakurai, T. and Shinozaki, K. 2013. Large-scale collection and analysis of full-length cDNAs from *Brachypodium distachyon* and integration with Pooideae sequence resources. *PLoS ONE*, 8:e75265.

Morrell, P.L. and Clegg, M.T. 2011. *Hordeum*. In: Kole, C., (ed.) *Wild crop relatives: genomic and breeding resources.* Springer-Verlag, Berlin, Germany. pp 309–319.

Norton, J.D. and Granberry, D.M. 1980. Characteristics of progeny from interspecific cross of *Cucumis melo* L. with *C. metuliferus* E. Mey. *J. Amer. Soc. Hort. Sci.*, 105: 174–180.

Oliver, R.E., Stack, R.W., Miller, J.D. and Cai, X. 2007. Reaction of wild emmer wheat accessions to *Fusarium* head blight. *Crop Sci.*, 47: 893–897.

Pakniyat, H. and Namayandeh, A. 2007. Salt tolerance associations with RAPD markers in *Hordeum vulgare* L. and *H. spontaneum* C. Koch. *Pak. J. Biol. Sci.,* 10: 1317–1320.

Pan, R.S. and More, T.A. 1996. Screening of melon (*Cucumis melo* L.) germplasm for multiple disease resistance. *Euphytica,* 88: 125–128.

Pilowsky, M. and Cohen, S. 1990. Tolerance to tomato yellow leaf curl virus derived from *Lycopersicon peruvianum. Plant Dis.,* 74: 248–250

Pink, D.A.C., Kift, N.B., Ellis, P.R., McClemant, S.J., Lynn, J. and Tatchell, G.M. 2003.Genetic control of resistance to aphid *Brevicoryne brassicae* in the wild species *Brassica fruticulosa. Plant breed.,* 122: 24–29.

Pitchaimuthu, M., Souravi, K., Ganeshan, G., Kumar, G.S and Pushpalatha, R. 2013. Identification of sources of resistance to powdery and downy mildew diseases in cucumber (*Cucumis sativus* L.) *Pest Manag. Hort. Ecosyst.,* 18 (1): 105–107.

Pradhan, A.K., Mukhopadhyay, A. and Pental, D. 1991. Identification of the putative cytoplasmic donor of a CMS system in *Brassica juncea. Plant Breed.,* 106 (3): 204–208.

Prescott-Allen, C. and Prescott-Allen, R. 1987. The first resource: wild species in the North American economy. *Brittonia,* 39: 427–427.

Prescott-Allen, R. and Prescott-Allen, C. 1988. Genes from the wild: Using wild Genetic Resources for Food and Raw Materials. International Institute of Environment and Development, Earthscan Publications, London.

Prischmann, D.A., Dashiell, K.E., Schneider, D.J. and Eubanks M.W. 2009. Evaluating *Tripsacum*-introgressed maize germplasm after infestation with western corn rootworms (Coleoptera: Chrysomelidae). *J. Appl. Entomol.,* 133: 10–20.

Pugalendhi, L., Veeraragavathatham, D., Natarjan, S. and Praneetha, S. 2010. Utilizing wild relative ((*Solanum viarum*) as resistant source to shoot and fruit borer in brinjal (*Solanum melongena* L.). *Elect. J. Plant Breed.,* 1 (4): 643–648.

Qi, L.L., Pumphrey, M.O., Friebe, B., Chen, P.D. and Gill, B.S. 2008. Molecular cytogenetic characterization of alien introgressions with gene *Fhb3* for resistance to *Fusarium* head blight disease of wheat. *Theor. Appl. Genet.,* 117: 1155–1166.

Qi, X., Li, M.W., Xie, M., Liu, X., Ni, M., Shao, G., Song, C., Yim, A.K.Y., Tao, Y. and Wong, F.L. *et al.* 2014. Identification of a novel salt tolerance gene in wild soybean by whole-genome sequencing. *Nat. Commun.,* 5: 4340.

Rajamony, L., More, T.A., Seshadri, V.S. and Varma, A. 1990. Reaction of muskmelon collections to cucumber green mottle mosaic virus. *J. Phytopathol.,* 129: 237–244

Rana, R.S., Thomas, T.A. and Arora, R.K. 1991. Plant genetic resources activities in okra: an Indian perspective. Report of an international workshop on okra genetic resources. International Board for Plant Genetic Resources, Rome. International Crop Network Series. 5.

Rick, C. and Chetelat, R. 1995. Utilization of related wild species for tomato improvement. *Acta Hort.*, 412: 21–38.

Rick, C.M. 1986. Germplasm resources in the wild tomato species In: El-Bollagy, A.S. and Pearson, A.R. (eds.) Symposium Tomato Production in Arid Land, Cario. pp 7.

Samarajeewa, P. and Rathnayaka, R. 2004. Disease resistance and genetic variation of wild relatives of okra (*Abelmoschus esculentus* L.). *Ann. Sri Lanka Depart. Agric.*, 6: 167–176.

Sandhu, J.S., Gupta, S.K., Singh, G., Sharma, Y.R. and Bains, T.S. 2006. Interspecific hybridization between *Cicer arietinum* and *Cicer pinnatifidum* for improvement of yield and other traits. In: 4th International Food Legumes Research Conference, New Delhi, India. pp 192.

Sarla, N. and Mallikarjuna Swamy, B.P. 2005. *Oryza glaberrima*: A Source for the Improvement of *Oryza sativa*. *Current Sci.*, 89: 955–963.

Saxena, K.B., Faris, D.G. and Kumar, R.V. 1987. Relationship between seed size and protein content in newly developed high protein lines of pigeonpea. *Plant food Hum. Nutr.*, 36: 335–340.

Saxena, R.K., Cui, X., Thakur, V., Walter, B., Close, T.J. and Varshney, R.K. 2011. Single feature polymorphisms (SFPS) for drought tolerance in pigeonpea (*Cajanus* spp.). *Funct. Integr. Genom.*, 11: 651–657.

Schmalenbach, I., Körber, N. and Pillen, K. 2008. Selecting a set of wild barley introgression lines and verification of QTL effects for resistance to powdery mildew and leaf rust. *Theor. Appl. Genet.*, 117: 1093–1106.

Seckin, B., Turkan, I., Sekmen, A.H. and Ozfidan, C. 2010. The role of antioxidant defense systems at differential salt tolerance of *Hordeum marinum* Huds. (sea barley grass) and *Hordeum vulgare* L. (cultivated barley). *Environ. Exp. Bot.*, 69: 76–85.

Selvakumar, R., Verma, R.P.S., Sharan, M.S., Bhardwaj, S.C., Shekhawat, P.S., Meeta, M., Singh, D., Devlash, R., Karwasra, S.S., Jain, S.K. and Sharma, I. 2013. Identification of resistance sources to barley yellow rust (*Puccinia striiformis* f. sp. *hordei*) in India. *Indian J. Plant Genet. Resour.*, 26 (2): 128–131.

Shan, C. and Liang, Z. 2010. Jasmonic acid regulates ascorbate and glutathione metabolism in *Agropyron cristatum* leaves under water stress. *Plant Sci.*, 178: 130–139.

Sharma, S.K. and Rana, J.C. 2012. Strategies for the Conservation of Crops Wild Relatives– Indian Context. In: Sharma, A.K., Ray, D. and Ghosh, S.N. (eds.).

Biological Diversity - Origin, Evolution and Conservation. Viva Books Private Limited. New Delhi, pp. 433–468.

Shavrukov, Y., Gupta, N., Miyazaki, J., Baho, M., Chalmers, K., Tester, M., Langridge, P. and Collins, N. 2010. HvNax3-a locus controlling shoot sodium exclusion derived from wild barley (*Hordeum vulgare* ssp. *spontaneum*). *Funct. Integr. Genomics*, 10: 277–291.

Singh, B., Rai, M., Kalloo, G., Satpathy, S. and Pandey, K. 2006. Wild taxa of okra (*Abelmoschus* species): reservoir of genes for resistance to biotic stresses. *Acta Hortic.*, 752: 323–328.

Singh, J., Lal, C., Kumar, D., Khippal, A., Kumar, L., Kumar, V., Malik, R., Kumar, S., Kharub, A.S., Verma, R.P.S. and Sharma, I. 2016. Widening the Genetic Base of Indian Barley Through the Use of Exotics. *International Journal of Tropical Agriculture*, 34 (1): 85–94.

Singh, K., Chhuneja, P., Ghai, M., Kaur, S., Goel, R.K., Bains, N.S., Keller, B. and Dhaliwal, H.S. 2007. Molecular mapping of leaf and stripe rust resistance genes. In: Buck *et al.*, (eds.). *T. Monococcum* and their transfer to hexaploid wheat. Wheat production in stressed environments. Springer publication, pp 779–786.

Singh, M., Bisht, I.S., Dutta, M., Kumar, K., Basandrai, A.K., Kaur, L., Sirari, A., Rizvi, A., Khan, Z., Sarker, A. and Bansal, K.C. 2014. Characterization and evaluation of wild annual *Cicer* species for agro-morphological traits and major biotic stresses under North-western Indian conditions. *Crop Sci.*, 54: 229–239.

Singh, M., Rana, M.K., Kumar, K., Bisht, I.S., Dutta, M., Gautam, N.K., Sarker, A. and Bansal, K.C. 2013. Broadening the genetic base of lentil cultivars through inter-subspecific and interspecific crosses of *Lens* taxa. *Plant Breed.*, 132: 667–675.

Sodkiewicz, W., Strzembicka, A. and Apolinarska, B. 2008. Chromosomal location in triticale of leaf rust resistance genes introduced from *Triticum monococcum*. *Plant Breed.*, 127: 364–367.

Sohail, Q., Inoue, T., Tanaka, H., Eltayeb, A.E., Matsuoka, Y. and Tsujimoto, H. 2011. Applicability of *Aegilops tauschii* drought tolerance traits to breeding of hexaploid wheat. *Breed. Sci.*, 61: 347–357.

Swarbreck, S.M., Lindquist, A.E., Ackerly, D.D. and Andersen, G.L. 2011. Analysis of leaf and root transcriptomes of soil grown *Avena barbata* plants. *Plant Cell Physiol.*, 52: 317–332.

Tester, M. and Langridge, P. 2010. Breeding technologies to increase crop production in a changing world. *Science*, 327: 818–822.

Tullu, A., Bett, K., Banniza, S., Vail, S. and Vandenberg, A. 2013. Widening the genetic base of cultivated lentil through hybridization of *Lens culinaris* "Eston" and *L. ervoides* accession IG72815. Can. *J. Plant Sci.*, 93: 1037–1047.

Vanblaere, T., Szankowski, I., Schaart, J., Schouten, H., Flachowsky, H., Broggini, G.A.L. and Gessler, C. 2011. The development of a cisgenic apple plant. *J. Biotechnol.*, 154: 304–311.

Verelst, W., Bertolini, E., De Bodt, S., Vandepoele, K., Demeulenaere, M., Pè, M.E. and Inzé, D. 2013. Molecular and physiological analysis of growth-limiting drought stress in *Brachypodium distachyon* leaves. *Mol. Plant.*, 6: 311–322.

Verma, R.P.S., Singh, D.P., Selvakumar, R., Chand, R., Singh, V.K. and Singh, A.K. 2013. Resistance to spot blotch in barley. *Indian J. Plant Genet. Resour.*, 26 (3): 220–225.

Vierling, R. A. and Nguyen, H.T. 1992. Heat-shock protein gene expression in diploid wheat genotypes differing in thermal tolerance. *Crop Sci.*, 32: 370–377.

Vincent, H., Wiersema, J., Kell, S., Fielder, H., Dobbie, S., Castaneda Alvarez, N.P. and Maxted, N. 2013. A prioritized crop wild relative inventory to help underpin global food security. *Biol. Conserv.*, 167: 265–275.

Wang, Q.B., Zhang, Y.Y., Fang, Z.Y., Liu, Y.M., Yang, L.M. and Zhuang, M. 2012. Chloroplast and mitochondrial SSR help to distinguish allo-cytoplasmic male sterile types in cabbage (*Brassica oleracea* L. var. *capitata*). *Mol. Breed.*, 30: 709–716.

Winter, S.M., Shelp, B.J., Anderson, T.R., Welacky, T.W. and Rajcan, I. 2007. QTL associated with horizontal resistance to soybean cyst nematode in *Glycine soja* PI464925B. *Theor. Appl. Genet.*, 114: 461–472.

Wright, S.I., Bi, I.V., Schroeder, S.G., Yamasaki, M., Doebley, J.F., McMullen, M.D. and Gaut, B.S. 2005. The effects of artificial selection of the maize genome. *Science*, 308: 1310–1314.

Xianjun, P., Xingyong, M., Weihong, F., Man, S., Liqin, C., Alam, I., Lee, B.H., Dongmei, Q., Shihua, S. and Gongshe, L. 2011. Improved drought and salt tolerance of *Arabidopsis thaliana* by transgenic expression of a novel *DREB* gene from *Leymus chinensis*. *Plant Cell Rep.*, 30: 1493–1502.

Xu, X., Liu, X., Ge, S., Jensen, J. D., Hu, F. Y., Li, X. and Wang, W. 2012. Resequencing 50 accessions of cultivated and wild rice yields markers for identifying agronomically important genes. *Nature Biotechnol.*, 30: 105–111.

Yadav, J.R. and Kumar, J. 1999. Evaluation of exotic and indigenous barley accessions for resistance against Indian pathotypes of *Puccinia striiformis hordei*. Rachis Newsletter, 18 (2): 60–63.

Yang, W. and Francis, D.M. 2007. Genetics and breeding for resistance to bacterial diseases in tomato: prospects for marker-assisted selection. In: Razdan, M.K. and Mattoo, A.K. (eds.), *Genetic Improvement of Solanaceous Crops.* Tomato vol 2. Science Publishers, New Hampshiore, USA: pp 379–419.

Yao, G., Zhang, J., Yang, L., Xu, H., Jiang, Y., Xiong, L., Zhang, C., Zhang, Z., Ma, Z. and Sorrells, M.E. 2007. Genetic mapping of two powdery mildew resistance genes in einkorn (*Triticum monococcum* L.) accessions. *Theor. Appl. Genet.*, 114: 351–358.

Yumurtaci, A. 2015. Utilization of wild relatives of wheat, barley, maize and oat in developing abiotic and biotic stress tolerant new varieties. *Emir. J. Food Agric.*, 27 (1): 1–23.

Zamir, D., Michelson, I.E., Zakay, Y., Navot, N., Zeidan, M., Sarfatti, M., Eshed, Y., Harel, E., Pleban, T., van Oss, H., Kedar, N., Rabinowitch, H.D. and Czosnek, H. 1994. Mapping and introgression of a Tomato yellow leaf curl virus tolerance gene, *Ty-1*. *Theor. Appl. Genet.*, 88: 141–146.

Zeng, J., Cao, W., Hucl, P., Yang, Y., Xue, A., Chi, D. and Fedak, G. 2013. Molecular cytogenetic analysis of wheat-*Elymus repens* introgression lines with resistance to *Fusarium* head blight. *Genome*, 56: 75–82.

Zhang, H., Li, C., Davis, E.L., Wang, J., Griffin, J.D., Kofsky, J. and Song, B.H. 2016. Genome-wide association study of resistance to soybean Cyst nematode (*Heterodera glycines*) HG type 2.5.7 in wild soybean (*Glycine soja*). *Front. Plant Sci.*, 7: 1214.

Zhang, H., Mittal, N., Leamy, L.J., Barazani, O. and Song, B.H. 2017. Back into the wild-Apply untrapped genetic diversity of wild relatives for crop improvement. *Evol. Appl.*, 10(1): 5–24.

Zhou, Z.K., Jiang, Y., Wang, Z., Gou, Z.H., Lyu, J., Li, W.Y. and Tian, Z.X. 2015. Rese quencing 302 wild and cultivated accessions identifies genes related to domestication and improvement in soybean. *Nature Biotechnol.*, 33: 408–414.

3

Crop Biofortification for Malnutrition

Ramadoss Bharathi Raja[1*], Manu Pratap Gangola[1], Vidhya Venkatesan[2], Surinder Singh[1], and Hukkeri Shivappa[1]

[1]Department of Plant Sciences, University of Saskatchewan,
51 Campus Drive, Saskatoon, Saskatchewan S7N 5A8, Canada.
[2]Department of Plant Genetic Resources, Centre for Plant Breeding and Genetics,
Tamil Nadu Agricultural University, Coimbatore-641 003, Tamil Nadu, India.
*Corresponding author: bharathiraja_6@yahoo.co.in

Abstract

Malnutrition is mainly caused by nutritionally deficient or excessive diet thus induces health problems in humans. Malnutrition affects more than half of the world's population therefore, is one of the major problems worldwide. Many strategies including supplementation, food fortification, and dietary diversification have been successfully implemented to alleviate the impact of malnutrition in certain parts of the world. Crop biofortification targets to enhance the nutritional qualities of crop plants by improving nutrients' concentration or bioavailability. Crop biofortification is a realistic and cost-effective method of delivering nutrient to the target population that may have limited access or knowledge to diverse diets and other interventions. Crops can be biofortified using agronomic practices, conventional and transgenic breeding approaches. HarvestPlus program, Consultative Group on International Agricultural Research (CGIAR) institutes and its partners have demonstrated that more than 20 million people in developing countries are now growing and consuming biofortified staple crops such as rice, wheat, maize, common beans and cassava that assures improved health of the population by eradicating malnutrition.

Keywords: Biofortification, Malnutrition, Plant breeding, Genetic engineering, Nutritional/grain quality traits

Introduction

The term 'Food security' can be defined as providing adequate, non-toxic and nutritionally rich food to meet the dietary needs of an individual leading to a healthy life. To achieve food security for more than 9 billion people by 2050, is an eminent challenge for the farmers, agricultural scientists, and government as agricultural food production needs to increase by 70% worldwide and 100% in developing countries to meet the growing food demand of the increasing population (Godfray *et al.*, 2010). Malnutrition, one of major concerns of the food security especially in developing countries, refers to the inadequate or excessive energy intake. The malnutrition can be categorized in to two groups,(i) energy-, and (ii) micronutrient- malnutrition. Energy malnutrition prevails mainly due to improper proportion of energy in the diet. A balanced proportion of carbohydrate, protein, fat, vitamins and minerals in the daily diet is crucial for proper growth and development of an individual. People from the developed countries usually get surplus energy, attributed to high calorie food availability/consumption and improved economic status, that may lead to increased risk of diabetes and other cardio-vascular diseases. Conversely, people from the developing or poor countries acquire less energy from their diet due to lack of resources, awareness, and purchasing power, that can cause impaired growth/development, increased risk of diseases, and death or abnormalities in children. Micronutrient malnutrition has been described as a hidden form of malnutrition which has a significant adverse impact on individual's health, social well-being and national economic efficiency.

According to the estimates of Food and Agriculture Organization (FAO), about 5-10 million of preschool children in the developing countries suffer from malnutrition every year (FAO, 2014). According to the FAO estimates, about 805 million people were chronically malnourished during 2012-2014, that is more than 100 and 209 million downcompared to last decade and during 1990–1992, respectively (FAO, IFAD, and WFP 2014). The prevalence of the undernourishment has also declined from 18.7 % to 11.3% globally and from 23.4% to 13.5% for the developing countries during the corresponding period (Sharma *et al.*, 2016). About 63 (98.2%) developing countries has not even achieved the target of halving the number of undernourished people by 2015, set by the World Food Summit held at Rome (Italy) during1996 (FAO, 1996).

Nearly five billion people are affected by micronutrient deficiency; around two billion people each are affected by iron (WHO, 2015) and iodine deficiency (De Benoist *et al.*, 2008) whereas, about 17.3% of the population is affected by zinc deficiency (Wessells and Brown 2012). World Health Organization (WHO) estimated that 250 million preschool children were vitamin A deficient and a considerable proportion of pregnant women suffered from vitamin A deficiency during 2015 (WHO, 2015a).

Many strategies including supplementation, food fortification, and dietary diversification have been successfully implemented to reduce malnutrition in certain parts of the world. However, single intervention is unable to eliminate the global risk of malnutrition in all the population groups. Therefore, biofortification of crops along with existing intervention strategies are being followed to reduce the malnourishment throughout the life span of an individual in an economically accessible alternative across the globe (Meenakshi *et al.*, 2010). Consequently, the agricultural research programs, mainly focused to increase the global crop production during the last four decades, are now being directed to develop nutritionally superior crop varieties to alleviate the problem of hidden hunger or malnutrition through HarvestPlus program.

Crop biofortification is the process (agronomic, breeding or genetic engineering practices) of enhancing nutrients into food crops (Garcia-Casal *et al.*, 2016). Biofortification provides sustainable and long-term delivery of health benefits by providing per day adequacy of nutrient intake among consumers for their entire life period rather than higher supply of nutrients like in case of supplements or industrially fortified foods. Therefore, biofortification ensures the easy and economic accessibility of the impoverished people to the nutritious food. This strategy fortifies the micronutrient dense traits in the crop variety which is preferred by both farmers and consumers. Therefore, this chapter summarizes the process, methods, progresses, and challenges of crop biofortification to alleviate the malnutrition across the globe.

Important Considerations for Biofortification to Overcome Malnutrition

The important questions need to be asked before starting any biofortification program, are:

i) Is it possible to increase the micronutrient concentration to the target level?

ii) What will be the digestibility of the nutrient with increased concentration so that it can be easily absorbed and utilized by the digestive tract to improve human health and nutrition?

iii) Does the increased level of micronutrient in staple crop affect the consumer preference? (e.g. golden rice).

iv) Will farmers grow the biofortified varieties?

v) Will the consumer buy/eat them in sufficient quantities? Sometimes crop biofortification increases the cost of production due to increased fertilizer application.

These questions or considerations need to be addressed to ensure the success of any crop biofortification program.

Implementing Crop Biofortification for Malnutrition

The implementation of biofortification consists of three important activities: discovery, development and dissemination/delivery (Bouis *et al.*, 2011; Saltzman *et al.*, 2013). Each activity is further processed into three to four internal steps (Figure 1).

Discovery

The discovery phase is the first step in implementing crop biofortification. It mainly includes the identification of target population (with specific nutrient requirement), and staple crops of a region or country. Thereafter, target concentration of the nutrient should be decided and breeding strategies should be designed by consulting with the breeder(s). The target concentration of the nutrient is decided based on average food intake and pattern as influenced by socio-economic, cultural, environmental and dietary factors (Hotz and McClafferty, 2007; Grusakand Cakmak, 2009). In the next step, plant breeder(s) screen available crop varieties/cultivars/accessions in a global genepool to determine the genetic and phenotypic variability for the nutrient concentration. In absence of genetic variability for the trait of interest in the available genepool, mutation breeding and genetic transformation are the potential alternatives to develop genotypes with improved nutrients concentrations (Menkir, 2008; Ashok Kumar *et al.*, 2012; Velu*et al.*, 2012; Fageria *et al.*, 2012). However, ethical issues limiting the implementation and/or releasing of variety developed through these approaches are still under consideration in many of the developing countries (Raja *et al.*, 2016).

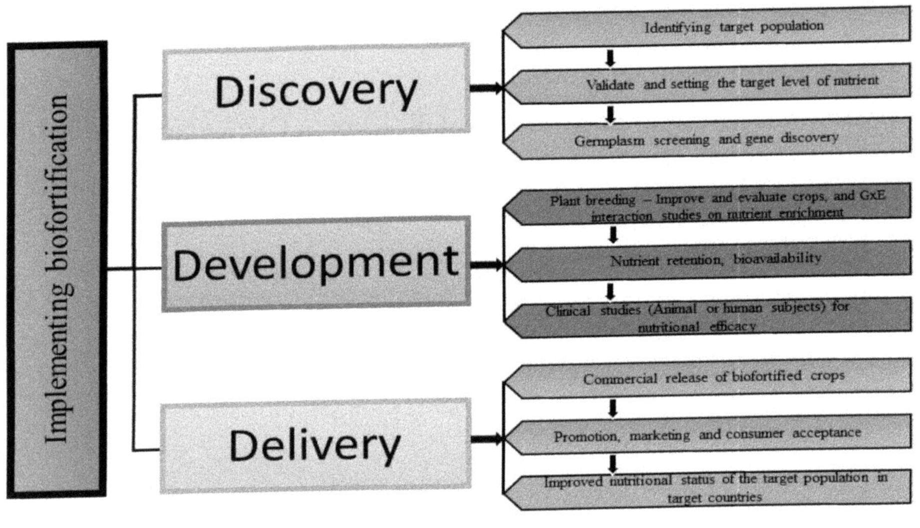

Fig 1. Workflow to implement crop biofortification (Bouis *et al.*, 2011; Saltzman *et al.*, 2013)

Development

This phase of implementing biofortification involves essential crop breeding activities. The breeding programs are initiated at national and international research institutes or/in partnership with private research centers to develop crop varieties with improved nutritional composition and better agronomic performance while enhancing or maintaining the traits affecting consumers acceptance. The ensued high yielding or nutritious genotypes are evaluated for their performances in multilocation trials during different years to assure their consistent performance in field conditions. The better performing genotypes are selected and processed to be released as a variety for a region or country. Thus, the complete development process can be attained usually in 6-10 years depending on the type of breeding approach followed (Bouis *et al.*, 2011; Saltzman *et al.*, 2013).

Nutrient retention and bioavailability in a crop is influenced by numerous factors like processing (e.g. hulling and milling rice), storage conditions/time and cooking methods. *In vitro* and *in vivo* (animal studies) experiments have also been performed to determine the relative absorption and bioavailability of the nutrients, and their interaction with other nutrients. The developed genotypes can also be compared with the current elite cultivars or varieties for their nutritional composition, agronomic performance, and digestibility to assure their superiority.

Dissemination

Based on impact studies, biofortified crops must be officially released to target countries and thereafter, to the target population. Dissemination involves consumer acceptance, varietal adaptation, and successful marketing strategies to maximize the adoption and consumption of biofortified crops to the target population. Finally, assessment of nutritional status of the target population in the target countries needs to be performed (Bouis *et al.*, 2011; Saltzman *et al.*, 2013).

Methods of Crop Biofortification

There are three non-mutually exclusive methods of crop biofortification which can be used to enhance the nutritional composition of the crop: (i) agronomic biofortification (application of fertilizer), (ii) classical plant breeding or molecular breeding/molecular marker-assisted breeding, and (iii) bio-engineering or genetic modification (including trans-genetic modification).

Agronomic Crop Biofortification

Agronomic crop biofortification is the application of a specific nutrient rich fertilizer to the crop canopy or soil to increase the micronutrient concentration in the edible part of the crop, consumption of which will ultimately increase the intake of essential nutrients by the consumers. Improved selenium (Se) content in the wheat (Lyons *et al.*, 2005) and rice (Chen *et al.*, 2002) has been achieved using Se rich fertilizer. Likewise, zinc (Zn), iodine (I) and cobalt (Co)concentrations can also be modulated utilizing agronomic biofortification. Success of this method mainly depends on the uptake and translocation of the nutrient into the crop from either natural (soil) or artificial (fertilizer/foliar spray) sources (Singh *et al.*, 2016a).

Conventional/Classical Plant Breeding

Conventional plant breeding method of biofortification uses deliberate interbreeding of closely or distantly related species to develop new varieties or hybrids with desirable level of nutrients or micronutrient without compensating their agronomic performances and consumer preferences (Khush *et al.*, 2012).

Different methods of plant breeding have been used in classical plant breeding to develop a variety with specific end use or desired traits. Conventional breeding programs to improve nutritional traits mainly involves,(a) evaluation of germplasm for nutritional traits, (b) inheritance pattern and choice of breeding method, (c) selection of suitable genotypes for crossing program, (d) evaluations in F_1 and segregating generations, (e) backcross method, or mutation breeding if required followed by (f) screening of selected lines for nutritional traits. Different breeding methods can be used based on the available genetic variability, nature of gene action and type of procedure to be used for evaluating the genotypes for nutritive traits.

Molecular Breeding/Marker Assisted Breeding

In molecular or marker-assisted breeding (MAB), DNA markers substitutes phenotypic selection/classical breeding to accelerate the selection process and release of improved cultivars. Molecular and conventional breeding are not mutually exclusive, instead both are complementary to each other in most of the breeding programs (Xu, 2010). Molecular breeding is more popular and has been effectively used to enhance the efficiency of classical breeding process by applying modern genomic tools and resources. Most of the molecular breeding strategies relies on the DNA markers that help breeder(s) to choose genotypes with desirable traits within a brief time. Various molecular marker techniques, currently available to identify thepolymorphism at DNA level, have been grouped into two predominant approaches: (1) Non-PCR based approaches, e.g. RFLP (Restricted Fragment Length Polymorphism), and (2) PCR based approaches,e.g. RAPD (Random Amplified Polymorphic DNA), AFLP (Amplified Fragment Length Polymorphism), SSR (Simple Sequence Repeats), SNP (Single Nucleotide Polymorphism) and so on (Acquaah, 2012). These molecular markers have been extensively utilized to improve the nutrient status of the crop plants. Other approaches like genotyping by sequencing (GBS), Targeting Induced Local Lesions IN Genomes (TILLING), Ecotype Targeting Induced Local Lesions IN Genomes (EcoTILLING), multi-parent advanced generation intercross (MAGIC) population have also successfully employed to identify novel markers/allele for nutrient enhancement in various crops. These well-known technologies should be followed in biofortification programs particularly to map/isolate the genomic region associated with high mineral/micronutrient accumulation, nutrient uptake, nutrients bioavailability, and reduced anti nutritional factors in the grain. Identified genomic regions can be effectively utilized as a functional marker to accelerate the breeding process to develop nutritionally superior crop varieties.

Genetic Modification / Transgenic

Transgenic or genetic modification is the process of introducing foreign gene, called transgene, into a crop species to modulate the expression of the gene (transgene) regulating the nutrients concentrations and compositions. This can be achieved by using liposomes, enzymes, plasmid vectors, viral vectors, pronuclear injection, protoplast fusion and ballistic DNA injection (Ganeshan and Chibbar, 2010). Golden Rice (Enhanced beta-carotene content) is an example of a genetically modified biofortified crop (Ye *et al.*, 2000). This was achieved by transferring gene from the daffodil to the recipient organism rice (Burkhardt *et al.*, 1997). Genetic modification method of biofortification has two important benefits over conventional biofortification: (a) it takes less time to produce a crop variety with stable expression of trait of interest with some exception, and (b) this method provides flexibility to transfer specific genes to any crop species (Ganeshan and Chibbar, 2010).

Status of Biofortified Crops For Various Nutritional Traits

Combination of conventional and molecular approaches have been successfully utilized to develop crops varieties with increased nutritional density. However, in some cases individual approaches have also been successful in developing suitable varieties. The progress in key crops has been reviewed and summarized in Table 1.

Iron

Iron (Fe) as an electron donor and acceptor participates in various important body functions. Fe supports the heme complex formation which is an essential component of hemoglobin, myoglobin and the catalytic center of cytochromes (Singh *et al.*, 2016). Therefore, Fe is crucial for oxygen transport and energy metabolism in human system. Iron deficiency leads to iron deficiency anemia (IDA) that affect at least two billion people globally. IDA affects 50% of the pregnant women and 40% of the preschool children in the developing countries. Fe has also been associated with brain development, structure and function. Apart from this, 0.2% of deaths and 0.5% of DALY (Disability Adjusted Life Years) have been imputed to IDA alone in children below 5 years. This can be overcome by increasing the Fe intake in the diet. Most of the developing countries relies on their cereal staples that are either low in Fe or contain antinutritional factor like phytate which limits its bioavailability (Zimmermann *et al.*, 2004).

Table 1. list of biofortified crops released in different countries (Saltzman *et al.*, 2013)

S.No	Crop	Target Nutrient	Target country	Release year
1.	Rice	Zinc (Iron)	Bangladesh, India Brazil	2013 2014
		Provitamin A Carotenoids*	Philippines, Bangladesh, Indonesia, India	2013
		Iron*	Bangladesh, India	2022
		Iron	China	2010
2.	Maize	Provitamin A Carotenoids	Zambia	2012
			Nigeria	2012
			Brazil	2013
			China	2015
			India	Unknown

S.No	Crop	Target Nutrient	Target country	Release year
3.	Wheat	Zinc (Iron)	India, Pakistan	2013
			China	2011
			Brazil	2016
4.	Sorghum	Zinc, Iron	India	2015
		Provitamin A Carotenoids*	Kenya, Burkina Faso, Nigeria	2018
5.	Pearl millet	Iron (Zinc)	India	2012
6.	Lentil	Iron, Zinc	Nepal, Bangladesh, Ethiopia, India, Syria	2012
7.	Cowpea	Iron, Zinc	India	2008
			Brazil	2008
8.	Bean	Iron (Zinc)	Rwanda, DR Congo	2012
			Brazil	2008
9.	Cassava	Provitamin A	DR Congo	2008
		Carotenoids	Nigeria	2011
			Brazil	2009
		Provitamin A Carotenoids Iron*	Nigeria, Kenya	2017
10.	Irish potato	Provitamin A	Rwanda, Ethiopia	-
11.	Sweet potato	Carotenoids	Uganda Mozambique Brazil China	2007 2002 2009 2010

** Denotes transgenic variety*

Rice

In recent years, conventional breeding approach have been mainly used to increase the Fe concentration in important staple crops. The rice variety IR 68144-2B-2-2-3, developed through conventional breeding, contains7–13 μg of Fe/g (Graham *et al.*, 1999). Both traditional and improved rice varieties contacting high Fe concentrations (6.6 μg/g to 16.7 μg/g) have been identified (Nachimuthu *et al.*, 2014). Rice genotypes with high Fe concentration were crossed with location specific high yielding rice varieties to produce progeny with both high yield, and increased level of iron and zinc (Khush *et al.*, 2012). Rice lines developed through transgenic approach using *Phaseolous vulgaris* ferritin gene under the control of endosperm specific rice glutelin promoter showed two times increase (37 μg/g) in iron content (Vasconcelos *et al.*, 2003). However, ethical issues with transgenic is still under consideration in most of the developing countries restricting their adoption for commercial cultivation (Vasconcelos *et al.*, 2003).

Pearl Millet, Cowpea and Lentil

The International Crops Research Institute for the Semi-Arid Tropics (ICRISAT) released pearl millet variety ICTP8203 (Dhanshakti) with increased Fe concentration of 107 ppm, the desired target of the breeding program (Cherian, 2014). Four early maturing high iron and zinc cowpea varieties, Pant Lobia-1 (82 ppm Fe and 40 ppm Zn), Pant Lobia-2 (100 ppm Fe and 37 ppm Zn), Pant Lobia-3 (67 ppm Fe and 38 ppm Zn), and Pant Lobia-4 (51 ppm Fe and 36 ppm Zn) were released in India during 2008, 2010, 2013 and 2014, respectively (Singh, 2014).

International Center for Agricultural Research in the Dry Areas (ICARDA) is leading the lentil program to develop high Fe-Zn containing varieties (Sarker, 2014). Large panel of lentil germplasms which includes breeding lines (more than 1,600), landraces, and released varieties have been analyzed for Fe and Zn concentrations. Fe concentration ranged from 42–132 ppm whereas Zn concentration varied from 23 to 78 ppm. Three varieties, L4704 (125 ppm Fe and 74 ppm Zn), ILL 7723 (83 ppm Fe and 61.5 ppm Zn), and Barimasur-7 (81 ppm Fe) were released during 2012-2014 in India, Nepal and Bangladesh, respectively containing more than the target Fe and Zn concentrations (Sarker, 2014).

Microbiological intervention using plant growth promoting rhizobacteria (PGPR) has been used as a strategy to improve the iron uptake from the rhizosphere soil by plants. Among PGPR, siderophore assisted *pseudomonas fluorescent* was successful in mobilizing iron from the soil to the plants (Prasanna *et al.*, 2016). Biofortification through PGPR is also a viable option to be considered as a possible supplementary measure along with breeding approaches that can increase the micronutrient concentration in crops especially cereals, and improve the crop yield by ameliorating the soil health (Prasanna *et al.*, 2016).

Zinc

Zinc (Zn) is an essential micronutrient for plants and humans as it plays pivotal role in protein, nucleic acid, carbohydrate, and lipid metabolism. Zn is an important cofactor for many enzymes (WHO/FAO 1998). In addition, Zn is also critical for the transcription and the coordination of other biological processes. Zn deficiency is one of the most serious problems in human nutrition. The abnormalities caused by Zn deficiency include retarded growth, depressed immune function, anorexia, skeletal abnormalities, diarrhea, alopecia and dermatitis. Children are more sensitive to Zn deficiency. Therefore, Zn deficiency is listed as a major risk factor for human health and cause of death globally (Cakmak, 2008).

Zn is generally low in the global food supply therefore, a greater proportion of the national population is at risk of Zn deficiency which ranges from 1-13% in North America-Europe to about 68-95% in South- and Southeast- Asia, Eastern Mediterranean and Africa. About 50% of the total world population is at risk of low Zn intake (Brown *et al.*, 2001).

Rice

Gregorio *et al.* (2000) screened International Rice Research Institute (IRRI) rice germplasm bank for high zinc concventration. Zn concentrations ranged from 15.3–58.4 mg/kg among a subset of 1138 samples. Traditional varieties, Jalmagna and Zuchen, contained almost twice as much iron and 50% more zinc compared to widely grown varieties, IR36 and IR64. Many aromatic rice varieties such as Basmati 370 from India, Pakistan and Azucena from the Philippines also showed consistently higher zinc concentration. Rice with enhanced level of zinc for Bangladesh and India have been developed by IRRI and the Bangladesh Rice Research Institute (BRRI). The final breeding target was set at 32 ppm, and varieties with more than 75% of the target concentration with better agronomic performances were released for commercial cultivation in Bangladesh and India.

Wheat

Crop biofortification of high-Zn wheat for Pakistan and India is led by International Maize and Wheat Improvement Center (CIMMYT) and released for commercial cultivation in 2013-2014. A wheat variety with Zn concentration more than the target concentration of 44 ppm, was released in China during 2011 (www.harvestplus-china.org).

Provitamin A

Plant carotenoids are the primary dietary source of provitamin A worldwide, with β-carotene as the most well-known provitamin A carotenoid. Other forms of carotenoid include α-carotene and β-cryptoxanthin. Provitamin A deficiency has been associated with color blindness, decreased reproduction, reduced bone growth, and insensitive immune response that may lead to increased risk of disease and death in humans especially children and pregnant women. About 140 million children and 7 million pregnant women are deficient for provitamin A in Africa and South/Southeast Asia, with 2,50,000-5,00,000 becoming blind every year, half of which die within 12 months of losing their sight. About 90% of these worlds annual death occurs mostly in middle income countries (Jones *et al.*, 2003). Human body can synthesize vitamin A, if the precursor compound β-carotene (also known as pro vitamin A) is present which can be obtained from many crops but cereal grains. Among the cereal grains, rice is a staple food of more than half of the world's population (Boonyaves *et al.*, 2017). Therefore, a precise metabolic step was introduced in rice to facilitate β-carotene biosynthesis (Ye *et al.*, 2000).

Rice

The first step was taken bydeveloping golden rice by Prof. I. Potrykus, in Taipei 309 rice variety and Dr. P. Bayer, using known precursor of β-carotene available in rice endosperm, geranyl geranyl pyrophosphate (GGPP) (Potrykus, 2001). Potrykus and his colleagues have modified the rice genome through genetic engineering and introduced two genes from daffodil (Burkhardt *et al.*, 1997) and one from *Erwinia* to complete the biochemical pathways (Ye *et al.*, 2000). GGPP, in a phytoene synthase (from daffodil) catalyzed reaction, synthesizes phytoene that is catalyzed into lycopene and β-carotene in two consecutive reactions catalyzed by phytoene desaturase (from *Erwinia*) and lycopene cyclase (from daffodil), respectively. This work was targeted to provide at least 2μg of Vitamin A/g of rice grain but, it has provided only 1.6 μg/g of rice. This rice variety with enhanced vitamin A concentration was termed as "Golden Rice 1". However, the vitamin A concentration in golden rice 1 was not sufficient to meet the daily requirement of a healthy person. Therefore, golden rice 2 (Syngenta 2), with 23-fold increased vitamin A content (37 μg/g of rice) using phytoene synthase gene from corn, has been developed (Paine *et al.*, 2005). Developing countries are now using golden rice for the development of local varieties with increased β-carotene by conventional breeding methods, because of ethical issues related to genetically modified crops.

Two yellow, or *amarillo* rice line that may supply the missing genes for provitamin A bio synthesis was naturally discovered in IRRI(Graham and Rosser, 2000). These two lines showed varying level of β-carotene, due to carotenoid decay in storage. Exploiting this kind of germplasm lines will help to develop the rice with higher β-carotene without addressing GMO issues, especially in developing countries.

Orange sweet potato (OSP)

The International Potato Center (CIP), and National Agriculture Research and Extension System (NARES) scientists adopted conventional breeding strategy to identify orange sweet potato with increased provitamin A (60 ppm). Further, cooking, sensory and agronomic parameters were improved to meet the local needs of the growers and consumers. The bioavailability of provitamin A is higher in OSP and showed significant increase across age groups (Haskell *etal.*,2004; Low*etal.*,2007; Van Jaarsveld *et al.*,2006). High provitamin A OSP varieties have been released in Mozambique and Uganda in 2002 and 2007, respectively. Sweet potato for Profit and Health Initiative (SPHI) was launched by CIP to deliver OSP to African countries so that it can be available to about 10 million households by 2020 to eradicate vitamin A malnutrition.

Maize

CIMMYT and International Institute of Tropical Agriculture (IITA) in collaboration with NARES directed provitamin A maize breeding program for the benefit of Southern Africa. Exploration of germplasm studied discovered temperate maize with the target level (15 ppm) of provitamin A which was used to introgress into tropical cultivar. Maize varieties that can provide 25% of the Estimated Average Requirement (EAR) were released in Zambia and Nigeria (Dhliwayo*et al.*, 2014).

Cassava

Provitamin A cassava is being developed for Nigeria and the Democratic Republic of the Congo (DRC). Conventional breeding strategy was used to identify the cassava line in South America. Further, it was improved for cassava mosaic disease and abiotic stresses by IITA and the International Center for Tropical

Agriculture (CIAT) to be suitable for Africa. Three cassava varieties, TMS 01/1371, TMS 01/1412, and TMS 01/1368 with sufficient provitamin A (25% of the EAR) for preschool children and women were released for Nigeria in 2011. Similarly, variety (I011661) with same level of provitamin A was released in DRC in 2008 (Kulakow and Peter, 2014).

Sorghum

Genetically modified sorghum for increased level of provitamin A (up to 21 ppm), reduced phytate (35–80%), and an improved protein profile through transgenic breeding was led by Africa HarvestPlus and Pioneer international. Transgenic lines are currently under evaluation and expected to be released for commercial cultivation in 2018 (Saltzman *et al.*, 2013).

Banana

Bioversity International has continued to work on vitamin A banana/plantain. About 400 germplasm accessions from different genome sub group were screened through conventional breeding for provitamin A carotenoids (pVACs) in fruit of nutritionally rich Musa cultivars to identify the genes involved in biogenesis of carotenoid pathways. Banana/ plantain varieties with 20 ppm of provitamin A have been released in Nigeria, Ivory Coast, Cameroon, Burundi, and DRC by IITA and Bioversity International (Ekesa, 2014).

National Agricultural Research Organization of Uganda and Queensland University of Technology are developing transgenic provitamin A banana for Uganda with up to 20 ppm provitamin A. Initial trials were positive and released for commercial cultivation in 2021 (https://www.sciencedaily.com/releases/2017/07/170707095806.htm).

Limitation in Crop Biofortification

Yield is the Primary Concern

Increased yields and better adaptation to various biotic and abiotic stresses are among the priority area of crop improvement. Yield potential and stability unquestionably deserve high priority however enough attention needs to be paid to improve the nutritional traits of the plant produce for maximum utilization or end use.

Breeding for Nutritional Traits Should Safeguard the Economic Interests of Consumer and Farmers

The producer, the retailer, the miller, the baker and the consumer interpret quality in their own way. Within market class, the breeder must consider development of cultivars with quality characteristics of the grain or other economic parts of the plant that will produce a superior product, which should acceptable to the consumers and without compromising the yield or economic benefit of the farmers.

Sometimes there is no Premium Price for Improved Products with Nutritional Traits

Most of the high lysine or high protein maize lines show poor grain filling which ultimately resulted in lower yield potential. For every 1% increase in protein content approximately 11% more nitrogen is needed. Therefore, high lysine or high protein cultivars of maize are not likely to be popular among farmers unless quality is taken into consideration for pricing of the produce.

Lack of High through Put Efficient Screening Methods or Techniques

Many biochemical procedures for quality analysis are tedious and less throughput and time consuming. Screening for large number of germplasm line or mutants we need rapid and reliable method of estimation to speed up the screening process. Some of the cost effective and rapid method developed for screening the germplasm lines are Cut Grain Dip (CGD) method of amylose in rice (Agasimani *et al.*, 2013; Raja *et al.*, 2017), FAD method of cyanide estimation in sorghum (Reddy *et al.*, 2016), Rapid screening method for grain Iron and zinc content in pearl millet (Velu *et al.*, 2006; Velu *et al.*, 2008).

Negative Correlation of Grain Yield and Protein Content

This is common observation in plant breeding, especially in cereals, that as the grain protein increases, the yield in plants decreases. The analysis of wheat evolution reveals a change in grain biochemical composition. Increases in wheat yield have led to a reduction in the starch and protein content (Triboi *et al.*, 2006). This fact clearly observed in maize population carrying *opaque*-2 and *floury*-2 genes yielding 8-10 % less than normal maize lines (Prasanna *et al.*,2001). Similar fact was also observed in sorghum (Liang *et al.*, 1969). Therefore, is a failure to develop genotype having both high yield and protein.

Quality Traits are Highly Influenced by Environment

Most of the nutritional traits are highly influenced by growing environmental condition. Studies by Kumar *et al.* (2010) showed significant effect of genotype × environment on sucrose, raffinose and stachyose concentration in seven soybean genotypes. Tahir *et al.* (2011) reported significant effect of genotype, environment and their interaction on sucrose and raffinose family oligosaccharides (RFOs) content in lentil seeds. Recently Gangola *et al.* (2013) showed significant effect of genotype and growing environment on concentration of myo-inositol, galactinol, glucose, fructose, sucrose, raffinose, stachyose, verbascose and total RFO in both desi and kabuligenotypes.

Prospects

The quantum jump in the production and productivity has reached in all the crops. Now, we are in the era of functional or quality foods. The need of the hour is the improvement of the nutritional or grain quality traits. High throughput phenotyping coupled with molecular genetics offers new frontiers for improving nutritional traits through biofortification. The conventional plant breeding for quality improvement would continue to play its role, recent improvement in molecular biology, next generation sequencing (NGS), and recombinant DNA technology offers greater potential for rapid progress in improving nutritional traits of major staple crops to reduce the global burden of malnutrition.

References

Acquaah G. Principles of Plant Genetics and Breeding. Wiley-Blackwell Publication. ISBN: 978-1-4051-3646-4; 2012. http://onlinelibrary.wiley.com/book/10.1002/9781118313718.

Agasimani, S., Selvakumar, G., Joel, A.J. and Ganesh Ram, S. 2013. A simple and rapid single kernel screening method to estimate amylose content in rice grains.*Phytochem Anal*, 24(6): 569-573.

Ashok Kumar, A., Reddy, B.V., Ramaiah, B., Sahrawat, K.L. and Pfeiffer, W.H. 2012. Genetic variability and character association for grain iron and zinc contents in sorghum germplasm accessions and commercial cultivars. *Eur J Plant Sci Biotechnol*, 6(1): 1-5.

Boonyaves, K., Wu, T.Y.,Gruissem, W. and Bhullar, N.K. 2017. Enhanced Grain Iron Levels in Rice Expressing an iron-regulated metal transporter, nicotianamine synthase, and ferritin Gene Cassette. *Front Plant Sci*, 8.

Bouis, H.E., Hotz, C., McClafferty, B., Meenakshi, J.V. and Pfeiffer, W.H. 2011. Biofortification: a new tool to reduce micronutrient malnutrition. *Food Nutr Bull*,32: S31-S40.

Brown, K.H., Wuehler, S.E. and Peerson, J.M. 2001. The importance of zinc in human nutrition and estimation of the global prevalence of zinc deficiency. *Food Nutr Bull*, 22(2): 113-125.

Burkhardt, P.K., Beyer, P., Wünn, J., Klöti, A., Armstrong, G.A., Schledz, M., Lintig, J. and Potrykus, I. 1997. Transgenic rice (*Oryza sativa*) endosperm expressing daffodil (*Narcissus pseudonarcissus*) phytoene synthase accumulates phytoene, a key intermediate of provitamin A biosynthesis. *Plant J*, 11(5): 1071-1078.

Cakmak, I. 2008. Enrichment of cereal grains with zinc: agronomic or genetic biofortification?*Plant Soil*, 302(1-2): 1-17.

Chen, L., Yang, F., Xu, J., Hu, Y., Hu, Q., Zhang, Y. and Pan, G. 2002. Determination of selenium concentration of rice in China and effect of fertilization of selenite and selenate on selenium content of rice. *J Agric Food Chem*, 50(18): 5128-5130.

Cherian, B. 2014. Delivery of iron pearl millet in India. Biofortification Progress Brief 30. Washington, D.C.: International Food Policy Research Institute (IFPRI).

De Benoist, B., McLean, E., Andersson, M. and Rogers, L. 2008. Iodine deficiency in 2007: global progress since 2003. *Food Nutr Bull*, 29(3): 195-202.

Dhliwayo, Thanda; Palacios, Natalia; Babu, Raman; San Vincente, Felix; Pixley, Kevin; Menkir, Abebe; Maziya-Dixon, Bussie; Alamu, Oladeji and Rocheford, Torbert. 2014. Vitamin A maize. Biofortification Progress Brief 5. Washington, D.C.: International Food Policy Research Institute (IFPRI).

Ekesa, B. 2014. Vitamin A banana/plantain. Biofortification Progress Brief 8. Washington, D.C.: International Food Policy Research Institute (IFPRI).

Fageria, N.K., Moraes, M.F., Ferreira, E.P.B. and Knupp, A.M. 2012. Biofortification of trace elements in food crops for human health. *Commun Soil Sci Plant Anal*, 43(3): 556-570.

FAO, 1996. Rome declaration on World Food Security and World Food Summit plan of action. World FoodSummit 13–17 Nov 1996. FAO, Rome.

FAO, I., 2014. WFP. The State of Food Insecurity in the World 2014 strengthening the enabling environment for food security and nutrition. FAO, Rome.

FAO, IFAD, WFP. (2014) The state of food insecurity in the world 2014: strengthening the enabling environment for food security and nutrition. Food and Agricultural Organization of the United Nations, Rome.

Ganeshan, S. and Chibbar, R.N. 2010. Gene Transfer Methods. In Transgenic Crop Plants (pp. 57-83). Springer Berlin Heidelberg.

Gangola, M.P., Khedikar, Y.P., Gaur, P.M., Båga, M. and Chibbar, R.N. 2013. Genotype and growing environment interaction shows a positive correlation between substrates of raffinose family oligosaccharides (RFO) biosynthesis and their accumulation in chickpea (*Cicer arietinum* L.) seeds. *J Agric Food Chem*, 61(20): 4943-4952.

Garcia-Casal, M.N., Peña-Rosas, J.P., Pachón, H., De-Regil, L.M., Centeno Tablante, E. and Flores-Urrutia, M.C. 2016. Staple crops biofortified with increased micronutrient content: effects on vitamin and mineral status, as well as health and cognitive function in the general population. The Cochrane Library.

Godfray, H.C.J., Beddington, J.R., Crute, I.R., Haddad, L., Lawrence, D., Muir, J.F., Pretty, J., Robinson, S., Thomas, S.M. and Toulmin, C. 2010. Food security: the challenge of feeding 9 billion people. *Science*, 327(5967): 812-818.

Graham, R., Senadhira, D., Beebe, S., Iglesias, C. and Monasterio, I. 1999. Breeding for micronutrient density in edible portions of staple food crops: conventional approaches. *Field Crops Res*, 60(1): 57-80.

Graham, R.D. and Rosser, J.M. 2000. Carotenoids in staple foods: their potential to improve human nutrition. *Food Nutr Bull*, 21(4): 404-409.

Gregorio, G.B., Senadhira, D., Htut, H. and Graham, R.D. 2000. Breeding for trace mineral density in rice. *Food Nutr Bull*, 21(4): 382-386.

Grusak, M.A. and Cakmak, I. 2009. 12 Methods to improve the crop-delivery of minerals to humans and livestock. *Plant Nutritional Genomics*, 265.

Haskell, M.J., Jamil, K.M., Hassan, F., Peerson, J.M., Hossain, M.I., Fuchs, G.J. and Brown, K.H. 2004. Daily consumption of Indian spinach (*Basella alba*) or sweet potatoes has a positive effect on total-body vitamin A stores in Bangladeshi men. *Am J ClinNutr*, 80(3): 705-714.

Hotz, C. and McClafferty, B. 2007. From harvest to health: challenges for developing biofortified staple foods and determining their impact on micronutrient status. *Food Nutr Bull*, 28(2): 271-279.

Jones, G., Steketee, R.W., Black, R.E., Bhutta, Z.A., Morris, S.S. and Bellagio Child Survival Study Group. 2003. How many child deaths can we prevent this year?.*The Lancet*, 362(9377): 65-71.

Khush, G.S., Lee, S., Cho, J.I. and Jeon, J.S. 2012. Biofortification of crops for reducing malnutrition. *Plant Biotechnol Rep*, 6(3): 195-202.

Kulakow, Peter. 2014. Vitamin A cassava. Biofortification Progress Brief 6. Washington, D.C.: International Food Policy Research Institute (IFPRI).

Kumar, V., Rani, A., Goyal, L., Dixit, A.K., Manjaya, J.G., Dev, J. and Swamy, M. 2010. Sucrose and raffinose family oligosaccharides (RFOs) in soybean seeds as influenced by genotype and growing location. *J Agric Food Chem*, 58(8): 5081-5085.

Liang, G. H. L., Overley, C. B and Casal, A. J. 1969. Inter-relationships among agronomic characters in grain sorg-hum. *Crop Sci*, 9:299-302

Low, J.W., Arimond, M., Osman, N., Cunguara, B., Zano, F. and Tschirley, D. 2007. A food-based approach introducing orange-fleshed sweet potatoes increased vitamin A intake and serum retinol concentrations in young children in rural Mozambique. *J Nutr*, 137(5): 1320-1327.

Lyons, G., Ortiz-Monasterio, I., Stangoulis, J. and Graham, R. 2005. Selenium concentration in wheat grain: Is there sufficient genotypic variation to use in breeding?.*Plant Soil*, 269(1-2): 369-380.

Meenakshi, J.V., Johnson, N.L., Manyong, V.M., DeGroote, H., Javelosa, J., Yanggen, D.R., Naher, F., Gonzalez, C., García, J. and Meng, E. 2010. How cost-effective is biofortification in combating micronutrient malnutrition? An ex ante assessment. *World Dev*, 38(1): 64-75.

Menkir, A. 2008. Genetic variation for grain mineral content in tropical-adapted maize inbred lines. *Food Chem*, 110(2): 454-464.

Nachimuthu, V.V., Robin, S., Sudhakar, D., Rajeswari, S., Raveendran, M., Subramanian, K.S., Tannidi, S. and Pandian, B.A. 2014. Genotypic variation for micronutrient content in traditional and improved rice lines and its role in biofortification programme. *Indian J Sci Technol*, 7(9): 1414-1425.

Paine, J.A., Shipton, C.A., Chaggar, S., Howells, R.M., Kennedy, M.J., Vernon, G., Wright, S.Y., Hinchliffe, E., Adams, J.L., Silverstone, A.L. and Drake, R. 2005. Improving the nutritional value of Golden Rice through increased pro-vitamin A content. *Nat Biotechnol,* 23(4): 482-487.

Potrykus, I. 2001. Golden rice and beyond. *Plant Physiol,* 125(3): 1157-1161.

Prasanna, B.M., Vasal, S.K., Kassahun, B. and Singh, N.N. 2001. Quality protein maize. *Curr Sci,*1308-1319.

Prasanna, R., Nain, L., Rana, A. and Shivay, Y.S. 2016. Biofortification with microorganisms: present status and future challenges. In Biofortification of Food Crops (pp. 249-262). Springer India.

Raja, R.B., Agasimani, S., Anusheela, V., Thiruvengadam, V., Chibbar, R.N. and Ram, S.G. 2016. TILLING and EcoTILLING for Discovery of Induced and Natural Variations in Sorghum Genome. In The Sorghum Genome (pp. 257-267). Springer International Publishing.

Raja, R.B., Anusheela, V., Agasimani, S., Jaiswal, S., Thiruvengadam, V., Chibbar, R.N. and Ram, S.G. 2017. Validation and Applicability of Single Kernel-Based Cut Grain Dip Method for Amylose Determination in Rice. *Food Anal Method,* 10(2): 442-448.

Reddy, R.H., Karthikeyan, B.J., Agasimani, S., Raja, R.B., Thiruvengadam, V. and Ram, S.G. 2016. Rapid screening assay for precise and reliable estimation of cyanide content in sorghum. *Aust J Crop Sci,* 10(10): 1388.

Saltzman, A., Birol, E., Bouis, H.E., Boy, E., De Moura, F.F., Islam, Y. and Pfeiffer, W.H. 2013. Biofortification: progress toward a more nourishing future.*Glob Food Sec,* 2(1): 9-17.

Sarker, A. 2014. Iron and zinc lentils. Biofortification Progress Brief 9. Washington, D.C.: International Food Policy Research Institute (IFPRI).

Sharma, P., Dwivedi, S. and Singh, D. 2016. Global Poverty, Hunger, and Malnutrition: A Situational Analysis. In Biofortification of Food Crops (pp. 19-30). Springer India.

Singh, B.B. 2014. Iron cowpea. Biofortification Progress Brief 11. Washington, D.C.: International Food Policy Research Institute (IFPRI).

Singh, U., Praharaj, C.S., Chaturvedi, S.K. and Bohra, A. 2016. Biofortification: Introduction, Approaches, Limitations, and Challenges. In Biofortification of Food Crops (pp. 3-18). Springer India.

Singh, U., Praharaj, C.S., Singh, S.S. and Singh, N.P. eds. 2016a. Biofortification of Food Crops. Springer.

Tahir, M., Vandenberg, A. and Chibbar, R.N. 2011. Influence of environment on seed soluble carbohydrates in selected lentil cultivars. *J Food Compos Anal,* 24(4): 596-602.

Triboi, E., Martre, P., Girousse, C., Ravel, C. and Triboi-Blondel, A.M. 2006. Unravelling environmental and genetic relationships between grain yield and nitrogen concentration for wheat. *Eur J Agron*, 25(2):108-118.

Van Jaarsveld, P.J., Harmse, E., Nestel, P. and Rodriguez-Amaya, D.B. 2006. Retention of β-carotene in boiled, mashed orange-fleshed sweet potato. *J Food Compos Anal*, 19(4):321-329.

Vasconcelos, M., Datta, K., Oliva, N., Khalekuzzaman, M., Torrizo, L., Krishnan, S., Oliveira, M., Goto, F. and Datta, S.K. 2003. Enhanced iron and zinc accumulation in transgenic rice with the ferritin gene. *Plant Sci*, 164(3): 371-378.

Velu, G., Bhattacharjee, R., Rai, K.N.,Sahrawat, K.L. and Longvah, T. 2008. A simple and rapid screening method for grain zinc content in pearl millet. *J of SAT Agric Res*, 6:1-4.

Velu, G., Kulkarni, V.N., Muralidharan, V., Rai, K.N., Longvah, T., Sahrawat, K.L. and Raveendran, T.S. 2006. A rapid screening method for grain iron content in pearl millet. *Int Sorghum Millets Newsletter*, 47:158-161.

Velu, G., Singh, R.P., Huerta-Espino, J., Peña, R.J., Arun, B., Mahendru-Singh, A., Mujahid, M.Y., Sohu, V.S., Mavi, G.S., Crossa, J. and Alvarado, G. 2012. Performance of biofortified spring wheat genotypes in target environments for grain zinc and iron concentrations. *Field Crops Res*, 137: 261-267.

Wessells, K.R. and Brown, K.H. 2012. Estimating the global prevalence of zinc deficiency: results based on zinc availability in national food supplies and the prevalence of stunting. *PloS one*, 7(11):p.e50568.

WHO, 2015 Micronutrient deficiencies: iron deficiency anaemia. World Health Organization, Geneva. http://www.who.int/nutrition/topics/ida/en/. Accessed1,July 2017.

WHO 2015a Micronutrient deficiencies: vitamin A deficiency. World Health Organization, Geneva.http://www.who.int/nutrition/topics/vad/en/. Accessed 1 July 2015.

WHO/FAO (1998) Vitamin and mineral requirements inhuman nutrition: report of a joint FAO/WHO expert consultation, 2nd edn. Bangkok. (http://www.who.int/iris/handle/10665/42716)

Xu, Y. 2010. Molecular plant breeding. Cabi.

Ye, X., Al-Babili, S., Klöti, A., Zhang, J., Lucca, P., Beyer, P. and Potrykus, I. 2000. Engineering the provitamin A (β-carotene) biosynthetic pathway into (carotenoid-free) rice endosperm. *Science*, 287(5451): 303-305.

Zimmermann, M.B., Wegmueller, R., Zeder, C., Chaouki, N., Biebinger, R., Hurrell, R.F. and Windhab, E. 2004. Triple fortification of salt with microcapsules of iodine, iron, and vitamin A. *Am J ClinNutr*, 80(5):1283-1290.

https://www.sciencedaily.com/releases/2017/07/170707095806.htm

www.harvestplus-china.org

4

Peri-Urban and Organic Agriculture: Thrust areas of Farming

Jyoti Yadav[1*], Swati Sindhu[2] and Jayant Yadav[3]

[1]Department of Zoology, CCS Haryana Agricultural University, Hisar - 125004, India
[2]Department of Microbiology, CCS Haryana Agricultural University, Hisar - 125004, India
[3]Department of Entomology, CCS Haryana Agricultural University, Hisar - 125004, India
*Corresponding author: yadavjyoti694@gmail.com

Abstract

Peri urban agriculture has emerged in a supporting role for mitigating the nutritional needs especially in urban areas. By ensuring the multiple use of land, it ensures the availability of perishable products like meat, dairy, mushroom, etc. However, the problems of waste management, zoonosis and pollution increased with the increased peri urban agricultural activities. Organic agriculture on other hand ensures the agricultural production by minimizing the environmental degradation. Composting, biogas production, vermicomposting offers a solution for the waste generated out of animal husbandry or agricultural activities. While for pest management, the use of biopesticides and botanicals are encouraged. The use of microbes for enhancing soil fertility instead of using chemical fertilizers can not only prevented the environmental pollution but also strengthens the biodiversity of that particular area.

Keywords: Peri Urban, Organic, Vermi-composting, Biocontrol, Biopesticides, Biofertiliser

Introduction

The agricultural activities taking place in urban periphery is referred to as peri-urban agriculture. Due to population influx from both urban and rural side leads to the hiking of land prices, leading to the emergence of multiple land use over a period of time. Thus, such agricultural systems are primarily sustained by more intensive production in smaller scale. More perishable crops and animal production (meat and dairy products) form the major concerns of far enterprises of peri urban spaces. But, the factors like size, capital input, available technologies, market orientation, and water availability particularly determine the peri urban agricultural activities (Danso, 2002). Globally, 15-20% of the total consumed food is produced by urban and peri urban agriculture (Corbould, 2013).According to FAO (2006), about 11 million residents are involved in peri urban agriculture that includes intensive production of high value products like milk and vegetables that contributes to the substantial food security to urban areas nearby. Agricultural productive systems can be specialised (involves the production of single crop or animal), mixed (having two main crops or animals) or hybrid (combines two activities like crop production and animal husbandry) (Vagneron *et al.*, 2002). However, lack of proper waste management can lead to the problems of pollution. Landless farmers usually dump the manure as a waste and this unattended manure is washed away to water bodies, thereafter resulting in health hazards, environmental degradation, including eutrophication. Other wastes are normally dumped either in controlled or uncontrolled city landfills and dumpsites (Komakech *et al.*, 2014). Such dumpsites effects the environment negatively, as the greenhouse gases such as methane are produced which are almost 25 times stronger as compared to carbon dioxide (IPCC, 2007). Organic agriculture presents viable solution for reducing and managing the waste. The techniques like vermicompost and compost help in turning the organic waste into nutrient rich manure while the use of biofertilisers, biopesticides and biocontrol agents protect the environmental stability. Integrated form of agriculture is also successful in mitigating waste reducing needs. The recycled organic waste like crop residues can directly be used as animal feed. As the demand for food increases, the agricultural production needs to be intensified; ultimately increasing the demand of fertilizers. If the methods of vermicomposting and composting are adopted by the farmers locally, sustainable development can be ensured. Alternative value chains, comprising the production of animal feed by earthworms and larva result in production of organic fertilisers as well as protein rich animal feed. Thus, organic agriculture not only decreases environmental and health hazards but also provides high value products such as biogas, compost, feed and fertilizers. A comprehensive review of peri urban and organic agriculture will provide an evolved path to precisely attain sustainable development.

Potentials of Peri Urban Agriculture

a) ***Ensuring urban food security:*** Reduction in the cost of supplying and distribution of food can be ensured by promoting agriculture in peri urban spaces. Alongwith, the benefit of nutritional enhancement, timely and even distribution of food are also ensured (Nugent, 2001). The various activities involved in peri urban agriculture are enlisted below.

Urban horticulture: Characterised by high turnover, high resource utilisation, high and good quality yield and production of more than one crop in one season is ideal for prevailing in peri urban areas (Gallaher *et al.*, 2013). Horticultural cropping systems in peri urban areas may also be linked to aquaculture by using aquaponics to ensure the production of animal protein too (Asp and Alsanius, 2014).

Aquaculture: As stated by Bostock *et al.* (2010), aquaculture supplies a total of 50% of fish for human consumption globally. In peri urban areas, aquaculture is done in waste water that contains domestic waste having night soil (used to fertilise the pond after preliminary treatment). This process involves decomposition processes in which bacteria, phytoplankton or zooplankton and detritivores find their place in micro aquatic food web (Bunting, 2004). Waste water use for aquaculture offers viable option in water scarce areas (Tenkorang *et al.*, 2012). In addition, instead of discharging nutrients in environment, they are recycled in aquaculture, thereby reducing the incidences of eutrophication (Liu *et al.*, 2010). However, fish culture in contaminated water may contain pathogens and harmful chemicals (AsemHiablie *et al.*, 2013). However, the guidelines for use and wastewater treatment for reuse in aquaculture as well as agriculture has been stated by WHO (2006) and The Hyderabad Declaration, (2002).

Animal husbandry: Animal husbandry especially for dairy and meat products has great potential to contribute towards food security and family income in peri urban and urban areas. Cattle keeping in cities id more profitable and market oriented as compared to rural cattle farming (Ayenew *et al.*, 2011). Poultry is also an alternative with low input of feed ensuring high output of egg and meat (Berg *et al.*, 2014). Pig farming for pork has been highly recommended in tropics to eliminate the animal protein deficiency. People in peri urban areas are generally involved with pig farming, marketing live pigs, slaughtering and selling pork. Limitations of animal husbandry include land shortage for cattle grazing, food, manure management and market access. Endemic animal diseases along with zoonotic diseases affect animals as well as humans especially where the animal density is higher are also a major concern (Kagira *et al.*, 2010).

b) Economic Development in Peri Urban Areas

Peri urban agriculture adds to the family income from the sales of surplus and by decreased household expenditures as the food is grown by their own. In addition, the processing, manufacturing and sales of the agricultural inputs like compost, earthworms, bio-fertilizers and transportation services also enhances the growth of micro enterprises (Bon *et al.*, 2010).

c) Ensuring Environmental Sustainability

By reducing the transportation inputs and making available the fresh food for urban areas peri urban agriculture aids in reducing the ecological footprint of city. In addition, the low input techniques such as compost, vermicompost and vermiculture help in waste management generated from urban households (Konijnendijk, 2004 and Yadav *et al.*, 2017). Peri urban spaces when used for agriculture pave pathway for greener environment, thereby supports biodiversity and improve the micro environment (dust reduction and shade) locally.

Disadvantages of Peri Urban Agriculture

- Lack of irrigation facilities may lead to the use of water from polluted streams of untreated sewage water and can lead to the contamination of crops with pathogenic organisms.
- Tick borne diseases and mosquito breeding are the major resultants of peri urban agricultural activities.
- Contamination due to pesticides' residue and excessive use of fertilizers of only contaminates the environment but also spoils the health of animals nearby.
- Peri urban areas are exposed to heavy metal contamination, which may contaminate the crops and biomagnifies at different trophic levels in a food chain.
- The disease transfer from the domestic animals to people (also referred as zoonosis) is a primary concern of peri urban agriculture.
- Nutrient rich manure and fertilizers when added to land in higher amounts, runs away and contaminate water bodies and may result in eutrophication.

Table 1: Comparison Between Rural and Peri Urban Agricultural Set up

	Rural agriculture	Urban and peri-agriculture (UPA)
Farm types	Conventional; farms consists of interdependent subunits	Unconventional; more specialized and independent subunits
Livelihood	Agriculture is full time job and is the source of primary income	Farming is usually secondary livelihood; farmers often work on a part-time basis only
Farmer type	Usually 'born farmers'; Strong traditional knowledge	Normally they are beginners and have weak traditional knowledge
Products	Mainly staple crops and animal husbandry	Perishable products, such as green vegetables, dairy products, mushrooms, ornamental plants, fish etc.
Crop season	Seasonal periods	Year-round growing of crops (irrigated)

	Rural agriculture	**Urban and peri-agriculture (UPA)**
Production factors	Low land price; lower costs of labor; high costs of commercial inputs; variable cost of water	High land price, land scarcity; higher costs of labor; lower costs of commercial inputs; high cost of clean water; availability of low-cost organic wastes and wastewater
Environmental Context	Relatively stable; land and water resources rarely polluted	Fragile; land and water resources often more polluted
Research and extension services	More likely (but declining)	Hardly available, but individuals may gain direct access to libraries, research organizations, market information, etc.
Market	Distant markets; marketing through chain; low degree of local processing	Closeness to markets; direct marketing to customers possible; higher degree of local processing (including street foods)
Land security	Relatively high	Insecure; often informal use of public land; competitive land uses

Source: De Zeeuw, 2004.

Composting: Small scale composting is an efficient low input management strategy marked with low levels of greenhouse gases emissions (Ermolaev *et al.*, 2014). FAO model "Save and Grow" for sustainable mode of crop production also recommends the use of compost for improvising soil quality (Hoornweg and Munro, 2008). Composting may be defined as the process of biological decomposition where organic substrates are stabilised in biologically produced thermophilic conditions, with a final stable product ideal for storage and land application. Organic residues like kitchen waste, garbage, molasses and agricultural wastes can be used as substrate for composting. Composting process involves four phases as described below:

a) First phase involves the proliferation of diverse mesophilic bacteria and fungus and degradation process by microbes if followed by the increase in temperature to about 45°C. As the temperature rises the vegetative cells and hyphae will lyse and die but heat resistant spores will survive only.

b) During second phase, as the temperature reaches to 70-80°C the population of thermophilic bacteria, actinomycetes and fungus population increases.

c) Third phase often referred as stationary phase marks no significant temperature change due to the fact that microbial heat production and heat dissipation balance out each other and hence no marked difference in microbial community is observed.

d) Gradual decline in temperature marks the start of maturation phase of composting. This phase is characterised by the increase in mesophilic microbial population which ultimately increase the rate of decomposition process.

Vermicomposting: On one hand, vermicomposting offers eco-friendly, self-sustainable and low input technique to combat the issues of managing waste that has been generated as the by-products of peri urban agriculture by using earthworms while on other hand, it promotes organic agriculture. Earthworms have the capacity to degrade the organic waste into nutrient rich valuable vermicast that has the potentials as plant fertilizers. About 3,627 earthworm species of terrestrial earthworms has been recognised throughout the world (Samaranayke *et al.*, 2010). *Eisenia fetida, Lumbricus rubellus, Eudrillus euginea, Perionyx excavatus* and *Eisenia andrei*are among the most cultured species for vermicomposting (Yadav and Gupta, 2017). Earthworms ingest and digest the organic waste and then excrete the vermicast. The feeding activity of earthworm result in improvising the soil physical (soil porosity, soil aggregates) and chemical characteristics (decreased carbon, increased N, P and K).The microbes present in worms gut offers more hostile environment for rapid degradation of organic waste. Worm population (400-500g/sq. m) in 1m× 1m × 0.3 m bin can recover upto 50 Kg of vermicompost in 6-8 weeks.Chauhan *et al.* (2010) has documented *Eisenia fetida* as the most efficient species for vermicomposting as compared to *Eudrilus eugeniae* and *Perionyx excavatus*.

Methodology of Vermicomposting:

Organic matter comprising kitchen waste, garden waste (Wani *et al.*, 2013), wheat straw, rice straw etc. alongwith cowdung forms the suitable substrate for vermicomposting (Yadav *et al.*, 2017). The raw organic waste is then pre-digested for 15 days to avoid the incidences of increased temperature due to microbial digestion. The earthworms are then released into pre- digested waste to start the process of vermicomposting. Moisture (60%- 70%), proper aeration, temperature and pH (7-8) would be maintained to facilitate the growth of worms. High levels of oily substances and citrus must be avoided. The vermibeds should either be made in shadow or should be covered with gunny bags.

Table 2: Comparison between the nutritional components of compost and vermicompost

Parameter	Compost	Vermicompost
pH	8.66	8.41
Conductivity	2.89 Mmho	3.21 Mmho
TDS	1932 mg/l	2260 mg/l
Sodium	85 mg/l	92 mg/l
Potassium	1520 mg/l	155 mg/l
Chloride	14 mg	15 mg
Nitrate	1.506	1.494
Calcium	11 mg	15 mg
Magnesium	6 mg	10 mg

Source: Khan and Ishaq, 2011

Biofertilisers

Biofertilisers combat the problems pertaining to pollution as an alternative to chemical and conventional fertilizers. Various studies conducted worldwide on biofertilizers showed that rhizobia nodulate legumes like beans, soybean, chickpea, pigeon pea can fix 50-500 kg atmospheric N/hectare under favourable environmental conditions (Mazid and Khan, 2014). Biofertilisers can be defined as the living carrier based microbial inoculants which applied to seed, soil or plants which enhance plant growth and productivity. It is estimated that by 2020 to nutrient requirement of total population will be 28.8 million tonnes, while only 21.6 million tones will be available with a loss of about 7.2 million tones (Mishra D.J. *et al.*, 2013). Biofertilisers not only enhance nutrient uptake by their mobilization from organic and chemical sources but also improve soil structural frame, enhance growth of beneficial microflora, suppression of soil borne pathogens, increase formation of soil aggregates which helps to protect from soil erosion. Biofertilisers'production has some limitations like selection of appropriate and efficient strains of bacteria, suitable carrier and viability of microbial cultures during transportation and storage over a period of time (Mishra and Dash, 2014). Their composition is highly fluctuated. Lab grown cultures having beneficial plant growth promoting properties are not always able to grow well in field conditions due to harsh environmental conditions. Due to the involvement of living microorganisms, strict safety regulations must be followed to avoid biohazard, hence making marketing process difficult.

Microbial biofertilizers are mainly of three types:

1. Nitrogen Fixing Bacteria
2. Phosphate Solubilizers
3. Zinc Solubilizers

Nitrogen Fixing Bacteria

Rhizobium fixes nitrogen in symbiotically with legumes like chickpea, mungbean, black gram etc. Only a compatible strain forms nodule on a plant, these relations are species specific like a particular strain of rhizobia form symbiotic association only with specific legume plants not with all plants according to their cross-inoculation groups. *Rhizobium melilotti* for leucerne, *Rhizobium phaseoli* for green gram, black gram; *Rhizobium japonicum* for soyabean; *Rhizobium leguminoserum* for pea, lentil; *Rhizobium lupine* for chickpea and *Rhizobium trifoli* for berseem. Rhizobium can also fix nitrogen in nonlegume parasponia. Rhizobial population is directly proportional to the legumes in the soil. Restoration of their population can be done with artificial inoculation. *Azospirillum* fixes about 20-40 kg/ha and produce plant growth promoting substances. *A.lipoferum and A.brasilense*are dominant for use as a biofertiliser. *Azospirillum* form symbiotic association with many plants like maize, sugarcane, sorghum, pearl millet, so recommended as a biofertiliser for these crops. *Azotobacter chroococcum*is the most commonly used species in arable soils. The number of *Azotobacter* is low in rhizospheric and uncultivated soils due to lack of organic matter and presence of antagonistic microorganisms in soil. This bacterium also produces anti-fungal antibiotics which are inhibitory to several pathogenic fungi.

Zinc Solubilizing Bacteria

Zn has many roles in biological systems, integrity of biological membranes, biomass production, chlorophyll formation, nodulation, carbohydrate synthesis, enhanced stress tolerance and reproductive processes(Gandhi *et al.*, 2014). Zinc is required for the formation of phytohormones like auxins, cytokinins which are critically required for plant growth (Hussain *et al.*, 2015).The zinc solubilizing bacteria can be isolated from soil samples of rhizospheric crop by enrichment culture technique.The zinc can be solubilized by many microbes like *Bacillus subtilis and Thiobacillus thioxidans* which can be used as bio-fertilizers in combination with insoluble cheaper zinc compounds like zinc oxide, zinc carbonate and zinc sulphide.

Phosphate Solubilizing Bacteria

Phosphorus is second important nutrient for plants after nitrogen. Normally in soil phosphorus is found in unavailable forms, bound with other atoms depending on acidic or alkaline conditions or pH which can be available back to plants by some bacteria which can solubilize phosphate through production of organic acids. Strains of *Pseudomonas, Bacillus* are promiscuous phosphate solubilizing bacteria which are better alternative of chemical phosphatic fertilizers.

Plant Growth Promoting Rhizobacteria (PGPR)

PGPRs are those bacteria which colonize rhizosphere rigorously, thereby enhancing the plant growth and productivity either directly or indirectly by producing different plant growth promoting substances. PGPR can enhance yield, nitrogen uptake and root growth. They also act as bicontrol agents by production of siderophores, antibiotics and bacteriocin activities. These bacteria have ability to produce hydrogen cyanide, anti-fungal agents and chitin degrading enzymes which inhibit fungal pathogens to grow. *Pseudomonas fluorescens*is one of the best studied plant growth promoting rhizobacteria, because it can antagonize many pathogens.

Table 3: Various PGPRs Used in Organic Agriculture

Sr. No.	Bacteria	Observations	Test plant	Reference
1.	*Serratia proteamaculus*	Positive effects on root diameter, stem Length, total leaf chlorophyll, yield	Tomato	Moustaine *et al.*, 2017
2.	*Pantoeaag glomerans*	Positive effects on stem length, total leaf chlorophyll	Tomato	Moustaine *et al.*, 2017

Sr. No.	Bacteria	Observations	Test plant	Reference
3.	*Bacillus cereus*	Positive effects on stem length, total leaf chlorophyll	tomato	Moustaine *et al.*, 2017
4.	*Pseudomonas putida* BA-8	Increased shoot length	Apple	karakurt and aslantas, 2010
5.	*Bacillus megatherium*	Inhibitory effects on mycelia growth of *fusarium oxysporum*	Soil	Shobha and kumudini, 2012
6.	*Bacillus pumilus* 8N-4	Increased crop yield	Wheat variety orkhon	Hafeez *et al.*, 2006
7.	*Pseudomonas flourescens* and *Azospirillum lipoferum*	Significant effect on yield	Rice	Khorshidi *et al.*, 2011

Biocontrol agents: Biological control may be defined as the control of one living organism with the help of another living organism which relies upon the action of natural enemies or bio-control agents such as predators, parasitoids and entomopathogens. It may also involve the application of microbes, their genes or products (secondary metabolites) that aids in preventing plant pathogens, thereby promoting plant growth. Predators generally are free living which tend to feed on smaller preys or require several preys to complete their life cycle e.g. Coccinellid beetles. While parasitoids are of same size or smaller as compared to their host and require only one host for its development into a free living adult e.g. *Trichogramma chilonis*. Entomo pathogens are disease causing organisms such as bacteria, viruses, fungi, nematodes and protozoa which either kill their host or malformed the future generations e.g. *Beauveria bassiana*. Bacteria *Bacillus thuringiensis*, fungi (*Trichoderma, Metarhiziuman isopliae, Beauveria bassiana, Beauveria brongniartii, Verticillium lecani, Clerodendron inerme*) and nematodes (*Steinernema spp., Heterorabditis spp.*) are the major entomopathogens which either kill their host or cause malformations in future generations. Predatory mites such as *Typhlodromuspyri, Kampimodromus aberrans, Amblyseius andersoni* and *Phytoseilus persimilis* has found to be successful in controlling pest mites.

Table 4: Successful Cases of Bio-Control In India

Crop	Insect pest	Natural enemy	Type	Imported from
Apple	San Jose scale, *Quadraspidiotus perniciosus*	*Aphytis diaspidis*	Parasitoid	USA
Citrus	Common mealybug, *Planococcus citri*	*Leptomastix dactylopoii*	Parasitoid	Trinidad, West Indies
Apple	San Jose scale, *Quadraspidiotus perniciosus*	*Encarsia perniciosi*	Parasitoid	China

Crop	Insect pest	Natural enemy	Type	Imported from
Apple	Wolly apple aphid, *Eriosoma lanigerum*	*Aphelinus mali*	Parasitoid	England
Castor	Castor semi-looper, *Achaea janata*	*Telenomus remus*	Parasitoid	NEW Guinea
Groundnut	*Oryctes rhinoceros*	*Platymeris laevicollis*	Predator	Zanzibar (Tanzania)
Citrus	Cottony cushion scale, *Icerya purchasi*	*Rodolia cardinalis*	Predator	USA
Papaya, Mango	Spiralling whitefly, *Aleurodicus dispersus*	*Encarsia guadeloupae*	Parasitoid	Within the country
Subabul	Subabulpsyllid, *Heteropsylla cabana*	*Curinus coeruleus*	Predator	Thailand
Fruit crops, Coffee	Mealybugs	*Cryptolaemus montrouzieri*	Predator	Australia
Citrus	*Planococcus citri* and *P. lilacinus*	*Leptomastix dactylopii*	Parasitoid	Trinidad, West Indies
Coconut	Coconut leaf eating caterpillar, *Opisina arenosella*	*Stomatomyia bezziana*	Parasitoid	Sri Lanka
Coffee	Berry borer, *Hypothenemus hampei*	*Phymastichus coffea*	Parasitoid	Columbia
Potato	Potato Tuber moth *Phthorimaea operculella*	*Bracon gelechiae*	Parasitoid	Canada Hawaii Peru
Cotton, Tomato	American bollworm, *Helicoverpa armigera*	*Eucelotoria bryani*	Parasitoid	USA
Cotton	Pink bollworm, *Pectinophora gossypiella*	*Bracon kirkpatricki*	Parasitoid	USA
Tobacco	Tobacco caterpillar, *Spodoptera litura*	*Steinernema carpocapsae*	Parasitoid	Czech Republic
Okra, Beans	Spider mites, *Tetranychus spp.*	*Amblyseius chilenensis*	Predator	USA UK
Vegetables, Ornamentals	Giant African snail, *Achatina fulica*	*Euglandina rosea*	Predator	Bermuda
Sugarcane	Sugarcane pyrilla, *Pyrilla perpusilla*	*Epiricania melanoleuca*	Parasitoid	Within the country
Sugarcane	Sugarcane top borer, *Scirpophaga excerptalis*	*Isotima javensis*	Parasitoid	Within the country
Sugarcane	Sugarcane stalk borer, *Chilo auricilius*	*Cotesia flavipes*	Parasitoid	Indonesia

Crop	Insect pest	Natural enemy	Type	Imported from
Sugarcane	Sugarcane inter node borer, *Chilo saccharipha-gus*indicus	*Diatraeophagas triatalis*	Parasitoid	Indonesia

Source: Prasad, 2014

Trichoderma: A Novel Biocontrol Fungus

This fungus produces various lytic enzymes which play important role in biocontrol by cell wall degradation, hyphal growth and stress tolerance and also have additional benefits like enhanced plant growth, reproduction, rhizosphere modification and defence system. *Trichoderma* based Biocontrol agents (BCAs) are about 60% of all fungal based BCAs and *T. harzianum* as an active agent of commercially available biopesticides (Lorito *et al.*, 2010). Various mechanisms through which *Trichoderma spp.* antagonize a broad range of plant pathogens include niche exclusion through nutrient competition, induced systemic resistance, plant growth promotion, antibiosis and direct mycoparasitism. This immense biocontrol potential responsible for effective biological control applications of different *Trichoderma* strains as an alternative method to chemical agents for the control of plant pathogens.*Trichoderma* strains have been found to be effective in controlling pineapple disease of sugarcane caused by *Ceratocystis paradoxa* (Rahman *et al.*, 2009). *T. virens* produce gliovirin and gliotoxin which are involved in synthesis of antibiosis making it efficient biocontrol agent and effective against *Pythium ultimum*in cotton (Howell, 1998). *T. viride* isolates produce volatile compounds like harzianic acid, alamethicins, tricholin, peptaibols, antibiotics, viridin, gliovirin, glisoprenins, heptelidic acid which prove to be harmful for pathogens (Raaijmakers *et al.*, 2009). Mycoparasitism, which is a negative interaction involving direct attack of one fungal (Biocontrol) species on another fungal (Pathogenic) species, has been found in *Trichoderma. Trichoderma* produces small sized iron chelating compounds called siderophores to mobilize iron from its surrounding environment. *Trichoderma* spp. competes with pythium for iron availability in soil and thus, is responsible for suppression of *Pythium*(Wagunde *et al.*, 2016). It is also worth noticing that *T. Harzianum* strain T-22 is the only microbe reported to induce systemic resistance to pathogens in model plants and also in tomatoes and maize (Yoshioka *et al.*, 2012, Saksirirat *et al.*, 2009). Induced systemic resistance is one of the most important mechanisms of biocontrol effects of *Trichoderma*(Harman, 2006). Many strains of *T. virens, T. asperellum, T. harzianum,* and *T. atroviride* activate some metabolic changes that enhance higher tolerance to many plant-pathogenic microbes including viruses.

Biopesticides: Biopesticides include a variety of microbial esticides, biochemicals derived either from micro-organisms or other natural sources, and processes involving the genetic modification of plants to express genes encoding insecticidal toxins. These mainly include microbial pesticides and plant extracts. They mainly act as antifeedants, inhibit oviposition, reduce fecundity, increase egg sterility or inhibit metamorphosis, thereby reducing the pest survival.

Microbial Control

The utilization of pathogen or micro-organisms for the management of pest populations is known as microbial control and the commercial formulations are known as microbial pesticides. Micro-organisms (pathogens) can be formulated for use as a pesticide for the control of pests. In recent years microbial control has become a significant weapon in the pest control. It has predominantly involved the artificial manipulation of pathogens *viz.*, fungi, bacteria, viruses, nematodes and protozoans, formulated as a spray or dust to suppress or threaten outbreaks of pest.

a) Bacteria: Although more than 100 bacteria have been identified as arthropod pathogens, only *Bacillus thuringiensis*has achieved commercial importance. Two other *Bacilli, B. popilliae* and *B. sphaericus* have been used in more specialized niches and likely to go further commercial development in future. There are about 70 registered *Bt* products with more than 450 formulations. Several commercial products of *Bt* such as Thuricide, Dipel, Trident, Condor and Biobit are marketed worldwide. Formulations based on*Bt* account for 90 per cent of the total biopesticide sales worldwide.

b) Fungi: Of over 800 species of fungi are known to be pathogenic to insects, only seven have been commercialized. The first mycoinsecticide to be registered was *Hirsutella thompsonii*produced by Abbott Labs under the trade name of Mycar. Currently the most widely used fungal insecticide is *B. bassiana* and there are three major strains being marketed in USA, Europe and South America. There are currently two registered fungal insecticides in the genus *Metarhizium* namely *M. anisopliae* and *M. flavoviride*.

c) Viruses: A number of viral pesticides have been registered for commercial use for control of lepidopteran pests of agricultural and horticultural crops. However the most successful use of viruses has been against the temperate forest pests. The first viral insecticide developed was the single nucleocapsidnucleopolyhedrovirus of *Helicoverpazea* (HzSNPV) registered in 1971 and known as Elcar which was effective against all the major *Helicoverpa/Heliothis* species. It provided efficient control on soyabean, sorghum, maize, tomato and chickpea. In India commercial formulations of NPVs and GV are widely available and used by farmers.

d) Nematodes: More than 300 nematode-insect relationships have been described to-date. However, mainly two families of nematodes, *Heterorhabditidae* and *Steinernematidae* have been extensively used to develop commercial formulations. In India, both are marketed by Bio-Sense Crop Protection, Mumbai. *Steinernema sp.* is commercially available for the control of American bollworm, pink bollworm, tobacco caterpillar, white grub, shoot borer, spotted bollworm and leaf eating caterpillar on crops like cotton, groundnut, sugarcane, rice, potato, tomato etc. *Heterorhabditis sp.* is available for the control of white grub, flea beetle, grey weevil and red palm weevil on groundnuts and coconut.

Plant Extracts:

There is no doubt botanical insecticides are an interesting alternative to insect pest control, and on the other hand only a few of the more than 250,000 plant species on our planet have been properly evaluated for this purpose. This means the potential for the future may be huge. In fact, plants like neem (*Azadirachta indica* J., Meliaceae), have shown excellent results and there already are commercial products in the market made from it. But one should not think success is at hand and botanical insecticides will replace all synthetic products. They are only alternatives that may be used in Integrated Pest Management programs and they should be used together with other available control measures. Promising pesticidal plants are: neem, chinaberry, chrysanthemum, tobacco, rotenone, pongram, custard apple, sabadilla, rynia and quassia. Sharma *et al.*, (2015) showed that biopesticides *viz.* Bio Magic (92.67%), Racer (91.90%), Pacer (91.50%), Mealikil (90.84%) were highly effective following Bio Power (87.53%) and Biocide Manic (85.8%) in reducing the population of whitefly over control. A 2% concentration of mineral oil + neem oil and mineral oil + *Pongamiaglabra* seed oil produced 95% and 93.33% mortality at 48 h after treatment under laboratory conditions. Whereas in pot culture, 2% concentrations of mineral oil + neem oil and mineral oil + *Pongamia glabra* seed oil were effective against *B. tabaci* with a mean population reduction of 81.83% and 81.52%, respectively (Chandra Shekhar*et al.*, 2015). Datura proved to be the most effective bringing about significant reduction in the whitefly population followed by neem oil. Garlic and eucalyptus also produced significant results compared to untreated check (Khan *et al.*, 2013). Ali *et al.*(2016) evaluated that Neem extract showed highest reduction percent (82.60%), followed by Tobacco extract (75.95%), Eucalyptus extract (73.93%) and lowest for untreated control (11.07%).

Botanical Pesticides

Neem: Neem, *Azadirachta indica* (Fam. Meliaceae) is indigenous to India and from centuries this tree is known for its medicinal and insecticidal values. Pradhan *et al.* (1962) successfully protected standing crop from locust attack at IARI, New Delhi, by spraying 0.001 per cent neem seed kernel extract. All the parts of Neem tree possess insecticidal activity but seed kernel is the most active part. Neem possesses bactericidal, nematicidal, fungicidal, molluscidial, diueretic and antiarthritic properties. Azadirachtin has systemic effects in certain crop plants, enhancing its efficacy and persistence. The repellent and antifeedant effects of neem have been reported against a wide range of insect pests. Concentrations ranging from 0.001

to 0.4 per cent of various neem seed kernel (NSK) extracts have generally been found to deter feeding of most of the insects. The growth inhibitory effects of neem derivatives result in various developmental defects. These defects reduce feeding and disturb the neuro-endocrine system of insects caused by azadirachtin. Neem products also affect insect vigour, longevity and fecundity. Azadirachtin the most important biologically active component of neem shows phagorepellent and toxic effects at 0.1 to 1000 ppm. It is safe to mammals and a 90 day oral feeding of rats with 10,000 ppm of azadirachtin did not show chronic toxicity.Commercialization of neem based pesticides is expanding at a rapid rate and over 100 commercial neem formulations are available. Formulated neem products are required to contain at least 1500 ppm of a.i. *azadirachtin* in kernel based formulations and 300 ppm in neem oil based ones. Recently neem formulations with 10000 and 50000 ppm of azadirachtin have been introduced.

Table 5: List of Neem-based Pesticides

Sr. No.	Formulations	Trade Name
1.	0.03 % aza	Bioneem, Jai Neem, Jawan, Limonool, Neem oil, Neemark, Neemgaurd, Nimbecidine
2.	0.05 % aza	Jeevan crop, Nim-76, Shaktiman,
3.	0.15% aza	Aza, Biopest, Fortuna Aza, Margocide-CK, Neem Gold, Neem Hit, Neem plus, Neemolin, Neempourn, Rakshak
4.	0.5 % aza	Achook, Neemarin
5.	1.0 % aza	Econeem, Neemarin
6.	5.0 % aza	Econeem plus, NeemAzal-F

Chrysanthemum: Pyrethrum derived from the dried flowers of *Chrysanthemum cinerariaefolium* has been used as an insecticide. Worldwide annual production of pyrethrum averages 30,000 tonnes. Pure pyrethrins are moderately toxic to mammals but technical grade pyrethrum is considerably less toxic. Pyrethrum is a highly effective insecticide against household insects like house flies, mosquitoes, fleas and lice. It is safe to mammals and is easily broken down to non- toxic metabolites. Pyrethrins act quickly on central nervous system of insect and cause a knockdown effect. The use of pyrethrum in agriculture is mainly restricted to some vegetables.

Rotenone Plants: Rotenone occur in the dried roots of tropical legumes, *Derris elliptica* (4-5 %) and *Lonchocarpusnicou* (8-10 %) belonging to family Fabaceae. Rotenone is commonly sold as dust containing 1 to 5 per cent active ingredients for home and garden use but liquid formulations used in agriculture may contain as much as 8 per cent rotenone and 15 per cent total rotenoids. It mainly acts as contact poison and respiratory inhibitor. However, its use has been discontinued due to high fish and mammalian toxicity and rapid degradation.

Conclusion

Organic agriculture in peri urban and rural areas offers solution for nutritional requirements as well as environmental sustainability. It may also be concluded that integrated farming with multiple use of land and resources including waste management offers the best output. However, awareness regarding use of organic mode of agriculture still needs to be created.

References

Ali, S.S., Ahmad, S., Ahmed, S.S., Rizwana, H., Siddiqui, S., Ali, S.S., Rattar, I.A. and Shah, M.A. 2016. Effect of biopesticides against sucking insect pests of brinjal crop under field conditions. *J. Basic and Appl.Sci.,* **12: 41-49.**

Asem-Hiablie, S., Church, C.D., Elliott, H.A., Shappell, N.W., Schoenfuss, H.L., Drechsel, P., Williams, C.F., Knopf, A.L. and Dabie, M.Y. 2013. Serum estrogenicity and biological responses in African catfish raised in wastewater ponds in Ghana. *Sci. Total Environment,* **463-464:** 1182-91.

Asp, H. and Alsanius, B. 2014. Potential of urban horticulture to secure food provisions in urban and peri-urban environments. SLU-Global Report. 33- 34.

Ayenew, Y. A., Wurzinger, M., Tegegne, A. and Zollitsch, W. 2011. Socioeconomic characteristics of urban and peri-urban dairy production systems in the North western Ethiopian highlands. *Trop. Anim. Health Prod.,* **43(6):**1145-1152.

Bapiri, A., Asgharzadeh, A., Mujallali, H., Khavazi, K. and Pazira, E. 2012. Evaluation of zinc solubilization potential by different strains of fluorescent *Pseudomonads. J. Appl. Sci. Environ. Manage,* **16(3):** 295-298.

Berg, M., Ikwap, K. and Fellstrom, C. 2014. Poultry – a major source of protein for the poor. SLU-Global Report, pp. 46-49.

Bostock, J., McAndrew, B., Richards, R., Jauncey, K., Telfer, T., Lorenzen, K., Little, D., Ross, L., Handisyde, N., Gatward, I. and Corner, R. 2010. Aquaculture: global status and trends. Philosophical Transactions of the Royal Society B: *Biological Sciences,* **365(1554):** 2897-2912.

Bunting, S. W. 2004. Wastewater aquaculture: perpetuating vulnerability or opportunity to enhance poor livelihoods. *Aquatic Resources, Culture and Development,* **1(1):** 51-75.

Chandra Shekhar K., Sridharan S. and Ramakrishnan N. 2015. Bioefficacy, phytotoxicity and biosafety of mineral oil on management of whitefly in okra. *Int. J. Agr. Sci.,* **21:** 28-35.

Chauhan, A., Kumar, S., Singh, A.P. and Gupta, M. 2010.Vermicomposting of vegetable wastes with cowdung using three earthworm species *Eisenia foetida, Eudriluseugeniae* and *Perionyx excavates. Nat. Sci.,* **8(1):** 33-43.

Corbould, C. 2013. Feeding in the Cities: Is urban agriculture the future of food security. Future Directions. Retrieved from www.futuredirections.org.au/publications.

Danso, G., Drechsel, P., Wiafe-Antwi, T. and Gyiele, L. 2002. Income of farming systems around Kumasi, Ghana. *Urban Agriculture Magazine,* **7:** 5-6.

De Bon, H., Parrot, L. and Moustier, P. 2010. Sustainable urban agriculture in developing countries: A review. *Agron. Sustain. Dev.,* **30(1):** 21-32.

De Zeeuw, H. 2004. Introduction to urban agriculture. Nairobi Course. Leusden, Urban Harvest, RUAF.

Dhaliwal, G.S., Singh, R. and Jindal, V. 2013. A textbook of integrated pest management. Kalyani publishers, 179-256.

Ermolaev, E., Sundberg, C., Pell, M. and Jönsson, H. 2014. Greenhouse gas emissions from home composting in practice. *Bioresour. Technol.,* **151:**174-182.

FAO. 2006. World agriculture: towards 2030/2050 - Prospects for food, nutrition, agriculture and major commodity groups. Interim report, Global Perspective Studies Unit. Rome, Italy: Food and Agriculture Organization of the United Nations.

G, Shobha. and B.S. Kumudini. 2012. Antagonistic effect of the newly isolated PGPR *Bacillus* spp. on *Fusarium oxysporum. Int. J. Appl. Sci Engineering Res,* **1:** 463-474.

Gallaher, C.M., Kerr, J.M., Njenga, M., Karanja, N.K. and WinklerPrins, A.M.G.A., 2013.Urban agriculture, social capital, and food security in the Kibera slums of Nairobi, Kenya. Agriculture and Human Values, 30: 389-404.

Gandhi, A., Muralidharan, G., Sudhakar, E. and Murugan, A. 2014. Screening for elite zinc solubilizing bacterial isolate from rice rhizosphere environment. *Int. J. Recent Sci. Res.* **5:** 2201-2204.

Hafeez, F.Y., Yasmin, S., Ariani, D., Mehboob-ur-Rahman, Z.Y. and Malik, K.A. 2006. Plant growth promoting bacteria as biofertilizer. *Agron Sust Develop,* **26:** 143–150.

Harman, G.E. 2006. Overview of mechanisms and uses of *Trichoderma*spp. *Phytopathol,* **96:**190-194.

Hoornweg, D. and Munro-Faure, P. 2008. Urban Agriculture For Sustainable Poverty Alleviation and Food Security. FAO Position paper on Urban Agriculture.

Howell, C.R. 1998. The role of antibiosis in biocontrol. In: *Trichoderma & Gliocladium.* Harman GE, Kubicek CP (Eds.). 2. Taylor & Francis, Padstow. 173-184.

Hussain, A., Arshad, M., Zahir. Z.A. and Asghar, M. 2015. Prospects of zinc solubilizing bacteria for enhancing growth of maize. *Pak. J. Agri. Sci,* **52(4):** 915-922.

IPCC. 2007. Climate Change 2007. The Physical Science Basis.Intergovernmental Panel on Climate Change. Cambridge University Press.

Kagira, J. M., Kanyari, P. W. N., Maingi, N., Githigia, S. M., Ng'ang'a, J. C. and Karuga, J. W. 2010. Characteristics of the smallholder free-range pig production system in western Kenya. *Trop. Anim. Health Prod.*, 42(5): 865-873.

Karakurt, H. and Aslantas, R. 2010. Effects of some plant growth promoting rhizobacteria (pgpr) strains on plant growth and leaf nutrient content of apple. *J. Fruit Ornamental Plant Res*, **18(1)** : 101-110.

Khan, A. and Ishaq, F. 2011.Chemical nutrient analysis of different composts (Vermicompost and Pitcompost) and their effect on the growth of vegetative crop *Pisumsativum. Asian J. Plant Sci Res,* 1(1): 116-130.

Khan, M.H., Ahmad, N., Rashdi, S.M.M., Rauf, I., Ismail, M. and Tofique, M. 2013. Management of sucking complex in bt cotton through the application of different plant products. *Pakhtunkhwa J. Life Science,* 1(1): 42-48.

Khorshidi, Y.R., Ardakani, M.R., Ramezanpour, M.R., Khavazi, K. and Zargari, K. 2011. Response of yield and yield components of rice (Oryza sativa L.) to *Pseudomonas fluorescens* and *Azospirillum lipoferum* under different nitrogen levels. *Am Eurasian J Agric Environ Sci,* **10(3):** 387–395.

Komakech, A., Banadda, N.E., Gebresenbet, G. and Vinnerås, B. 2014. Maps of animal urban agriculture in Kampala. *Agronomy for Sustainable Development,* **34(2):** 493-500.

Konijnendijk, C., Gauthier, M. &Veenhuizen, R. van. 2004. *Urban Agriculture Magazine,* Trees and cities, growing together. Leusden, RUAF.

Liu, J.U., Wang, Z.F. and Lin, W. 2010. De-eutrophication of effluent wastewater from fish aquaculture by using marine green alga Ulvapertusa.*Chinese Journal of Oceanology and Limnology,* **28:** 201-208.

Lorito, M., Woo S.L., Harman G.E. and Monte, E. 2010. Translational research on *Trichoderma*: from omics to the field. *Ann. Rev. Phytopathol.,* **48:** 395-417.

Mazid M. and Khan T.A. 2014. Future of Bio-fertilizers in Indian Agriculture: An Overview. *Int. J. Agric. Food Res,* **3(3):** 10-23.

Mishra D.J., Singh R., Mishra U.K. and Kumar S.S. 2012. Role of Bio-Fertilizer in Organic Agriculture: A Review. *Res. J. Recent Sci.,* **2:** 39-41.

Mishra P., and Dash, D. 2014. Rejuvenation of Biofertilizer for Sustainable Agriculture and Economic Development. *Consilience: The Journal of Sustainable Development,* **11:** 41-61.

Moustaine, M., Elkahkahi, R., Benbouazza, A., Benkirane, R. and Achbani, E.H. 2017. Effect of plant growth promoting rhizobacterial (PGPR) inoculation on growth in tomato (*Solanum lycopersicum*L.) and characterization for direct PGP abilities in Morocco. *Int. J. Environ. Agric. Biotechnol (IJEAB),* **2:** 590-596.

Nugent, R.A. 2001. Using economic analysis to measure the sustainability of urban and periurban agriculture: A comparison of cost-benefit and contingent valuation analyses. Presentation at workshop on Appropriate Methodologies in Urban Agriculture, Nairobi, Kenya.

Prasad, T.V. 2014. Handbook of entomology. New Vishal Publications. Pp. 173-229.

Raaijmakers, J.M., Paulitz, T.C., Steinberg. C., Alabouvette, C. and Moënne-Loccoz, Y. 2009. The rhizosphere: a playground and battle-field for soil-borne pathogens and beneficial microorganisms. *Plant Soil,* **321:** 341-361.

Rahman, M.A., Begum, M.F. and Alam M.F. 2009. Screening of *Trichoderma* isolates as a biological control agent against *Ceratocystis paradoxa* causing pineapple disease of Sugarcane. *Mycobiol,* **37:** 277-285.

Saksirirat, W., Chareerak, P. and Bunyatrachata, W. 2009. Induced systemic resistance of biocontrol fungus, Trichoderma spp. against bacterial and gray leaf spot in tomatoes. *Asian J. Food Agro-Industry,* **2:** 99-104.

Samaranayake, J. W. K. and Wijekoon, S. 2010. Effects of selected earthworms on soil fertility, plant growth and vermicomposting.*TARE,* **1:** 1-5.

Sharma, M., Budha, P.B. and Pradhan, S.B. 2015. Efficacy test of bio-pesticides against tobacco whitefly *Bemisia tabaci* (Gennadius, 1889) on tomato plants in Nepal. *JIST,* **20(2):** 11-17.

Tenkorang, A., Yeboah-Agyepong, M., Buamah, R., Agbo, N.W., Chaudhry R. and Murray, A. 2012. Promoting sustainable sanitation through wastewater-fed aquaculture: a case study from Ghana.*Water International,* **37(7):** 831–842.

The Hyderabad Declaration. 2002. The Hyderabad Declaration on wastewater use in Agriculture. A workshop – "Wastewater Use in Irrigated Agriculture: Confronting the livelihood and Environmental realities" Sponsored by the International Water management Institute (IWMI, based in Colombo, Sri Lanka) and the International Development Research centre (IDRC, based in Ottawa, Canada).

Vagneron I., Pages J. and Moustier P. 2003. Economic appraisal of profitability and sustainability of peri-urban agriculture in Bangkok. Rome, FAO / CIRAD.Pp. 38.

Waghund, R. R., Shelake, R.M. and Sabalpara, A.N. 2016. *Trichoderma*: A significant fungus for agriculture and environment. *African J. Agric. Res,* **11(22):** 1952-1965.

WHO. 2006. Guidelines for the safe use of wastewater, excreta and grey water. Wastewater and excreta use in aquaculture. Geneva (Switzerland), 3.

Yadav, J. and Gupta, R.K. 2017. Dynamics of nutrient profile during vermicomposting. *Eco. Env. & Cons,* **23 (1):** 516-521.

Yadav, J., Gupta, R.K. and Kumar D. 2016. Chances in C:N of different substrates during vermicomposting. *Eco. Env. & Cons,* **23 (1):** 368-372.

Yoshioka, Y., Ichikawa, H., Naznin, H.A., Kogure, A. and Hyakumachi, M. 2012. Systemic resistance induced in *Arabidopsis thaliana* by *Trichoderma asperellum* SKT-1, a microbial pesticide of seed-borne diseases of rice. *Pest Manage. Sci,* **68:** 60-66.

5

Bio-nanotechnology for Sustainable Agriculture Development

Gulab Singh[*] and Khushboo

Department of Biotechnology, Central University of Haryana, Jant Pali, Mahendergarh-123029

[*]Corresponding author: gulab24@gmail.com

Abstract

Nanoscience and nanotechnology are interesting areas of research for researcher in present century. Exhaustive research in nanotechnology provides ample opportunity in various fields like medicine, electronics, food and agriculture sector. Nanotechnology in agriculture began with the growing awareness about the conventional farming technologies that would neither be able to increase productivity any further nor restore ecosystems damaged by existing technologies such as long-term effects of farming with irrigation, fertilizers, and pesticides. The adequate uses and benefits of nanotechnology are plentiful. Bulky populations of developing countries are facing the daily food scarcity due to negative environmental impact and political instability. There is urgent need for the development of drought and pest resistant variety with high productivity. Nanotechnology has the prospective to elevate the agriculture and food sector with novel nanotools for rapid disease diagnostic, by enhancing the capacity of plants to absorb nutrients, smart delivery of nutrients, bioseparation of proteins, rapid sampling and analysis of biological and chemical contaminants etc.The eloquent interests of using nanotechnology in agriculture includes specific applications like nano-fertilizers and nano-pesticides increase the productivity without decontamination of soils, waters, and also protection against several insect pest and microbial diseases. Developing field of nanotechnology provide glimpses of potential applications in the agri-food sector, including feed and food ingredients, intelligent packaging and quick detection systems, food and water safety that could have significant impacts on the rural populations in developing countries. This chapter summaries the potential uses and benefits of nanotechnology in agriculture.

Keywords: Agriculture, nanotechnology, nanoparticles, nanopesticides, nanoencapsulation

Introduction

The word nanotechnology is normally used for materials having size range between 1 - 100 nm (Huang *et al.*, 2007). Nanotechnology is knocking at the door of brilliance with advantages like miniature size, large surface area and more reactive sites (Prasad *et al.*,2014). Nanotechnology is an innovative scientific access that involves the use of materials and equipment capable of conducting physical as well as chemical properties of a substance at molecular level. On the contrary, biotechnology is implicated with the knowledge and techniques of biology to manipulate molecular, genetic and cellular processes to develop products in diverse fields of science i.e. from medicine to agriculture (Fakruddin *et al.*, 2012). Agriculture is always the most extensive field because it provides the crude material for food and feed industries. Agriculture is the backbone of the developing countries like India, where more than 60% population is reliant on it for their existence (Brock *et al.*, 2011).

To sustain the increasing population and natural resources, the agriculture development should be economically viable, environment friendly and efficient (Mukhopadhyay, 2014). To eradicate the poverty and hunger from the developing countries, improvement of soil fertility can play a very important role (Campbell *et al.*, 2014). Nanotechnology has the potential to diversify the agricultural and food industry (Roco 2003; Kuzma and VerHage, 2006), with novel tools for genetic improvement (Kuzma, 2007; Scott, 2007), delivery of genes and drug molecules to specific sites at cellular levels (Maysinger, 2007), and nanoarray-based gene-technologies for gene expressions in plants under stress conditions (Walker, 2005). Nano-biotechnology has the feasibility to enhance the nutritional value, ability of plant to absorb nutrients and improve the systems for directing the environmental conditions (Tarafdar *et al.*, 2013). Nanotechnology can be beneficial for enhancing the performance of microarrays, development in early detection of pathogens, and contaminants in the food (Moraru *et al.*, 2003, Suman *et al.*, 2010). There are new challenges in the agricultural field including a growing demand for healthy and safe food, an increasing liability of disease and threats to agricultural production from changing climate conditions (Biswal *et al.*, 2012).

Without any confusion the sustainable growth of agriculture utterly depends on the nanotechnology. The developments of nano - devices and nano –materials have higher dynamics than other applications. Currently, in food and agriculture sector there is higher demand of nanomaterial especially for packaging (Scrinis and Lyons, 2007). The innovative properties of nanomaterials combined with inventive engineering would have novel applications for agriculture and food systems; and as nanotechnology progresses, might lead to idea of new materials and devices (Scott & Chen, 2013). A variety of materials are used to compose nano particles like metal oxides, ceramics, magnetic materials, quantum dots, lipids polymers, dendrimers and emulsions (Puoci *et al.*, 2008). The varying concentration of nano particles influences the processes like germination and growth of plant (Zheng *et al.*, 2005). Nanotechnology has the ability to change the whole scenario of the present agriculture and food industry with aid of novel tools for the treatment of plant diseases, accelerated pathogen detection kits and Nanobiosensor (Joseph and Morrison, 2006). Nanotechnology acts at same level with virus or disease causing particles and thus holds the potential of primitive detection and elimination. By coordinated sensing, monitoring and controlling system the farmer could identify the presence of disease, long before the symptoms develop and activate the bioactive systems (Prasad *et al.*, 2012). It is expected that in forthcoming days nanocatalysts will be available to raise the efficacy of commercially available pesticides and herbicides, allowing the lower dose to be used for crop plants (Joseph and Morrison, 2006).

The consumption of pesticides is about two million tonnes per year in whole world. Studies indicate that insect pests cause loss in crop production i.e. 25% in rice, 5-10% in wheat, 30% in pulses, 35% in oilseeds, 20% in sugar cane and 50% in cotton (Dhaliwal *et al.*, 2010). Careless use of pesticide enhance the pathogen and insect resistance, decrease soil biodiversity, causesbio magnifications of pesticide, decline in pollinators (De *et al.*, 2014). Formulation of nano chemical pesticides increase the agricultural yield using nanoparticles enclosed fertilizers for slow and constant release of nutrients and water (Tilman *et al.*, 2002). Some boon of the nanotechnology for agriculture sectors are nanoparticles mediated precise management of soil, nano-mediated pest control, diversification of farming practice and products etc. (Rai & Ingle 2012).

Nanotechnology will have an impact on every field of science by its ground breaking scientific innovations. The development of high-tech agricultural system will continue to have an excellent impact in main areas of food industry and agricultural practices as well as enhance the quality and quantity of the yields (Prasad *et al.*, 2012a). Apart from conventional role, future agriculture will have to counter the energy needs by entrapping solar energy and material manufacturing to sustain culture. Here an attempt is made to sum up the modern strategies of nanotechnology applications for sustainable agriculture development.

Applications of Nanotechnology in Agriculture

Nanofertilizers

Fertilizers plays important role in enhancement of agricultural yield (35-40%). Bio-fertilizers include fungal mycorrhizae, Rhizobium, Azotobacter, Azospirillum and blue green algae, bacteria like *Pseudomonas* and *Bacillus* (Wu *et al.*, 2005). Microorganisms convert complex organic matter into simple compounds to provide essential nutrients to plants, enhance soil fertility and increase crop yield (Jha and Prasad, 2006). Short shelf life, temperature sensitivity and storage desiccation are the problems associated with the usage of biofertilizers. Polymeric nanoparticles have been utilized for coating of bio-fertilizers to yield the formulations resistance to desiccation.

Nano-fertilizers are synthesized in order to enhance nutrient use efficiency of plant and nano-fertilizers are more competent than ordinary fertilizer (Liu *et al.*,

2006a). In current decade nano-fertilizers are freely available in the market, which may contain nano zinc, silica, iron and titanium dioxide, ZnCdSe/ZnS core shell QDs, InP/ZnS core shell QDs, Mn/ZnSe QDs, gold nanorods, core shell QDs. Zinc, iron and molybdenum deficiency has been reported as one of the major problems in alkaline nature of soils (Sadeghzadeh, 2013). Consequences of spraying of iron oxide nanoparticles on the response of wheat growth, yield and quality has been studied. Effect of manganese nanoparticles has been studied to increase the growth of mung bean (*Vigna radiata*) (Pradhan *et al.*, 2013). Improvement in the growth of mung bean and chickpea (Cicer arietinum) at lower concentration of zinc- oxide nanoparticles were observed (Mahajan *et al.*, 2011). Loss of nitrogen due to leaching and emission could be reduced by the use of nanofertilizers. Toxic effects of fertilizers could be decreased by the slow controlled release fertilizers and might improve the soil (Suman *et al.*, 2010).

Nanoparticles as Pesticides and Herbicides

Production of food and plant protection has an explored are in future through use of nanoparticles. It is obvious that insects and pests are the potent ones in agricultural fields. Synthetic agrochemicals plays essential role in the agricultural field, it has also developed resistance in the pests. Nanoparticles are obligatory for management of insect pest of modern agriculture. Thus, the nanoparticles can play effective role against insect pests and host pathogens (Khota *et al.*, 2012). Nanoparticle gene mediated DNA transfer can be used to deliver DNA and other desired chemicals into the plant tissue for the protection of plant. For the controlled release of water soluble pesticide porous hollow silica nanoparticles loaded with validamycin (pesticide) can be used as efficient delivery system. Activity of garlic essential oil (insecticide) against red flour beetle (Tribolium castaneum) has been enhanced by the polyethylene glycol-coated nanoparticles. Yang *et al.*, (2009) studied that the control efficacy of PEG coated nanoparticles was about 80% due to slow release of active components of nanoparticles. Applications of various types of nanoparticles were studied such as silver nanoparticles, aluminium oxide, zinc oxide and titanium dioxide in the control of rice weevil (caused by Sitophilusoryzae) and grasserie disease in silkworm (caused by Bombyxmori and baculovirus BmNPV (B. mori nuclear polyhedrosis virus) (Goswami *et al.*, 2010).

Barik *et al.*, (2008) studied the mechanism of control of insect pest using nano- silica as nano insecticide. For preventing death by dessication insect pest use a variety of cuticular lipids to defend their water barrier. But nano- silica particles cause the death of insect by being absorbed into the cuticular lipids by physiosorption. Treatment of leaves of mulberry (B. mori) with ethanolic suspension of hydrophobic alumino-silicate nanoparticle resulted in significant decrease in viral load.Amorphous silica nanoparticles cause more than 90% mortality of rice weevil *Sitophilus oryzae* as compared to bulk – sized silica (individual particles larger than 1.0μm). Significant mortality of two insect pest i.e. *Sitophilus oryzae* L. and *Rhyzopertha dominica* was reported after continuous exposure to nanostructured alumina-treated wheat for three days (Teodoro *et al.*, 2010). Thus, inorganic nanostructured alumina may provide a cheap and reliable substitute for control of insect pests as compared to commercially available insecticides.

Decline in crop yield to a greater extent is mainly because of weeds which use the nutrients in greater quantity as compared to crop plants. Mostly chemically synthesized herbicides are commercially available, they kill the weeds but in turn cause soil pollution, decrease soil fertility and damage crop plants as well. Being very small, nano-herbicides will blend with soil to evacuate weeds in an eco- friendly manner without leaving any toxic residues in soil and environment (Luque & Rubiales, 2009). Weeds endure and spread via tubers and deep roots. Target specific nanoherbicides have been invented for transport in roots of weeds. Nanoherbicides lead to death of weeds inhibit the metabolic pathway of weeds such as glycolysis by entering in root system. Thus, the efficacy of herbicides can be improved by the use of nanotechnology.

Nanocapsules for Efficient Delivery of Pesticides

Bhattacharyya *et al.*, (2016) reviewed that nanotechnology will improve the agriculture in near future in addition to pest management. Nano- encapsulation is one of the examples of this technology. Nano- encapsulation is a process in which chemical such as an insecticide is released slowly and efficiently to a particular host plant for insect pest control with enhanced solubility, specificity, permeability and stability. Nano-encapsulation includes nanoparticles in the form of pesticides that can be timely released and properly absorbed by the plants. Currently, nano-encapsulation is the most promising technology which led to decrease the dosage of pesticide and human beings exposure to them (Nuruzzaman *et al.*, 2016). In the present scenario, most of the leading chemical companies target on the formulation of nanoscale pesticide for the development of non- toxic and promising pesticide delivery system to reduce the negative impacts on the ecosystem (Bhattacharyya *et al.*, 2016).

Nanoparticles as Agrochemicals

To rectify the problem related with pesticides new strategies being used like agrochemicals. For the formulation of agrochemicals, several surfactants, polymers (organic), and metal nanoparticles (inorganic) in the nanometer size range are used. However, a significant proportion of agrochemicals do not reach their targets when applied traditionally because of many environmental factors like wind, sunlight, rain etc (Liu *et al.*, 2008). Advanced technologies like encapsulation and controlled release methods are used to modify the use of agrochemicals in crop production (Mulqueen, 2003). For more precise nozzle spray, nano-pesticides have been formulated with small size, improved absorption solubility and stability. Encapsulation of agrochemicals leads to improvement in bio-efficacy, protection from rapid degradation, controlled release and prolong duration of bioactive agents. The use of Bifenthrin (a synthetic pyrethroid insecticide) is enhanced by formulating nanoparticle suspension with the use of flash nano-precipitation (Liu *et al.*, 2008). To treat the plant pathogenic fungi two fungicides named as Chlorothalonil and Tebuconazole are used. Liu *et al.*, (2001) studied enhanced biological activity of these fungicides against *G. trabeum* when loaded with nanoparticles. Essential oils (substitute of conventional pesticides) have low mammalian and fish toxicity and also less persist in fresh water and soil. Chemical instability of these oils in the presence of air, light, moisture, high temperature cause rapid evaporation and degradation of some active components. Nano-formulation of essential oil could provide the protection to the environmental degradation. There should be ideal nano-formulation for effective and continuous controlled release of essential oil. Lai *et al.*, (2006) reported that Artemisia (essential oil) formulated with solid lipid nanoparticles exhibited higher physical stability and decreased rate of evaporation. Gonzalez-Melendi *et al.*, (2008) observed that iron nanoparticles treated with carbon had entered and translocated in whole plant (*Cucubita pepo*) through vascular system. Because of magnetic property of iron nanoparticle, they can be positioned in the tissue by applying a magnetic field gradient.

Nanobiosensors for Agriculture Sustainability

Sensors are sophisticated instruments which sense the physico- chemical and biological changes and transform that response into a signal which can be used by humans (NNCO, 2009).Nano biosensors can be productively used for sensing a wide array of agriculture like fertilizers, herbicide, pesticide, insecticide, pathogens, moisture, soil pH for enhancing crop productivity to assist in sustainable agriculture. Heavy use of organophosphates disrupt cholinesterase enzyme and cholinergic dysfunction resulted in serious risk to human health. In present scenario,accurate detection of low concentrations of organophosphates is done by the development of a variety of enzyme-based biosensors. The hydrophobic surface based poly (dimethylsiloxane)-poly (di-allydimethylemmonium) (PDMS)-PDDA/AuNPs/cholineoxidase/achebiosensorexhibit excellent stability and unique sensitivity to organophosphorous group of pesticides (detection limit 5×10^{-10} g/L) (Zhao *et al.*, 2009).

There is tremendous demand for rapid, sensitive and cost-effective nanobiosensor systems in crucial areas of human activity such as health care, agriculture, genome analysis, food and drink, process industries, environmental monitoring, defense, and security (Fogel and Limson, 2016). Atomic force microscopy based nanobiosensor functionalized with the acetolactate synthase enzyme was detected for herbicide metsulfuronmethyl (an acetolactate synthase inhibitor) through the procurement of force curves (da Silva *et al.*, 2013). Biosensors enhance the food safety of consumer via rapid detection of bacteria and viruses (Otles and Yalcin, 2010).

Nano sensors have the potential to indicate the nutrient and water status of the crop plants which in turn makes the farmers to apply nutrients or water according to requirement. For real time monitoring of soil conditions and crop growth the autonomous sensor linked to a global monitoring system. For monitoring agricultural pollutants and assessment of impacts on living matter or health, electrochemically functionalized single walled carbon nanotubes based nano sensors with metal/metal oxide nanoparticles or nanotubes for gases viz. ammonia, nitrogen oxides, hydrogen sulfide, sulfur dioxide and volatile organics are used (Sekhon, 2014). Toincrease the crop productivity and reduce land burden, nanosensors based on using electrochemically functionalized single walled carbon nanotubes with either metal nanoparticles or metal oxide nanoparticles, and metal oxide nanowires and nanotubes for gases such as ammonia, nitrogen oxides, hydrogen sulfide, sulfur dioxide, and volatile organics are used (Wanekaya, 2006). Finally, use of smart sensors leads to improve the productivity of crop by providing accurate information and helping farmers to make excellent decision.

A significant enhancement in crop yield with green revolution and new farming practices has constantly reduced the micronutrients of soil, water and fertility of soil (Alloway, 2008). The adequate use of agricultural natural property like water, nutrients and chemicals as nanosensors is user friendly. Nanomaterials linked with global positioning systems are used for satellite imaging of fields to detect the pests and facts of stress like drought. To find out the existence of plant viruses and level of soil nutrients,nanosensors are dispersed in the field (Ingale and Chaudhari, 2013). For economical, proficient, rapid effortless decoding grocery barcodes are used that may tag perhaps multiple pathogens in the field (Li *et al.*, 2005).

To detect pathogens associated with different plant diseases, Quantum dots-fluorescence resonance energy transfer based sensors have been developed. These sensors are used to detect various plant diseases like Witches' broom disease of lime caused by *Phytoplasmaaurantifolia,* Rhizomania disease in sugar beet caused by beet necrotic yellow vein virus(Rad *et al.*, 2012, Safarpour *et al.*, 2012). To detect organophosphorous pesticides, highly sensitive biosensor based on enzyme inhibition mechanism are used like the optical transducer of cadmium telluride semiconductor quantum dots integrated with acetyl cholinesterase enzyme with detection limits of 1.05×10^{-11} M for paraoxon and 4.47×10^{-12} M for parathion (Zheng *et al.*, 2011).There is no detectable toxicity for seed germination and seedling growth at low concentration of QDs. So, QDs can be used for live imaging in plant root systems to certify known physiological processes (Hu *et al.*, 2010; Das *et al.*, 2015).QDs can be used for in situ hybridization of several plant chromosomes but are less sensitive than conventional system (Muller *et al.*, 2006).

Nanomaterials: Antimicrobial Agent for Plant Disease Control

Some of the metal nanoparticles with antimicrobial activity that have entered into the arena of controlling plant diseases and investigated by some researchers include carbon, silver, silica and alumino- silicates. The use of nanoparticles of silver as antimicrobial agents is increasing day by day as the technology is advancing because silver exhibit different modes of inhibitory action to microorganisms (Young , 2009). Park *et al.* (2006) studied the effectiveness of nanosized silica–silver particles to get rid of plant pathogenic fungi viz. *Botrytis cinerea, Rhizoctonia solani, Colletotrichum gloeosporioides, Magnaporthe grisea* and *Pythium ultimum.* Silver nanoparticles have the efficacy to alter many biochemical processes in microorganisms including expression of proteins responsible of ATP production

(Pal *et al.*, 2007). Jo *et al.* (2009) studied the antifungal activity of nano-size silver colloidal solution against plant pathogenic fungi like *B.sorokiniana* and *M. grisea.* Similarly, Kim *et al.* (2012) observed the most significant inhibition of plant pathogenic fungi on potato dextrose agar (PDA) and 100 ppm of silver nanoparticles.

Zinc oxide and magnesium oxide nanoparticles are effective and attractive antibacterial agent because of enhanced dispensability, optical transparency, and smoothness (Shah and Towkeer, 2010). Mixture of Copper nanoparticles and soda lime glass powder shows efficient antimicrobial activity against gram-positive, gram-negative bacteria and fungi (Esteban-Tejeda *et al.*, 2009). In forthcoming days, nanoscale devices might be used to diagnose plant health issues before visible to farmer.

Nanotechnology in Irrigation Water Filtration and Retention

The emerging trends in nanotechnology which will be profitable for the farmers all over the world include several nanomaterials which are economically effective in cleansing of irrigation water. Membrane filters constructed from carbon nanotubes, nanoporous ceramics and nanoparticles with magnetic property are used for filteration of irrigation water instead of using chemicals UV light treatment (Hillie and Hlophe, 2007). Nanofiltration is used to remove solids including bacteria, parasites while magnetic nanoparticles can be used to remove nanocrystals and arsenic from water (Yavuz *et al.*, 2006). The local brackish water is treated with nanofiltration membranes to gain high quality desalinated water. Desalinated water decrease the irrigation requirement of the crop upto 25% and in few cases crop yield also increases.

Because of porous and capillary suction properties, Zeolites (naturally occurring crystalline aluminium silicates) are used in water infilteration and retention in soil. It is an excellent remedy for non-wetting sands. Water retention capacity of sandy soil and porosity of clay soil can be significantly increased by the use of zeolites. Ultimately, the crop production can be increased in drought prone areas (Prasad, *et al.* 2014).

Nanotechnology in Organic Farming

The main aim of organic farming is to increase the crop yield with less use of fertilizers, pesticides, herbicides etc. To determine growing efficiency of the crops and highly localized environmental conditions, organic farming makes the use of computers, GPS system and remote sensing devices. Nanotechnology has banned in Canada for organic food production. An amendment was supplemented to Canada's national organic rules banning nanotechnology as a "Prohibited Substance or Method" (The Organic and Non-GMO Report, 2010).

Nanotechnology and Agribusiness

The global economic report says that the estimated market of agriculture ranged between US$ 20.7 billion- US$ 0.98 trillion in 2010. Till 2020, agribusiness will be more than US$ 3.4 trillion in the market. In the present scenario, USA is at top with 3.7billion USD investment through the National Nanotechnology Initiative (NNI) (Hirsh *et al.*, 2014, Banterle *et al.* 2014). Japan and the European Union are at second and third position with contribution of US$ 750 million and US$ 1.2 billion, respectively per year (Sodano *et al.*, 2014). At present, more than 400 companies are active in the field of nanotechnology and this figure is expected to increase to more than 1000 in upcoming years.

Public Acceptance of Nanotechnology

In upcoming years nanotechnology appears to be quite interesting and effective in agricultural field. Nanoparticles practically come in direct and indirect contact of everyone because of rapid distribution in food products. It is probable that merger of nanotech and biotech may cause unknown changes in soil, biodiversity and the environment. The nanomaterials have the ability to infiltrate in human body but there is no real information about the effects that they may have. In a public opinion survey, respondents considered nanotechnology as relatively neutral, less risky, and more beneficial than a number of other technologies, such as genetically modified organisms, pesticides, chemical disinfectants, and human genetic engineering (Currall *et al.*, 2006). Nanomaterials might be responsible for causing harm because of ambiguous chemical properties. We are having little understanding about fate, transport and behavior of engineered nanoparticles in the environment. Scientist are blazing a stream for a novel technology and looking at every possible route to improve upon current methods (Bernhardt *et al.*, 2010). Extensive studies are needed to understand the mechanism of nanoparticles toxicity and their impacts on natural environment.

Limitations of Nanotechnology

The main aspect on nanotechnology is to improve the quality and nutritional value of global food. In spite of potential benefits, a very less is known about the safety facets of nanotechnology. A number of researches have directed to estimate the toxicity of nanoparticles frequently used in industry (Rana and Kalaichelvan, 2013, Du *et al.*, 2017). The cytotoxic effect of metal nanoparticles is definitely depending on the charge at the membrane surface and structure of targeted cell-wall. The electrostatic interaction of nanoparticles with the membrane and their accumulation in cytoplasm may be accredited for toxicity (Aziz *et al.*, 2015, 2016). When photochemically active nanoparticles like TiO2, ZnO, SiO2 and fullerenes are exposed to light , they form superoxide radicals in the presence of oxygen (Hoffmann *et al.*, 2007). When the organisms are simultaneously exposed to photochemically active nanoparticles and UV light, the cells respond to oxidative stress (reactive oxygen species) by enhancing the number of protective enzymes (Vannini *et al.*, 2014). Thus, generation of ROS can be assigned in determination of toxicity of nanoparticles.

Nanoparticles enter in plant cellular system and translocate in the shoot, accumulate in edible tissues. Nanoparticles affect the rate of transpiration, respiration, photosynthesis and interfere with translocation of food material (Shweta *et al.*, 2016). The plants having exposure to cerium oxide nanoparticles show elevated cerium content, indicating that cerium oxide nanoparticles entered in roots and translocated to shoot and edible tissues (Wang *et al.*, 2012). Use of nanosilver as an antibacterial agent is being debated in whole world because of potential risks and benefits. (Boholm & Arvidsson, 2014).The only possibility for more production of agricultural crops is the use of nanoparticles which are devoid of any toxic nanocomposite. There should be a routine check up to maintain an eco- friendly environment after the introduction of engineered nanoparticles (either chemical or green) in the agricultural field.

Future Perspective

Currently, the research and development agenda of nanotechnology have the potential to change the agriculture production by allowing better management to aid the society(Sugunan and Dutta, 2004).The population is increasing at a very fast rate globally. Therefore, it is essential to organize the recent technology in food industry. Thus, the forthcoming nanotechnology will possesses vey exclusive properties from field to table: crop production, rapid testing related to control of pests by use of agrochemicals, testing cross contamination of agriculture and food products, improvement of food texture, intelligent feed, packaging and labeling etc. A lot of agriculture and food system applications need more attention in near future researches.

Processes of biosynthesis of nanoparticles engage more researchers to go for future evolution in area of electrochemical sensor, biosensors,pharmaceuticals and agriculture. Nanotechnology will revolutionize agriculture in near future by the use of nanosensors for detection of aquatic toxins, improved decontamination and recycling of heavy metals, nanoparticles as a novel photocatalyst for environmental catalysis. By the production of new pesticides, insecticides and insect repellents, green revolution would be speed up over the upcoming decades (Owolade *et al.* 2008).In long term way, Nanotechnology may provide ingenious and cost effective development routes for human nutrition worldwide.

Conclusion

Nanotechnology with the unique properties has the potential to revolutionize the current technologies used in diverse fields including agriculture. The diverse opportunities for the application of nanotechnology play an important role in food production and are immense for agriculture. With the help of controlled delivery, nanoparticles have given auspicious outcomes for plant disease resistance, enhanced plant growth and nutrition via site specific delivery of fertilizers and other essential nutrients. Although, a lot of knowledge is available about the individual nanomaterials, yet the noxious level of many nanoparticles is still ambiguous. So, the use of nanomaterials is limited. Nanotechnology can implement green, efficient and ecofriendly approach for insect pest management in agriculture.

Nanomaterials are like 'magic bullets' embarked with herbicides, fungicides, nutrients, fertilizers or nucleic acids and release their charge to the specific and desired part of plant to enact the appropriate result. However, introduction of any new technology always has social and ethical problems associated with it. To incredibly amplify the advancement of novel applications in all fields, it is very much essential to inform common man about the benefits of nanotechnology. Taking lessons from unconstructive response of public towards genetically modified crops, it is crucial to explain the positive and negative aspects of this highly promising technology. The government should consider relevant labeling and should set down regulations to increase public acceptance. There is a horrible call to hack the ragged outline existing among the society, common man and flowering scientific notions. If we succeed in overcoming this, then an amazing, bright and auspicious future will be at the door step of society.

References

Alloway, B.J. 2008. Micronutrients and crop production: an introduction, in: B.J. Alloway (Ed.), Micronutrient Deficiencies in Global Crop Production, Springer, pp. 1–39.

Aziz, N., Faraz, M., Pandey, R., Sakir, M., Fatma, T., Varma, A.*et al.* 2015. Facile algae-derived route to biogenic silver nanoparticles: synthesis, antibacterial and photocatalytic properties. *Langmuir,* 31: 11605–11612.

Banterle, A., Cavaliere, A., Carraresi, L., and Stranieri, S. 2014. Food SMEs face increasing competition in the EU market: marketing management capability is a tool for becoming a price maker. *Agribusiness* 30, 113–131.

Barik, T. K., Sahu, B., and Swain, V. 2008.Nano-silica—from medicine to pest control. *Parasitol Res,* 103: 253–258

Bernhardt, E.S., Colman, B.P., Hochella, M.F.Jr *et al.* 2010. An ecological perspective on nanomaterial impacts in the environment. *J Environ Qual.,* 39: 1–12

Bhattacharyya, A., Duraisamy, P., Govindarajan, M., Buhroo, A.A., and Prasad, R. 2016. "Nano-biofungicides: emerging trend in insect pest control," in Advances and Applications through Fungal Nanobiotechnology, ed. R. Prasad (Cham: Springer International Publishing), 307–319.

Biswal, S.K., Nayak, A.K., Parida, U.K., and Nayak, P.L.2012.Applications of nanotechnology in agriculture and food sciences.*Int. J. Sci. Innovat. Discov.* 2(1): 21 – 36.

Boholm, M. and Arvidsson, R. 2014. Controversy over antibacterial silver: implications for environmental and sustainability assessments. *J Clean Prod.* 68:135–143.

Brock, D.A., Douglas, T.E., Queller, D. C. and Strassmann, J.E. 2011. Primitive agriculture in a social amoeba. *Nature,* 469: 393- 396.

Campbell, B. M., Thornton, P., Zougmoré, R., vanAsten, P. and Lipper, L. 2014. Sustainable intensification: what is its role in climate smart agriculture? *Curr. Opin.Environ. Sustain.* 8: 39–43.

Currall, S. C., King, E. B., Lane, N., Madera, J. and Turner, S.2006. What drives public acceptance of nanotechnology? *Nat Nanotechnol,* 1:153–155.

Da Silva, A. C. N., Deda, D. K., Da Róz, A. L., Prado, R. A., Carvalho, C.C., Viviani, V., and Leite, F. L. 2013.Nanobiosensorsbasedon chemically modified AFM probes: a useful tool for metsulfuron-methyl detection. *Sensors(Basel),* 13: 1477–1489.

Das, S., Wolfson, B.P., Tetard, L., Tharkur, J., Bazata, J., and Santra, S.2015. Effect of N-acetyl cysteine coated CdS:Mn/ZnS quantum dots on seed germination and seedling growth of snow pea (Pisumsativum L.): imaging and spectroscopic studies. *Environ. Sci.,* 2: 203–212.

Dhaliwal, G.S., Jindal, V., and Dhawan, A. K. 2010. Insect pest problems and crop losses: changing trends. *Indian J Ecol,* 37(1): 1–7.

Du, W., Tan, W., Peralta-Videa, J.R., Gardea-Torresdey, J. L., Ji, R., and Yin .2017. Interaction of metal oxide nanoparticles with higher terrestrial plants: Physiological and biochemical aspects. *Plant Physiol. Biochem.,* 110: 210–225.

Esteban-Tejeda, L., Malpartida, F., Esteban-Cubillo, A., Pecharromán, C., andMoya, J.S. 2009.Antibacterial and antifungal activity of a soda-lime glass containing copper nanoparticles. *Nanotechnology,* 20: 505701

Fakruddin, Md., Hossain, Z., and Afroz, H. .2012. Prospects and applications of nanobiotechnology: a medical perspective. *J. Nanobiotechnology,* 10:31.

Fogel, R. and Limson, J. 2016. Developing biosensors in developing countries: South Africa asa case study. *Biosensors*6(1) 5.

Gonzalez-Melendi, P., Fernandez-Pacheco, R., Coronado, M. J., Corredor, E., Testillano, P.S., Risueno, M.C., Marquina, C., Ibarra, M. R., Rubiales, D. and Perez-de-luque, A. 2008. Nanoparticles as smart treatment-delivery systems in plants: Assessment of different techniques of microscopy for their visualization in plant tissues. *Ann. Bot.* 101:187–195.

Goswami, A., Roy, I., Sengupta, S. and Debnath, N. 2010. Novel applications of solid and liquid formulations of nanoparticles against insect pests and pathogens.*Thin Solid Films* 519:1252–1257

Hirsh, S., Schiefer, J., Gschwandtner, A. and Hartmann, M. 2014.The determinants of firm profitability differences in EU food processing, *J. Agric. Econ.,* 65: 703–721.

Hoffmann. M., Holtze, E. M. and Wiesner, M. R. 2007."Reactive oxygen species generation on nanoparticulate material," in Environmental Nanotechnology.

Applications and Impacts of Nanomaterials, eds M.R. Wiesner and J.Y. Bottero (New York, NY: McGraw Hill), 155–203.

Hu, Y., Li, J., Ma, L., Peng, Q., Feng, W. and Zhang, L. 2010. High efficiency transport of quantum dots into plant roots with the aid of silwet L-77.*Plant Physiol.Biochem.*48,703–709.

Huang, J., Li, Q., Sun, D., Lu, Y., Su, Y., Yang, X., Wang, H., Wang, Y., Shao, W., He, N., Hong, J. and Chen, C., 2007.Biosynthesis of silver and gold nanoparticles by novel sundried Cinnamomumcamphora leaf. *Nanotechnology*, 18: 105104.

Jha,M.N.,Prasad,A.N.,2006.Efficacyofnewinexpensivecyanobacterialbiofertilizer including its shelf-life.*World J. Microbiol. Biotechnol.* 22: 73–79.

Jo, Y. K., Kim, B.H. and Jung, G. 2009. Antifungal activity of silver ions and nanoparticles on phytopathogenic fungi. *Plant Dis*, 93: 1037–1043

Joseph, T. and Morrison, M., 2006. Nanoforum Report: Nanotechnology in Agriculture and Food. *European Nanotechnology Gateway*.

Khota, L. R., Sankarana, S., Majaa, J. M., Ehsania, R. and Schuster, E.W., 2012. Applications of nanomaterials in agricultural production and crop protection: a review. *CropProt.*, 35,64–70.

Kuzma, J., 2007. Moving forward responsibly: Oversight for the nanotechnology-biology interface.*J. Nanopart. Res.*, 9: 165–182.

Kuzma, J. and VerHage, P. .2006. Nanotechnology in agriculture and food production: Anticipated applications. Woodrow Wilson International Center for Scholars.Project on emerging nanotechnologies and the consortium on law, values and health and life sciences.

Liu, Y., Tong, Z. and Prud'homme, R. K. 2008. Stabilized polymeric nanoparticles for controlled and efficient release of bifenthrin. *Pest Manag. Sci.* 64: 808–812.

Liu, Y., Yan, L., Heiden, P. and Laks, P. 2001. Use of nanoparticles for controlled release of biocides in solid wood. *J. Appl. Polym. Sci.*, 79: 458–465.

Luque, A. P. and Rubialesm, D. 2009. Nanotechnology for parasitic plant control. *Pest Manag Sci.* 65: 540–545.

Mahajan, P., Dhoke, S. K. and Khanna, A. S. 2011. Effect of nano-ZnO particle suspension on growth of mung (Vignaradiata) and gram (Cicerarietinum) seedlings using plant agar method.*J. Nanotechnol.* 7 (2011) 1–7.

Maysinger, D. 2007. Nanoparticles and cells: Good companions and doomed partnerships. *Org. Biomol. Chem.*, 5:2335–2342.

Moraru, C., Panchapakesan, C., Huang, Q., Takhistov, P., Liu, S. and Kokini, J. 2003. Nanotechnology: A new frontier in food science. *Food Technol.*, 57:24–29.

Mukhopadhyay, S. S. 2014. Nanotechnology in agriculture: prospects and constraints. *Nanotechnol.Sci.Appl.* 7: 63–71.

Muller, F., Houben, A., Barker, P., Xiao, Y., Kas, J., andMelzer, M. 2006. Quantum dots a versatile tool in plant science. *J. Nanobiotechnol*, 4: 5.

Otles, S., and Yalcin, B. 2010.Nano-biosensors as new tool for detection of food quality and safety.*Log Forum*. 6: 67–70.

Owolade, O. F., Ogunletim D. O., and Adenekanm, M. O. 2008. Titanium dioxide affects disease development and yield of edible cowpea. *EJEAF Chem*. 7(50): 2942- 2947.

Pal, S., Tak, Y. K. and Song, J. M. 2007. Dose the antibacterial activity of silver nanoparticles depend on the shape of the nanoparticle? A study of gram negative bacterium *E.coli.Appl. Environ. Microbiol.,* 73: 1712-1720.

Park, H. J., Kimm S. H., Kimm H.J., and Choim, S. H. 2006. A new composition of nanosized silica-silver for control of various plant diseases. *Plant Pathol J.,* 22:295–302.

Pradhan, S., Patra, P., Das, S., Chandra, S., Mitra, S., and Dey, K.K. 2013. Photochemical modulation of biosafe manganese nanoparticles on Vignaradiata: a detailed molecular biochemical, and biophysical study. *Environ. Sci. Technol.,* 47: 13122–13131.

Prasad, R., Badge, U. S. and Varmam, A. 2012a. Intellectual property rights and agricultural biotechnology : an overview . *Afr. J. Biotechnol.,* 11(73): 13746-13752.

Prasad, R., Kumar, V., and Prasad K. S. 2014. Nanotechnology in sustainable agriculture:presentconcernsandfutureaspects. *Afr.J.Biotechnol.,* 13: 705–713.

Puoci, F., Lemma, F.,Spizzirri, U. G., Cirillo, G., Curcio, M. and Picci, N. 2008. Polymer in agriculture: a review.*Am. J. Agri. Biol. Sci.,* 3: 299–314.

Rad, F., Mohsenifar, A., Tabatabaei, M., Safarnejad, M. R., Shahryari, F., Safarpour, H., Foroutan, A., Mardi, M., Davoudi, D., Fotokian, M. 2012. Detection of CandidatusPhytoplasmaaurantifolia with a quantum dots fret-based biosensor. *J. Plant Pathol.,* 94: 525–534.

Rai, M. and Ingle, A. 2012. Role of nanotechnology in agriculture with special reference to management of insect pests. *Appl. Microbiol. Biotechnol.,* 94: 287–293.

Rana, S. and Kalaichelvan, P.T. 2013.Ecotoxicity of nanoparticles. ISRN *Toxicol.,*

Roco, M. C., 2003. Broader societal issues of nanotechnology. *J. Nanopart. Res.,* 5: 181–189.

Sadeghzadeh, B., 2013. A review of zinc nutrition and plant breeding. *J. Soil Sci. PlantNutr.,* 13: 905–927.

Safarpour,H., Safarnejad, M. R., Tabatabaei, M., Mohsenifar, A., Rad, F., Basirat, M., Shahryari, F. and Hasanzadeh, F. 2012. Development of a quantum dots FRET-based biosensor for efficient detection of Polymyxabetae. *Can. J. Plant Pathol.,* 34: 507–515.

Scott, N. and Chen, H. 2013. Nanoscale science and engineering for agriculture and food systems. *Industrial Biotechnology,* 9: 17–18.

Scott, N. R. 2007. Nanoscience in veterinary medicine. *Vet. Res. Commun*, 31: 139–144.

Scrinis, G., and Lyons, K. 2007. The emerging nano—corporate paradigm: nanotechnology and the transformation of nature, food and agri- food systems. *Int. J. Sociol. Food Agric.* 15: 22-44.

Sekhon, B. S. 2014. Nanotechnology in agri-food production: an overview. *Nanotechnol. Sci. Appl.*, 7: 31–53.

Shah, M. A., and Towkeer, A. 2010.Principles of nanosciences and nanotechnology. Narosa Publishing House, New Delhi

Sodano, V., and Verneau, F. 2014. Competition policy and food sector in the European Union. *J. Int. Food Agribusiness Mark.*, 26: 155–172.

Sugunan, A., and Dutta, J. 2008. Nanotechnology Vol. 2: Environmental Aspects (Krug Harald (ed.), wiley –VCH ,Weinheim.

Suman, P. R., Jain, V. K., and Varma, A. 2010. Role of nanomaterials in symbiotic fungus growth enhancement. *Curr. Sci.*, 99: 1189-1191.

Tarafdar, J. C., Sharma, S., and Raliya, R. 2013. Nanotechnology: Interdisciplinary science of applications. *Afr. J.Biotechnol.*, 12(3): 219-226.

Teodoro, S., Micaela, B., and David, K. W. 2010. Novel use of nano-structured alumina as an insecticide.*Pest ManagSci*, 66(6): 577–579.

Tilman, D., Cassman, K. G., Matson, P. A., Naylor, R., and Polasky, S. 2002. Agricultural sustainability and intensive production practices. *Nature.*, 418: 671–677.

Tripathi, D. K., Singh, S., Singh, S., Dubey, N. K. and Chauhan, D. K. 2016. Impact of nanoparticles on photosynthesis: challenges and opportunities.Mater. *Focus*, 5: 405–411.

Vannini,C., Domingo, G., Onelli, E., De Mattia, F., Bruni, I., and Marsoni, M. 2014.Phytotoxic and genotoxic effects of silver nanoparticles exposure on germinating wheat seedlings.*J. Plant Physiol.*, 171: 1142–1148.

Walker, L. 2005. In: Nanotechnology for agriculture, food and the environment. Presentation at nanotechnology biology interface: Exploring models for oversight, University of Minnesota, University of Minnesota, September 15.

Wanekya, A. K., Chen, W., Myung, N.V. and Mulchandani, A. 2006. Nanowire-based electrochemical biosensors. *Electroanalysis*. 18(6): 533–550.

Wang, Q., Ma, X., Zhang, W., Pei, H., Chen, Y. 2012. The impact of cerium oxide nanoparticles on tomato (Solanumlycopersicum L.) and its implications for food safety.*Metallomics*, 4(10):1105–1112.

Wu, S. C., Cao, Z. H., Li, Z. G., Cheung, K. C. and Wong, M. H. 2005. Effects of biofertilizer containing N-fixer, P and K solubilizers and AM fungi on maize growth: a greenhouse trial, Geoderma 125: 155–166.

Yang, F. L., Lim X. G., Zhum F., and Leim C. L. 2009.Structural characterization of nanoparticles loaded with garlic essential oil and their insecticidal activity against Triboliumcastaneum (Herbst) (Coleoptera: Tenebrionidae). *J Agric Food Chem* 57(21): 10156–10162.

Yang, F. L., Li, X. G., Zhu, F. and Lei, C. L. 2009. Structural characterization of nanoparticles loaded with garlic essential oil and their insecticidal activity against Triboliumcastaneum (Herbst) (Coleoptera: Tenebrionidae). *J Agric Food Chem* 57(21): 10156–10162.

Young, K. J. 2009. Antifungal activity of silver ions and nanoparticles on phytopathogenic fungi. *Plant Dis.* 93(10): 1037-1043.

Zhao, W., Ge, P., Xu, J. and Chen, H. 2009. Selective detection of hypertoxic organophosphates pesticides via PDMS based acetylcholinesterase-inhibition biosensor. *Environ. Sci. Technol.,* 43: 6724–6729.

Zheng, L., Hong, F., Lu, S., Liu, C. 2005. Effect of nano-TiO(2) on strength ofnaturally aged seeds and growth of spinach. *Biol. Trace Elem. Res.* 104: 83–91.

Zheng, Z., Zhou, Y., Li, X., Liu, S., Tang, Z. 2011. Highly-sensitive organophosphorous pesticide biosensors based on nanostructured films of acetylcholinesterase and CdTe quantum dots, *Biosens.Bioelectron.,* 26: 3081–3085.

6

The Potential of Underutilized Plants-Current and Future Perspective

Sanjay Yadav[1*], Pawan Kumar Yadav[2] and O.P. Bishnoi[3]

[1*]District Consultant, NPPCF, Civil Surgeon Office, Narnaul

[2]Lecturer (Biology), Govt. Senior Secondry School, Ferozepur Jhirka, Nuh, India

[3]Department of Genetics and Plant Breeding, CCS HAU, Hisar 125004 Haryana, India

*Correspondence author: sanju15583@gmail.com

Abstract

Underutilized and Neglected plants are lesser-known plant species in terms of marketing and research and are well adapted to marginal and stress conditions. They are important not only for their significant contribution to human nutrition, but also because they could be used to prepare various value- added products. These products are different from those of mainstream commodities. Underutilized plants bear a number of benefits as they are rich source of micronutrients, vitamins, as well as health-promoting secondary plant metabolites. Therefore, agricultural and horticultural research should develop strategies not only to produce more food, but also to improve access to more nutritious food. Thus the present chapter is focusing to evaluate Underutilized and Neglected plants as important "Food for 21st century" for improving the nutritional status as well as increasing resilience of agro- and horti-food systems.

Keywords: Malnutrition, ethnobotany, underutilized, anti-microbial, phytochemicals.

Introduction

Since ancient times plants have been playing an important vital role for human health. With the increasing population pressure worldwide, around 40000 to 100000 plant species have been used for food, fodder, medicine, fuel, cultural and industrial purposes and out of these only 30 species are used to meet 90% world's food requirements (Abbasi *et al.*, 2013). Due to rapid rise in global population, feeding becomes major problem. The 20th Century witnessed the undertaking of systematic collection to rescue the genetic resources of staple crops (Pistorius., 1997) whereas the 21st Century has started with the awareness on the need to rescue and improve the use of those potentially important species which are left aside by research, technology, marketing systems as well as conservation.

Malnutrition, hunger, poor health and even starvation are still the world's greatest challenges. Malnutrition is defined as deficiency of nutrition due to not ingestion of proper amount of nutrients or not eating enough food and consuming nutrient poor food in respect to daily nutritional requirements (Blossner and de Onis, 2005). Malnutrition reduces physical and mental development during childhood and weakened health and productivity (Triantaphylides *et al.*, 2009). Malnutrition and diseases are closely associated to each other. In developing countries like India there has been a dramatic increase of chronic diseases such as cancer, diabetes and cardiovascular diseases. According to World Food Programme, poor nutrition causes nearly half of deaths in children below five years in developing countries. Incidences of these diet related diseases are high in low income communities where foods of animal origin are unaffordable (Yang *et al.*, 2009). Various neglected or underutilized crops and plants can offer an alternative source of nutrients, vitamins and health promoting plant secondary metabolites. Therefore mainstreaming these crops into local food system will help in alleviating malnutrition especially in rural communities (Baldermann *et al.*, 2016).

Plant species which are neglected or underutilized can contribute significantly to meet requirements of healthy diet, appropriate food, energy requirements, nutrition and income generation (Kumar *et al.*, 2014). Underutilized crops are plant species that are used traditionally for their food, fiber, fodder, oil and medicinal properties, but have yet to be adopted by large scale agriculturalists. In general underutilized/neglected plant species are those that occur as life support species in extreme environmental conditions and threatened habitats having genetic tolerance to survive under harsh conditions (Monika, 2014) The Underutilized foods can also be defined as "The foods which are less available, less utilized, rarely used and region specific" (William and Haq. 2002).

Underutilized/Neglected plants have poor shelf-life, unrecognized nutritional value, poor consumer awareness and reputational problems, therefore also called as poor people's food. These are also referred as minor, orphan, neglected, underutilized, underexploited, alternative, local, traditional marginal crops and have been included world-wide plans of action, after having successfully raised the interest among decision makers. As a result of peoples demand for food changes, many formerly neglected crops are now globally became significant crops such as soybean, oil palm and kiwi fruit.

Underutilized plant species have a distinctive past in terms of their use and value, but their use is currently limited relative to their economic potential (Gruer *et al.,* 2006). Ethnobotanical surveys reveal that in India, there are a number of underutilized/neglected plants which are valuable from medicinal and nutritional point of view.

Criteria to Become Underutilized/Neglected Crop

Any plant species to be considered as underutilized food crop, it must have following features-

- It must have ethnobotanical and scientific proof of food value.
- These crops should have indigenous uses in localized area.
- Should also involved in folklore practices.
- Crop must have been cultivated in past or being cultivated in specific geographical area.
- Should have weak or no formal seed supply system.
- Currently cultivated in less area then other conventional crops.
- Should have good nutritional and therapeutic value with little attention of farmers and scientific world.

Need to Promote Underutilized/Neglected Plants?

Increasing population pressure and over-dependence on a few plant species for food, fodder and medicine may exacerbates many acute difficulties with regard to food security, nutrition, health, ecosystem sustainability and cultural integrity (Jaenicke and Hoschle-Zeledon, 2006). Agricultural development and food security depend, in part, on our ability to broaden the range of agricultural and forestry species in an effective and sustainable way. This means finding ways to protect and enhance locally important species so that they can be deployed more widely in agriculture and environmental management. Current global and socioeconomic changes enjoin us to act now because, as crop populations and knowledge of their uses disappear, the continuing neglect and under- use of many valuable agricultural species is eroding their genetic base. Therefore the developing countries like India are being encouraging their scientific community, policy makers and non government organizations to contribute to creation of new opportunities to rescue the genetic base of many neglected or underutilized plants to maximize their contribution for global food security (ICUC, 2006).

Potential of Underutilized/Neglected Plants

Globally there are large area of marginal and wasteland which is not suitable for the cultivation of staple crops due to their poor fertility and water resources.

Most of underutilized/neglected plants are tolerant to harsh conditions and suitable to grow on marginal or wastelands (Hedge, 2002). Worldwide almost all regions and food cultures have examples of nutritionally rich and culturally valued species that suffer neglect and may be falling into disuse. Focusing attention on such neglected plant species is an effective way to maintain healthy and diverse diet to combat micronutrient deficiencies mainly among rural people in developing countries. Many underutilized fruits or crops are very rich in nutrition but unfortunately they are not popular among farmers for cultivation. Most of these plants have very good potential to improve economic and food security in developing countries like India that include-

- Helpful in improving physical and economic status of poor people.
- It provides the ways to reduce the risk of over-dependency on limited number of major staple food crops.
- A way to preserve and celebrate cultural and dietary diversity.
- Way to increase sustainability of agriculture and food quality.
- Use of Marginal and wastelands for agriculture to meet ever increasing food demand (Mayes *et al.*, 2011).

Apart from being the house of nutrients, these crops are evolved with very important genetic pool for resistance to biotic and abiotic stresses.

Some Underutilized/ Neglected Plants of North-West India

There are many underutilized/neglected plants in India and majority of them are not well documented (Singh *et al.*, 2012). The ethnobotanical data for all the underutilized species is still to be explored. Some of the Underutilized/ Neglected plants are documented in Table.

Nutitional and Medicinal Value of Neglected/Underutilized Plants

Traditional medical knowledge of medicinal plants and their use by indigenous cultures are not only useful for conservation of cultural traditions and biodiversity but also for community healthcare and drug development in present and future. Demand for medicinal plants is increasing in both developed and developing countries due to growing recognition of natural products being non narcotic, having no side effects and easily available at affordable price. The poor people of country are mostly malnourished along with multiple nutrient-deficiency disorders due to ignorance about importance of fruits and vegetables in their diets. Rural area is considered to be rich in biodiversity which is not harnessed fully. Plants which are neither grown commercially on large scale nor traded widely are considered as underutilized plants. They are cultivated and consumed locally and easier to grow in adverse conditions. These underutilized/ neglected plants are rich of source of carbohydrates, fats, proteins, energy, vitamins, minerals and dietary fiber. So, exploitation of such plants can become a solution to prevent and cure various diseases like kwashiorkor, night blindness, anemia, diabetes, cancer etc. It is also an established fact that seasonal, locally available, and cheap fruits and vegetables can also keep the population healthy and nutritionally secure rather than costly off-seasonal ones. Some of the underutilized/neglected plants found in North-West India which used as raw or processed preparations are described below:

Chenopodium album L.

Chenopodium album is popularly called Bathua and found abundantly in the winter season.

List of Some important Underutilized/Neglected Plants of India.

Sr. No.	Common Name	Botanical Name	Family	Medicinal/Nutritional Value
1	Ber	*Ziziphus jujube* Lam.	*Rhamnaceae*	Paste is applied to treat ringworm and fruits are eaten raw as they are very nutritious.
2	Lasora	*Cordia dichotoma* Frost.	*Boraginaceae*	Plant is very useful in ulcers, headache and urinary diseases.
3	Gondhani	*Cordia rothii* Roem. & Schult.	*Boraginaceae*	Root, bark and leaves are used to cure malaria, intestinal disorders and conjunctivitis.
4	Barbanta	*Ficus religiosa* L.	*Moraceae*	Fruits are known as Barbanta by local people. Leaves fruits and bark are used to relieve vomiting and hiccup.
5	Kair	*Capparis decidua* (Roth.)	*Capparaceae*	Whole plant decoction is used to treat scurvy and fruits are used in preparation of pickle.
6	Sangar	*Prosopis cineraria* L.	*Mimosaceae*	Stem bark is used to treat intestinal worms. Fruits which are known as sangar in regional language are eaten raw or as a vegetable.
7	Karonda	*Carissa carandas* L.	*Apocynaceae*	Fruit is rich source of iron and vitamin C. Mature fruits is harvested for pickles, jam, syrup, chutney and also used for treatment of anemia.
8	Bel	*Aegle marmelos* L. (Corr.)	*Rutaceae*	Fruit pulp is used in diarrhea and dysentery.
9	Bathua	*Chenopodium album* L.	*Chenopodiaceae*	Whole plant is used as Vegetable and to treat Diarrhoea.
10	Choulai	*Amaranthus spinosus* L.	*Amaranthaceae*	Root juice is applied in Itching and leaves are used as vegetable.

Sr. No.	Common Name	Botanical Name	Family	Medicinal/Nutritional Value
11	Jangli Karela	*Momordica charantia* L.	*Cucurbitaceae*	Used for the treatment of diabetes, cough, respiratory diseases, skin diseases, wounds, ulcer and rheumatism.
12	Imli	*Tamarindus indica* L.	*Caesalpiniaceae*	Bark, fruits and seeds are used as blood purifier, cholera and snakebite.
13	Khareti	*Sida acuta Brum*. F.	*Malvaceae*	Root is used for itching.
14	Jangli sarsoan	*Sisymbrium irio* L.	*Cruciferae*	Immature leaves are stimulant and also used in the treatment of asthma and throat infections.
15	Garmunda	*Citrullus colocynthis* L.	*Cucurbitaceae*	Fruits, seeds and root is used in respiratory, intestinal and urinary problems.
16	Peel	*Salvadora oleoides* Decne.	*Salvadoraceae*	Bark, fruits and leaves are used to relive pain in piles and skin diseases.
17	Gokharu	*Tribulus terrestris* L.	*Zygophyllaceae*	Whole plant is used in Fever, Sterility and skin disease.
18	Jawar	*Sorghum bicolor* (L) Moench	Poaceae	Sorghum is used in traditional medicine to cure anemia, cancer, and a variety of infectious diseases.
19	Kachari	*Cucumis melo* L. var.agrestis	*Cucurbitaceae*	Fruits and seeds are use used raw and have digestive properties.
20	Saatthi	*Boerhaavia diffusa* L.	*Nyctaginaceae*	Roots are used for fomentation to alleviate pain and swelling. Fresh root juice used to mitigate ailments of eye. Locally used as vegetable.

This plant is a rich source of oxalic acid, protein, vitamin A, calcium, phosphorus, and potassium. It has many medicinal and nutritional properties like antirheumatic, contraceptive, laxative etc. *Chenopodium album* used in the treatment of urinary problems, bug bites, sunstroke and skin problems. The leaves and young shoots are eaten as a leaf vegetable and also used in dishes such as soups, curries, and paratha (Poonia and Upadhayay, 2015).

Ziziphus jujube Lam.

Ziziphus fruits and seeds are delicious and an effective herbal remedy used in traditional medicine system they have anti-fungal, anti-bacterial, anti-ulcer, anti-inflammatory, antispastic, antinephritic, cardiotonic, antioxidant and wound healing properties. Being mucilaginous, the fruits are very soothing to the throat and decoction used to treat sore throats. Plant is good source of saponins, triterpenoids and alkaloids. The roots are used in the treatment of dyspepsia, fever and wound healing and also promote the growth of hairs. (Preeti. 2014).

Cordia Dicotoma

Various parts of the plant are used medicinally. It is highly esteemed for treating coughs and diseases of the chest and also has anti-inflammatory effects. Mucilage is used to treat ulcers. Juice of the leaves is applied as a poultice to treat migraine. The immature fruits are used to prepare pickle, vegetable and fodder (Kuppast and Vasudev, 2006).

Cordia rothii Roem. & Schult

Fruits and leaves are used to treat mouth ulcers. This plant is effective in heart ailments and cardiotonic activity in Saurashtra region of Gujarat. It is a rich source of phytochemicals like alkaloids, glycoalkaloids, coumarins and inorganic salts (Chauhan, and Chavan, 2009).

Ficus religiosa L.

It is commonly known as pepal which sheltered the Buddha as he divined the "Truths." It is used traditionally to treat about 50 types of disorders, as it contains antiulcer, antibacterial, antidiabetic, antioxidant, hypolipidimic, hypoglycemic, wound healing, immunomodulatory and in the treatment of gonorrhea and skin diseases. This plant contained appreciable amounts of phytochemicals like phenols, tannins, steroids, alkaloids, flavonoids and vitamins. The fruits are a rich source of vitamins and some essential amino acids such as isoleucine, and phenylalanine. Leaves contain carbohydrate, protein, lipid, calcium, sodium, potassium, and phosphorus (Kaur *et al.*, 2011).

Capparis Decidua

Capparis decidua is used in Unani system of medicine to cure ailments like scurvy, heart disease, improves appetite, rheumatism, cough, asthma, boils, swellings and as antidote to poison. It is very efficacious in relieving toothache and pyorrhoea when chewed and decoction of ground stems and leaves is used to kill intestinal worms. The fruits are used to prepare pickle and vegetable for diabetic patients. In Rajputana, the plant is a wholesome fodder for camels. Various biochemical compounds like alkaloids, phenols, sterols or glycosides are present in considerable amount (Joseph and Jini, 2011).

Prosopis cineraria (L.)

Prosopis cineraria (L.) is an herbal plant used traditionally as fodder and fruits are used as vegetable. It is also used for the treatment of various ailments like leprosy, dysentery, asthma, dyspepsia and earache. Pharmacological activities like analgesic, antipyretic, antihyperglycemic, antioxidant, antihypercholesterolemic, antitumor have been also reported from different plant extracts. Various phytoconstituents like tannins, steroids, flavone derivatives and alkaloids has been isolated from the plant (Garg and Mittal, 2013).

Aegle marmelos L.

According to Vedic literature *Aegle marmelos* is extensively used for the treatment of various diseases like jaundice, constipation, chronic diarrhea, dysentery, stomachache, fever, asthma, inflammations, acute bronchitis, snakebite, abdominal discomfort, acidity, burning sensation, epilepsy, indigestion, leprosy, spermatorrhoea, leucoderma, eye disorders, ulcers, nausea, sores, swelling, thyroid disorders, and upper respiratory tract infections. Phytochemicas like alkaloids, glycosides, terpenoids, saponins, tannins, flavonoids and steroids are found in good amount. Fruits possess high nutritional value and used to make juice, jam, sirup, jelly, toffee and other products. Pulp contains water, sugars, protein, fiber, fat, calcium, phosphorus, potassium, Iron, minerals and vitamins (Sekar *et al.*, 2011).

Amaranthus spinosus L

Traditionally it is used in folklore system to treat abscesses, burns, wounds, inflammation, monorrhagia, gonorrhoea and eczema. Juice of *Amaranthus spinosus* is used to prevent swelling around stomach. Boiled leaves without salt consumed for 2-3 days to cure jaundice and stomach ache. Root paste in equal volume of honey is used to control vomiting. *Amaranthus spinosus* contains medicinal properties like antioxidant, anti-inflammatory, anti-malarial, anti-tumour, anti-nociceptive, antibacterial, anti-helmintic and anti-diabetic. Active constituents like alkaloids, flavonoids, glycosides, phenolic acids, steroids, amino acids, terpenoids, lipids, saponins, b-sitosterol, stigmasterol, linoleic acid, rutin, catechuic tannins and carotenoids are present in good amount (Rai *et al.*, 2014).

Momordica charantia L.

Momordica charantia also commonly known as bitter melon. It is an alternative therapy that is used for lowering blood glucose levels in patients with diabetes mellitus. Components of bitter melon extract appear to have structural similarities to animal insulin. It contains antiviral, hypoglycemic and antineoplastic activities (Basch *et al.*, 2003).

Tamarindus indica L.

It belongs to the family *Leguminosae* and used extensively in traditional medicine. Preferably it is used for abdominal pain, diarrhea and dysentery, some bacterial infections and parasitic infestations, wound healing, constipation and inflammation. Pulp is used in foods for flavour confections, curries and sauces. Tamarind is a fair source of Vitamin C, B and minerals like P, K, Ca and Mg. It is a rich source of phytochemicals, and hence the plant is reported to possess antidiabetic, antimicrobial, antivenomic, antioxidant, antimalarial, cardioprotective, hepatoprotective and antiasthmatic activity (Kuru, 2014).

Sida acuta Brum. F.

Sida acuta is widely used for spiritual practices. The plant may be used alone or in combination with other plants according to the diseases or to the healers. It may be administrated orally route in the case of fever or by external application of the paste directly on the skin for skin diseases or snake bites (Karou 2007).

Citrullus colocynthis L.

Citrullus colocynthis has been reported to possess a wide range of traditional medicinal uses including asthma, bronchitis, jaundice, diabetes, leprosy, common cold, cough, joint pain, cancer, toothache and wound healing. It also protect from different microbial infections. The plant was also shown to be rich in nutritional value with high protein contents and excellent biological activities, which include antidiabetic, antioxidant, cytotoxic, insecticidal, antimicrobial and anti-inflammatory. Several bioactive chemical are isolated such as, glycosides, flavonoids, alkaloids, fatty acids and essential oils (Abdullah *et al.*, 2014).

Salvadora oleoides Decne.

Salvadora oleoides Decne. is commonly known as meetha jal in India. It is used in the treatment of various ailments like piles, tumors, bronchitis, cough rheumatism, fever, conjunctivitis and ulcer. The major phytochemicals of this plant are sterols like beta-sitosterol and their glucosides, flavonoids, dihydroisocoumarin, terpenoids like-methoxy-4-vinylphenol and cis-3-hexenyl benzoate. It also possess anti hypoglycemic, hypolipidemic, analgesic, and antimicrobial activity (Garg *et al.*, 2014).

Tribulus terrestris L.

It is an herb and used by our rishis to treat various body disorders. It bears various medicinal properties. It is commonly known as Gokhshura. The leaves are useful in affection of urinary calculi, stem is astringent and its infusion is useful in gonorrhea. Literature also reveals that it have stimulating effect on sexual functions by stimulating effect on spermatogenesis. Phytochemistry of this plant shows the presence of chlorgenin, gitogenin, kaempferol 3- glucoside, kaempferol 3 – rutinoside, diosgenin glycosides, β-sitosterol, stigmasterol, diosgenin and neotigogenin (Ukani *et al.*, 1997).

Cucumis melo var. agrestis

Cucumis melo var. agrestis, is an annual creeper, is available all over India and well known for its antioxidant property. The fruits can be used as a cooling light cleanser or moisturiser for the skin. They are also used to treat burns and abrasions. Seed is antitussive, digestive, febrifuge and vermifuge. Phytochemical investigation reveals that carbohydrates, tannins, flavonoids, alkaloids, saponins, steroids & triterpinoids and glycosides were present in both the extracts (Alagar *et al.*, 2015).

Boerhaavia diffusa L.

Boerhaavia diffusa is one of the most renowned medicinal plant used to treat large number of human ailments as mentioned in Ayurveda, Charaka Samhita, and Sushrita Samhita. The Plant have numerous medicinal properties and are used by endemic and tribal people of India. Various phytochemicals such as flavonoids, alkaloids, steroids, triterpenoids, lipids, lignins, carbohydrates, proteins, and glycoproteins. have been isolated and studied in detail for their biological activities such as Anti-bacterial, Anti-nociceptive, hepato-protective, hypo-glycemic, anti-estrogenic, anti-inflammatory, anti-stress, dyspepsia, abdominal pain, inflammation and jaundice (Mahesh *et al.*, 2012).

Carissa carandas L.

Carissa carandas is a useful food and medicinal plant of India which is used as a traditional medicinal plant. Ethnopharmacological ly it shows anti-cancer, anti-oxidant, analgesic, anti-inflammatory, anti-ulcer, cardiovascular, anti-nociceptive, anti-diabetic, antimicrobial and cytotoxic activities. The major bioactive constituents are alkaloids, flavonoids, saponins, glycosides, triterpenoids, phenolic compounds and tannins (Singh and Uppal., 2015).

Advantages of Underutilized Plants

World population is facing rapid urbanization and around 50% population of developed and developing countries are resides in cities. This rapid urbanization and simultaneous growing world population creating challenge of food supply to consumers. In India there is lack of policies of authentication and documentation of neglected/underutilized crops. Government should take into account the underutilized crops by considering the ethnobotanical knowledge available in literature. Indigenous knowledge must be tapped and combined from various local communities which will be merged with scientific world to create new food opportunities. These underutilized/neglected plants exhibit a number of benefits such as-

Extra Income for Rural Community- Growing demand from consumers for novel food items may create new market for Neglected and Underutilized species which can generate extra income for community.

Ecosystem Stability- Modern changing global environment led to a growing interest in crops/plants species which are well adopted for such as adverse conditions.

Maintaining Cultural Diversity- Plants are used in local culture and traditions from ancient times. Many neglected and underutilized plants play an important role in keeping cultural diversity alive.

Food Security and Nutrition- Many neglected and underutilized plant species are nutritionally rich providing important vitamins, proteins and minerals. Further neglect and genetic erosion of such plant species can have immediate consequences on the nutritional status and food security around the globe

Various Processed Food Items from Underutilized/Neglected Plants.

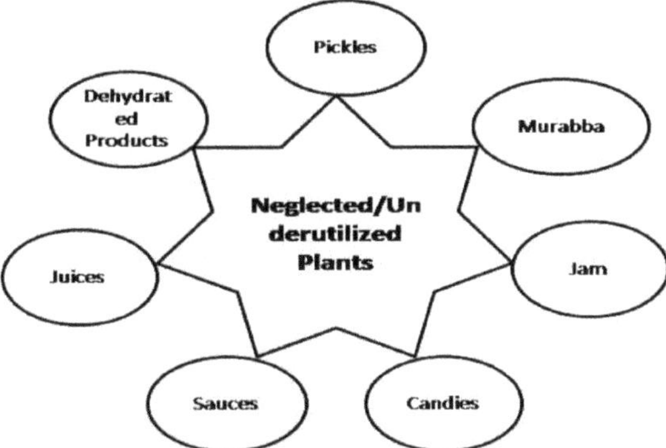

Various pharmacological activities showed by Underutilized/Neglected plants.

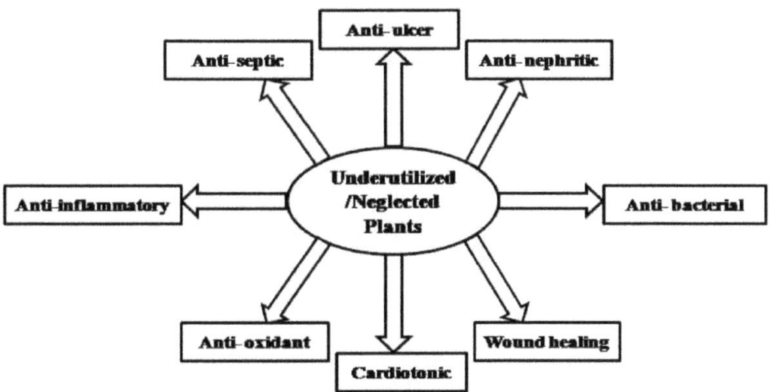

Conclusion

In present global scenario many underutilized/ Neglected plants which were widely grown earlier but today falling into disuse due to a variety of economic, cultural, agronomic and genetic factors. They can be used to prepare various useful products. These products are nutritious good antioxidants and high in fiber contents. These are different from the mainstream commodities. Limited and seasonal availability and difficulties in harvesting/collection make these commodities costly. High cost prevents the sale of products in local market. Generally available commodities and products are cheaper. In developing countries like India the demand for fruit and vegetable preserves and preparations is increasing day by day. But due to small scale production, marketing and sold locally the products of underutilized plants could not provide proper income to poor. Demand for underutilized fruit/vegetable products, market share, and profits could be increased. Awareness on benefits of these products and market promotion is necessary.

Refrences

Abbasi, A.M., Khan, M.A., Shah, M.H., Shah, M.M., Pervez, A. and Ahmad, A. 2013. Ethno botanical appraisal and cultural values of medicinally important wild edible vegetables of lesser Himalayas, *Pakistan Journal of Ethno biology and Ethno Medicine*. 1: 9-13.

Abdullah, I.H., Hussain, A.R., Munavvar, Z.A.S., Shahzad A.S.C., Satyajit, D.S. and Anwar, H.G. 2014. *Citrullus colocynthis* (L.) Schrad (bitter apple fruit): A review of its phytochemistry, pharmacology, traditional uses and nutritional potential. *J.Ethnopharmacology*. 155: 54-66.

Alagar, R .M., Sahithi, G., Vasanthi, R., David Banji, K.N.V. and Rao, S. D. 2015. Study of phytochemical and antioxidant activity of *Cucumis melo* var. agrestis fruit. *Journal of Pharmacognosy and Phytochemistry*. 4 (2): 303-306.

Baldermann, S., Frede, K., Klopsch, R., Neugart, S., Neumann, A., Ngwene, B., Schroter, D., Schweigert, F.J., Wiesner, M. and Schreiner, M. 2016. Are Neglected Plants are the Food for the Future. Critical Reviews in Plant Sciences. 35 (2): 106-119.

Basch, E., Gabardi, S. and Ulbricht, C. 2003. Bitter melon (*Momordica charantia*): A review of efficacy and safety. *American Journal of Health-System Pharmacy*. 60 (4): 356-359.

Blossner, M. and De Onis, M. 2005. Malnutrition: Quantifying the health impact at national and local levels. In: WHO (Ed.), WHO Environmental Burden of Disease Series (Vol. 12, pp. 43), WHO, Geneva.

Chauhan, M. G. and Chavan, S S. 2009. Pharmacognosy and biological activity of *Cordia rothii* Roem. & Schult. Bark. *Indian Journal Of Traditional Knowledge*. 8: pp. 598-601.

Garg, A. and Mittal, S.K. 2013. Review on *Prosopis cineraria*: A potential herb of Thar desert. *Drug Invention Today*, 5: 60-65.

Hegde, N.G. 2002. Promotion of underutilised fruit crops. In: Fruits for the future in Asia. Haq, N. and Hughes, A. (eds). Proceedings of the Regional Consultation Meeting, Bangkok, Thailand. International Centre for Underutilised Crops, University of Southampton, Southampton, UK, pp. 45-53.

ICUC. 2006. Annual Report 2005-2006 (PDF online reproduction). Colombo, Sri Lanka: International Centre for Underutilized Crops. ISBN 955-1560-03-5. ISSN 1800-2315.

Jaenicke, H. and Hoschle-Zeledon, I. 2006. Strategic Framework for Underutilized Plant Species Research and Development: With Reference to Asia and the Pacific and Sub-Saharan Africa. International Center for Underutilized Crops, Sri lanka and Global Facilitation Unit for Underutilized Plants, Rome, Italy, p. 33.

Joseph, B. and Jini, D. 2011. Medicinal Potency of *Capparis deciduas*- A Harsh Terrain Plant. *Research Journal of Phytochemistry*, 5: pp. 1-13.

Karou, S.D., Nadembega, W.M.C., Ilboudo, D.P., Ouermi, D. and Simpore, J. 2007. *Sida acuta* Burm. f.: A medicinal plant with numerous potencies. *African Journal of Biotechnology*. 6 (25): 2953-2959.

Kaur, A., Rana, A. C., Tiwari, V., Sharma, R. and Kumar, S. 2011. Review on Ethanomedicinal and Pharmacological Properties of *Ficus religiosa*. *Journal of Applied Pharmaceutical Science*, 1: pp. 6-11.

Kumar, D., Dwivedi, S.V., Kumar, S., Ahmed, F., Bhardwaj, R. K., Thakur, K.S. and Thakur, P. 2014. Potential and biodiversity conservation strategies of underutilized or indigenous vegetables in Himachal Pradesh. *International Journal of Agricultural Sciences*. 10:459-462.

Kuppast, J and Vasudev, N. 2006. Wound healing activity of *Cordia dichotoma* fruits. *Natural Product Radiance*, 5: 99.

Kuru, P. 2014. *Tamarindus indica* and its health related effects. *Asian Pacific Journal of Tropical Biomedicine*. 4 (9): 676-681.

Mahesh, A.R., Kumar, H., Ranganath, M.K. and Devkar, R.A. 2012. Detail Study on *Boerhaavia Diffusa* Plant for its Medicinal Importance- A Review. *Research Journal of Pharmaceutical Sciences*. 1 (1): 28-36.

Mayes, S., Massawe, P.G., Alderson, J.A., Roberts, S.N., Azam- Ali and Hermann, M. 2011. The potential for underutilized crops to improve security of food production. *Journal of Experimental Botany*, 1-5.

Pistorius, R. 1997. Scientists, Plants and Politics – A history of the Plant Genetic Resources Movement. International Plant Genetic Resources Institute, Rome, Italy. 134.

Poonia, A. and Upadhayay, A. 2015. *Chenopodium album* Linn: review of nutritive value and biological properties. *J. Food Sci Technol*. 52 (7): 3977–3985.

Preeti, and Tripathi, S. 2014. *Ziziphus jujube*: A Phytopharmacological Review. *International Journal of Research and Development in Pharmacy and Life Sciences*. 3: pp. 959-966.

Rai, P.K., Jindal, S., Gupta, N. and Rana, R. 2014. An Inside Review of *Amaranthus spinosus* L.: A Potential Medicinal Plant of India. *Int. J. Res. Pharmacy Chemistry*. 4(3): 643-653.

Sekar, D.K., Kumar, G., Karthik, L. and Rao, B.K.V. 2011. A review on pharmacological and phytochemical properties of *Aegle marmelos* (L.). *Asian Journal of Plant Science and Research*, 2: 8-17.

Singh, A. and Uppal, G.K. 2015. A Review on *Carissa carandas* – Phytochemistry, Ethnopharmacology and Micropropagation as Conservation strategy. *Asian Journal of Pharmaceutical and Clinical Research.* 8 (3): 26-30.

Singh, S.J., Batra, V. K., Singh, S. K. and Singh T. J. 2012. Diversity of underutilized vegetable crops species in North- East India with special reference to Manipur: A review. *NeBIO*. 3 (2): 87-95.

Thakur, M. 2014.Underutilized food crops: treasure for the future India. *Food Science Research Journal*, 5: 174-183.

Triantaphylides, C. and Havaux, M. 2009. Singlet oxygen in plants: Production, detoxification and signaling. *Trends Plant Sci*. 14: 219–228.

Ukani, M.D., Nanavati, D.D and Mehta, N.K. 1997. A Review on the Ayurvedic Herb *Tribulus terrestris* L.. *Ancient Science of Life*. 17 (2): 144-150.

Williams, J.T. and Haq, N. 2002. Global research on underutilized crops - an assessment of current activities and proposals for enhanced cooperation. Southampton, UK: International Centre for Underutilized Crops. ISBN 92-9043 545-3.

Yang, R.Y. and Keding, G.B. 2009. Nutritional contributions of important african indigenous vegetables. African indigenous vegetables in urban agriculture (pp. 105–144). Earthscan, London.

7

Biotechnology: A Potential Tool for Upliftment of Economic Status of Farmers

Subhash Kajla, Renu Singh, Anil K. Poonia and Ram C. Yadav*

Centre for Plant Biotechnology, CCS Haryana Agricultural University Campus, Hisar, India

**Corresponding author: rcyadavbiotech@yahoo.com*

Abstract

Modern Biotechnology is considered as a powerful tool and leads to a great evolution in crop improvement in terms of production, quality and safety, besides preserving the environment. The techniques like Plant Tissue Culture and Molecular Biology have contributed a lot in genetic improvement, mass multiplication, testing of genetic stability/diversity and production of disease free quality planting material of various crops at global scale. Other tissue culture techniques like embryo rescue, protoplast fusion, anther culture, ovary culture, somatic embryogenesis, and synthetic seeds were also being used widely for development of new varieties and conservation of existing germplasm. The tissue culture industry witnesses a gross installed production capacity of millions of plantlets per annum in a short spell of time. Biological investigation comprises use of different markers used for phenotypic, chemotypic and genotypic characterization. Genotypic characterization involves the use of DNA markers, which include Random Amplified Polymorphic DNA (RAPD), Amplified Fragment Length Polymorphism (AFLP), microsatellites etc. are used in breeding programmes. Each of these markers has a specific merits and demerits. Use of molecular markers, in association with linkage maps and genomics, offers quantitative potential to breeders for genetic improvement more precisely and rapidly as compared to other traditional and conventional methods. Genetic Transformation is the most recent technique to enhance genetic variability, available within a crop. Organisms produced from this process are commonly called Genetically Modified Organisms (GMOs). This is a promising tool for the development of new varieties with specific traits which are not present within the crop gene pool. The potential of biotechnological techniques has reduced pesticide use and has reduced environmental footprint associated with pesticide use by 20%. Similarly in the period from 1996 to 2013, cultivation of *Bt* maize caused a huge reduction of 72 million kg in insecticides. The adoption of this technology has resulted in significant socioeconomic benefits and advantages of these techniques for enhancing the income of farmers.

Keyword: Biotechnology, Transformation, Genetic fidelity and RAPD

Introduction

The positive correspondence between agricultural biotechnology and food security has for long been established. Early in the 54th UN general assembly session declared the application of agricultural biotechnology in developing countries provides viable opportunities for improving productivity and increasing production capacity in the agricultural sector (Mugabe, 2000; Santaniello, 2005). It was also pointed out in connection to the green revolution that "agricultural biotechnology is the major technological innovation to be made available to farmers after the end of the green revolution". These positions at once identify and recommend the application of biotechnology in agriculture as an apparatus for addressing food insecurity in developing countries. Asserting the linkage between agricultural biotechnology and food security, it was pointed out that agricultural biotechnology promises to play a critical role in improving agricultural productivity and reducing the environmental impact of agriculture leading to agricultural sustainability and food security in many regions of the world. Indeed, agricultural biotechnology has enlarged and increases the abilities of science to overcome various genetic and environmental constraints which impose serious limitations on the capacities of crops and animals to yield their optimum outputs. Biotechnological approaches, specifically, plant tissue culture, molecular biology and now genetic transformation, are found to have good potential as a supplement to traditional agriculture for multiplication and genetic enhancement of the crops with improved food, fibres and fuel. Using the tools of agricultural biotechnology, plant breeders can select single genes that produce a desired trait and can transfer them into another plant easily by overcoming the genetic barrier between these plants. Thus, through agricultural biotechnology, the yields of particular plants can be increased, their nutrients content and useful biochemical may be improved faster and in a much easier way than as compared to other traditional plant breeding methods. Applying the techniques of agricultural biotechnology in combination with other scientific techniques the insert and innate productivity potentials of some of our crops can be harnessed and unleashed to aid the nation as a whole to realize its dream of food security. Plant tissue culture techniques offer an integrated approach for the mass production and multiplication of true to type and disease-free plants within a small spell of time. Use of molecular markers for improved selection in plant breeding and genetic engineering has also proved very effective. The present chapter includes the biotechnological techniques which have been accepted as an important tool for direct application of different biotechnological techniques in the field of agriculture development. The adoption of these technologies has resulted in significant social and economic benefits to the farmers and proved with vast potential in agriculture sector. Besides three important pillars of food security will be discussed briefly in this chapter.

The term "Biotechnology" has been used to refer to many biological processes that produce useful products, including some ancient ones such as fermentation in beer, wine and cheese (Coombs, 1992; Zaid *et al.*, 1999). Most frequently used term to refer about the natural processes of DNA replication, breakage, ligation, and repair that has made possible a deeper understanding of the mechanics of cell biology and the hereditary process (Mc Couch, 2001). The chapter describes DNA based molecular techniques used to modify the genetic composition of agriculturally useful plants. Traditional or conventional methods of modifying the genetic composition of plants, still widely used alone and in conjunction with DNA based methods for crop improvement has been modified by transfer of DNA from one organism to another using DNA based techniques, i.e., not breeding, are referred as transgenic, genetically engineered, or rDNA (recombinant DNA). These terms are preferred to genetically modified organisms (GMOs) because the genetic composition of various agricultural crops has been modified. (Ruttan, 2004). Biotechnology has a series of powerful tools in addition to genetic engineering that are useful for changing the genetic composition of plants, including those mentioned below (Toenniessen *et al.*, 2003; Naylor *et al.*, 2004). The biotechnological tools may likely to have great impacts in crop improvement, where even small changes like change in colour, aroma quality and postharvest behaviour would make significant commercial impacts. Genetic transformation, micropropagation, plant tissue culture *in vitro* conservation of germplasm, biofertilizers, biopesticides, post-harvest biotechnology, synthetic seed technology and virus cleaning through STG techniques are important areas in biotechnology. In our country, micropropagation of different crops has established and *in vitro* protocols have been successfully applied in various medicinal, ornamental, horticultural, forest, and other food crops. Systematic efforts in the field of plant research and development focused at commercialization of product and process may go in a long way in promotion of agricultural productivity in the country for enhancing India's capability to compete globally (Sengar and Chaudhary, 2009)

Biotechnology has provided a number of specific agronomic traits and overcome a number of production barriers for farmers. This led to improved productivity and profitability for millions to the adopting farmers who have applied the biotechnological techniques. Agricultural biotechnology resulted in significant benefits for farmers like including increased crop yield. The implementation of effective and successful agricultural development strategies ensure food security in Indian Subcontinent represents and proved one of the most crucial issues of the 21st century. Following are some important issues to target in order to increase food security using biotechnology:

1. Ensuring Food Availability:

Certain techniques of agricultural biotechnology can be applied in the field of food production:

a) **Increasing per Seed Yield in different Crops**

Use of techniques of genetic transformation, through which a few genes can be transferred from one crop to another, and allowed to be incorporated into the gene of the recipient plant, may lead to develop the high yielding, short maturing and pest resistant varieties of particular crops. This technique when adopted can be used to transfer genes that have the ability to confer high yields, into some of our major staple crops like maize, sorghum, peas etc. Through this process, genes that confer early maturity or enhanced growth can be imparted into these crops thereby obtaining new cultivars that are both high yielding and early maturing. These cultivars will enhance output in terms of yield from individual plants and the total production per acre or hectare. When short maturing varieties of these staple crops become available, the limitation imposed by crop production seasons and exacerbated presently by climate change and total absence of irrigation facilities, will be surmounted. This will result in two or even three harvest of these staple crops in one farming season, ensuring their availability in greater quantity. Again by adding genes responsible for pest and disease resistance to these to conventional crops will help them to resist against pest and diseases, all the crops losses in the field incurred due to pest and diseases attack will be curtailed, these crops will grow healthy from infancy to maturity and be harvested intact, and that harvest will certainly be greater. So will be more production from the same area.

b) **Multiplying the Planting Materials for Farmers**

Using the technique of plant tissue culture, healthy planting materials can be produced especially in case of some of vegetative propagating crops in India that require high volume of materials for planting. For example, sugar cane, aloevera, strawberries, brahmi, potatoes cuttings, pineapple etc. can be multiplied into millions through tissue culture. This will substantially increase the volume of disease free planting materials and the hectares of these crops under cultivation thus increasing the quantity of these crops harvested and available for food. Centre for Plant Biotechnology, Hisar, Department of Science and Technology Haryana state, uses this technique to propagate sugarcane, strawberries, aloevera, bamboo, brahmi, stevia and provided lakhs of plantlets to farmers of Haryana and other states in the country. This feature can be replicated for other crops, thereby eliminating or reducing shortage of planting materials and boosting food production (Herdt, 1995).

c) **Increasing the area of Land Under Cultivation**

Also through the process of gene transfer, crops that can withstand biotic and abiotic stresses can be developed. These crops are enabled genetically to grow successfully under extreme soil conditions in harsh environmental conditions. This will significantly increase the area of land under cultivation since land areas that are hitherto not cultivable due to drought and excessive water logging will now be able to support and sustain crop production. With the incorporation of these non-cultivable lands into the areas under cultivation a greater quantity of crops can be raised and more food can be produced.

2. Enhancing/Improving Nutritional Qualities

Agricultural biotechnology techniques can also be used to enhance the nutritional qualities of some of our staple crops. Through the techniques of gene transfer the nutrient composition of some of our crops (cereals and legumes) can be altered to give a higher nutritive content and value. Also, some vital nutrients not readily available from the staple diets of Indian households can be inserted into the staple foods. This will provide highly nutritious foodstuff containing the appropriate minerals, vitamins, and hormones which are capable of providing high level of nutrition to individuals to keep them in good health. This will lead ultimately to drastically reducing the menace of malnutrition caused by food insecurity as individuals will be able to deploy their human and material endowments in this direction. Also, less money will be spent by government on individual's health bills as good health from good nutrition will keep people from failing ill often. Moreover, absenteeism from work due to hospital visits and admissions, loss of family breadwinners due to ill health will decline drastically and individual's households' and national expenditures on health care will be curtailed and the money realized conserved and used for other developments. Improving the taste, texture and appearance of some of our crops some techniques of agricultural biotechnology can be used to slow down the process of spoilage so that fruits and other perishable foods seventy percent (70%) of which are spoiled in India today, can be preserved for a long time and yet be fit for consumption by the end user even long after harvest. As such these food items arrive at the end users table without any alteration in taste, texture and appearance. This enables crops that are ready for harvest in one of the seasons of the year and are scarce off season, to be preserved in taste, texture and appearance and made available experienced during scarcity. This makes food stuff available and accessible to more people all year round. Also, certain foods produced in the southern part of India and are not available in the north can conveniently be transported without alteration to their taste, texture and appearance to meet the demand in the north. While certain foods produced in the north and therefore not available in the south can also be transported with their taste, texture and appearance intact to the south to meet the demand for such products there. This will promote better food distribution and increase the availability and accessibility of these foods and enhance their utilization, thus contributing to food security. Again improving the taste, texture and appearance of food through the delayed spoilage techniques of agricultural biotechnology can significantly decrease the level and amount of damage caused to crop produce and product, while in the store, thereby increasing food quantity and availability, preserving nutritional content, and enhancing food utilization. Further, with the availability of a variety of food stuff to people throughout the seasons of the year, coupled with other food materials not native to particular regions being available there, Indians now have a wide amplitude to choose from an array of food materials to meet their dietary needs. This enables people to eat what their own choice food, as it is typical under a condition of food security, and not just what is available.

3. Reduced Dependence on Agrochemicals

The potentials of agricultural biotechnology in producing crops that need less herbicide application than normal and are resistant to specific pest and disease whose infestation reduces crop yields and quality has been established. Through some techniques of agricultural biotechnology crops have been genetically engineered to acquire the ability to withstand high biotic pressure from weeds and pests, with less use of herbicides. Herbicide tolerant crops that have the potentials of withstanding the application of broad spectrum herbicides without phyto-toxicity risks have been developed. Thus reducing the frequent use of herbicide application during the crop seasons, assuring less crop injury and increasing yields and this principle can be applied on crops like maize, sorghum, millet, cowpea, groundnut and sweet potato whose yield and quality are seriously hampered by weeds and pests. Another principle of agricultural biotechnology has been used to engineer crop plants to develop resistance to specific insects that tends to cause damage to it. Such crops were made to produce certain chemicals which if ingested by the attacking insect in the process of feeding, causes the insect to stop feeding and thereafter dies. This ensures that the yields from these crops which would have been reduced due to pest, weeds and disease infections are saved for harvest. The yields are also free from damage so food availability is enhanced.

Various Biotechnological techniques useful to enhance the income and improve the economic status of farmers:

The various biotechnological techniques given below have proved useful in agricultural biotechnology for crop improvement and responsible for enhancing the income and uplifting the economic status of farmers includes:

1. Plant Tissue Culture

The technique is also known as *in vitro* plant production and defined as the production of all types of plant cells, tissues and organs under aseptic conditions. Tissue culture manipulates cells, anthers, pollen grains, or other tissues; so they live for extended periods under laboratory conditions or become whole, living, growing organisms; genetically engineered cells may be converted into genetically engineered organisms through tissue culture. (Bhojwani and Razdan, 1983; George and Sherrington, 1984; Pierik, 1987). Plant tissue culture has played an important role in the manipulation of plants for improved agronomic performance and an integral part of molecular approaches to plant improvement. It also acts as an intermediary whereby advances made by the molecular biologists in gene isolation and modification are transferred to plant cells. Some of the techniques that are more approachable and found to be applied directly in plant propagation and genetic improvement of plants are (i) micropropagation, (ii) meristem culture, (iii) somatic embryogenesis, (iv) somaclonal variation, (v) embryo culture, (vi) in vitro selection, (vii) anther culture, and (viii) protoplast culture. The most applied among the above mentioned is the technique of micropropagation, which has revolutionized the modern agriculture industry.

2. Micropropagation

Micropropagation is the application of tissue culture technique to the propagation of plants starting with very small parts grown aseptically in a test tube or other suitable containers (Hartman *et al* 1993). Micropropagation is one of the key tools of plant biotechnology that has been extensively exploited to meet the growing demands for elite planting material. Farmers have demand at large scale for disease free plants of superior quality of various ornamental, horticultural, floricultural and agroforestry plants. Micropropagation is a good enough technique to fulfil the demands of farmers. As a result, several plant tissue culture laboratories have come up worldwide, and more so in India. The demand for micropropagated plants in agriculture is increasing by the day, since the traditional methods of propagation do not yield sufficient quantity and quality of planting material in some crops. It is important to note that the demand for some crops like banana, papaya, guava, pineapple, strawberry, sugarcane, potato, turmeric, ginger, cardamom, vanilla and ornamentals like anthurium's, orchids, chrysanthemums, rose, lily, and gerberas are on the rise in different states in the country. Small quantities of medicinal plants like Aloevera, brahmi, Chlorophytum, Mulethi, Gloriosa and forestry crops like Bamboo, Teak, Eucalyptus, Sandal, are also produced and consumed in the domestic market. Various micropropagation protocol have been developed at Centre for Plant Biotechnology, Hisar, Haryana for commercial production of various ornamental, horticultural, floricultural and other crop plants.

The growth in demand for tissue culture banana has increased at a high rate of 25-30% and a similar trend for other crops is observed particularly for that of sugarcane due to the introduction of ethanol blended petrol. It can be noted that there is growing awareness of superiority of tissue cultured plants, and demand for crops like banana, grapes, papaya, ginger, turmeric, cardamom, vanilla, potato, Jatropha is increasing. Major Pot plants and landscaping ornamentals like Ficus, Spathiphyllums, Syngoniums, Philodendrons, Nerium, Alpenia, Yucca, Cordylines, Pulcherrima, Sansevieria, Gerbera, Anthuriums, Rose, Statis, Lilies, Alstromeria etc. are routinely produced by various plant tissue culture laboratories in India. The tissue culture laboratory can also be used to produce biofertilisers like rhizobium, azotobacter, azospirillum, phosphate solubilising bacteria culture as well as mushroom spawn culture that indirectly contribute to the agricultural sector. Most of the plants are conventionally propagated through cutting, grafting, stooling or air layering, but these methods are time consuming (Chandra *et al.*, 2004). Micropropagation protocol of sugarcane COH-119 was developed for commercial production of sugarcane (Kajla *et al.*, 2012). The planting of extensive new orchards of vegetatively propagated clones of some tropical fruits has sometimes been limited by pathogens (Litz and Jaiswal, 1991). Commercial production of fruit crops using biotechnology is very successful (Kajla *et al.*, 2014a). In majority of trees, propagation by root cutting is often characterized by a rapid loss of rooting capacity of the cutting with increasing age of parent plant (Thorpe *et al.*, 1991). Clonal propagation using cell, tissue and organ culture techniques have considerable potential for the improvement of economically important trees within a limited time frame (Giri *et al.*, 2004; Singh *et al.*, 2004).

Somatic embryogenesis plays an important role in clonal propagation. When integrated with conventional breeding programs and molecular and cell biological techniques, somatic embryogenesis provides a valuable tool to enhance the pace of genetic improvement of commercial crop species (Stasolla and Yeung, 2003). As compared to organogenesis, somatic embryogenesis provides an ideal experimental process for investigation of plant differentiation as well as a mechanism for expression of totipotency in plant cells (Litz and Gray, 1992). Therefore, it is very important that superior germplasm of crops with respect to its vitamin content, disease and other abiotic stress tolerant plants, low seed content and longer shelf life is multiplied at large-scale through somatic embryogenesis. Alginate encapsulation of somatic embryos or non-embryogenic vegetative propagules to produce synthetic seeds could possibly be utilized as means for germplasm storage and transportation of elite germplasm (Rai *et al.*, 2009).

Embryo rescue is one of the earliest and successful forms of *in vitro* culture techniques used to assist in the development of plant embryos that might not survive to become viable plants (Sage, 2010). Embryo rescue plays an important role in modern plant breeding, allowing the development of many interspecific and intergeneric food and ornamental plant crop hybrids. This technique nourishes the immature or fragile embryo, thus allowing it the chance to survive. Plant embryos are multicellular structures that have the potential to develop into a new plant. The most widely used embryo rescue procedure is referred to as embryo culture, and involves excising plant embryos and placing them onto media culture (Miyajuma, 2006). Embryo rescue is used to create interspecific and intergeneric crosses that would normally produce seeds which are aborted. Interspecific incompatibility in plants can occur for many reasons, but most often embryo abortion occurs (Reed, 2005). In plant breeding, wide hybridization crosses can result in small shrunken seeds which indicate that fertilization has occurred, however the seed fails to develop. Many times, remote hybridizations fail to undergo normal sexual reproduction, thus embryo rescue can assist in circumventing this problem. (Bridgen, 1994) Embryo rescue places embryos containing transferred genes into tissue culture to complete their development into whole organisms. Embryo rescue is often used to facilitate "wide crossing" by producing whole plants from embryos that are the result of crossing two plants that would not normally produce offspring. (Singh and Shetty, 2011)

Molecular Markers

Molecular markers, useful for plant genome analysis, have now become an important tool in crop improvement (Keswat and Basanta, 2009). The development and use of molecular markers for the detection and exploitation of DNA polymorphism is one of the most significant developments in the field of molecular genetics. Molecular markers are segment of DNA sequence of the genome that can differentiate two or more genotypes and follow the Mendelian pattern of inheritance. Molecular markers have been introduced over last two decades, which has revolutionized the scenario of agriculture. Since the publication of restriction fragment length polymorphism (RFLP); first DNA marker techniques three decades ago, many DNA based marker techniques have been developed which includes RAPD, SSR, ISSR, CAPs etc. DNA based markers have found large scale applications in crop improvement programs, because of several advantages over morphological and biochemical markers. Molecular markers are environmentally neutral, highly abundant, and independent of tissue and stage of the plants. Conventional breeding programs rely on visible traits to select improved varieties however; MAS rely on identifying marker DNA sequences that are inherited alongside a desired trait during the first few generations. Although there is no marker technique which is without any limitation but an ideal molecular marker should have following features: 1) technically should be simple and easy to perform 2) follow co-dominant inheritance 3) require no prior sequence information about genome 4) reproducible and 5) generate more number of markers per reaction. However, most of the currently used molecular markers are random DNA markers (RDMs) derived from unknown regions of the genome, limiting their use in trait specific crop improvements. Therefore, the focus in the recent years has been to develop markers which are derived from the within the gene regions making them more useful for crop improvement. These markers are referred as gene targeted markers (GTMs) and those gene targeted markers which are derived from the polymorphic site within the genes having effect on phenotypes of the plants are referred as functional markers (FMs).

Clonal identifications are traditionally based on various morphological characters; however, morphological characters may not be reliable to discriminate between closely related genotypes (Chandra *et al.*, 2005). Most of the cultivars grown on commercial scale are seedling selections from the well-known parent cultivars (Jaiswal and Amin, 1992). A close genetic relationship among cultivars, somatic mutations and changes due to environmental alterations can create problems in correct identification of germplasm. In recent years, different molecular markers (RAPD, RFLP, AFLP, SSRs, ISSR and VNTRs) have been employed for the investigations of cultivar origin and taxonomic relationships of several plant species. Genetic Diversity assessment has been successfully done in banana (Choudhary *et al.*, 2015). Detection of genetic variation is also important for micropropagation and *in vitro* germplasm conservation to eliminate undesirable somaclonal variations. Molecular characterization of the taxon will increase the knowledge about the available germplasm. In addition, identification of superior genotype of crops can pave the way for mass propagation and conservation through tissue culture.

The maintenance of genetic integrity of micropropagated plants with respect to the mother plant is the most vital consideration for upholding certain horticultural traits using elite genotype over natural seedlings. Under the influence of several factors such as the species, donor genotype, explants type, composition of the culture medium, conditions of the physical culture and the duration between successive subcultures there are chances of getting somaclonal variations amongst the regenerants (Larkin and Scowcroft, 1981) and they may be heritable (Jain, 2001). Thus, there arises the necessity to evaluate whether the regenerants are genetically identical with that of the mother plant. In recent years, molecular markers, viz., random amplified polymorphic DNA (RAPD), inter simple sequence repeat (ISSR), have been successfully applied to establish genetic stability or variability in various plant species. Rapid plant regeneration and molecular assessment of genetic stability using ISSR and RAPD marker in Grand Nane G-9 variety of banana was reported (Kajla *et al.,* 2014b). Genetic stability of a superior cultivar is important for the in vitro conservation and transport of elite genotype. However, Genomics applied to whole genomes of species together with other biological data about the species to understand what DNA confers what traits in the organisms. Similarly, proteomics analyses the proteins in a tissue to identify the gene expression in that tissue to understand the specific function of proteins encoded by particular genes. Both, along with metabolomics (metabolites) and phenomics (phenotypes), are subcategories of bioinformatics.

3. Genetic Engineering

Genetic engineering, also called genetic modification, is the direct manipulation of an organism's genome using biotechnology. In this technique inserts fragments of DNA into chromosomes of cells and then uses tissue culture to regenerate the cells into a whole organism with a different genetic composition from the original cells. This is also known as rDNA technology; it produces transgenic organisms. Tremendous studies have been made in crop improvement programs through genetic engineering by transferring genes of interest across taxonomic barriers from bacteria, fungi, viruses, insects and even mammals. The three widely adopted techniques of gene transfer includes, direct gene transfer through protoplast/cells, Agrobacterium mediated transformation and variety independent biolistic method of gene delivery. Transgenic plants have been generated from *Actinidia deliciosa* (kiwi fruit) to *Zea mays* (maize) including legumes, cereals, fruits, vegetables, ornamental plants and medicinal plants and have been reviewed extensively (Armstrong, 1999; Giri and Narasu, 2000; Altman, 2004; Bhojwani and Rajdan, 2005; Singh *et al.,* 2014). However, Agrobacterium-mediated transformation offers several advantages over direct gene transfer methodologies (particle bombardment, electroporation etc), such as the possibility to transfer only one or few copies of DNA fragments carrying the genes of interest at higher efficiencies with lower cost and the transfer of very large DNA fragments with minimal rearrangement (Hansen and Wright, 1999; Shibata and Liu, 2000).

The adoption of GMO technology has resulted in significant socio-economic benefits and advantages for farmers in developing and developed countries. The income and productivity from the GM crops have enabled farmers to switch to more sustainable farming practices (Brookes and Barfoot, 2011; Brookes and Barfoot, 2012). The four main crops in which GM traits have been commercialised worldwide are soybeans, maize, cotton and canola. They account for almost half of current global GM crop plantings, most of which are in the US, Canada, Asian countries and South and Central America. In particular, crops with herbicide tolerant and insect resistance traits have significantly improved global productivity and profitability for farmers. In recent years, herbicide tolerant and insect resistance crops boosted farm incomes by $18.8 billion. On a global productivity level, GM soy beans alone, contributed 122 million tonnes to global soy bean production. GM cotton has made a significant contribution to the income gains in developing countries coming from GM crops (James, 1999; James, 2011). Nevertheless, over the past 17 years, biotech crops have overcome a number of production constraints for many farmers worldwide. As GM technology has evidently improved productivity and profitability levels for farmers in developing and developed countries, it should be considered an important tool towards a more efficient and sustainable global agriculture.

Important Benefits of Genetic Engineering In Agriculture

Transgenic organisms can offer much wider range of benefits beyond those that emerged from innovations in conventional or traditional agricultural biotechnology. Following are examples of benefits resulting from applying this technique currently available in the field of agricultural biotechnology.

a. Increased Crop Production

Biotechnology has proved helpful tool to increase crop productivity by introducing such qualities as disease resistance and salt resistance to the crops. Now, researchers can select genes for these characters from other species and transfer them to important crops. Similar examples come from dry climates, where crops must use water as efficiently as possible. Genes from naturally drought-resistant plants can be used to increase drought tolerance in many crop varieties.

b. Enhanced Crop Protection

Crops such as corn, cotton, and potato have been successfully transformed through genetic engineering to make a protein that kills certain insects when they feed on the plants certain pests, not just the part of the plant to which *Bt* insecticide has been applied. So, farmers can use crop-protection technologies because these provide cost-effective solutions to pest problems which, if left uncontrolled, would severely lower the yields. In these cases, yields increase as the new technology provides more effective control. In other cases, a new technology is adopted because it is less expensive than a current technology with equivalent control. There are cases in which new technology is not adopted because for one reason or another it is not competitive with the existing technology. For example, organic farmers apply *Bt* as an insecticide to control insect pests in their crops, yet they may consider transgenic *Bt* crops to be unacceptable (Choudhary *et al.*, 2010).

Improvements in Food Processing Like Improved Nutritional Value

Genetic engineering has suggested new options for improving the nutritional value, colour, taste, flavor, and texture of foods. Transgenic crops in development include soybeans with higher protein content, potatoes with more nutritionally available starch and an improved amino acid content, beans with more essential amino acids, and rice with the ability produce beta-carotene, a precursor of vitamin A, to help prevent blindness in people who have nutritionally inadequate diets.

c. Better Flavor

Flavor can be altered by enhancing the activity of plant enzymes that transform aroma precursors into flavoring compounds. Transgenic peppers and melons with improved flavor are currently in field trials.

d. Fresher Produce

Genetic engineering can result in improved keeping properties to make transport of fresh produce easier, giving consumers access to nutritionally valuable whole foods and preventing decay, damage, and loss of nutrients. Transgenic tomatoes with delayed softening can be vine-ripened and still be shipped without bruising. Research is under way to make similar modifications to broccoli, celery, carrots, melons, and raspberry. The shelf life of some processed foods such as peanuts has also been improved by using ingredients that have had their fatty acid profile modified.

e. Environmental Benefits

When genetic engineering results in reduced pesticide dependence, we have less pesticide residues on foods, we reduce pesticide leaching into groundwater, and we minimize farm worker exposure to hazardous products. With *Bt* cotton's resistance to three major pests, the transgenic variety now represents half of the U.S. cotton crop and has thereby reduced total world insecticide use by 15 percent! Also, according to the U.S. Food and Drug Administration (FDA), "increases in adoption of herbicide-tolerant soybeans were associated with small increases in yields and variable profits but significant decreases in herbicide use".

f. Benefits For Developing Countries

Genetic engineering technologies can help to improve health conditions in less developed countries. Researchers from the Swiss Federal Institute of Technology's Institute for Plant Sciences inserted genes from a daffodil and a bacterium into rice plants to produce "Golden rice," which has sufficient beta-carotene to meet total vitamin A requirements in developing countries with rice-based diets. This crop has potential to significantly improve vitamin uptake in poverty-stricken areas where vitamin supplements are costly and difficult to distribute and vitamin A deficiency leads to blindness in children.

Agricultural Biotechnology and Economic Empowerment in India: Beyond agricultural benefits, agricultural biotechnology could offer several economic benefits to individuals and the nation at large. As described earlier, agricultural biotechnology allows the farmers to increase crop yields, through the development of high yielding, quick maturing crop varieties. This will lead to increase in food production which makes for greater availability of food for individual, household and national consumptions. When this happens individuals will feed well on adequate and nutritious food and become healthy. These healthy individuals can contribute their energies and labours through gainful employment in other sectors including agriculture for a free or a wage, thus gaining economic empowerment out of poverty and disease. Again individuals, households and national expenses on food will be minimized and then invested into other economic ventures for greater profits, thus bringing about financial independent and economic empowerment. Also with the abundance of healthy individuals, government and personal expenses on health care will be reduced and the money saved can be used to get other products to enhance people's quality of life for meaningful contribution to nation development. Through decreased agro-chemicals usage yield of crops will also increase thereby lowering production costs with higher agricultural productivity and income. This will yield lower production cost and food prices and improve nutrition leading to conservation of funds.

Conclusion

Modern Biotechnology has tremendous potential for crop improvement and in meeting food demand and environmental challenges, particularly where traditional breeding techniques have not been able to solve these specific problems. Therefore application of biotechnological tools is urgently required for multiplication, conservation and genetic enhancement of critical genotypes of plants. In-vitro regeneration is useful for multiplying and conserving the species and will be able to develop and use superior varieties that can perform under severe biotic and abiotic stresses. The aim of approaching biotechnological technique for enhancing yields – so that farmers can get more out of each seed. More out of each seed means more profitability, or a way to meet on farm needs, or a new way to diversify their farming operations. Besides all this the following recommendations were made; (1) More seminars, workshops and talks should be organized for scientist, policy, makers, farmers and agricultural extension agents on the gains that could be made from using genetically engineered crops to familiarize them with the benefits of embracing such technology. So as to engender a wide acceptance and participation in the use of this technology in crop production in India by them. (2) Government should buy biotechnologically engineered crops and distribute them to farmers to promote its full incorporation into the Indian farmer's crop production practices. (3) A public-private partnership deal should be made to enable massive funding to develop centers of excellence in agricultural biotechnology where special research can be undertaken to explore and use the genetic potentials of some of our staple crops to boost their production capacities.

References

Altman, A. 2004. Agriculture Biotechnology. Marcen Dekkr, USA.

Armstrong, C.L. 1999. The first decade of Maize transformation: A review and future perspective. *Maydica*, 44: 101-109.

Bhojwani, S.S. and Razdan, M.K. 1983. In: Plant Tissue Culture: Theory and Practice. Elsevier, Amsterdam.

Bhojwani, S.S. and Razdan, M.R. 2005. Studies in Plant Science 5. Plant Tissue Culture: Theory and Practice, a revised edition. Elsevier, Amsterdam Netherlands.

Bridgen, M.P. 1994. "A Review of Plant Embryo Culture". *Hort. Science*, 29: 1243–1246.

Brookes, G. and Barfoot, P. 2012. Forthcoming GM Crops: Global socio-economic and environmental impacts 1996-2010, PG Economics Ltd, Dorchester, UK.

Brookes, G. and Barfoot, P. 2011. GM crops: global socio-economic and environmental impacts 1996-2009.http://www.pgeconomics.co.uk/page/29/ sustainable,-profitable and productive agriculture-continues-to-be-boosted-by-the contribution-of biotech-crops.

Chandra, R., Bajpai, A., Gupta, S. and Tiwari, R.K. 2004. Embryogenesis and plant regeneration from mesocarp of *Psidium guajava* L. (guava). *Indian J. Biotech.* 3: 246-248.

Chandra, R., Mishra, M., Abida, M. and Singh, D.B. 2005. Paclobutrazol mediated somatic embryogenesis in guava. In: Abstract of 1st International Guava Symposium, CISH, Lucknow, India, pp 34.

Coombs, J.M. 1992. Macmillan Dictionary of Biotechnology. Hants, UK: Macmillan.

Choudhary, B. and Guar, K. 2010. "Bt Cotton in India: A Country Profile." http://www.isaaa.org/resources/publications/biotech_crop_profiles/bt_cotton_in_india-a_ country_profile/download/Bt_Cotton_in_India-A_Country_Profile.pdf.

Choudhary, D., Kajla, S., Poonia, A.K., Brar, B., Surekha and Duhan, J.S. 2015. Molecular assessment of genetic stability using ISSR and RAPD markers in *in vitro* multiplied copies of commercial banana cv. Robusta. *Indian Journal of Biotechnology*, 14: 420-424.

Singh, G. and Shetty, S. 2011. Impact of Tissue Culture on Agriculture in India. *Biotechnol. Bioinf. Bioeng.* 1(3): 279-288.

George, E.F. and Sherrington, P.D. 1984. In: Plant propagation by Tissue culture. Exegetics Ltd, Basingstoke, UK.

Giri, A. and Narasu, M.L. 2000. Transgenic hairy roots: recent trends and applications. *Biotechnology Advances*, 18: 1-22.

Giri, C., Shyamkumar, B. and Anjaneyulu, C. 2004. Progress in tissue culture, genetic transformation and applications of biotechnology to trees: an overview. Trees – Structure and Function. 18: 115–135.

Hansen, G. and Wright, M.S. 1999. Recent advances in the transformation of plants. *Trends in Plant Sciences*, 4: 226-231.

Herdt, R.W. 1995. The potential role of biotechnology in solving food production and environmental problems in developing countries. In Agriculture and Environment: Bridging Food Production and Environmental Protection in Developing Countries, ed. ASR Juo, RD Freed.

Jaiswal, V.S. and Amin, M.N. 1992. Guava and jackfruit. In: Hammerschlag, F.A. and Litz, R.E. (ed) Biotechnology of perennial fruit crops, Biotechnology in agriculture 8, CAB International, Wallingford, UK, pp 421-431.

Jain, M. 2001. Tissue culture–derived variation in crop improvement. *Euphytica,* 118: 153–166.

James, C. 1999. Global Review of Commercialized Transgenic Crops. ISAAA Briefs, No. 12, International Service for the Acquisition of Agri-biotech Applications, Ithaca, NY.

James, C. 2011. Global Status of Commercialized Biotech/GM Crops: 2011. ISAAA Brief No. 43. ISAAA: Ithaca, NY. http://www.isaaa.org/resources/publications/briefs/43/ executive summary/default.asp

Kajla, S., Goyal, S.C., Poonia A.K., Sehrawat, A.R. and Dhawan, A.K. 2012. An efficient protocol for large scale production and effect of various chemicals and growth substances on micropropagation of elite sugarcane cultivar CoH-119. Crop Science and Technology for food security, bioenergy and sustainbility. In: Behl, R.K., Bona, L., Pauk, J., Merbach, W. and Veha, A. (ed) Agro-bios International, Jodhpur, pp 45-54.

Kajla, S., Choudhary, D., Poonia, A.K. and Duhan, J.S. 2014b. Rapid Plant Regeneration and Molecular Assessment of Genetic Stability Using ISSR and RAPD Markers in Commercial Banana Cv. Grand Naine (G-9). *Journal of Advances in Biotechnology*, 4(3): 393-403.

Kajla, S., Poonia, A.K., Kharb, P. and Duhan, J.S. 2014a. Role of Biotechnology for Commercial Production of Fruit Crops. In: Salar, R.K., Gahlawat, S.K., Siwach, P. and Duhan, J.S. (ed) Biotechnology: Prospects and Applications. Springer New Delhi, Heidelberg, New York, Dordrecht London. ISBN 978-81-322-1682-7 ISBN 978-81-322-1683-4 (eBook), pp 27-38. DOI 10.1007/978-81-322-1683-4.

Keswat, M.P. and Basanta, D.K. 2009. Molecular markers: It's application in crop improvement. *Journal of crop science and technology*, 12 (4): 169-181.

Larkin, P.J. and Scowcroft, S.C. 1981. Somaclonal variation – a novel source of variability from cell culture for plant improvement. *Theor. Appl. Genet.*, 60: 197–214.

Litz, R.E. and Gray, D.J. 1992. Organogenesis and somatic embryogenesis. In: Hammerschlag, F.A., Litz, R.E. and Gray, D.J. (ed) Biotechnology of Pernnial Fruit Crops. CAB International, Wallingford, UK, pp 3–34.

Litz, R.E. and Jaiswal, V.S. 1991. Micropropagation of tropical and subtropical fruits. In: Debergh, P.C. and Zimmerman, R.H. (ed) Micropropagation. Kluwer Academic Publishers, Dordrecht, pp 247–263.

Mc Couch, S.R. 2001. Is biotechnology an answer? In Who Will Be Fed in the 21st Century? In: Wiebe, K., Ballenger, N., Pinstrup-Andersen, P. (ed). Washington, DC: Int. Food Policy Res. Inst./Econ.Res. Serv./Am. Agric. Econ. Assoc., pp 29–40.

Miyajuma, D. 2006. "Ovules that failed to form seeds in zinnia (*Zinnia violacec* Cav)". *Sci Hortic*, 107 (2): 176–182.

Mugabe, J. 2000. Biotechnology in Developing Countries and Countries with Economies in Transition: Strategic Capacity Building Considerations. Background Paper Prepared for the United Nations Conference on Trade & Development (UNCTAD) Geneva, Switzerland April 5, 5: 15-20.

Naylor, R.L., Falcon, W.P., Goodman, R.M., Jahn, M.M., Sengooba, T. *et al.* 2004. Biotechnology in the developing world: a case for increased investments in orphan crops. *Food Policy*, 29: 15–44.

Pierik, R.L.M. 1987. *In vitro* culture of higher plants. Martinus Nijhoff Publishers, Dordrecht, Netherlands.

Rai, M.K., Asthana, P., Singh, S.K., Jaiswal, V.S. and Jaiswal, U. 2009. The encapsulation technology in fruit plants – A review. *Biotechnology Advances*, 27: 671-679.

Reed, S. 2005. Plant Development and Biotechnology. In: Robert, N., Trigiano and Dennis, J. Gray, (ed) CRC Press, pp 235–239.

Ruttan, V.W. 2004. Controversy about agricultural technology lessons from the green revolution. *Int. J. Biotechnol*, 6: 43–54.

Sage, T.L., Strumas, F., Cole, W.W. and Barret, S. 2010. "Embryo rescue and plant regeneration following interspecific crosses in the genus *Hylocereus* (Cactaceae)". *Euphytica*, 174: 73–82.

Sengar, R.S. and Chaudhary, R. 2009. Role of Biotechnology in the Development of Horticultural Crops 2(1): 1-10.

Shibata, D. and Liu, Y.G. 2000. Agrobacterium-mediated plant transformation with large DNA fragments. *Trends in Plant Sciences*, 5: 354-357.

Santaniello, V. 2005. Agricultural Biotechnology: Implications for Food Security Agricultural Economies 2005, 32: 1.

Singh, M., Jaiswal, U. and Jaiswal, V.S. 2004. *In vitro* regeneration and improvement in tropical fruit trees: an assessment. In: Srivastava, P.S., Narula, A. and Srivastava, S. (ed) Plant biotechnology and molecular markers. Anamanya Publishers, New Delhi, India, pp 228-243.

Singh, R., Kumar, R., Heusden, A.W., Yadav, R.C. and Visser, R.G.F. 2014. Genetic improvement of mungbean (*Vigna radiata* L): Necessity to increase the levels of the micronutrients iron and zinc: a review. *Journal of Current Research in Sciences*, 2 (1): 1-11.

Stasolla, C. and Yeung, E.C. 2003. Recent advances in conifer somatic embryogenesis: improving somatic embryo quality. *Plant Cell Tissue Organ Culture*, 74: 15–35.

Thorpe, T.A., Harpy, I.S. and Kumar, P.P. 1991. Application of micropropagation to forestry. In: Debergh, P.C. and Zimmerman, R.H. (ed) Micropropagation. Kluwer Academic Publishers, Dordrecht, pp 311–336.

Toenniessen, G.H., O'Toole, J.C. and DeVries, J. 2003. Advances in plant biotechnology and its adoption in developing countries. *Curr. Opin*, 6: 191–198.

Zaid, A., Hughes, H.G., Porceddu, E. and Nicholas, F.W. 1999. Glossary of biotechnology and genetic engineering. Rep. 7, FAO, Rome.

8

Value Addition to Oilseeds for Their Sustainable Production

G. Nagaraj

ICAR-Indian Institute of Oilseeds Research, Rajendranagar, Hyderabad - 500 030, India
Corresponding author: guttarla@hotmail.com

Abstract

Oilseeds are highly useful commodities in the human food chain as also in the industrial sector. Edible seeds furnish around 500 calories of energy. They are rich sources of fats, proteins, sugars, essential fatty acids, amino acids, minerals and vitamins. They also contain health promoting phyto-chemicals and hence have gained importance as neutraceuticals. This is against the staple food grains like rice, wheat or millets which serve mainly as energy sources. The non edible oilseeds and their constituents have gained importance as raw material for the industry. A large number of high value oleochemicals like lubricants, cosmetics, etc. can be prepared from the oils. The derivative oleochemicals are costlier and have greater potential. The oils can be converted to environment friendly biodiesel. The fat free oilcakes are being utilized as value added protein rich foods. The inferior grade cakes find wide applications as feed ingredients for the cattle and poultry and thus indirectly help in enhancing milk and meat availability to the human population. Proteins from toxic castor seeds and rapeseed can be converted to fibre or utilized as organic manure. Thus development of costly and valuable products from oilseeds can make them more remunerative to the producer, namely, the farmer and thus can help in their sustainable production.

Keywords: Oilseeds, Value added products, Oilcakes, Nutrients, Neutraceuticals, Oleochemicals, Food products, Feeds, Fibre, Fuel, and Fodder

Introduction

Vegetable oils are important energy sources in the human food chain as also in the non-food industrial sector. The edible vegetable oils are important as energy as well as nutritious components to the human beings. Cultivated oilseeds and some tree borne seeds are the major sources of oils. There is a high degree of variation in annual production of these oilseeds due to their cultivation under low and uncertain rainfall and input starved conditions along with poor crop management.

About 64% of the area under oilseeds in India is rain-fed. Drought from uncertain rainfall and the consequent water deficit are common occurrence in the dry-lands. The production of oilseeds directly varies according to the rainfall pattern. Currently oilseeds are cultivated in an area of 270 lakh ha. The oilseeds production in India stands at around 300 lakh tones. At an average recovery of 30% oil from the oilseeds the production of vegetable oils stands at 90-100 lakh tonnes. Since this quantity is insufficient to meet the requirements of the country, an expenditure of around Rs. 60,000 crores is being spent to import additional quantity of edible oils which is a big drain on the foreign exchange component of the country. The requirement of edible oils will further increase due to increase in population as well as improvements in standards of living.

India is the fifth largest producer of oilseeds in the world. It occupies 15% of world area and accounts for only 9% oilseeds and 11% of edible oils production of the world. Within the country oilseeds occupy 14% of the gross cropped area and accounts for 1.4% of the gross domestic product (GDP). The oilseeds also account for 8% of the value of all agricultural products. It is encouraging to note that India accounts for 15% of the total export earnings. It is thus evident that oilseeds are a major export earner despite their low yields and non remunerative production.

About 14 million farmers are engaged in cultivation of oilseeds in India. Another one million people are involved in the post harvest processing like oil extraction etc. Oilseeds cultivation encounter a multitude of problems. There are a lot of uncertainties in oilseeds production. They are cultivated in less fertile soils by marginal and poor farmers mostly under rain-fed conditions. Hence the yields are low at around 500 to 1000kg/ha.

Low yields under resource poor conditions are a big deterrent to the oilseeds cultivation in India. In the international market edible oils are available at a cheaper rate encouraging imports and simultaneously discouraging oilseeds cultivation within the country. Also policy decisions by the Government of India are favoring cultivation of food crops and pulses. All these factors have made the cultivation of oilseeds non-remunerative.

There is thus an imperative need to make oilseeds cultivation more remunerative. This is possible only if value added products are made from oilseeds. The importance of oilseeds as energy and nutrition rich grains is well known. They are important sources of essential nutrients like linoleic and linolenic acids (essential fatty acids), amino acids and health promoting phytochemicals (neutraceuticals) in addition to their richness in oils and proteins. In general oilseeds can be considered as natural nutri - nuggets. They also serve as raw materials to the industry for manufacture of lubricants, cosmetics, synthetic fibre and biodiesel etc. Hence there is need to utilize oilseeds in the manufacture of high value food and non-food products to make them more remunerative.

There are only a few alternatives before the country to increase production of oilseeds. One of them is to increase area which may not be practicable. The other alternative is to increase the irrigated area under oilseeds cultivation. This aspect also may not gain priority as food grains and fibre crops will always have more importance. Increased productivity and remunerative prices will certainly boost up oilseeds production. In this connection oilseeds being sources of proteins and energy rich fats apart from industrially valuable phytochemicals and neutraceuticals have a great potential for value added products. This step will ensure higher prices and help in their increased production. The importance of in cultivated oilseeds with respect to value addition and related sustainable production is discussed in this chapter.

Oilseeds- Natural Nutri-Nuggets

Commonly consumed staple food grains like rice, wheat and millets are rich sources of carbohydrates and hence are useful as energy sources. Oilseeds on the other hand are rich in oils, protein, minerals, sugars, vitamins and many health promoting phytochemicals. Hence they serve as balanced food constituents in addition to being sources of neutraceuticals. The oilseeds differ with respect to their proximate composition as well as minor components.

Oilseeds mainly contain oil which is a concentrated source of energy. They also contain proteins, carbohydrates, minerals and useful vitamins. All these together make the oilseeds as good sources of energy for the human being and animals. Oilseeds provide 500-600 calories of energy per 100g. Groundnut, sesame, soybean, sunflower kernels, and niger are directly edible. Minimal processing like dehulling, roasting, cooking etc. makes them more acceptable and hence are widely consumed as food items all over the world. Other oilseeds like safflower, linseed and rapeseed-mustard also are consumed to a limited extent as part of food items.

Among edible oilseeds, groundnut and sesame contain higher oil and proteins followed by niger, rapeseed and sunflower while soybean and safflower contain lower levels of oil. However, soybean serves as the richest source of protein. In general, all the oilseeds contain 10-25% sugars and 3-5% minerals. Oilseeds also contain sufficient levels of vitamins. The main vitamins that are present are vitamin B complex group, and vitamin E (tocopherols). Vitamins A, C and D are present at very low levels. Compared to cereals, oilseeds serve not only as concentrated sources of energy, but also as balanced source of most of the food nutrients.

Table 1. Proximate Composition of Oilseeds(/100G)

Oilseed	Protein	Oil	Sugars	Minerals	Energy (Joules)
Groundnut	23-24	47-51	6-12	2-3	2700
Rapeseed	25-30	38-43	15-20	5-9	2500
Safflower	12-15	28-32	10-15	3-7	2000
Sunflower	14-19	28-35	16-27	2-3	2200
Sesame	20-22	48-55	14-16	5-7	2700
Soybean	30-40	14-23	25-30	3-5	2000
Niger	10-30	30-43	7-8	3-4	2200
Castor	18-24	40-55	5-7	2-3	2500
Linseed	20-22	30-35	25-28	2-3	2300

Groundnut, rapeseed, sesame and soybean are rich in proteins. Groundnut rapeseed ,sesame and castor are rich in oils. Groundnut and sesame seeds are edible in their raw or roasted forms and hence they are highly useful at household level for direct consumption and for preparation of simple food products. Soybean though richer in proteins needs to be processed for its utilization. Niger seed with sufficient levels of oil and protein can be utilized after roasting. The seed coat is richer in crude fibre. Sunflower with moderate levels of protein and oil needs to be decorticated before its consumption. Roasted linseed also serves as a highly nutritious oil and protein source. Other oilseeds like rapeseed, safflower and castor contain some antinutrients in addition to being good sources of oil and protein and cannot be consumed directly. They need to be treated to remove their antinutrients before their utilization.

All the oilseed cakes after removal of oil (richer in protein and sugars) are more useful. At present they find widespread use as animal feeds. The seeds or the cakes are good sources of minerals. They have energy values of around 2000- 2700 joules per 100g. The whole seeds as well as oils are rich in vitamins. Hence the edible seeds serve as wholesome foods. Their consumption directly or in combination with other regular foods will enhance the nutritive value of foods. In view of the richness in almost all the nutrients the edible oilseeds and their products can be considered as natural nutri-nuggets.

The quality of a protein depends on the amino acid composition. The amino acid composition of the oilseeds is presented in the Table 2. In general all the oilseeds are rich in all the major amino acids. A few of the oilseeds are deficient in essential and sulfur containing amino acids except sesame and niger. Consumption of two or three oilseeds or their cakes will balance the amino acid requirement.

Table 2: Essential Amino Acid Composition (%) of Oilseed Proteins

Amino acid	Ground nut	Rape seed	Safflow- er	Sun flower	Sesa- me	Soy bean	Ni- ger	Cas- tor	lin- seed
Valine	4.3	5.6	2.4	5.0	2.4	4.6	4.2	-	5.1
Leucine	6.3	7.8	2.7	7.3	4.6	7.8	5.4	3.0	5.8
Isoleucine	3.4	4.3	1.7	3.7	2.6	4.6	3.8	4.9	5.0
Threonine	2.5	5.0	1.4	4.7	1.9	3.9	2.6	4.9	1.6
Methionine	1.3	0.4	0.7	-	1.6	1.1	2.2	0.6	1.6
Lysine	5.9	5.4	1.3	3.5	-	6.4	3.4	2.0	3.7
Histidine	2.2	3.5	1.0	2.8	1.2	2.6	2.6	-	-

The fatty acid composition of the oils derived from the cultivated oilseeds is presented in the table 3. In general oilseeds are rich in fatty acids like oleic acid, linoleic acid and linolenic acids. The latter two fatty acids are essential and need to be consumed through food by the humans. Oleic acid is neutral and is a useful fatty acid. Oils rich in mono unsaturated fatty acids (MUFA)like oleic and erucic acids are stable and can be used for repeated fat frying and for storage up to one year or more. Saturated fatty acids(SFA) like palmitic, stearic and arachidic acids are also present in some of the oils. They also impart stability to oils. They are not useful from the point of cardiac health. Linoleic and linolenic acids have two and three double bonds and are called poly unsaturated fatty acids (MUFA). They serve as precursors of hormones and are essential for human nutrition.

A good edible oil should have equal proportions (1:1:1) of the three types of fatty acids namely, SFA, MUFA and PUFA. In reality, none of the natural oils do contain the fatty acids in this proportion. Hence it is better to blend the oils to obtain this combination of fatty acids. Otherwise, it is sufficient if we consume two or more oils to balance the fatty acid needs. The presence of some minor constituents like phytosterols and tocopherols influence the cooking and nutritional quality of edible oils. These are discussed separately.

Table 3: Fatty acids of oilseeds

Oilseed/ oil	Palmitic 16:0	Oleic 18:1	Linoleic 18:2	Linolenic 18:3	Erucic 22:1	ricinoleic
Groundnut	13	47	30			
Rapeseed	4	19	15	6	50	
Safflower	6	8	76			
Sunflower	5	30	55			
Sesame	10	40	45			
Soybean	10	25	50			
Niger	8	25	60			
Castor	1	3	4			88
Linseed	6	16	15			

The oilseeds are very good sources of mineral elements required by the human beings. Nutrients like Ca, Mg, K, P, S, Fe, Mn, etc. are present in sufficient quantities. Similarly the whole seeds or the oils are rich sources of vitamins like A, B group and E(tocopherols). In view of the presence of all the nutrients, oilseeds are considered better food sources than cereals. Hence their consumption directly or through value added products need to be encouraged or popularized. Such a step would provide better price to oilseeds and help production on a larger scale. The consumption of oilseeds benefits the community as a whole as they are wholesome foods.

Table 4: Mineral Content and Composition of Oilseeds. mg/100g

Oilseeds	Mineral %	P	K	Ca	Mg	S	Zn	Mn	Cu	Fe	B
Groundnut	2-3	400	700	55	220	300	4.0	2.0	1.0	4.5	1.5
Rapeseed	5-9	100	125	66	50	90	1.0	1.0	-	1.5	1.0
Sunflower	2-3	100	200	60	40	-	8.0	5.0	3.0	4.0	1.5
Sesame	5-7	700	400	100	20	-	6.0	2.0	5.0	2.0	1.5
Soybean	3-5	550	350	220	230	-	-	-	2.4	-	2.0

Table 5: Vitamin content of oilseeds. mg/100g oil

Vitamin	Groundnut	Rapeseed	Sunflower	soybean
Thiamine	1.0	0.5	3.7	14.0
Riboflavin	0.1	0.6	0.4	0.2
Pyridoxine	0.3	0.7	-	0.6
Niacin	15.0	16	31	24
Choline	170	670	-	340
Folic acid	0.3	2.3	-	0.3
Pantothene	2.7	1.0	4.5	1.2
VitaminE	4.0	1.4	5.0	9.0

Nutritional Quality of Vegetable Oils

Fats improve the taste and flavour of foods. Fatty foods satiate a hungry person fast because fats remain in the stomach for a longer period than either protein or carbohydrate (sugar) foods. Fats also carry essential vitamins like A, D, E and K. They also provide the body with essential fatty acids. Thus, fats have a very important role to play in human nutrition. Vegetable oils are the major source of fats in our diet. The cultivated oilseeds like groundnut, rapeseed, sesame, sunflower, safflower, soybean and niger are the major sources of edible oils to the mankind. The quality and composition of these oils vary as also the preference of people.

Fats and oils are chemically known as `triglycerides` because every molecule of fat has one molecule of glycerol linked up with three molecules of fatty acids, either the same acid or three different fatty acids. The fatty acids are either saturated or unsaturated. Both kinds contain molecules made up of varying numbers of carbon, hydrogen and oxygen only. The saturated fatty acids (SFA) are so called because no more hydrogen atoms can be linked directly to any of the carbon atom in them. The unsaturated fatty acids however, have one (monounsaturated fatty acids-MUFA) or two or more (polyunsaturated fatty acids-PUFA) carbon atoms which can still link up with hydrogen. Some of the polyunsaturated fatty acids like linoleic acids are not synthesized in our body and need to be supplied through the food. Hence, they are called **essential fatty acids** (EFA). Absence of these fatty acids in the diet results in hair fall, eczema like skin eruptions, slow healing of wounds and dry, scaly skin.

Functions of dietary lipids: Fat is an important component of human diet and fulfills several nutritional functions. It is a concentrated source of energy and helps to increase the calorie density of diets. Fat is more important in the diet of young children. It ensures adequate intake of energy without making the diet bulky. It is a source of essential fatty acids (EFAs), namely, linoleic and alpha linolenic acids. It is a carrier of fat soluble vitamins and facilitates their absorption and mobilization. It also contributes to flavour and palatability of the diet.

Fatty acids and nutrition: The fatty acids present in the fats determine their value as cooking media. Oils rich in saturated fatty acids and MUFA are stable and can be stored for longer periods (6 to 12 months). They can also withstand repeated deep frying. Oils rich in PUFA are more nutritious, due to their ability to supply EFA. A premium food fat should have equal proportions of saturated fatty acid, MUFA and PUFA.

Oils have nearly equal amounts of the three categories of fatty acids and hence can be considered as good food fats. Safflower, sunflower, niger and soybean containing more of polyunsaturated oils are rich sources of EFA. Sesame, groundnut and rapeseed oils are balanced with respect to keeping quality and nutritional quality. They have the SFA, MUFA and PUFA in roughly equal proportions.

Cooking and frying medium: A good cooking oil should be bland without any taste. It should not be spoiled on frying for a long time. A golden brown colour is appetizingly attractive. The frying medium should not easily turn rancid or get smoked on heating. MUFA fats serve as good frying media with reasonably good keeping and nutritional qualities. Fats like safflower and sunflower cannot be stored over longer periods. Soybean oil loses its original flavour after one deep frying.

When the free fatty acid content of fats is above 2%, their quality is considered to be poor. Repeated deep frying and long periods of storage increase the fat acidity which in turn decreases the smoking point. Such a deterioration in quality is more in the case of PUFA oils. Fishy odour develops in these oils due to oxidation. However, vegetable oils contain natural antioxidants like Vitamin E which to some extent prevent the oxidation. Partial hydrogenation or addition of antioxidants are some of the techniques employed to avoid rancidity development in high PUFA oils. It is also the basis of vanaspati industry where unsaturation is reduced through the addition of hydrogen. Rancid food is unpalatable and can even be harmful. Therefore, in case of children who depend mainly on one type of food, care should be taken to feed only fresh fats instead of rancid ones.

Essential fatty acids and prostaglandins:EFA deficiency in humans results in abnormal skin conditions such as scaliness and dermatitis, increased water loss, reduced regeneration of tissues, increased susceptibility to infection and increase in the ratio of 20:3 to 20:4 acids (the first number indicates the carbon atoms and the second one the double bonds in the fatty acid) in the serum phospholipids. These effects are probably due to the lack of prostaglandins produced from specific C_{20} polyene acids. EFA deficiency can usually be cured by linoleic acid because this acid which cannot be produced in animals and is entirely of plant origin can be metabolized to the required C_{20} and C_{22} polyene acids. Such acids are preferentially incorporated into cell membrane phospholipids where they have an important structural and functional role. The recommended daily intake of EFA for adult humans is at least 3% of energy requirement or 2-10g per day of linoleic acid.

Natural prostaglandins (PG) are produced in many animal tissues from the 20:3, 20:4 and 20:5 polyene acids which furnish the PG1, and PG3 series of compounds respectively. The several members of each series vary in the degree of oxygenation. All prostaglandins are derivatives of the cyclic C_{20} acid, namely, prostanoic acid. These compounds are not stored in the mammalian tissue, but are elaborated in response to various stimuli. Prostaglandins are used in child birth to induce parturition. They are used for treatment of gastric ulcers and bronchial asthma and they inhibit platelet aggregation.

Fats/Fatty acids and health: Dietary fat is related to the level of blood cholesterol, the main culprit linked to the cardiovascular diseases. Human beings with high blood cholesterol levels (260mg/ 100 ml) have double the risk of developing a heart disease than a person with normal cholesterol level (<200 mg/100 ml). Cholesterol is carried in the blood by either low density lipoprotein (LDL) or high density lipoprotein (HDL). It is the LDL which is found to have links with the incidence of heart attacks. On the other hand the HDL removes cholesterol from arteries and reduces the risk. Saturated fats increase the level of LDL in the blood. But PUFAs increase HDL and thus are beneficial. Even MUFAs have been shown to decrease cholesterol and LDL. Therefore, lower intake of saturated fat reduces cholesterol and thus reduces the chances of developing arterial thrombosis, malfunctioning of heart, increase in blood pressure and onset of diabetes in adults. Higher levels of EFA/PUFA in the diet have been reported to hasten ageing and are implicated in tumor production due to their ability to produce free radicals- very highly reactive oxygen or hydrogen atoms.

Minor fat components: Fats and oils also contain endogenous components referred to as unsaponifiable matter (non glyceride fraction, NGF). The components of NGF have hypocholesterolemic and/ or antioxidant effects. The latter property is useful in imparting stability to the oil and serving as free radical scavengers. Phytosterols are present in vegetable oils in free and esterified forms. The chemical structure of phytosterols is similar to cholesterol, differing only in the side chain. Phytosterols mimic and compete with cholesterol at the site of absorption and reduce absorption of dietary cholesterol.

Different isomers of tocopherols like α, β, γ and δ exist in oils. The δ isomers exert the highest and α isomers exert the least antioxidant effects. Tocopherols and tocotrienols capture and destroy damaging free radicals that have been suggested to play a role in cellular aging, atherosclerosis and cancer. δ tocotrienol has been shown to be effective in inhibiting breast and liver cancer cells. Tocotrienols have a role in molecular signaling. Being free radical scavengers, tocopherols, reduce the carcinogenic properties of PUFA. An adult human needs to consume 8 to 10 mg of tocopherols daily.

Sesame oil has lignans (sesamin, sesamolin and sesamol) which are powerful antioxidants that make the polyunsatured oil remain fairly resistant to oxidation and thermal decomposition. Sesame lignans inhibit endogenous synthesis of cholesterol and have antioxidant effects. Further, combination of sesame lignans and tocopherols have higher hypocholesterolemic and antioxidant effects as compared to the effects of lignans alone. Almost all vegetable oils contain phospholipids. Even squalene is present in some oils. All these constituents have anti oxidant properties.

Soybean oil provides isoflavones which have several health benefits. Natural phenols, either simple phenolic acids and their esters or more complex molecules such as flavones, and flavonoids in vegetable oils have antioxidant activity. Groundnut contains resveratrol at 200-250 µg/100g of seeds. This compound has a role in reducing heart disease risk. Resveratrol is present in red wine at a level of 800 µg/glass.

Table 6: Minor components in vegetable oils:

Diacylglycerols	0-8.0%
Monoacylglycerols	0-0.2%
Free fatty acids	0.2-9.5%
Phospholipids	0.01-0.1%
Sterols	0.4-2.0%
Tocopherols	150-2000 ppm
Tocotrienols	0-1500 ppm
Phenolics	0-50 ppm
Chlorophylls and derivatives	0-2.0 ppm
Carotenoids	0-500 ppm

Recommendations for dietary fat: Initially, FAO had recommended equal levels of SFA, MUFA, and PUFA out of a total 30 % energy from fats and oils. The current recommendations for safe range of intake of dietary fat for optimal health and prevention of diet-related chronic diseases are as follows (% energy): total fat 20-30, SFA<10, LA(linoleic acid) 6-8, ALNA (alpha linolenic acid)1-2.5, LA/ALNA 5-10, MUFA 10-12, cholesterol <200 mg/day and adequate antioxidants.

The ratio of PUFA/SFA should be 1 for an edible oil to be considered good. Groundnut, mustard, sesame and soybean oils have values between 1 and 3. Other oils like sunflower, safflower and linseed are either very rich sources of PUFA resulting in very high PUFA/SFA(P/S) ratios. Blending of oils can be done to obtain a balance of fatty acids as per the recommendation. Use of two or more oils, depending on the purpose, to offset their imbalance in fatty acid composition is another way of obtaining balanced nutrition. Saturated fats have a sound relationship to coronary heart disease. Hence a reduction in saturated fats can be suggested for certain individuals. It is difficult to cure coronary heart disease by altering the fat content of the diet. The principle "A little fat will do good" should be kept in mind and practiced.

The available edible oils within the country have all the qualities to serve as cooking as well as health promoting oils. Any deficiencies can be covered up through techniques like blending and or use of two or more oils in our day to day life. Keeping this aspect in view, imports of oils should be regulated or almost can be avoided to conserve the country's foreign exchange reserves. Such a step also will go a long way in encouraging our indigenous farmers in cultivation of oilseeds.

Oleo-Chemicals

Oils and fats have traditionally been utilized as food. However, they have more varied uses in the non-food sector. The oldest use has been as lamp oil and lubricant. In recent times they are greatly being used in the manufacture of paints, lacquers and soaps. Fats, fatty acids and their derivatives also find use in detergents, cosmetics, plastics and as rubber additive. Their wide use and adaptability is presently under check because of the availability of cheaper priced petroleum derivatives.

Vegetable oils have some advantages over petro-products. They are a renewable resource, they are biodegradable (environment friendly), they have excellent technical properties and they cause fewer medical problems (less carcinogenic) and allergies. The main disadvantages of vegetable oils are that they are high priced, they have limited oxidative stability, they have high viscosity and they are too reactive.

Oleo-chemicals: Oleochemicals are the chemical substances which are produced from oils and fats. Oleo-chemicals find diverse uses. All of them have a common functionality namely the processes at the surface. The long hydrocarbon chain favours the oily or hydrophobic phase, while the carboxyl group favours the aqueous or hydrophilic phase. In most of the situations, it is the carboxyl group that is modified through oleo-chemical reactions to fulfill specific functions, keeping intact the hydrophobic chain.

The oleochemicals derived from oils and fats can be divided into two categories. The chemicals which can be obtained by carrying out unit operations on oils and fats are called 'basic oleochemicals'. Other chemicals which are produced by derivative operations on oils and fats and/or basic oleochemicals are known as 'oleochemical derivatives'. The fatty acid molecules present in oils and fats are the 'building blocks' of these oleochemicals.

There are many derivatives of basic oleochemicals. The major derivatives of basic oleochemicals, fatty acids, fatty alcohols, fatty acid methyl esters, fatty amines and glycerol are the major oleochemicals from which several derivatives are made. Some derivatives can be made from more than one starting oleochemical.

Fat splitting: Oleo-chemistry emphasizes alkene, ester, hydroxyl and carboxyl reactivities. Industrial value of a vegetable oil generally depends on its content of specific fatty acid and the ease with which it can be modified. The general chain length of fatty acids ranges from C:12 to C:22 with saturated, mono, di or tri unsaturated acids. The main exception is ricinoleic acid which has a hydroxyl group in its molecule.

Triglycerides (oils and fats) are split into fatty acids and glycerol by 'fat splitting'. The mixture of fatty acids is distilled to get colourless, odourless, distilled grade fatty acids. By fractionation, the fatty acid mixture can be separated into fatty acids of different chain lengths. The glycerol-water mixture is passed through a series of evaporators to get glycerol of various grades, including chemically pure (CP) glycerine. Fatty acids and fatty acid methyl esters can be converted into fatty alcohols.

Glycerol and derivatives: Glycerol is an important by-product of soap industry. It has wide uses, in pharmaceuticals, food emulsifiers, resins, cellulose, polyols, polyurethane, tobacco products, explosives etc. Glycerol directly is used as a humectant in various foods and tobacco. Triacetyl glycerol (triacetin) finds use in the production of cigarette filters. It is also used in the manufacture of alkyd resins (esterification of dibasic acids with glycerol). Partial fatty acid esters of glycerol are used as emulsifiers, since the free hydroxyl group confers hydrophilicity, for example glycerol mono stearate. Ethoxylated monoacyl glycerols are used in shampoos and foam bath preparations, transparent creams, and in suntan lotions. Glycerol monoesters (ex: monolaurin) find use in cosmetics, and as food preservative. At present glycerol is in abundant supply.

The consumption of glycerine is maximum (31.5%) in the manufacture of drugs and personal care products followed by tobacco/triacetin, alkyd resins and polyether polyols. More than 90% of all triacetin is used in cigarette papers and filters. Miscellaneous applications also consume significantly large quantities of glycerine amounting to 16 % share of the total market.

Fatty acids and derivatives: Saponification of storage lipids produces fatty acids. Hydrolysis based on counter current splitting at high pressure is the method widely used by the industry. The resulting fatty acids are fractionated to C:8/10, C:12/14, C:16/18 etc. or of any other defined chain length. Saturated fatty acids are separated through crystallization. For example stearin fraction crystallises first which can easily be separated through centrifugation from the liquid olein fraction.

The carboxyl group is the reactive portion in a fatty acid. Compounds like methyl esters, alcohols, amides, amines and quaternary ammonium compounds can be manufactured. Soaps are nothing but alkali metal salts of fatty acids and are utilized as such. Various grades of soaps are possible depending upon the proportion of lauric and stearic acids. Fatty acids find use as polishes, buffing compounds, household and industrial cleaning agents. They also find use in the preparation of high grade candles and rubbers for tyres. Metallic stearates are used in the preparation of moulds in the manufacture of plastics. They serve as lubricants for coated papers. In food and pharmaceutical industry, metallic stearates are used to coat tablets or spices to increase their flow and prevent sticking. Other applications of fatty acids are in the field of cosmetics, greases and varnishes. The applications of some fatty acid groups belonging to cultivated vegetable oils are presented below.

Oleic-linoleic acid group: The oils rich in oleic acid are more stable and free from any serious tendency towards flavour reversion. For the same reason they find much use in soap making and produce firm stable soaps of desired properties. The vegetable fats from this group also find extensive applications in cosmetic formulations and are good ingredients imparting emolliency and good skin compatibility. Sesame oil, apart from being a cooking and salad oil, finds use as a soap fat, in pharmaceuticals and as a synergist for insecticides. Other low grade edible oils also can similarly be utilized for industrial purposes

Linolenic acid group: The oils from the group are distinguished by the presence of large proportion of linolenic acid, in addition to oleic and linoleic acids. The important members from this group include linseed, soybeen and rapeseed. The oils from this group have drying properties and find applications in paint, varnish and allied products.

Erucic and behenic acids: Oleochemicals based on the long chain C22 fatty acids erucic and behenic acids have special properties and find increasing applications. The *Cruciferae* family of plants, (rape, mustard, *crambe*) provide the most economical and readily available sources of erucic acid. Behenic acid is produced today by the hydrogenation of high erucic acid rapeseed (HEAR) oil. A major high value application of erucic acid can be in the manufacture of Nylon 13 and Nylon 13, 13. Other applications of erucic acid/benenic acid based oleochemicals would be in surfactants and detergents, plastics and plastic additives, photographic and recording materials, food and food additives, cosmetics and personal care products, ink additives, paper, textiles, lubricants and fuel oil additives.

Hydroxy acid group: Castor oil is the most unusual product, among all the vegetable oils. It is, by far more versatile due to presence of ricinoleic acid, which constitutes about 90% of the oil. Castor has a wide diversity of commercial applications, all directly related to the unique hydroxy fatty acid namely ricinoleic acid. Castor oil is non-toxic, renewable and biodegradable. It's oil is used in medicines as purgative and also as hair oil. Castor oil, a viscous liquid finds direct application in lubricating greases, polishes and plastics. Fully hydrogenated castor oil is a substitute for 'Carnauba Wax'. By catalytic dehydration process it can be converted to conjugated acid oil and this dehydrated castor oil is extensively used in protective coating industry.

Hydroxyl group can be acetylated or alkoxylated. The unsaturation can be altered by hydroxylation or expoxidation. The hydroxyl group can be removed by dehydration to increase the unsaturation of the compound, which can further be made to yield drying compounds and polymers. The hydroxyl position is so reactive that the molecule can be split at that point by high temperature pyrolyis and by caustic fusion to yield useful products of short chain length. Castor oil is used in the manufacture of cleavage products like sebacic acid, undecylenic acid and heptaldehyde and nylon through dimerization, polymerization etc.

Castor oil on sulfonation yields sulfated castor oil, known as 'Turkey Red Oil' which is used in cotton dyeing and printing industry to give lusture and brightness. On alkaline fusion, castor oil yields, sebacic acid and 2-octanol (caprylic alcohol). Sebacic acid is used as plasticiser and it's alcohol is used a as defoaming agent and as a solvent in plastic and lacquer industry.

Pyrolytic decomposition of castor oil at temperatures above 250 ^{0}C yield, undecylenic acid and heptaldehyde. The undecylenic acid finds wide use as a fungicide and bactericide. Heptadehyde is used as raw material for a host of perfumery chemicals like alpha amyl cinnamic aldehyde, methyl heptine carbonate etc.

When dehyrated, castor oil is converted into quick-drying oil. It is used extensively in paints and varnishes. It's water resistant qualities make it ideal for coating fabrics and for protecting coverings, insulation, food containers and gums. Sebacic acid, the castor oil product, is the basic ingredient in the production of nylon and other synthetic resins and fibres. The superior oiliness of castor oil and its ability to cling to very hot moving parts make it an outstanding lubricant. Castrol-R, racing motor oil, for high-speed automobile and motor cycle engines is made from castor oil.

Fatty acid methyl esters(biodiesel): Fatty Acid Methyl Esters (FAME) are derived from the reaction of fatty acids with methanol or direct methanolysis of triglycerides. The advantages of methyl esters are that they can easily be transported, they can be distilled since they have lower boiling points, and they have lower corrosive activity than free fatty acids. FAME serve as intermediates in the manufacture of alkanolamides, sucrose esters and methyl ester sulfonates. Some of these compounds perform well as detergents which are cheaper and biodegradable. In recent years FAME have acquired importance as biodiesel. These are mostly prepared from non edible oils to substitute the declining stocks of the non renewable fossil fuels, namely, petroleum products. In recent years, soybean and rapeseed oils have been used to manufacture bio-diesel. Jatropha and karanj are also being explored for this purpose.

Fatty acid methyl esters apart from biodiesel, do not have many direct applications. They are used as intermediates to produce a number of olecochemicals. Fatty alcohols are the major derivatives of fatty acid methyl esters.

Preparation of Biodiesel: Take 200 ml methanol and add 3.5 g sodium hydroxide. This reaction produces sodium methoxide. Mix the methanol and sodium hydroxide until the sodium hydroxide has completely dissolved (2 minutes), then add 1 liter of vegetable oil to this mixture. Shake continuously. The mixture separates out into layers. The bottom layer will be glycerine. The top layer is the biodiesel. Allow at least a couple of hours for the mixture to fully separate. The top layer is the diesel fuel. Normally one can use pure biodiesel or a mixture of biodiesel and petroleum diesel as a fuel in any unmodified diesel engine. If the engine is run at a temperature lower than 13°C, one should mix biodiesel with petroleum diesel. A 50:50 mixture will work for cold weather. At higher temperatures pure biodiesel can be used without any problem.

Biodiesel Stability and Shelf Life: The chemical stability of biodiesel depends on the oil from which it is derived. Biodiesel from oils that naturally contain the antioxidant tocopherol or vitamin E (Ex. rapeseed oil) remain stable longer than biodiesel from other types of vegetable oils. Biodiesel prepared from highly unsaturated lipids like soya oil or sunflower oil may not remain stable for longer periods. Temperature also affects fuel stability in that excessive temperatures may denature the fuel.

Properties of biodiesel: Biodiesel has promising lubricating properties and cetane ratings compared to low sulfur diesel fuels. Depending on the engine, this might include high pressure injection pumps, pump injectors (also called *unit injectors*) and fuel injectors. The calorific value of biodiesel is about 37.27 MJ/kg. This is 9% lower than regular petrodiesel. Variations in biodiesel energy density is more dependent on the feedstock used than the production process. It has been claimed biodiesel gives better lubricity and more complete combustion thus increasing the engine energy output and partially compensating for the higher energy density of petrodiesel.

The color of biodiesel ranges from golden to dark brown, depending on the production method. It is slightly miscible with water, has a high boiling point and low vapor pressure. The flash point of biodiesel (>130 °C, >266 °F) is significantly higher than that of petroleum diesel (64 °C, 147 °F) or gasoline (−45 °C, -52 °F). Biodiesel has a density of ~ 0.88 g/cm³, higher than petrodiesel (~ 0.85 g/cm³). Biodiesel contains virtually no sulfur.

Fatty alcohols and derivatives: Hydrogenation under high pressure of fatty acids yield fatty alcohols. More than 60% of fatty alcohols are derived from petro-chemicals while the rest are from fatty acids. Fatty alcohols are environment friendly and are utilized as plasticizers, lubricants, textile auxiliaries and cosmetics. Detergents like alcoholic ethoxylates, fatty alcohol ether sulfates and alkyl sulfates are made from fatty alcohols. These detergents are easily biodegradable and are efficient at low temperatures, have good water solubility and better hard water performance. Alkyl polyglycosides (derived from reaction with starch etc.) have excellent lather properties and are used in hair shampoos and food technology.

Fatty amines and derivatives: Fatty amines are the most important nitrogen containing oleochemicals. Fatty amines are prepared by converting the fatty acids to their nitrites and their hydrogenation to the amines. They can also be prepared by special hydrogenation in the presence of NH_3 of FAME. Different types of amines, namely, primary, secondary, tertiary and quaternary amines are possible. These compounds are useful as bacteriostats, textile and paper softeners, laundry detergents, emulsifiers, organophilic clays, viscosifiers in paints and varnishes, additives to lubricants, corrosion inhibitors etc. Oleoamines have manifold uses, in petroleum exploration industry, as well as drilling auxiliaries, and as anticorrosion agents. Tertiary and quaternary ammonium compounds function as flotation agents in metallurgy. Fatty amine oxides (eg. lauramine oxides) are shampoo ingredients, which clean the hair and give body and sheen and to make the hair more manageable. Amines and their derivatives can be used as intermediates in chemical processes. The major appplications are in laundry and household products.

Fatty amides: Fatty amides are obtained through reaction of fatty acids with anhydrous NH_3. Primary amides ($RCONH_2$) like erucamides are useful as slip or anti-blocking agents in the production of polyolefin films and as internal and external lubrication additives. Monosubstituted amides (R`CONHR) or secondary amides are useful internal lubricants. They are used to reduce cohesive forces of polymer chain and improve flow of polymers. Alkanolamides are surface active compounds and are incorporated into shampoos and light duty detergents. Sulfosuccinate monoesters of fatty amides are useful as hair shampoos, carpet cleaners and rust prevention agents.

General applications of vegetable oils: The main products that are derived from oils and fats and their utilization in various walks of our life are as follows:

a) **Soaps:** Saponified fatty acids have been the simplest surface active agents since ancient times. Various proportions of C_{12}-C_{14} and C_{16}-C_{18} fatty acids are used to obtain a soap with good lather, water solubility and detergency. Though stearin from palm oil is available at reasonable rates, it does not produce white soap and hence is utilized in low quality laundry soaps.

b) **Cosmetics and pharmaceuticals:** Oils like coconut, seasame, groundnut and castor oils have medicinal uses. They are also utilized in making specialty soaps like face and shaving creams and hair lotions. Vegetable oils function as emollients in creams and lotions. They serve as carriers of vitamins. Lecithins are excellent emulsifiers.

c) **Plastics and polymers:** Esterification of dibasic acids with polyhydroxy alcohols produces alkyd resins which dissolve in common solvents and represent tough and flexible polymers suitable for protective coatings. Urethane resins are made by esterifying di-isocyanates with dibasic acids. Epoxy resins are made from poly-epoxide and diphenolic compounds.

d) **Surface coatings and inks:** Drying and semi-drying oils like linseed and soybean oils as well as heat polymerized and bodied oils are used in the manufacture of paints, enamels, varnishes and other protective and decorative coatings. Vegetable oils or fatty acids sometimes function as plasticizers rather than coatings. They are useful as educts for alkyd resins and reactive diluents.

   Recently vegetable oil based printing inks have gained more importance. Because of different printing types and surface properties of paper, plastic films or metallic foils, a great diversity of inks is required to fill the needs of the printing trade. Inks used in offset or letter printing are pastes consisting of pigments, pigment carrier, or binding agent (alkyd resins). Heat bodied or blown soybean or linseed oils with varying degrees of viscosity are substituted for petro-products in the manufacture of inks.

e) **Lubricants and hydraulic oils:** The basic function of a lubricant is to prevent moving surfaces from coming into direct contact with each other. Vegetable oils exhibit better adhesion properties than mineral oils. Undesirable frictional wear and temperature rise are also lower with plant oils. They have higher viscosity indices. However, they are less stable than mineral oils. Sometimes hydrolysis products like fatty acids may cause corrosion.

   Rapeseed and castor oils are widely used as lubricating oils. The popular 'castrol' is a special blend of lubricating oil containing castor oil. Some of the applications of vegetable oils include heavy duty lubricants, hydraulic oils and automobile brake fluids. At present the main disadvantage of vegetable oils is their higher cost.

f) **Oiled cloth, linoleum and rubber additive:** Oiled cloth is manufactured by impregnating cotton cloth with drying oils, paints, filler material and varnish or wax. The popular floor covering linoleum is prepared from drying oils, filler compounds (rosin, alkyds), pigments and ground cork or wood. Factice, the vulcanized oil is a rubber substitute. Along with rubber, it improves the strength of rubber.

g) **Metal treatment:** Core oils are used as binders for sand to make a mold in which castings are made. The cores are baked at 200-230^{0}C, at which temperature, the oils polymerize and get hardened. Oils also find application in quenching of metals, tinning, and rust proofing of metals etc.

h) **Textile and leather industry:** Surface active compounds prepared by treatment of oils with sulfuric acid have been very important in imparting desirable softness or stiffness to fabrics. Turkey red oil is an important oil utilized in textile dyeing. Vegetable oils are used in fat liquoring of leather after chrome tanning treatment.

Table 7: Major applications of oleochemicals

Oleochemicals	Applications
Fatty Acids & Derivatives:	Plasticizers, lubricants, surfacants, washing & cleaning agents soaps, Alkyd resins, dyes, textiles, leather, paper, rubber, paints and varnishes. Food additive, surface coatings, inks, oilfield chemicals
Fatty Alcohols & Derivatives:	Washing & cleaning agents, surfactants, cosmetics, textiles, leather, paper, mineral oil additives.
Fatty Acid Methyl Esters	Cosmetics, surfactants, washing & cleaning agents. Bio-diesel
Fatty Amines & Derivatives	Fabric conditioners, mining, biocides, road making, textiles, fibres, mineral oil additives.
Glycerol and its Derivatives	Cosmetics, toothpastes, pharmaceuticals, foodstuffs, lacquers, plastics, synthetic resins, tobacco, explosives, cellulose processing, dyes.
Drying Oils	Lacquers, varnishes, paints, Linoleum
Neutral oils	Soaps, cosmetics, lubricating oils.

At present castor and linseed serve as industrial oil sources with castor being India's monopoly. Mustard oil having high erucic acid has great potential as source of industrial oil. Similarly all other low grade edible oils can find varied uses in the industry. Foregoing information has already made it evident that costly oleochemicals and derivatives can be manufactured from the vegetable oils. The cost of some of the chemicals are anywhere around 200 times or more than the raw material like oilseeds or oils. Hence development of oleochemical industry not only supports our country in sustainable production but also helps earn much needed foreign exchange.

Oilseed Based Food Products

Groundnut, sesame, soybean and niger seeds and their products are being utilised in the preparation of household food items. Processing like dehulling, roasting or boiling etc. makes these oilseeds more flavorful and tasty food items apart from increasing the nutritional value. All these oilseeds and their defatted flours serve as good energy and nutrition sources. Oilseeds and nuts are more popular as snack food items all over the world. Protein concentrates and protein isolates can be manufactured from them. Milk and butter substitutes, *chikkis, laddus, chutneys,* dry powders, fermented and coagulated products, fortified foods like biscuits and breads, meat analogues and texturised products can be made either at home or by the industry which are presently in good demand. The oil rich foods are more satiating, help in delayed hunger and are more balanced with respect to nutritional constituents.

Groundnut based foods: Groundnuts can be used as component of any food. The peanuts are unique in that they find a wide range of uses. The kernels are used after roasting, frying, salting, or boiling. They are used in a variety of dishes and in the preparation of confectionery products. Peanuts eaten with some jaggery form a cheap, tasty and nutritious food. They can be mixed with rice wheat, sorghum, maize etc. and can be eaten as part of the foods prepared. Groundnuts can also be used in soups or stews. They are broken up first and boiled in the soup, together with meat, fish and or vegetables. They are consumed after first roasting and then pounding into a paste. This may be either a coarse or a fine paste, similar to the peanut butter. Groundnut stews are popular in many countries. **Roasted groundnuts:** Roasting of groundnuts frequently with a little table salt is a very common practice all over the world. Groundnuts are roasted either by applying dry heat or in some vegetable oil. Dry roasted groundnuts are used in the preparation of butter, confectionery, or bakery products. Roasting reduces moisture content, develops a pleasant flavor, and makes the meal more acceptable for consumption. The reduction in moisture content during roasting of groundnuts prevent molding and reduces staleness and rancidity. Proteins are denatured during roasting. Excessive heating during roasting lowers the nutritional quality of the protein.

Amino acids such as glutamic acid, aspartic acid, phenylalanine, and histidine react with sugars to contribute to the flavor quality of roasted groundnuts. These amino acids are released from a large peptide, while glucose and fructose are released from sucrose. Roasting decreases the methionine-rich proteins and aggregates arachin proteins.

Cake and meal: Groundnut cake (defatted), when powdered, gives the meal. Full-fat meal is also obtained by direct powdering of groundnuts. The defatted peanut meal contains about 43 to 65% protein, 6 to 20% fat, (depending upon the method of oil extraction), and some vitamins of the B-complex. The defatted groundnut meal has higher PER, NPU, and digestibility than the full-fat groundnut meal. The chemical score of groundnut flours can be improved by blending with other plant proteins. These blends could be used in bakery products. Heat-processed groundnut meal is lower in lysine, methioine, and threonine. Autoclaving and oven heating of groundnut cake reduces the protein digestibility, PER and available lysine.

Protein Isolates: Production of groundnut protein in the form of concentrates and isolates for human consumption has become a reality. They have a lot of uses in the preparation of many types of food products. They impart beneficial characteristics to foods in which they are incorporated. Isolated protein (90-95%) from groundnut is a sophisticated product. It offers immense possibilities in the development of protein enriched foods and meat analogues including milk-like products. The isolates can be textured and flavored to meet a wide variety of consumer preferences. The biggest advantage is that it is bland and does not interfere with the natural flavors of the foods enriched with it.

Presoaking of split groundnut kernels in 4% NaCl solution overnight followed by hot-water washing helps in washing out the water solubles and flavor components. The protein isolate thus obtained has high solubility, white color and is free from a nutty flavor. Groundnut protein isolates and oil have scope to be used in the manufacture of cheese analogues. Groundnut proteins have been used to extend meat products without adverse effects.

Groundnut (peanut) butter: Manufacture of groundnut butter is gaining importance all over the world. It is prepared by grinding roasted and blanched kernels to which 1 to 4% common salt is added. Groundnut butter flavor is a major factor in determining the acceptability of the product. Groundnut butter contains 25% protein, 50% fat, 23% carbohydrates and 4% minerals. It is rich in calcium, iron, phosphorus, sodium and potassium. It also has good levels of vitamin A, thiamine, riboflavin and niacin. The calorific value of groundnut butter is around 590 cal/100g. This is similar to an item called chutney made in India. The only difference is that chillies, tamarind, salt and spices are added to the chutney.

Fermented products: Groundnut cake or meal can be used for human consumption after partial hydrolysis of protein by the process of fermentation using certain molds. Such products are readily digestible, tasty, and nutritious, "*Oncom*" is a fermented groundnut press cake and is a popular product in Indonesia. "*Oncom*" contains 3-9% oil, 20-30% protein and up to 22% carbohydrates.

Groundnut milk: Good-quality peanut kernels are cleaned of impurities and roasted lightly to facilitate removal of red skin. The testa or skin can be removed in a blanching machine. The germ is separated by sieving and the spoilt seeds by hand picking. The cleaned kernels are ground to a smooth paste. The paste is mixed with 7 times the weight of water in a blender. Calcium hydroxide solution is added till the pH of the milk is adjusted to 6.8. A mixture of disodium phosphate and acid potassium phosphate having pH 7.0 is added to stabilize the milk. The milk is filtered through a fine cloth and fortified with vitamins A, D, riboflavin, folic acid, vitamin B12 and minerals (calcium and iron salts). Cane sugar is added at 7% level. The milk is homogenized, steamed, bottled and kept in a refrigerator till distributed.

This is a milk-like product or a milk analogue called "miltone" in India. It consists of groundnut milk extended with buffalo milk. It has the same nutritive value as that of milk and can be used as its substitute, flavoured beverage or for making curds by lactic fermentation. During fermentation, large molecular weight globulins, such as arachin, are hydrolyzed to smaller components. As a result of protein breakdown, the nitrogen solubility profiles of groundnut protein at various pH values change drastically. Thiamin, riboflavin, and niacin levels are increased in fermented groundnut flour. There is degradation of phytic acid during the preparation of oncom. Protein efficiency ratios (PER) of fermented groundnuts are however, not increased over the non-fermented materials.

Infant and weanling foods: Milk, the ideal food for babies, infants and other vulnerable groups of population is an expensive item. A nutritionally sound infant food based on protein isolate and skim milk powder has been developed. This spray dried product with 26% protein is easily dispersible in water and can be fed to a baby from the bottle. Two third of its protein is derived from the groundnut protein isolate and one-third from milk. At 15% protein level in the diet, the PER of this product is about the same as that of milk. Spray dried infant foods based on groundnut protein isolate, soya flour, and skim milk powder, have also been developed with added hydrogenated fat and hydrolyzed starch. The PER of this food is 2.7 and when fortified with methionine is 3.0 or equivalent to milk.

Composite flours: Groundnuts flour can be used for the preparation of composite flours. Such composite flours have a higher protein content and better nutritive value. Whole wheat and maize flours are supplemented with 10, 20, and 30% levels of a 1:1 mixture of groundnut and chickpea flours. Supplementation increases the protein content of the wheat and maize blends by 20 to 61%. A significant increase in other proximate constituents as well as lysine also takes place. The chemical score of wheat flour increases from 53 to 72 and that of maize flour from 49 to 71 with 30% groundnut chickpea flour. Biological evaluation of wheat breads at 10% protein level and maize breads at 8% protein level in the diet has shown a significant improvement in the protein quality of breads as judged by gain in body weight, protein efficiency ratio, net protein utilization, biological value, and utilizable protein content. A supplementation level of 20% is adequate to achieve the desired nutritive benefits

Mysore flour: It is made of 75% of edible grade tapioca flour and 25% low fat edible grade groundnut flour in India. This product has 12-13% protein and could be used to prepare a number of traditional foods such as *chapati, parotha, puri* and several other savoury and sweet dishes. It helps to upgrade the nutritive value of tapioca.

Pausthik atta (nutritious flour): Two formulations of this product are prepared. Formula-A has 75% whole wheat flour, 17% tapioca flour and 8% groundnut flour. This is more suitable for use in areas where indigenously grown tapioca flour is available at low price. Formula-B consists of 5-10% edible grade low fat groundnut flour added to whole wheat flour. Both the formulae have very good acceptance. A vast variety of enriched flours can be developed and used for several purposes.

Traditional products: Groundnuts are utilized for preparation of products like *laddu* and *chikki* in India. For preparation of *laddu*, peanut kernels are roasted and decorticated. The separated cotyledons are mixed with thick and hot jaggery syrup. Small portions of the mixture are pressed with the hand to obtain balls or *laddus* of about 3 to 5 cm diameter. *Laddus* may also be prepared from roasted and decorticated groundnut kernels after coarse grinding to obtain particles of about 1 to 5 mm diameter which are mixed with a small quantity (2:1 or 3:1) of crushed jaggery. Small portions of this mixture are pressed between the hands to obtain balls of about 3 to 5 cm diameter. *Chikki* is a very popular product being liked and utilized all over India. It is prepared by mixing roasted and decorticated peanut cotyledons with a hot slurry of jaggery. The mixture is spread on the tray or table or any surface with a uniform thickness of about 1.0 to 1.5cm. After cooling, the material is cut into small pieces and packed.

Soy food items: Soybeans being very rich in protein have received lot of attention for preparation of many foods. A large number of traditional technologies have been developed to make fermented products and textured from soya. The problem with most soy based products is their strong flavour which is preferred in the far eastern cultures but not in South Asia. Lactic fermentation minimises this flavour.

Soy flour: Full-fat soy flour comes from whole, dehulled soybeans that have been ground into a fine powder; it retains the fat naturally found in soybeans. As the name implies, defatted flour has the fat removed before grinding. Although soy flour is not a common ingredient in home baking, it is used extensively by the food industry. Soy flour is approximately 50% protein on a dry weight basis.

Full fat soy flour: One of the simplest ways of using soybean in human diet is in the form of full fat soy flour in combination with wheat, sorghum, maize or rice flour. 10% addition of soy flour is recommended with cereals to start with. Preparation and use of recipes from soy-cereal blended flour does not involve any change in the traditional food habits of the people. The process adapted/developed for the production of full fat soy flour consists of cleaning, dehulling, soaking/steeping of soy-splits in bean to water (containing 0.5-1% $NaHCO_3$, w/v) ratio of 1:3 followed by immersion cooking in water for 15-20 min, drying of cooked splits to about 8% moisture. The splits are then milled to obtain soy fluor. The soy flour thus produced contains about 20% oil, 40% protein with PER value of about 2.0 with water absorption capacity of about 200%. Urease and trypsin inhibitor activities are almost nil and the soy flour can be stored safely up to 3-months. Organoleptic and sensory evaluation experiments on use of full fat soy flour in various soy based food recipes have shown encouraging results.

Partially defatted soy flour: Cleaned and dehulled soybeans are conditioned with steam for 15-18 min at atmospheric pressure. They are flaked followed by oil expelling which is done in three passes. Soy flakes/cakes are powdered, cooled and packed. The partially defatted soy flour has about 5-6% oil. The cake obtained is of edible grade. Flour is creamy yellow. It offers potential for producing oil as well as food grade partially defatted soy flour.

Soy*paneer*: It is a coagulated and pressed soy protein. Milk-*paneer* is very popular among high-income group urban people for making vegetarain-curry and *Paneer Pakoda*. However, the cost of milk-paneer is very high and hence, it is out of the economic reach of poorer sections of people. In this context, soypaneer is an appropriate alternative to milk-paneer as the cost of the former is only one-third the cost of milk paneer and gives all the nutritional advantages and psychological satisfaction of eating a *paneer*-dish.

The process adapted/developed for the production of soy paneer consists of cleaning and grading, dehulling, soaking of soy splits in water (containing 1% $NaHCO_3$, w/v basis) with bean to water ratio of 1:3 at room temperature for 4 hours or for one hour at 60°C, grinding of soaked and thoroughly washed splits using hot water (bean to water ratio = 1:8), boiling of soy puree for 15-20 min, and then removing the roughage through filtering, cooling the milk thus obtained to about 70°C, addition of 0.02 M $CaSO_4/MgSO_4$ and stirring the mixture for 2 or 3 times, allowing it to settle for 30 min in order to get protein coagulated, filtering of proteinate complex through muslin cloth lined paneer forming boxes, pressing the soy paneer at 10 g/cm^2 for about an hour, removing the soy paneer blocks to cold water and keeping it there after washing for about 30 min, and then storing the paneer in water till use.

1.5-2 kg soy *paneer* can be obtained from one kg cleaned and dry soybean. Soy paneer at about 75% moisture content contains approximately 12% protein, 7% oil, 4% carbohydrate and 2% ash. In an organoleptic assessment of soy paneer, the mean score values of taste panel judges for overall acceptability of soy paneer is 7.5 as compared to 9 for milk-paneer. Paneer prepared out of soy-animal milk blends at 25, 50 and 75% levels has shown encouraging results in overcoming the initial hesitation of people in accepting soy paneer.

Soy dal: Vegetarians generally consume cereals along with cooked pulse in a soup/ broth form. Appropriate fortification of these pulses with soybean can provide a better protein with all essential amino-acids at low cost. Experimental results have shown that blending of soy *dal* with other pulses namely pigeon pea, green gram, black gram and lentil upto 40-50% and pressure cooking provides a well cooked, dispersed and organoleptically acceptable dal containing about 32% protein.

Soy flakes: It is another product which can be used along with rice, maize or sorghum flakes. The process adapted/developed for making soy flakes consists of cleaning of raw soybean to remove impurities, brokens and shrivelled grain, dehulling and splitting of cleaned soybeans, blanching of soydal for 1 hour and then drying to 25-30% moisture, flaking of the semi-dried soy dal, drying of flakes to 8% moisture, grading and packaging.

Soy flakes thus obtained contains approximately 44% protein and 19% oil with no trypsin inhibitors. Storage of soy flakes at 6-8% moisture in polyethylene bags or metallic containers is safe for at least 6 months. Soy flakes/grits could be added to daily dietary preparations like curries, dal, etc. at 10-15% level. The broken soy flakes are good food ingredients in preparation of rice based fermented products like *Idli* and *Dosa*.

Soy-protein ingredient products: Soy-protein ingredient products are commercially available sources of soy protein that are used in the preparation of other food- stuffs. Soybeans are first dehulled, flaked, and defatted to make "white flakes". These white flakes are made into defatted grits or flours, soy concentrates, or have their flavor compounds and flatulence-producing sugars extracted to produce isolated soy protein.

Soy protein has become an ingredient of many common foods, including salad dressings, cream soups, and candy coatings. In addition to the superior nutrition provided by the consumption of soy proteins, they have functional properties that make them important ingredients in many processed foods. For example, soy proteins absorb water and fat, act as an emulsifier, aid in whipping (aeration), and improve texture in foods to which they are added.

Soy protein concentrates: Soy protein concentrates are made by removing a portion of the carbohydrates from defatted and dehulled soybeans. Alcohol extraction is the method most commonly used to manufacture soy protein concentrates; it results in the loss of isoflavones and saponins, which are alcohol soluble. Soy protein concentrates retain some of the fiber in the original soybean and are about 70% protein on a moisture-free basis. They are most often used as a functional or nutritional ingredient in a wide variety of food products.

Isolated soy protein: The most concentrated source of soy protein is isolated soy protein (sometimes called soy protein isolate), which is usually at least 90% protein on a dry-weight basis. Most isolated soy protein is manufactured by water extraction of protein from defatted and dehulled soybeans. Since extensive water washing can remove isoflavones and other compounds, processing must be carefully controlled to retain these naturally occurring components. Isolated soy protein is heat treated during processing to ensure inactivation of trypsin inhibitors. It is used as a source of high-quality protein and serves as a functional ingredient in many familiar food products.

Textured vegetable protein and meat extenders: Soy protein isolate has been used as a meat substitute and to prepare this, the soybean meal is suspended in dilute sodium hydroxide and centrifuged. The meat extender may now be prepared in several ways. The 2 main manufacturing processes consist of either obtaining an alkaline dispersion of the protein, which is then mixed with colour and flavour and steam extruded, or the dispersion is mixed with gum, colour and flavour and extruded through fine dyes of 0.18 mm diameter. The extruded material is stretched and layered on reciprocating tables, coagulated with live steam and thus given a fibrous, more meat-like structure, in contrast to the rather amorphous material obtained with the first method. These meat substitutes are used extensively by mixing the with commercial meat products in the manufacture of sausages and hamburgers.

Fabricated products made with spun protein analogues, even at 50% level, are very similar to their all meat counterparts and have high acceptability. The flavour and texture problems associated with many other meat extenders which restrict their use to 10-30% can be overcome by use of spun protein analogues. These spun protein analogues are very adaptable to applications with sectioned and formed products and in some cases actually improve product quality.

Extruded products: Extrusion cooking, a process of high temperature, high pressure short-time cooking under controlled conditions of moisture, is particularly suited for high-protein low-fat materials like solvent extracted soy flour or groundnut flour.

Milk substitutes and infant foods: Milk prepared from soybean has long been used in China and other Asian countries for feeding of infants. Preparation of soybean milk without any beany odour can be carried out using the following steps: (i) cleanining and dehulling of soybean in a mill; (ii) soaking dehulled split beans in water for 5-10 hr; (iii) steaming the soaked beans for 30 min to destroy lipoxidase, trypsin inhibitors and other anti-nutrients; (iv) grinding the cooked beans in boiling water; (v) the ground mass is then filtered and sugar, minerals and vitamins are added to the filtrate; and (vi) the final product is then homogenized, steamed and bottled.

Soymilk can also be prepared from commercially available defatted flour. The defatted flour is suspended in hot water (80°C), blended and the pH of the suspension is adjusted to 10.0 to solubilize proteins. The suspension is centrifuged and the filtrate's pH is then adjusted to 7.0-7.4 and centrifuged again. The protein content in the filtrate is adjusted with water and the mixture is blended with sugar and fat, homogenized, bottled and sterilized. Flavoured milks can be prepared from soy milk using sweeteners and suitable colouring and flavouring agents. Dried soymilk powders have also been prepared. Protein concentrates and isolates can also be used to prepare soymilk.

Soya snack: Cereal based snack foods are quite popular all over the world. Addition of soybean improves the nutritional value. A process developed for making soy-based snack food consists of cleaning of cereals and treated soybean splits and grinding of the soy-cereal mixture into flour. The dough made from fluor is extruded (through a form of extruder) at about 100°C, semi-drying the extrudate, flaking, drying of flakes to 8% moisture and packaging. The flakes thus prepared from soy-wheat/sorghum/maize (30:70) have approximately 20% protein, 7% oil, 55% carbohydrates.

Soy-badi: It is a dehydrated and spherical shaped pulse based vegetable prepared in villages for house-hold consumption. It is generally made using green gram pulse along with other ingredients such as spices, salt etc. The process of soy-badi making consists of cleaning of soybean, dehulling and splitting, soaking of soy splits in water containing 1% $NaHCO_3$ for 4 h at a bean-water ratio of 1:3, grinding of soaked soy dal to make paste, addition of salt, spices, etc. as per taste, cooking for about 30 min, forming badi and then drying to 6-8% moisture. The soy-badi thus made has about 43% protein, 19% oil and negligible trypsin inhibitor which is further eliminated during cooking of the badi.

Rapeseed and Mustard Products: There is a great potential for preparation of high quality oilseed protein concentrates from rapeseed- mustard. The iso-thyocyanates are not desirable in the protein meal. Glucosinolates and their breakdown products from mustard and rapeseed cause enlargement of thyroid, adrenal gland, kidney and liver haemorrage. Hence it is better to remove these anti nutrients, prior to utilization of the meal. They can be removed by methods such as extraction by suitable solvents, enzymatic decomposition, treatment with chemicals etc. such techniques when utilized will make available, a million tonnes of high quality protein for human consumption.

Rape seed meal can be utilized in preparation of protein concentrates, isolates and textured products. The nutritional quality of glucosinolate free rapeseed protein is comparable to that of casein with a Protein Efficiency Ratio (PER) of 2.4 and Net Protein Utilisation (NPU) value of 65. The following foods can be prepared using mustard protein.

1. Using mustard protein concentrate (MPC) protein rich biscuits with 15% protein can be prepared. The biscuits have desirable texture and taste. It has a NPR value of 3.64.

2. High protein candies incorporating MPC can also be prepared. This type of product finds ready acceptance with children and provide them protein supplement.

3. Weaning foods: MPC can also be used for preparing weaning foods containing mixtures of cereals and pulses. Germinated cereals and pulses can be used to prepare such foods. The *in vitro* digestibility of protein of such products are 67 for under-germinated and 65 for germinated.

Sesame based foods: Sesame protein is a rich source of sulphur amino-acids in which most other vegetable proteins are deficient. It is an oilseed of choice with many uses. Nearly 20% of sesame seeds produced is consumed directly in confections and other foods. Production and use of sesame protein presents several problems but they have now been solved. It's husk is hard to remove and has undigestible fibre. Besides, the oxalic acid content of the husk is also quite high. The protein concentrate made form dehusked sesame is an excellent supplement of sulphur amino-acids for incorporation with other oilseed concentrates and grain legumes to make foods with high PER, especially for consumption by children.

Sesame seed kernels are a nourishing food that heals. The seeds are digestive, rejuvenative, anti-aging and rich source of quality oil (50%), protein (25%) and corbohydrates (15%). Seeds provide 640 calories/100g energy and very rich in vitamins E, A, B complex and minerals like calcium, phosphorus, iron, copper, magnesium, zinc and potassium. Sesame seeds contain extraordinary quantities of methionine and tryptophan amino acids, which are essential for healhy liver and kidneys and utilization of B complex. Linoleic acid; methionine and tryptophan, minerals and vitamins which are deficient in general in most vegetable items, are abundant in sesame seeds. This unique composition coupled with the balanced proportion of saturated, mono-unsaturated and poly-unsaturated fatty acids including linoleic acid and tocopherol makes sesame seed a nearly perfect food.

Sesame has very little possibilities of causing any allergy and therefore it is the best alternative source of food to the people with milk allergies. The husk is also very rich in minerals and vitamin B1, but these are unfortunately non-extractable by human digestive system due to the presence of anti-nutritional factors, oxalic and phytic acids. Oxalic acid binds calcium to form non-available calcium oxalate. Therefore, dehulling is preferred before direct consumption of sesame seeds.

Protein concentrates and isolates: Sesame is widely processed into several high-protein products such as flakes, flours, protein concentrates and isolates. The protein concentrate contains more protein (about 70%) than the flour while protein isolate contains about 90% or more protein. The defatted flour and isolates prepared from sesame do not contain any undesirable pigments, off-flavors or toxins. The protein isolate contains very high levels of protein and free from oil, ash, crude fiber and phytate phosphorus with very low levels of nitrogen-free extract.

Functional properties : The functional properties such as water absorption, fat or oil absorption, emulsification, foaming and viscosity are important in determining the applications of food products in various food systems. Sesame flour and protein concentrate exhibit less water absorption than soy flour. Sesame products show high fat absorption than soy products. The emulsifying capacity and emulsion stability of sesame products were comparable to those of soy products. The foam expansion and foam stability are also higher in sesame products.

Sesame *chikkies* and *laddus*: Sesame seeds are used in the manufacture of traditional confections such as halva, *laddu* and *chikki*. They are also eaten whole after roasting. A confection called *laddu* is prepared from roasted groundnuts and sesame seeds by pounding with jaggery in the proportion of 2:1:2. Small balls are prepared by hand. *Laddus* are also prepared from sesame seeds by mixing them with hot jaggery syrup. *Chikki* is another confection popular in India. It is prepared by pouring sesame seeds in boiling jaggery solution to obtain thick slurry. The slurry is spread uniformly on a metallic sheet or table and cut into small rectangular pieces.

In India *laddus* (balls) of sesame and jaggery are prepared mostly at household level. These *laddus* keep people healthy, young and energetic and protect from cold and cold related ailments. Sesame seeds in combination with jaggery, cow's ghee and other ingredients complement and supplement each other to produce synergistic effect on human body to overcome debility and revitalize with the dilapidated system. Besides improving general health and stamina, this preparation acts as laxative for bowel movement, lubricates the joints to relieve rheumatic pain, induces resistance to cold and reduces other problems associated with old age. If taken twice daily in the morning and evening with tea/milk, the *laddus* reduce the process of aging and removes the debility.

Ready-to-eat foods: Minor millet fluors like pearlmillet, sorghum and finger millet(ragi) are used to prepare ready to eat foods in Indian households. Roasted bajra (pearl millet) flour (60g) is mixed with roasted green gram dhal (15g), roasted groundnut (10g) and sesame seeds (5g). The mixture is pounded/ ground to obtain a fine powder. It is mixed with boiling water or milk to the desired thickness. Sugar or salt are added to taste. Ragi/ sorghum instant foods are prepared in the same way by replacing bajra flour.

Sesame seeds are utilized in the production of tehineh (sesame butter) and halwah (halva). Tehineh is made from a paste of dehulled roasted seeds similar to peanut butter. Halwah is a sweet made up of tehineh, sugar, citric acid and *saponaria officinalis* root extract. Tehineh is added to a variety of foods, bread and bakery products .

Sesame seeds and kernels are used in commercial bakeries for the preparation of quick breads, rolls, crackers, coffee cakes, pies and pastry products. The seeds are lightly roasted and used in salads and salad dressings . Cholam (sorghum) porridge is prepared by mixing roasted flours of cholam (50g), green gram dhal (15g), and sesame cake (25g) with hot water. The butter is poured in boiling jaggery solution while stirring and bioled for 10-15 min.

Sesame flour has been used as a methionine supplement in the preparation of fermented foods, namely, vada and dosa. The sesame flour is used to replace 5-20% of rice-blackgram flour. Sesame meal is used as a vegetable in preparation of curries in the Eastern part of India. Sesame flour has been used in the preparation of bread and cookies. When used as a part of replacement of wheat flour, sesame flour performs better than the sunflower flour. High-protein biscuits can be prepared by mixing wheat flour with roasted Bengal gram and roasted sesame flour to prepare the dough. Sesame fluor can be used in formulating high protein beverages. A nutritious beverage could be prepared using 70% soy protein and 30% sesame protein.

Sunflower Based Foods: The whole seeds and kernels of the non-oilseed-type sunflowers are used for human consumption after roasting, similar to roasted peanuts. Roasted and salted sunflower seed, either dehulled or whole, is a popular snack food all over the world. The broken or whole kernels with or without roasting can be used in a variety of bakery products, salads, candies, and some food dishes. The kernels, with or without roasting, are used for sprinkling on syrup or pancakes and waffles, blended with honey, butter and salt to make a spread, or used as a fondue dip.

Confection: Sunflower seed is consumed as a snack food. Confectionery sunflower (generally bolder seeds) is consumed in-shell or dehulled. Kernel is used in a broad variety of snacks and food proudcts such as breads or as part of nuts and dried fruit mixes.Confectionery sunflower seeds are larger than oil varieties, normally black with white stripes and approximately five eighths of an inch long. The heavy hull accounts for 50% of the weight of the seed and is loosely fixed to the kernel inside. The seeds (in shell) are roasted, salted and packaged for human consumption and are classified as either large or jumbo. Medium-sized kernels are also packaged for human consumption. They are mostly utilized by the bakery industry. The black-shelled oilseed variety is richer in oil and are used for the production of sunflower oil.

In shell seeds are normally roasted and seasoned. They are eaten as a snack by cracking the shell with one's teeth, discarding the hull and eating the delicious morsel within. The shells are removed mechanically and the resulting kernels are roasted and utilized directly as snack food and or incorporated as components of food items to improve their taste and nutrition.

Sunflower seeds are a good source of vitamin E, mainly alpha-tocopherol. A one-ounce serving of sunflower seeds contains a whopping 76% of the Recommended Dietary Allowance for vitamin E. Sunflower seeds provide vitamin E, along with healthful, unsaturated fat. They are also good sources of the antioxidants, copper and selenium, which work in conjunction with vitamin E. Sunflower seeds provide six grams of protein and two grams of fiber per serving, and are power-packed with folate, pantothenic acid, vitamin B6, manganese, iron and zinc.

Some of these nutrients have been shown to offer a variety of potential health benefits, including protection against heart disease and cancer, as well as a role in memory and cognitive functions. Sunflower seeds contain high levels of the phytochemicals like choline, betaine, and phenolic acids in addition to tocopherols. The kernel is a good source of lignans and arginine. Due to the high levels of phytochemicals sunflower kernel is considered a powerhouse of health benefits.

A mixture of sunflower meal, corn, roasted chickpea flour and sesame in the ratio of 65:15:10:10 has been used in the school lunch programme in India. The mixture is effective in increasing the heights of growing children. Sunflower meal (defatted) and protein products can be utilized in the preparation of bread, cakes, cookies, muffins and pies as a partial substitute for wheat flour or nuts. The concentration of sunflower above 15% results in poor loaf volume and dark color. Autoclaving of flour, roasting of kernels or the use of commercial dough conditioners help restore bread texture and volume.

Addition of 5-10% sunflower protein concentrate to wheat flour can increase the protein content and improve the amino acid balance of bread without changing the crumb color. Sunflower meal and protein concentrates have been extruded to form meat-like, chewy, texturized vegetable protein particles. Sunflower protein isolate incorporated in batter/breading system can impart its antioxidant effect to the meat of the fried beef patties .

Sunbutter-a peanut butter alternative: Some humans are allergic to peanuts and hence there is need for alternatives to peanut butter. Sunflower seeds are a popular alternative for preparing peanut butter. Sunbutter looks like peanut butter and has a mild and distinctive sunflower seed flavor.

Safflower based foods: Safflower is an important traditional oilseed in many parts of the world especially the developing countries. It is a rich source of poly-unsaturated fatty acids and proteins along with useful minerals and vitamins.

Safflower seeds are utilized as snack food. Fried seeds with hulls are eaten like groundnuts. The kernels are a better option. The seeds are generally disliked because they contain phenolics which impart bitter taste. However, those accustomed easily relish them. They are especially useful as neutaceuticals because of the high PUFA, proteins, minerals and useful vitamins present in them. . Roasted seeds, generally mixed with chickpeas, barley or wheat, are eaten as a snack food. The ground kernels are mixed with sesame and eaten. In India the seeds are used to make a spicy item called *chutney*.

A paste of seeds is useful in hastening curd formation. In Ethiopia, finely powdred safflower kernels are mixed with water to prepare a drink called 'fitfit', which is used on fast-days or mixed with 'teff' bread and spices to form a porridge. Crushed dried seeds are also used to grease the baking plate for the 'teff' flatbread.

Utilization of cake and flour: The meal obtained from completely decorticated seeds is light in color, bitter in taste and contains about 60 to 67% proteins. The removal of bitter compounds by methanol extraction further increases the protein content upto 70%. Such meals have promise in human foods in the form of high-protein flours, protein isolates and concentrates, in a variety of bakery products and in weaning or traditional foods as a protein supplement.

The defatted meal can be further pulverized to flour or used to prepare protein isolates and concentrates. This facilitates the removal of hull or fiber. A process has been developed for preparing a high-protein (41 to 45%) meal which consists of impacting a meal and classifying the material on the basis of particle size. A low-fiber, high-protein flour and edible concentrate have been prepared by screening and aspirating cracked safflower seeds, extracting the oil, milling and sieving. A light-colored, bitter-flavored flour contains more fiber than the recommended 5% maximum for human consumption. The protein concentrate has been prepared by extracting the bitter and cathartic glucosides from the flour with 70 to 80% ethanol.

The edible safflower flour (70% proteins) can be incorporated into meat-like patties and bread at a level of 5%. Due to the presence of bitter-flavored compounds, the utilization of defatted flour in various processed foods has remained limited to the experimental stage. The defatted flour holds promise as a protein supplement in a variety of processed foods such as bakery products, weaning foods and traditional cereal-legume-based products.

Protein isolates: The safflower protein isolates show inferior foaming capacity and stability, and whipping capacity as compared to other oilseed protein isolates. Heating treatments decrease whipping and water-holding capacities of protein isolates. Isolates prepared by the micellization procedure show superior functional properties than those obtained by isoelectric precipitation. However, the recovery of proteins is lower in the former process. Protein isolates prepared from debittered meal can be used to fortify bread, pasta and nutritional drinks. Among amino acids, lysine is limiting, while methionine and isoleucine are on the border line.

Niger based foods: Niger seeds are edible. Their seeds fried in ghee can accompany other foods to improve their palatability. They are parched and mixed with pulses as snack food. A spicy preparation called chutney, prepared by mixing red chilli powder with roasted and pounded niger seeds is a common preparation in some parts of rural India .

Niger seeds are parched and ground with water to make a drink. Other minor uses include mixing the parched seed with pulses as a snack food, and making a stew or "wot" with ground seed and spices. Niger flour is also baked into a bread or sprinkled on bread prior to consumption. The seeds have good market and they are utilized as bird feed in some countries.

Niger cake in addition to being rich in proteins, is also rich in soluble sugars. Niger cake and its protein are balanced with respect to essential amino acids. Its quality is similar to that of sesame cake. Hence, attempts are needed for wider utilization of niger cake as a food ingredient. Some valuable work has been done on its protein. Protein from dehusked seed is quite rich in sulphur amino-acids, but deficient in lysine. When the deoiled meal is mixed with grain legumes or cereals, it gives a PER quite comparable to casein.

Development of over the counter food products from oilseeds has great potential. Such products not only improve the nutritional quality of the products, but also increase the value of oilseeds. Simple processes like dehulling and or roasting and boiling more than doubles oilseed value. Other products like protein concentrates and isolates serve as high value foods essential to the elders and children. Hence oilseed food product development again has a special role in sustainable oilseed production.

Linseed In Health or Functional Foods

Linseed because of it's richness in linolenic acid, directly serves as a neutraceutical and also indirectly through other foods. Linseed oil apart from being rich in 18:3 fatty acid is low in saturated fatty acids, especially, palmitic acid (about 5%). Thus, it certainly can serve as a good salad oil, if not as a frying oil. Various branded products with either linseed oil or the seed or its cake as constituents in different proportions are already available in the market. More products are likely, in view of its various health benefits. It is anyway, already a part of the folk Indian medicine and is widely being used by rural people in India.

Seeing the possible role of flax-seed in the treatment of some cancers, particularly mammary cancer in rats, some poultry farmers are feeding chicken, feeds rich in flaxseed for boosting the omega-3 fatty acid content of eggs (ω–3 rich) of the chickens, These functional foods have gained lot of importance in the market in recent times. Similarly attempts are being made to develop ω–3 rich beef by providing compound feeds rich in linseed to ruminant animals.

Oilseed Extractions

The production of oilseeds is adding a new dimension to our food and feed front by making available protein-rich oil cakes. These oilseed meals have traditionally been utilized as feed for cattle, fish and poultry and the inedible cakes as organic manure. After processing the oilseeds for the oil through expeller and the solvent extraction mills, the residue, namely oil meal or cake is treated as a by-product by the oil industry. Most of the time the cakes are handled carelessly and under unhygienic conditions. Such products lose their commercial value and importance.

Some oil cakes like groundnut and sesame obtained from *ghanies* and small scale extraction units in rural areas are being utilized as components of dietary items. The advent of modern vegetable oil processing mills have rendered the residue as by-product suitable only as livestock feed. The cakes are richer than the oilseeds in proteins, carbohydrates, minerals and even vitamins.Primarily oilseed cakes are solely used for manuring and as a wholesome feedstock material for poultry, swines, milch cattle etc., However, a very little (about 5-10%) protein of it is used for food uses. It is also a fact that the cakes of all oilseeds are not edible unless properly processed.

Oilcakes in food products: The oil cakes are by-products of the oil milling industry. Upto 60-70% of the seeds return back as oilcakes from oil industry. These cakes are rich in proteins (20-50%), carbohydrates (20-30%) and minerals (5-8%). As long as the oil extraction units have been in the village level *ghanies,* some oil cakes like groundnut and sesame have been utilised as components of dietary items. Hygienically processed cakes can be utilized as protein rich flours which can be incorporated as part of the household foods and snack food items. They are potentially used as a raw material in many food industries. They can serve as milk and meat analogues.

Proteins are necessary for physical and mental growth as well as for the maintenance of good health and vigour. Oilseeds and oil cakes serve as good suppliers of essential amino acids. Sesame protein is the best followed by niger, groundnut and soybean. Oil cakes can be utilized in the production of low cost protein flours. Value added products like protein hydrolysates (amino acids), protein concentrates and protein isolates can also be manufactured from oil cakes.

Fabricated foods: Texturized vegetable proteins (meat analogues), through extrusion cooking may be prepared along with other cereals and pulses. Other foods such as simulated dairy analogues (milk, infant foods, coffee whiteners, ice creams, malted foods), bakery and confectionery products (biscuits and toffees), fermented foods like Miso, Natto etc., may also be prepared.

Simple mixes: The oil cakes can be ground into flour and mixed with cereals and pulses to prepare various energy rich foods like Balahar and Incaparina. In this case, oil cakes with different fat contents may be used to meet the nutritional requirement of different categories of people. Protein isolates and protein concentrates can be prepared and used for various functional and nutritional properties in food formulations. They contain about 90% and 70% proteins, respectively. They are used in infant formulae, medical foods for oral and external feeding, weight-loss products and body building supplements, as coffee creamers, protein fortified juices and other acid beverages, baked food and confections, dairy analogues and non diary sour creams and frozen desserts.

Quality of oilseed meals: The protein content of the oilseed meals range from 20-50%. Oilseed extractions can be placed in three categories based on their protein content. High protein meals include those of soybean, groundnut and sesame with 40-50% protein, meals of niger, rapeseed-mustard and linseed have 30-40% protein while the others have lower protein contents of 20-30%. Proteins are necessary for physical and mental growth as well as for the maintenance of good health and vigour. An adult human being needs to consume 50g of protein per day. Animal protein is costly and hence oilseed extractions can play a major role in reducing the protein malnutrition in the developing countries.

The protein of oilseeds are very good sources of essential amino acids required in our food. Soybean and rapeseed are rich in lysine. But soybean is deficient in sulphur containing amino acids namely cysteine and methionine apart from trytophan. Sesame protein is rich in methionine. Hence, sesame cake can be blended with other oilseed cakes to improve their nutritional quality. Oil meals are also rich in carbohydrates and minerals. Sugar content of the meals range between 26 and 30%. They are less harmful to diabetic patients. The mineral contents range from 4 to 10%. They are rich in calcium, phosphorus, magnesium and iron which are essential nutritional constituents for the human body.

Undesirable factors and their elimination: Oilseed meals contain some toxins and anti-nutritional factors like polyphenols, protease inhibitors, oxalates, phytates, glucosinolates and aflatoxin. These reduce the nutritional quality of the meal apart from causing illness and other health problems. Such low grade cakes can be improved with some treatments. The cakes can be treated with ammonia gas to improve their texture, detoxify aflatoxin and allergen and enrich them with nitrogen (protein). Dehulling is another simple process to upgrade the meal quality by reducing the crude fibre. Steaming is useful in removing toxins from rapeseed, castor and linseed meals. There are a variety of simple techniques to improve the cake quality from feed to food and inedible cake to animal feed. Such improved cakes and those processed under hygienic conditions apart form being useful as food and feed can earn more foreign exchange to the producing countries.

The undesirable factors mainly associated with husks or hulls may be eliminated through dehulling/dehusking prior to crushing. The off-flavours such as the musty odour of mustard meal, nutty/beany flavour of soybean may be eliminated by certain pre-treatments like soaking in alkaline solutions or extractions with ethanol/isopropanol. These treatments may add to the cost of the whole system but yet yield acceptable edible grade oil cakes. Antinutritional factors are eliminated through thermal processing either dry heat/moist heat or supersaturated steaming. Biotechnology is also helping in breeding oilseed crops without these undesirable constituents. Mention may be made about groundnuts tolerant/ resistant to *A.flavus*, rapeseed varieties without glucosinolates and soybeans without trypsin which are being developed and released in several countries.

Quality of cake as affected by the method of extraction: The method of extraction of oil affects the quality of the resultant oilcakes. There are different methods of oil extraction namely traditional *ghanies*, mechanical expeller, solvent extraction and supercritical CO_2 extraction systems. The oilcakes obtained from the mechanical expression contain about 8-10% oil. There is less thermal destruction of desirable nutrients. However, unhygienic conditions exist in these operations which most of the times yield non-edible oilseed cakes. But when the oilseeds are subjected to solvent extraction, less than 1% oil is left in the cake. Thus, by adopting different extraction methods oilcakes containing varied amount of fats, are obtained which may meet the requirement of different groups of people.

Feed uses of oilseed meals: Now-a-days a large number of feed formulations include oilseed cakes as concentrates. The good quality protein improves the egg-laying capacity in chickens, milk yield in cattle, quality of fish catches and meat production.

Groundnut meal: Groundnut meal has been a major source of vegetable protein in animal diets all over the world.. Solvent extracted groundnut meal contains 43 to 48% crude protein. It is an excellent source of arginine but deficient in lysine, cystine and methionine. Supplementation of synthetic lysine to groundnut meal can make it an ideal protein source in feeds. However, it has been seen that groundnut meal is associated with mycotoxins like aflatoxins. Groundnut seed coat also contains tannins which lower the quality of protein compared to soybean meal.

Soybean meal: Soybean meal contains 40 to 48% protein. The amino acid profile of soybean meal is well balanced and is rich in lysine but deficient in methionine content. Soybean seeds have certain anti-nutritional factors like protease inhibitors, goitrogens and estrogenic factors that can affect the production performance of chicken and other animals. But they are all thermo-labile and can be destroyed by roasting/heating or autoclaving. An adequate balance of amino acids by substituting methionine to soybean meal can successfully replace even fish meal.

Mustard meal: Mustard meal has an average crude protein value of 37% with more lysine and methionine contents than groundnut meal. It's usage in animal feeds is limited due to the presence of deleterious factors like glucosinolates, tannins, erucic acid, high crude fibre etc. Glucosinolates are potential giotrogens which cause enlargement of thyroid gland, consequently reducing the performance of layers and broilers. The toxin sinapin affects palatability of feeds and imparts a fishy flavour. The process of solvent extraction after initial pressing of the seed destroys the goitrogens and isothiocyanates and reduces the toxic effets of the cake. The tannins can be checked to some extent by addition of calcium hydroxide and sodium carbonate in dry form in the animal diets.

Sesame meal: Sesame meal contains 40% protein and 8% crude fiber. The protein is rich in arginine, leucine and methionine but low in lysine. Since the sesame meal is deficient in lysine its combination with soybean meal is useful. Sesame meal has high calcium content but its availability is questionable because of phytate content in the hulls of the seed. The phytic acid reduces the availability of calcium and zinc from the diet. Oxalate is another antinutrient in sesame having a role similar to that of phytate. Nevertheless the diet can be included in animal feeds in combination with other lysine rich oil cakes.

Sunflower meal: The sunflower meal contains 30-40% of protein with variable levels of crude fiber depending upon the level of dehulling of the seed. The net protein utilization of sunflower meal is higher than groundnut cake, probably due to higher methionine and cystine content. Sunflower meal is generally not recommended beyond 20% in the diets due to its high crude fibre content and also due to the presence of polyphenolic compounds (esp. chlorogenic acid) which inhibit the activity of hydrolytic enzymes. To overcome this problem, supplementation of additional methionine and choline in the animal diet is suggested.

Castor seed meal: The protein content of castor seed meal varies from 21 to 48% depending upon the extent of decortication. It has an ideal amino acid profile with moderately high cystine, mithionine and isoleucine. But its anti-nutritional substances ricin, ricinine and the allergens restrict its use in animal rations even at low levels of inclusion. Processing of castor meal by heating at 80°C temperature in the presence of 6N ammonia or an acid or alkali reduces the toxins substantially. Detoxified meal needs to be effectively corrected for lysine. It's net protein utilization value is higher than other oil seed meals used in poultry.

Linseed cake: Linseed meal contains 25-30% protein and is a good cattle feed. The cake is rich in minerals like P, Ca, and K and is also utilized as an organic manure. The cake contains a glucoside linamarin. On hydrolysis it liberates HCN which is poisonous to animals. The amino acid conposition is similar to that of any other oilseed. It is rich in lysine and poor in methionine. However, linseed protein quality is better than that of rapeseed proteins.

If the toxic factor linamanin is removed, the linseed meal serves as a source of very good quality protein. It has a lot of uses in the industry as a source of mucilage and protein. Linseed cake is useful 1) as a filter aid in the processing of uranium, iron and copper sulphide and in the setting of cement slurry and kaolin clay, 2) for the preparation of plywood glues by blending with phenolic resins (4:6), 3) as a filler in thermosetting resins because it imparts improved water resistance, 4) as a corrosion inhibitor, 5) as a binding material in water paints, and 6) for stabilizing mud plaster.

Other Applications: Some of the undesirable factors that are isolated from the oil cakes such as glucosinolates from mustard have commercial applications like insecticides. Incorporation of soy proteins in diet therapy protects us from breast cancer, colon cancer and heart attacks. Most oilseeds function as neutraceuticals. Ricinine and ricin from castor have insecticidal and medicinal values. Hence utilization of various oilseeds based on their composition needs careful attention.

Linseed mucilage: Linseed mucilage is prepared from aqueous extract of seed (soaking in water for 24 hrs) by precipitation. It is a white fibrous mass and is friable when totally dry and gets dissolved in water giving high viscosity solution. It is used a) in cosmetic and pharmaceutical industries as a demulcent b) in food products as a water soluble emulsifying agent as well as thickner and binder c) as a substitute for acacia gum in stabilization of emulsion and d) as a useful base for eye ointment .

Oilseed extractions or de-oiled cakes being richer in protein need to be utilized on a larger scale. Development of infant and weanling foods will certainly enhance their value. Non-food grade cakes serve as poultry, fish and cattle feed. Through this route again the cakes help in increasing meat and milk production. Artificial fibre can also be manufactured from oilseed proteins. Low grade and toxic cakes serve as organic manure and can help increase not only crop production but also organic food which is in great demand.

Fodder, Fibre and Fuel

In addition to the utility of seed and its components, the crop or the residues also have varied uses. For example the groundnut haulms serve as highly nutritious cattle feed and thus helping in increased milk production. The stalks and stems mostly serve as fuel in the rural areas. However, they have potential as source of fibre. Linseed is one such popular fibre crop. In addition, plant parts like leaves, roots etc. have some medicinal properties and need to be utilized widely.

Flax also serves as a source of fibre which could be spun and woven into cloth. The fibre which is composed mainly of cellulose is stronger, more durable and more resistant to moisture than cotton or wool. Therefore, it is used in gloves, foot wear, netting and sports gear. It is also used in paper manufacture and textile industry. Flax is the oldest textile fibre. It is one of the most natural and most eco-friendly of all the textile fibres. Flax fibre has more strength, fineness and durability than cotton fibre. It is soft, lustrous and flexible and has high water absorbency. It can be blended with cotton, hemp and jute. Blending of flax with cotton will enhance not only its textile qualities but also save cotton for other uses. It also mixes well with wool and tussar silk. It is used for the manufacture of linen stitching including handkerchiefs, table and bed linens, dress inter lining, men's suitings, towels, tents, drapperies, upholstery, covers, surgical dressings, fire fighting hoses and water holding bags.

The dry scutched fibre on spinning produces rough and thick yarn which can be utilized for making strong ropes, which are damp resistant. The tensile strength is comparable to other fibre producing materials. The linseed fibre twines can be used for making fishing nets, wrapping purposes, house decoration purposes, etc. The yarn of linseed fiber can be utilized for the manufacture of rough textiles such as carpets, blankets, mats, mattresses etc,. This yarn can be a substitute for cotton in making tents.

Linseed-jute (Linju) and flax-cotton (Linco) blended fabrics show better quality as furnishing materials than cent-per-cent jute or cotton fabrics. With the introduction of linseed fibre for fabrics (pre treatment of linseed fibre), substitute of cotton for the manufacture of fine textiles like suiting and shirting, bed sheets, bed cover, curtains, etc. is possible. The linco-fabrics can supplement the existing cloth production as well as generate employment especially in linseed growing areas.

Conclusion

It is thus evident that oilseeds and their products have versatile uses in all the essential sectors for the human being and his dependant animals. They can serve as food, feed, fuel and fibre. They are already regular components of the day to-day life of mankind. Oleochemicals like lubricants, paints, cosmetics etc. will enhance the value and importance of oilseeds. These chemicals are environment friendly and have least side effects. As demand and research progresses more uses especially in the form of neutraceuticals will expand. The oilseeds have great potential and their sustainable and enhanced production will play a great role in the survival of mankind and his dependant animals. Let us hope opulence in oilseeds will usher in large number of benefits to the living beings and the environment on the earth.

References

Achaya, K.T. 1990. Oilseeds and oilmilling in India- A Cultural and Historical Survey. Oxford IBH. New Delhi

Murphy, D.J. 1994. Designer Oilcrops: Breeding, Processing and Biotechnology. Weinheim. London

Nagaraj, G. 2009. Oilseeds: Properties, Processing, Products and Procedures. New India Publishing Agency, New Delhi.

Nagaraj, G. 2015. Agricultural Plant Biochemistry. New India Publishing Agency, New Delhi.

Nagaraj, G. 2016. Plant Lipids. Aavishkar Publishers, Distributors, Jaipur.

9

Improvement and Cultivation of Medicinal and Underutilized Plants

Nidhi Bhakuni[1*] and Rahul Kaushik[2]

[1]DSB Campus, Kumaun University Nainital-263002

[2]Supercomputing Facility for Bioinformatics and Computational Biology, Indian institute of Technology Delhi, Hauzkhas, New Delhi: 110016

*Corresponding author: nidhi.kunainital@gmail.com

Abstract

Certain plants species are very crucial for their socio-economic and medicinal values. Among these species, a certain percentage of plants are prioritized for their ethno-traditional aspects by the local communities and thus are well protected by local communities. However, other plant species with medicinal and economical values needs immediate attention because of their extraction and deforestation. For harnessing their medicinal and economical properties to an optimum level, general awareness among local communities and a large-scale cultivation needs to be encouraged. The traditional uses of medicinal plants through the existing knowledge having a great significance for human health and socio-economic development. The mainstay of this chapter is to explore the possibilities of encouraging the cultivation of medicinal and underutilized plant for their optimal use. Further, this chapter suggests different prospects for improvement and commercialization of various product obtained from these plants. Various products derived from medicinal plant which may be very beneficial at industrial level for preparation of various medicines and some underutilized crop plants are also explored.

Keywords: Medicinal plants, ethno-medicinal, socio- economic, deforestation.

Introduction

The forest has been the source of valuable medicinal plants (Kumar *et al.*, 2011; Adnan and Holscher, 2012; FAO, 1997). With passage of time, human comprehended the preventive and curative properties of plants and started using them for health care. India has one of the oldest, richest and most diverse cultural traditions called folk tradition associated with use of medicinal plants based on indigenous knowledge skill and culture practices (Tewari *et al.*, 2014; Chauhan *et al.*, 2017). The medicinal and underutilized plants help in serving crucial purposes, mostly in isolated and traditional communities. These plants are on a radar of researchers, farmers and consumers for their nutritional and economical values (Agrawal *et al.*, 2012; Cleveland and Murray, 1997). Accumulating human pressure on natural resources and environmental fluctuations have necessitated the optimal use of medicinal and underutilized plants for uplifting rural livelihoods and combating environmental deprivation (Hummer *et al.*, 2012, Setyawan, 2009; Batugal *et al.*, 2004; *Zuhud et al.*, 1994). The medicinal plants can be categorized into herbs, shrubs and trees depending upon their life cycle (Antony and Thomas, 2011; Rana *et al.*, 2013; Sekar, 2012). Similarly, the underutilized plants can be trees with high economical values, indigenous crops with better environmental adaptability, flowering and aroma producing herbs and shrubs etc. (Bussann *et al.*, 2007; Hastorf, 1998).Among medicinal and underutilized plant, there are certain plant species which have huge medical perspectives and yet to be utilized at commercial levels which may offer profitable income opportunities to alleviate poverty in developing countries (Dhar *et al.*, 2002; Sharma and Joshi, 2010; Dafni *et al.*, 2006). Some of these plant species may provide tremendous nutritional values to exterminate the micronutrient malnutrition, which is affecting a huge population worldwide (Dangwal *et al.*, 2011; Misra *et al.*, 2004; Hefferon, 2015).

India is rich in its medicinal value from the ancient time with traditional knowledge system, which deals with various important aspects and health issues of the local people (Sarkar *et al.*, 2015; Chhibber, 2008).According to a report from World Health Organization, around 80% people of the world rely on traditional medicine for their primary healthcare and Himalayan biodiversity has been the reservoir of enormous natural resources of medicinal wealth (WHO, 2007; Spiertz, 2013). In this chapter, we will explore various medicinal and underutilized plants along with their geographical distribution in India, medicinal and economic importance. Further, we will discuss the plausibility of optimizing their cultivation at commercial levels. The initial part of the chapter enlists the medicinal plants, which are being used by local communities along with their medicinal uses and geographical distribution. Further, the chapter accounts for the medicinal and underutilized plants, which yields some chemicals or products to be used directly at industrial levels.

Medicinal and Underutilized Plants

In this section, we accounts for some of the medicinal and underutilized plants at commercial levels. The information about their medicinal uses has been known to local communities for years and is being passed on to next generation locally. Here, we describe some of the plants with medicinal uses and geographical distribution.

Rhododendron arboretum

Vernacular Name: Buransh

Family: Ericaceae

Part Used: Flower

Medicinal Importance: Flower juice added with sugar is useful during fever, headache, stomachache, and as freshener. It also helpful in heart and respiratory disorders.

Geographical Distribution: It is distributed at1800 - 2500 m elevation. The plant grows of up to 20 m height and binds of 15 to 20 bell-shaped flowers.

Viburnum cotinifolium

Vernacular Name: Goniya

Family: Adoxaceae

Part Used: Tree bark

Medicinal Importance: Decoction of bark used in hepatic and digestive troubles.

Geographical Distribution: It is found in the Himalayas, from Kashmir to Bhutan, at altitudes of 2100 to 3600 m. It is an short tree or shrub grow up to 3 m. Young leaves are downy but wear smooth, remaining grey and woolly beneath, ovate in shape and flower-clusters appear in May and June upon short woolly stalks, the small white flowers flushed with pink.

Myrica esculenta

Vernacular Name: Kafal

Family: Myricaceae

Part Used: Fruit

Medicinal Importance: Fruit juice mixed with salt is used for curing headache and body pain and also used as tonic.

Geographical Distribution: In India, it is widely distributed in Jammu & Kashmir, Himachal Pradesh, Uttar Pradesh, Sikkim, Arunachal Pradesh, Assam, Meghalaya, Nagaland, Manipur and Mizoram between the altitudes ranging from 1000-2300 m. Tree of medium height, 20-25 feet. Bark is soft and brittle. Fruit is a globose, succulent drupe, with a bard endocarp.

Symplocos chinensis

Vernacular Name: Loadh

Family: Symplocaceae

Part Used: Tree bark

Medicinal Importance: Bark juice beneficial as brain boosting tonic.

Geographical Distribution: Its altitudinal range varies from 1200 to 2000 m. Trees grow up to 10 m high, thick bark have greyish in color and smooth surface. Fruit ovoid, glabrous, shallowly furrowed, yellow and stone woody. Seeds varies from 1 to 2 and oblong.

Prunus cerasoides

Vernacular Name: Padam

Family: Rosaceae

Part Used: Bark and leaves

Medicinal Importance: Bark, and leaves in internal injury, rheumatic pain.

Geographical Distribution: It is distributed in the temperate forests from 1200 – 2400 m altitude. It has glossy, ringed bark and grows up to 30 m in height. The flower are pink in color and blooming in autumn and winter.

Pyrus pashia

Vernacular Name: Mehal

Family: Rosaceae

Part Used: Fruit

Medicinal Importance: Fruit of mehal mixed with salt dropped into infected eyes.

Geographical Distribution: Distribution of Mehal across the Himalayas and commonly occurs in mid-hill regions from the Caucasus to the Himalaya, between 750 and 2600 m above sea-level. The average tree height is 6 to 10 m. The leaves of a mature tree are characterized as simple, long-pointed, toothed, hairless and shining with an ovate shape.Fruit is edible and categorized as being pome.

Berberis asiatica

Vernacular Name: Kilnori

Family: Berberidaceae

Part Used: Root

Medicinal Importance: Roots decoction mixed with rose water is useful during eye infection. The roots are also helpful in treating ulcers, urethral discharges, ophthalmic, jaundice and fevers.

Geographical Distribution: Shrubberies, grassy and rocky slopes up to 2500 m. Found in heavy shade, on north-facing slopes. The shoots growth 1 to 2 mm, leaves on long shoots are non-photosynthetic, developed into one to three or more spines. The fruit is a small berry 5 to15 mm long, ripening red or dark blue, often with a pink or violet waxy surface bloom.

Rubus ellipticus

Vernacular Name: Hisalu

Family: Rosaceae

Part Used: Root and fruit

(Source: Forest & Kim Starr, *Plants of Hawaii)*

Medicinal Importance: Root is used for controlling blood pressure and diarrhea in human and haematurea in animals. The fruits help blood purification and very effective for heart patients.

Geographical Distribution: Asian species of thorny fruiting shrub. It is found around 1200 m to 2500m elevation in Himalayan region. The golden Himalayan raspberry is a large shrub with stout stems that can grow to up to 4.5 meters and flowering between the months of February and April.

Allium sativum

Vernacular Name: Garlic

Family: Rosaceae

Part Used: Root

Medicinal Importance: It is helpful of hypertension and prevent tick bites. The bulbs are used to cure sicknesses related to the heart, liver, digestion, diabetes and rheumatism.

Geographical Distribution: It is an annual herb and grows in open, dry lands at altitude ranging from 600 to 2000 m. It needs moist cool climate and well-drained fertile loamy soil.

Aloe barbedensis

Vernacular Name: Ghrtkumari

Family: Asphodelaceae

Part Used: Leaves

Medicinal Importance: Its leaf gel is widely used in folk medicine to treat ailments of the eyes, skin, liver, spleen, chronic constipation, indigestion, peptic ulcers and acid reflux.

Geographical Distribution: A perennial and stem less herb that grows up to an altitude of 1200 m. It flourishes in sunny warm climate, sandy and well-drained soil. It propagates via its root suckers.

Catharanthus roseus

Vernacular Name: Sadabahar

Family: Apocynaceae

Part Used: Roots and leaf

Medicinal Importance: Its roots and leaf extracts are used for preparing medicines to cure acute leukemia, cancer, sedation, blood pressure and diabetes.

Geographical distribution: It is an herb grows in mild, tropical climate up to an altitude of 1500 m. It requires moist, irrigated lands and light, well drained sandy loam soil. However, being a hardy plant, it grows in most soils. It is propagated through its seeds.

Centella asiatica

Vernacular Name: Brahmi

Family: Apiaceae

Part Used: Leaves

Medicinal Importance: It is helpful brain fever, apply paste of green leaves on forehead during fever and helping to retard the aging process.

Geographical Distribution: It is native to wetlands in Asia, cultivated in north-eastern Himalayan region. The flowers are white or pinkish to red in color, born in small, rounded bunches (umbels) near the surface of the soil.

Cuminum cyminum

Vernacular Name: Jeera

Family: Apiaceae

Part Used: Seeds

Medicinal Importance: It is helpful in treating minor digestive conditions, coughs, pain and rotten teeth. These seeds are familiar food flavoring agents especially in tropical Asia.

Geographical Distribution: It is distributed along Mediterranean coast, India, China, and Iran. In Himalayan region it found in 1600 to 2500m elevation. Annual herb, flower present in umbels.

Mentha piperitia

Vernacular Name: Pudina

Family: Lamiaceae

Part Used: Leaves

Medicinal Importance: Its leaf extracts are helpful as antiseptic, preservative, anti-spasmodic, diuretic, choleric, carminative and anti-bacterial, anti-fungal agents.

Geographical distribution: It is a perennial aromatic herb which grows in tropical, sub-tropical and temperate climate at 270-1500 m elevation. It needs well drained loamy, silt loam to clay loam soil and suitable rainfall through the growing stage.

Zingiber officinale

Vernacular Name: Adrak

Family: Zingiberaceae

Part Used: Rhizome

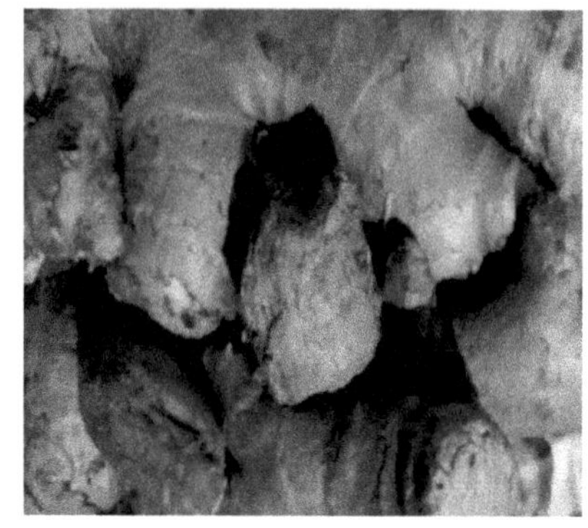

Medicinal Importance: Its rhizome is used to treat stomach pain, flatulence, dyspepsia, skin diseases, inflammations, swellings, vomiting, and indigestion, lack of appetite, sore throat, diabetes, and diarrhea. It also has anti-oxidant properties.

Geographical distribution: An herb with aromatic rhizomes grows well in warm, humid climates up to altitudes of 1800 m. It requires heavy laterite loam, clay loam or black soil. It is propagated through rhizomes and has a sprouting rate of 100 per cent.

Murraya koeningii

Vernacular Name: Kaddi patta

Family: Rutaceae

Part Used: Leaves,bark and root

(Source: Pinterest)

Medicinal Importance: The leaves, bark and roots are used to treat diseases such as jaundice, diarrhea, dysentery, piles, skin allergy, insect bites, scorpion stings and diabetes. The leaves are also act as a culinary agent.

Geographical distribution: It is usually a shrub that nurtures in plains in tropical climate up to altitudes of 1500 m. It requires red sandy loam soil with good drainage.

Ocimum sanctum

Vernacular Name: Tulsi

Family: Lamiaceae

Part Used: Leaves bark and root

Medicinal Importance: The leaves, roots and seeds are helpful in curing fever, cold, cough, chronic skin diseases, vomiting, bronchitis, diabetes, digestive disorders, genitourinary disorders, sores, inflammations, sprains, burns, bruises and psoriasis. It is regarded as a sacred plant.

Geographical Importance: It is an aromatic perennial herb grows at altitudes up to 2000 m. and needs well drained rich loam, saline or alkaline soil.

Potentilla fulgens

Vernacular Name: Bajardanti

Family: Rosaceae

Part Used: Root

(Source: Himalayanagro.in)

Medicinal Importance: It is used as an herbal remedy for diarrhea, inflammation and gastrointestinal disorders.

Geographical distribution: It is herbaceous perennial, distributed at 1600 to 4800 m elevation in Himalayan region. It has dry and inedible fruit.

Cynadon dactylon

Vernacular Name: Doob

Family: Poaceae

Part Used: Whole plant

Medicinal Importance: It is useful in external injury, extracts helpful in curing Jaundice treatment. The extracts also useful in treating the infected eyes.

Geographical distribution: It is sort of omnipresent plant and distributed all over the world between about 30° S and 30° N latitude and it needs 625 to 1750 mm yearly rainfall. It flourishes in warm climate. It may grow upto 40 cm length. **Medicinal use**: Leaves used in high fever, stomachache.

Ajuga parviflora

Vernacular Name: Ratpatia

Family: Lamiaceae

Part Used: Leaves

Geographical distribution: It is distributed along the Himalaya Kashmir to Nepal, at altitudes of 600-1500 m. It is a perennial herb, unbranched densely sheltered with the long villous hair.

Polygonum nepalense

Vernacular Name: Raunne

Family: Polygonaceae

Part Used: Root

Medicinal Importance: Juice of the root extract used in treatment of fevers.

Geographical distribution: Common in damp shaded situations, 1700 - 2000 m in Himalayan ranges. Flowering period June to September.

In Table 1, we have enlisted some of the plants and the products derived from them for medicinal purpose at industrial levels. Cultivation of these plants at large scale may be very helpful for local community in their socio-economic development.

Table 1: Certain plants with their chemical derivatives and respective medicinal uses.

Species	Chemical Derivatives	Used in Disease Condition
Curcuma longa	Curcumin	Cancer, acute Kidney injury
Camillia sinensis	Epigallocateclin	Alzheimer disease, Muscular dystrophy
Gassypium hirsutum	Gossypol	Leukemia.
Allium cepa	Quercetin	Pulmonary disease.

Species	Chemical Derivatives	Used in Disease Condition
Vitis vinifera	Reveratrol	Vascular system injuries
Artemisia annua	Artemisinin	Antimalarial drug
Caphalotaxus harringtonii	Harringtonine,	Anti- tumor, cancer drugs
Colchicum autumnate	Colchicine	Mediterranean fever, anti-leukemic
Papaver somniferum	Maorphine, Thebaine	Cerebral and myocardial ischemia
Datura metel	Scopolamine	Motion sickness
Ephedra sinica	Ephedrine	Bronchial asthma
Liquidambar styracifkua	Shikimic acid	Precursor for the antiviral drug

Underutilized Crop Plants

In this section, we accounts for some the crop plants which have high commercial values but yet to be cultivated at large scale because of lack of awareness and proper guidance.

Amaranthus Species (Amaranth)

Utility: As source of grain, vegetable and fodder, High carotene, 10-20% proteins, less nitrates and oxalates, folic acid, iron

Description: It protein contents constitutes all essential amino acids which makes it suitable for nutritional requirement. High carotene and iron content is useful in curing vitamin A deficiencies. Presence of folic acid is helpful in blood hemoglobin formation. It is major food component in Himalayan regions. In Himalayan regions, mainly three species of *Amaranthus* are cultivated namely, *A.hypochondriacus, A. caudatus*and *A. cruentus*.

Fagophyrum Species (Buckwheat)

Utility: Used as a leafy vegetable, extraction of rutin from green leaves and flowers used in medicine and making good quality honey.

Description: An important short duration crop cultivated at higher altitudes beyond 1600 m which accounts for 90% of cultivated land at higher Himalayan regions.

Chenopodium Species (Chenopod)

Utility: A vital vegetable and grain crop cultivated in Himalayan regions and consumed in combination with other cereals. Higher nutritional values then wheat, rice, barley and maize. Grains are used in local alcoholic beverages preparation.

Description: Mainly four speicies of Chenopodium are cultivated in Himalayan regions viz. *C. album, C. quinoa, C. nutaliae* and *C. pallidicaule*. Chenoodium is suited to mixed cropping practices.

Vigna umbellate (Rice Bean)

Utility: Used as food, fodder, green manure and a cover crop. Dry seeds are edible. Pods and leaves are used as vegetable. Rich source of proteins.

Description: Considered as drought tolerant plant, usually cultivated in lowland humid tropics. Its distribution is restricted to tribal regions of north-eastern hills, Eastern and Western Ghats. It is good sourceof protein, calcium and phosphorus.

Psophocarpus tetragonolobus (Winged Bean)

Utility: A rich source of proteins with grain and root being the edible parts. The plant is used as fodder and green manure. The young pods, leaves, sprouts and flowers are used as vegetable and seed oil for cooking. Seeds are rich in the antioxidants and tocopherol.

Description: It is mainly grown as a courtyard or a garden crop in Southeast Asia. In India, it is restricted to humid, subtropicalparts of the north-eastern region, Bengal, Bihar and Western Ghats.

Conclusion

It is imperious that agricultural productivity needs to been riched via use of alternate plant species which are underutilized. Also, the knowledge and utilization of medicinal plants is restricted to local communities which needs to be made widespread for combating hunger and poverty. The optimal use of above discussed medicinal and underutilized plant at a large scale may prove very beneficial economically as well as health wise. Encouraging commercial crops among the farmers may help in raising their economic status. For promoting the cultivation of underutilized plants, we need to have suitable policies for relieving poverty and sustainability of natural resources. The foremost bottlenecks in promoting underutilized plants and medicinal plants are the unavailability of knowledge and proper guidance along with unavailability of sufficient amount of quality seed or planting material.

Emerging suitable outline and skills for the cultivation of medicinal and underutilized plants is a crucial factor to safeguard an uninterrupted supply of medicinal plants for the pharmacological industry and to pause the deprivation of the natural resource base. The participatory research agenda may be of great help to the farmers, scientists and policy makers in the perspective of socio-economic growth for viable livelihoods of the rural communities.

References

Adnan M. and Holscher D. 2012. Diversity of medicinal plants amongs different forest-use types of the Pakistani Himalaya. *Econ Bot*, 66(4): 344-356.

Agrawal, A., Cashore, B., Hardi, R., Shepherd, G., Catherine, B. and Daniel, M. 2013. Economic contribution of forest: United Nation Forum on forest. 10, pp3-6.

Antony, R. and Thomas, R. 2011. A mini review on medicinal properties of the resurrecting plant Selaginella bryopteris (Sanjeevani). *Int J Pharm Life Sci,* 2(7): 933-939.

Batugal, P.A., Kanniah, J., Young, L.S. and Oliver, J.T. 2004. Medicinal plants research in Asia, Vol 1: The framework and project work plans. International Plant Genetic Resources Institute Regional Office for Asia, the Pacific and Oceania (IPGRI-APO). Serdang, Selangor DE, Malaysia. Pp 56-89.

Bussmann, R.W., Sharon, D. and Lopez, A. 2007. Blending traditional western medicine: medicinal plant use among patients at clinica Anticona in El Povennir, Peru. *Ethnobot Res App,* 5: 185-199.

Chauhan, P.S., Bisht, S. and Ahmed, S. 2017. Traditional ad entho-botanical use of medicinal trees in district Tehri Garhwal (Western Himalaya). *J Ayu Her Med.* 7(1):2442-2448.

Chhibber, B. 2008. Indian culture heritage and environmental conservation through traditional Knowledge. *Mainstream* 40(25): 6-8.

Cleveland, D.A. and Murray, S.C. 1997. The world's crop genetics resources and the right of indigenous farmer's. *Curr Anthropol,* 38(4):477-516.

Dafni, A., Lev, E., Beckmann, S. and Eichberger, C. 2006. Ritual Plants of Muslim graveyards in northern Isreal. *J Ethnobiol Ethnomed,* 2: 38-46.

Dahr, U., Manjkhola, S., Joshi, M., Bhatt, A., Bisht, A.K. and Joshi, M. 2012. Status and future strategy for development of medicinal plants sector in Uttaranchal, India. *Curr Sci,* 83(8): 956-963.

Dangwal, L.R., Rana, C.S. and Sharma A. 2011. Ethno-medicinal plants from transitional zone of Nanda Devi biosphere reserve, District Chamoli, Uttarakhand (India). *Indian J Nat Prod Resour.* 2(1): 116-120.

Hastorf, C.A. 1998. The Cultural Life of Early Domestic Plant Use. *Antiquity* 72:773-782.

Hefferon, K.L. 2015. Nutritionally enhanced food crops; progress and perspectives. *Int J Mol Sci,* 16: 3895-3914.

Hummee, K.E., Pomper, K.W., Postman, J. and Graham, G.J. 2012. Emerging fruit crops. ML Badenes and DH Byrne (eds), Fruit breeding, Handbook of plant Breeding 8. DOI 10.1007/978-1-4419-0763-7_4@springer+Business.

Kumar, M., Sheikh, M.A. and Bussmann, R.W. 2011. Ethnomedicinal and ecological status of plants in Garhwal Himalaya, India. *J Ethnobiol Ethnomed,* 7:32.

Misra, B.K., Sharma, R.K. and Nagarajan, 2004. Plant breeding: A component of public health strategy. *Curr Sci.* 86(9): 1210-1215.

Rana, C.S., Tewari, J.K. and Gairola, S. 2013. Faith herbal knowledge document of Nanda devi Biospehere reserve Uttarakhand, India. *Indian J Trad Know,* 12(2): 308-314.

Sarkar P., Kumar LDH, Dhubhal C, Panigarhi SS. and Choudhary R. 2015. Traditional and ayurvedic foods of Indian Origin. *J Ethnic Food.* 2: 97-109.

Sekar, K.C. 2012. Invasive alien plants of Indian Himalayan region diversity and implication. *Am J Plant Sci,* 3: 177-184.

Seters, A.P.V.1997. Forest based medicines in traditional and cosmopolitan health care. Medicinal plants for forest conservation and health care. Bodeker G., Bhat KKS., Burley J., Vantomme, eds Non-wood forest products, 2: 5-11.

Setyawan, A. 2009. Traditionally utilization of *Selaginella;* field research and literature review. *Bioscience,* 1(3): 146-158.

Sharma, V and Joshi, B.D. 2010. Role of Sacred plants in religion and health care system of local people of Almora district of Uttarakhand state (India). *Academic Arena,* 2(6): 19-22.

Spiertz, H. 2013. Challenges for crop production research in improving land use, productivity and sustainability. *Sustainability.* 5: 1632-1644.

Tewari, S., Paliwal, A.K. and Joshi, B. 2014. Medicinal use of some common plants among people of Garur block of District Bageshwar, Uttarakhand, India. *Octa J Biosci.* 2(1):32-35.

World health organization. 2007. Traditional medicine. pp22.

Zuhud, E.A.M., Ekarelawan and Riswan S. 1994. Indonesian tropical forests as a source of medicinal plant genetic diversity. In: Zuhud EAM, Hary (eds). Preservation of the utilization of medicinal plant diversity of Indonesian tropical forests. Faculty of Forestry IPB-LATIN. Bogor. [Ind onesia].

10

Role of Post Harvest Technology of perishables to Feed Growing Population

Suman Bala[1*] and Jitender Kumar[2]

[1]Centre for Plant Biotechnology, Hisar- 125 001, India
[2]Department of Botany and Plant Physiology, CCS HAU, Hisar
*Corresponding author: sumanmalika14@gmail.com

Abstract

Fruits and vegetables are living entities even after harvest, continue their metabolic activities even after harvest, and as results naturally get spoiled. This character puts fresh fruits and vegetables in the category of highly perishable commodities and 20–40% of horticultural produce are lost before they can be consumed, so there is need for maintaining adequate conditions during storage and transport that results in prolonging shelf life. We have a need to develop good systems of postharvest management and infrastructure for extending shelf-life of perishables. There are few proven methods and technologies used to slow down the metabolic changes for extending shelf life by some post harvest treatments & sorting, grading, precooling, suitable packaging, proper transportation, and maintenance of storage atmosphere.

Keywords: Perishables, packaging, shelf life, pre cooling and post harvest technology

Introduction

Fruits and vegetables are must important ones among the various field crops being a source of vitamins, minerals which cannot be provided by cereals and pulses. Moreover, as the India has buffer stock for the cereals so there is a need for diversion in agriculture towards fruits and vegetables production. In order to increase the economy of farmers and earn more export potential in term of foreign exchequer by exporting fruits and vegetables to many countries. India is blessed with wide agro-climatic conditions in different parts of country in different times of the year. So there is a large potential to grow various fruits and vegetables in the different parts of country throughout the year. However, fruits and vegetables are highly perishable as compared to cereals and pulses and they are liable to damage very fast as compared to other agriculture commodities. Fruits and vegetables being a living entities, have high rate of respiration and transpiration which reduces their shelf life, because of this reason government is unable to give minimum support price for these commodities and farmers are afraid of growing these commodities. Prices of these commodities sometimes are so low that farmers have to throw their produce on roads. So there is need to enhance the shelf-life of these commodities in order to boost the economy of farmers and to earn maximum foreign exchequer.

Horticultural commodities not only provide nutritional and healthy foods to human beings, but also generate a considerable good income for growers. However, horticultural crops typically have high moisture content, tender texture and high perishability. That is why if these commodities are not handled properly, a high-value nutritious product can rot in a matter of days or hours. The post harvest losses in fruits and vegetables vary according to the nature of commodities, production season and production areas. However, overall these losses are in the range of 20 to 40% in different fruits in terms of money it comes around 60 to 70 thousand crores of rupees per annum. That is why, these perishable commodities need very careful handling at every stage so that deterioration of produce is restricted as much as possible during the period between harvest and consumption (Dhatt and Mahajan, 2007).

Important sites of post-harvest losses: Important sites where post-harvest losses takes place are: Farmer's field (15-20%), packaging (15-20%), transportation (30-40%) and marketing (30-40%). Farmers and food sellers have been concerned about these losses since agriculture began. Cutting postharvest losses could, presumably, add a sizable quantity to the global food supply, thus reducing the need to intensify production in future.

Once harvested, vegetable and fruit are subject to the process of senescence. Numerous biochemical processes continuously change the original composition of the crop until it becomes unmarketable. The period during which consumption is considered acceptable is defined as the time of "postharvest shelf life". Postharvest shelf life is typically determined by overall appearance, taste, flavor and texture of the commodity

Post harvest technology: It is interdisciplinary branch of science for the protection, conservation, processing, packaging, distribution and marketing of agriculture produce to meet requirement of people. It helps to meet the requirement of growing population by making more nutritive items for low grade commodities by proper processing, fortification and utilization of waste into nutritive animal food.

The choice of technology package depends on circumstances, such as the scale of production, crop type, prevailing climatic conditions, and the farmer's affordability and willingness to provide basic infra-structure like storage, handling, grading, packing, transport and marketing facilities and technical expertise. This could be carried out by the public and private sectors. Various protocols have been standardized and available for adoption to get the best result, which will give economic benefits. Similarly, proper storage conditions, with suitable temperature and humidity are needed to extend the storage life. Greater emphasis need to be given on the training of farmers for post harvest handling of fruits and vegetable, creation of infrastructure for sorting, grading, packing and post harvest treatments.

Post harvest physiology: It is scientific study of physiology of living tissues of fruits and vegetables after they have been denied further nutrition from plant by picking. This process includes ripening, harvesting, transportation, handling, packaging and storage (post harvest chain). The most important goals of post-harvest physiology are keeping the product cool, to avoid moisture loss and slow down undesirable chemical changes, and avoiding physical damage such as bruising, to delay spoilage. Sanitation is also an important factor, to reduce the attack of pathogens that could be carried by fresh produce.

Ripening: Ripening is a complex physiological process resulting in softening, sweetening and increase in aroma compounds so that ripening fruits are ready to eat . The associated physiological or biochemical changes are increased rate of respiration and ethylene production, loss of chlorophyll and continued expansion of cells and conversion of complex metabolites into simple molecules. The various biochemical changes (Fig. 1) that take place during ripening are:

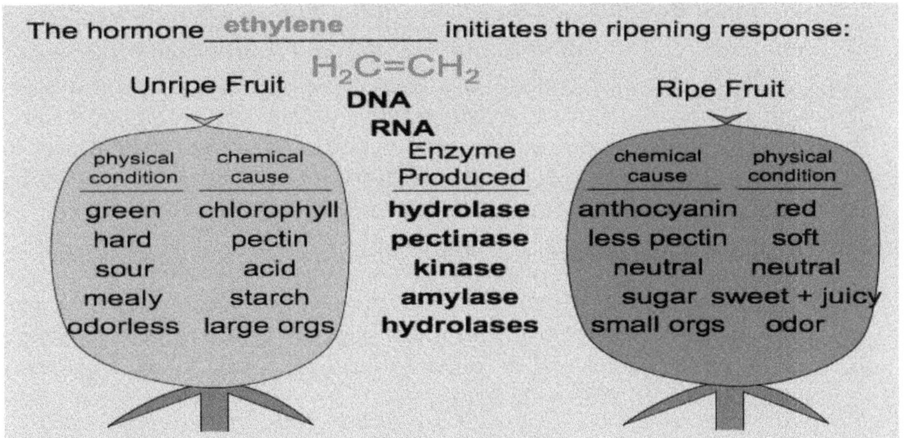

Critical factors contributing to postharvest losses: Postharvest loss occur because of a improper handling or bio-deterioration by microorganisms, insects, rodents or birds (Hodges *et al.*, 2011).

a) Internal factors:

Metabolic: These losses in perishables are due to higher rate of respiration, transpiration and senescence metabolism.

Mechanical: Losses due to improper handling, improper packaging and transportation causes bruising, cutting, breaking, impact wounding are mechanical injury.

Parasitic diseases: Losses due to the invasion of fungi, bacteria, insects and micro-organisms which attack fresh produce easily and spread quickly leads to parasitic losses.

Lack of market demand: Poor planning, inaccurate production and market information may lead to over production of certain fruits or vegetables which can't be sold in time. This situation occurs most frequently in areas where transportation and storage facilities are inadequate. Produce may lie rotting in production areas, if farmers are unable to transport it to people who need it in distant locations.

b) External Factors

These factors can be grouped into two primary categories: environmental factors and socio-economic patterns and trends.

i) Environmental factors

Temperature: Higher the temperature the shorter the storage life of horticultural products and the greater the amount of loss within a given time, as most factors that destroy the produce or lower its quality occur at a faster rate as the temperature increases (Atanda *et al.*, 2011).

Humidity: A moist comodities will give up moisture to the air while a dry comodities will absorb moisture from the air. Fresh horticultural products have high moisture content and need to be stored under conditions of high relative moisture loss and wilting (except for onions and garlic). Dried or dehydrated products need to be stored under conditions of low relative humidity in order to avoid adsorbing moisture to the point where mold growth occurs (Atanda *et al.*, 2011).

ii) Socio-economic factors:

Social trend such as urbanization has driven more and more people from rural area to large cities, resulting in a high demand for food products at urban centres, increasing the need for more efficient and extended food supply chains (Parfitt *et al.*, 2010).

Others: Various other reasons for losses include:

- Lack of clear concept of packing house operations.
- Lack of awareness among the growers, contractors and even the policy makers.
- Lack of infrastructure and Inadequate technical support.
- Inadequate post-harvest quality control.

- Absence of pre-cooling, cold chain and cold storage.
- Inadequate market facilities, market intelligence and market information service (MIS)

Research Activities and Infrastructure

Following steps have been taken to develop the technology for extending the shelf life of fruits.

A) Development of cultivars with larger shelf-life: Developing fruit and vegetable varieties with improved shelf-life so as to maintain the freshness and quality for a longer period has been a long cherished dream of horticulturists worldwide. Extension of shelf-life would enable the commodities to withstand the rigours of harvesting, transportation and storage, thus providing consumers with quality produce. However, the success in achieving this target has been meagre and only a few fruit and vegetable varieties with improved shelf-life are available. Different breeding methods have been recommended for improving shelf-life.

Domestication and introduction: Domestication involves bringing out the wild plant species under human management (Singh and Singh, 2011). Plant introduction intends to widen the genetic base of a plant species and can give rise to novel and potential genotypes which may be directly released into cultivation. A few fruit and vegetable cultivars such as Jonathan apple, Early Grande peach, Kinnow mandarin, Solo papaya and tomato varieties Roma and Labonita have been developed through introduction valued for their better shelf-life compared to other varieties in these crops.

Chance seedlings and clonal selections: Chance seedling is the name given to a fruit cultivar discovered by chance without any sustained breeding efforts. In fact, chance seedlings are the superior naturally occurring variants in a plant population. Mango varieties Alphonso, Banganpalli, and Dashehari, Ambri apple and guava varieties which originated as chance seedlings, are appreciated for their higher shelf-life. Selection is commonly used technique for the genetic improvement of fruits and vegetables. A number of fruit and vegetable cultivars developed through selection with better shelf life are given in table1.

Table 1: Selection made in fruits and vegetables with improved shelf-life

Crop	Variety	Parentage
Mango	Pusa Surya	Eldon
Papaya	Pusa Meiesty	Ranchi
Peach	TA 170	Flordasun
Guava	IIHR Sel-8	Allahabad Safeda
Bittergourd	Konkan Tara	Local Germplasm
Bottlegourd	Samrat	Local Germplasm

Crop	Variety	Parentage
Wintersquash	Arka Suryamukhi	Local Germplasm
Onion	RO 59	------
Muskmelon	Arka Rajhans	Rajasthan collection
Pumpkin	Arka Chandan	IIHR 105

Hybridization: Some hybrids show higher shelf-life and better fruit quality traits than parents (Singh and Singh, 2011). A few such hybrid fruit and vegetable varieties having better shelf-life are given in table 2.

Table 2: Fruit and vegetable hybrids with higher shelf-life

Crop	Variety	Parentage
Mango	Pusa Arunima	Amrapali × Sensetion
Guava	Hybrid 16-1	Apple Colour × Allahabad Safeda
Custard Apple	Arka Sahan	Island Gem × Mammoth
Papaya	Surya	Sunrise Solo × Pink Flesh Sweet
Persimmon	Kanshu	Shinshu × 18.4
Onion	Arka Niketan	------------------
Tomato	Dongnong 711	97-208 × 94-125
Watermelon	Arka Manik	IIHR 21 × Crimson Sweet KWMP-1 × KWMP

Mutation breeding: Spontaneous and induced mutations may result potentially novel genotypes with considerably higher shelf-life. Mutations provide a valuable source of variation in plant material from which the breeder can make selection. Mutants with increased sugar content and extended shelf-life have been reported in pear (Singh and Singh, 2011). The use of *in vitro* methodologies, the exploitation of somaclonal variation, and the setting up of early selection methods could make induced mutation techniques more reliable. Some authors have noted extended shelf-life in tomato by introgression of the *rin* gene and suggested that future research should attempt to reduce deleterious effects on flavour and colour components of ripening. The use of *rin* homologues to control shelf-life has promise in other species, including non-climacteric varieties, but flavour, colour and nutritional content should not be compromised.

Biotechnological approaches: With the advent of various biotechnology-based techniques antisense RNA technology and genetic transformation, it may now be possible to develop superior fruit and vegetable varieties with longer shelf-life (Surenderanathan, 2005). Various plant traits which need to be genetically modified for higher shelf-life include respiration, ethylene biosynthesis, and disease tolerance. Lower respiration and ethylene production rates, slower ripening, reduced browning, decreased chilling sensitivity and increased resistance to post-harvest rots may contribute to better keeping quality in fruits and vegetables. The importance of understanding the biochemical process of tissue softening and the use of such information for retarding the ripening process has been demonstrated. Modification of fruit ripening by suppressing expression of specific enzymes has been extensively studied and demonstrated in tomatoes. In tomato, plants have been engineered for reduced synthesis of polygalacturonase which is responsible for softening the cell wall. Using the antisense technology, antisense RNA for polygalacturonase (PGU) has been inserted (Singh and Singh, 2011). When normal pGU mRNA is synthesized, the antisense RNA binds to it and prevents synthesis of pGU. Antisense RNA against ACC synthase was also expressed in tomato plants and this inhibited ethylene biosynthesis by about 99.5% and markedly delayed fruit ripening. Expression of ACC synthase and ACC oxidase genes encoding the key enzymes required for ethylene biosynthesis can be specifically inhibited to constant ethylene production in ripening fruits. Expression of antisense RNA derived from the cDNA of LE-ACC2 gene resulted in almost complete inhibition of mRNA accumulation of both the ripening induced ACC synthase genes and the ethylene production in such transgenic fruits was reduced to very low levels. In Australia, transgenic work in pineapples has resulted in control of ethylene production and development of pineapples not susceptible to the postharvest disorder blackheart. In Brazil, transgenic passion fruit are being developed with an antisense construct to the ACC oxidase gene. Some other alternative strategies employed for delayed fruit ripening through reduction in ethylene biosynthesis include degradation of ACC by the bacterial enzyme, ACC deaminase and the formation of a conjugated derivative of N-malonyl ACC by the enzyme ACC N-malonyl transferase.

B) Cultural practices for better quality of fruits: Soil type, mulching, irrigation and fertilization influence the water and nutrient supply to the plant, which can affect the nutritional quality of the harvested plant part (Kader, 1988). The effect of fertilizers on the vitamin content of plants is less important than the effects of genotype and climatic conditions, but their influence on mineral content is more significant. For example, sulphur and selenium uptake influence the concentrations of organosulfur compound in *Allium* and *Brassica species* (Goldman *et al.*, 1999). High calcium content in fruits has been related to longer postharvest life as a result of reduced rates of respiration and ethylene production, delayed ripening, increased firmness and reduced incidence of physiological disorders and decay. In contrast, high nitrogen content is often associated with shorter postharvest-life due to increased susceptibility to mechanical damage, physiological disorders and decay. Increasing the nitrogen supply to citrus trees results in somewhat lower acidity and ascorbic acid content in citrus fruits, while increased potassium fertilization increases their acidity and ascorbic acid content (Buescher *et al.*, 1999). There are numerous physiological disorders associated with mineral deficiencies. For example, bitter pit of apples; blossom-end rot of tomatoes, peppers and watermelons; cork spot in apples and pears; and red blotch of lemons are associated with calcium deficiency in these fruits. Boron deficiency results in corking of apples, apricots and pears; lumpy rind of citrus fruits and cracking of apricots. Poor colour of stone fruits may be related to iron and/or zinc deficiencies. Excess sodium and/or chloride (due to salinity) results in reduced fruit size and higher solids content.

Severe water stress results in increased sunburns of fruits, irregular ripening of pears, tough and leathery texture in peaches, and incomplete kernel development in nuts. Moderate water stress reduces fruit size and increase contents of soluble solids, acidity and ascorbic acid. On the other hand, excess water supply to the plants results in cracking of fruits, excessive turgidity leading to increased susceptibility to physical damage, reduced firmness, delayed maturity and reduced soluble solids content.

Cultural practices such as pruning and thinning determine the crop load and fruit size which can influence nutritional composition of fruit. The use of pesticides and growth regulators does not directly influence fruit composition but may indirectly affect it due to delayed or accelerated fruit maturity.

C) Maturity Indices: Maturity at harvest is the most important factor that determines postharvest-life and final quality such as appearance, texture, flavor, nutritive value of fruit and vegetables. It is two types:

Physiological maturity: It is the stage when a fruit is capable of further development or ripening when it is harvested i.e. ready for eating or processing.

Horticultural maturity : It refers to the stage of development when plant and plant part possesses the pre-requisites for use by consumers for a particular purpose i.e. ready for harvest (Dhatt and Mahajan, 2007).

The necessity of shipping mature fruit-vegetables at long distances has often encouraged harvesting them early than ideal maturity, (Kader, 1995). Fruits harvested too early may lack flavor and may not ripen properly, while produce harvested too late may be fibrous or have very limited market life. Similarly, vegetables are harvested over a wide range of physiological stages, depending upon which part of the plant is used as food. For example, small or immature vegetables possess better texture and quality than mature or over-mature vegetables.

Types of indices and their components:

- Visual indices based on size, shape and colour of fruit
- Physical indices based on firmness and specific gravity of fruits
- Chemical measurement based on total soluble solids and titratable acidity of fruits
- Calculated indices based on calendar date and heat units of fruits.

D) Method and time of harvesting: Faulty harvesting techniques of fruits and vegetables leads to decay of fruits because it creates suitable medium for growth of micro flora at point of injury of fruits. There are basically three methods most commonly used for harvesting any fruits or vegetables.

- ◈ Harvesting individual perishable with hand by pulling or twisting the fruit pedicel.
- ◈ Harvesting individual fruits or fruit bunch or vegetable bunch with the help of fruit clippers/secateurs/scissors/ specially designed for harvesting.

Harvesting in noon hour is most harmful due to its maximum field heat present in fruit which reduces their shelf life. Fruits harvested at early in the morning and transported to pack house for sorting, grading and packing yield better quality and better shelf-life.

E) Packing house operations: Concept of packing station is completely lacking in India and packing is done mainly on farm in unhygienic manner. Mostly fruits are transported as such to mandis even without packing. Packing stations should have facilities for grading, sorting, pre-cooling and packaging in hygienic manner. This helps to reduce post harvest losses and better prices to farmers. Little attempts has been initiated by Government in this direction in 8th Five Year Plan by establishing NHB (National Horticulture Board). NHB has set up packing sheds, pre-cooling units, cold storage and special transport vehicle and many retail outlet in different parts of the country. Grading and postharvest treatments plant for Nagpur oranges has been installed at National Research Centre for citrus, Nagpur. Mechanical washing, waxing and grading unit has been installed at Banglore.

It is important to minimize mechanical damage by avoiding rough handling and bruising during the different steps of pack house operations. Secondly the pack house operations should be carried out in shaded area. Shade can be created using locally available materials like, shade cloth, woven mats, plastic tarps or a canvas sheet hung from temporary poles. Shade alone can reduce air temperatures surrounding the produce by 8–17 °C during summer season. The packing house operations include the following steps:

Dumping : The first step of handling is known as dumping. It should be done gently either using water or dry dumping. Wet dumping can be done by immersing the produce in water. It reduces mechanical injury, bruising, abrasions on the fruits, since water is more gentle on produce. The dry dumping is done by soft brushes fitted on the sloped ramp or moving conveyor belts. It will help in removing dust and dirt on the fruits.

Sorting and grading: This is one of the most important postharvest operations after harvesting. This is done primarily for quality packing and removal of diseased and defective produce from the lot. Proper sorting and grading gives assurance of quality produce (Dhatt and Mahajan, 2007). This is either done in the farmer's field or in the pack houses. Both manual and mechanical graders are used for grading. All round-shaped fruits and vegetables are easily graded by mechanical graders. Grading may be based on color, size, and extent of defects, while sorting is totally dependent on man power. Grading is done by simple to highly sophisticated graders. Today, many sophisticated graders are in use for fresh produce such as *GREEFA*.

Primary processing: Fresh fruits and vegetables are marketed in our country immediately after harvest without primary or minimal processing. So large amount of inedible parts are transported for marketing. These parts generate waste at that place and undue transportation and packing cost.

F) Pre-cooling : Pre-cooling for removing field heat from freshly harvested fruits reduces microbial activity, metabolic activity, respiration rates and ethylene production. This also decreases the ripening rate, diminishes water loss and decay, and thus, helps preserving quality and prolongs shelf life of the fruits (Ferreira *et al.*, 1994 and Reina *et al.*, 1995). The rate of pre-cooling of the fruits and vegetables immediately after harvest is a determining factor for their quality and durability in storage (Tonini *et al.*, 1979). It could be summarized the method of pre-cooling as follows (Shahi *et al.*, 2012).

Room cooling: It is low cost and slow method of cooling. In this method, produce is simply loaded into a cool room and cool air is allowed to circulate among the cartons, sacks, bins or bulk load.

Forced air cooling: Forced air-cooling is mostly used for wide range of horticultural produce. This is the fastest method of pre-cooling. Forced air-cooling pulls or pushes air through the vents/holes in storage containers. In this method uniform cooling of the produce can be achieved if the stacks of pallet bins are properly aligned. Cooling time depends on (i) the airflow (ii) the temperature difference between the produce and the cold air and (iii) produce diameter.

Hydro cooling : The use of cold water is an old and effective cooling method used for quickly cooling a wide range of fruits and vegetables before packaging. For the packed commodities it is less used because of difficulty in the movement of water through the containers and because of high cost involved in water tolerant containers. This method of cooling not only avoids water loss but may even add water to the commodity.

Vacuum cooling : Vacuum cooling takes place by water evaporation from the product at very low air pressure. In this method, air is pumped out from a larger steel chamber in which the produce is loaded for pre-cooling. Removal of air results in the reduction of pressure of the atmosphere around the produce, which further lowers, the boiling temperature of its water. As the pressure falls, the water boils quickly removing the heat from the produce.

Package icing : In some commodities, crushed or flaked ice is packed along with produce for fast cooling. However, as the ice comes in contact with the produce, it melts, and the cooling rate slows considerably. The ice keeps a high relative humidity around the product. Package ice may be finely crushed ice, flake ice or slurry of ice. Liquid icing distributes the ice throughout the container, achieving better contact with the product. Packaged icing can be used only with water tolerant, non-chilling sensitive products and with water tolerant packages such as waxed fiberboard, plastic or wood (Dhatt and Mahajan, 2007).

G) **Postharvest treatments**

The post-harvest treatments play an important role in extending the storage and marketable life of horticultural perishables. The most important postharvest treatments include:

Washing with chlorine solution : Chlorine treatment (100–150 ppm available chlorine) can be used in wash water to help control inoculums build up during packing operations.

Hormones to Extend the Shelf-Life of Fruits:

Gibberelins: Gibberelins are known as senescense delaying agents. Exogenous application of gibberelic (GA) acid delays changes such as degreening and synthesis of carotenoids. Ingle *et al.* (2001) revealed that foliar application of GA_3 @ 25 ppm increased the fruit weight, volume, total soluble solids (TSS), ascorbic acid, peel and yield over control in 'Nagpur' mandarin. Postharvest treatment of GA_3 (50-100ppm) reduces weight loss, delays colour development and fruit softening in Nagpur mandarin (Bastakoti and Gautam, 2007). Fruits of sapota are dipped in solution of GA_3 (300ppm) for 10 minutes enhances shelf life (8.75 days), delays ripening (7.50 days), retains fruit firmness (0.55kg/cm2) and reduces physiological loss in weight (PLW) stored in corrugated fibre boxes (CFB) at room temperature for 12 days (Patel and Katrodia, 1998).

Cytokinins: Cytokinin delays the senescence phenomena by delaying green colour degradation. PLW and rotting were observed to be minimum up to 60 days with the postharvest application of Kinetin (20ppm), while there were minimum with the same treatment when stored up to 28 days under ambient conditions. The maximum retention of Total Soluble Solids (TSS), Vitamin-C and juice content was observed with the postharvest application of Kinetin (20ppm) under the cold storage (4^0 C, 90-95% RH) and ambient conditions (6-21^0 C, 51-80% RH) in Kinnow mandarin. Benzyladenine (BA) (30mg/l) and Kinetin(10mg/L) applied at fruit setting stage has maximum TSS & total sugars respectively and acidity was maximum under BA (10mg/L) applied at fruit setting stage, stored under ambient conditions for 7 days in Kinnow mandarin (Khalid *et al.*, 2012).

Ethylene inhibitors: 1-MCP (1-methyl cyclopropene), AVG (Amenoethoxyvinyl gycine), silver nitrate, silver thiosulfate, cycloheximide, benzothiadiazole etc. are some of the chemicals which inhibit ethylene production and/or action during ripening and storage of fruits. 1- MCP is a synthetic cyclic olefin that inhibits ethylene by blocking access to the ethylene-binding receptor.

Calcium application : The post-harvest application of $CaCl_2$ or $Ca(NO_3)_2$ play an important role in enhancing the storage and marketable life of fruits by maintaining their firmness and quality. Calcium application delays aging or ripening, reduces postharvest decay, controls the development of many physiological disorders. Calcium infiltration reduces chilling injury and increase disease resistance in stored fruit.

Thermal treatments: Thermal treatments included (i) hot water treatment: Fruits may be dipped in hot water before marketing or storage to control various post-harvest diseases and improving peel color of the fruit. (ii) Vapor heat treatment (VHT): This treatment proved very effective in controlling infection of fruit flies in fruits after harvest. The boxes are stacked in a room, which are heated and humidified by injection of steam. The temperature and exposure time are adjusted to kill all stages of insects (egg, larva, pupa and adult), but fruit should not be damaged.

Fumigation: The fumigation of SO_2 is successfully used for controlling post- harvest diseases of grapes. This is achieved by placing the boxes of fruit in a gas tight room and introducing the gas from a cylinder to the appropriate concentration. However, special sodium metabisulphite pads are also available which can be packed into individual boxes of a fruit to give a slow release of SO_2. The primary function of treatment is to control the *Botrytis cinerea* . The SO_2 fumigation is also used to prevent discoloration of skin of litchis.

Irradiation: It helps to control insect infestation and extend product shelf life. Ionizing radiation can be applied to fresh fruits and vegetables to control micro-organisms and inhibit or prevent cell reproduction and some chemical changes. Gamma rays use the radiation given off by a radioactive substance (Cobalt-60 or Cesium-137). These particular substances do not give off any neutrons, which mean that they do not make anything around them radioactive. Irradiation of food takes place in a chamber with massive concrete walls that keep any rays from escaping. It can be applied by exposing the crop to radiations from radioisotopes (normally in the form of gamma-rays measured in Grays (Gy), Low Dose (< 1KGy) of gamma radiation is effective for disinfection of fruits for insects and larvae (< 1 K Gy), sprout inhibition in bulbs and tubers (0.03-0.20 K Gy) and delay of fruit ripening (0.25-0.75 K Gy). Medium Dose (1-10 K Gy) is effective for sterilization of herbs and spices (8-10 K Gy), reduction of pathogens i.e. Salmonella, Campylobacter, Escherichia or Vibro spp. (2.5-10 K Gy) and extension of shelf life of fruits and vegetables (1.5-3 K Gy). And high dose (> 10 K Gy) is effective for sterilization of packing material (10-25 K Gy). At commercial dose level, irradiation does not cause greater damage to nutritional and sensory quality . However, at higher dose level, chemical changes in the structure of amino acids and proteins cause changes in aroma and taste of food.

Irradiated foods are safe, wholesome and nutritious. Irradiation is endorsed by federal regulatory agencies and numerous national and international health organizations. The Food and Drug Administration (FDA) has approved the irradiation of meat and poultry, and recently fresh spinach and iceberg lettuce, and allows its use for a variety of other foods, including fresh fruits and vegetables and spices. Consumer education is needed to enhance the acceptance of irradiated foods by the public.

Ozone: It is the most effective natural effective natural bactericide of all the disinfecting agents. It is strong and ideal, germicide, sanitizer, sterilizer, vermicide, anti-microbial, bactericide, fungicide and detoxifying agent. Ozone oxidizes the metabolic products and neutrilizes the odours generated during the ripening in storage fruits. Growers and processors use ozonated water to wash fruits replacing chlorine.

Edible coatings: Edible coatings are defined as the thin layer of material which can be consumed and provide a barrier to oxygen, microbes of external source, moisture and solute movement for food. In edible coating a semi permeable barrier is provided and is aimed to extend shelf life by decreasing moisture and solute migration, gas exchange, oxidative reaction rates and respiration as well as to reduce physiological disorders on fresh cut fruits (Park, 1999). Different type of materials were used for coating and wrapping various fruits and vegetables to extend their shelf life, and this is eaten together with foods, with or without removal is considered an edible coating (Pavlath and Orts (2009). Edible coating or edible films provide shiny appearance to fruits and vegetables. Thickness of edible coating is generally less than 0·3 mm (Tharantharn, 2003). An edible coating protects outer membrane of fresh fruits and vegetables (Mohamed *et al.*, 2013). The edible coating serves as carrier of texture enhancer, antioxidants and it is used as a nutraceutical (Rojas-Grau *et al.*, 2008).

Classification of Edible Coatings

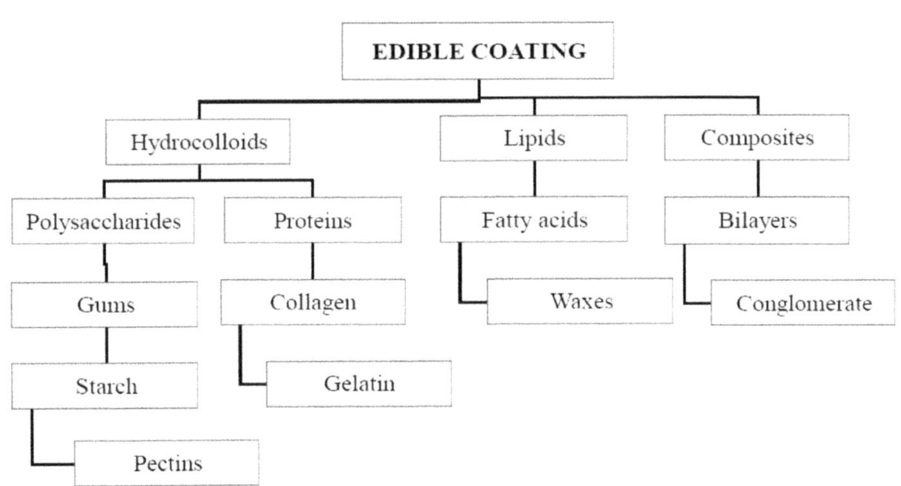

Fig. 2: Different types of Edible coating (Lin and Zhao, 2007).

Herbal Edible Coatings: A New Concept: Herbal edible coating is a new technique for food industry. It is made from herbs or combination of other edible coatings and herbs, most common herbs used in Edible coatings are such as *Aloe vera* gel, Neem, Lemon grass, Rosemary, Tulsi and Turmeric. Herbs have antimicrobial properties, it consists vitamins, antioxidants and essential minerals (Douglas *et al.*, 2005). *Aloe vera* gel is widely used in coating on Fruits and Vegetables, because of its antimicrobial property, it also reduces loss of moisture and water. Ginger essential oil, clove bud oil, turmeric neem extract, mint oil, other essential oil and extracts are also used in edible coating of fruits and vegetables. Herbs are natural source of vitamins, minerals, antioxidants, beneficial for health act as a nutraceutical and medicines (Chauhan *et al.*, 2014).

H) Packaging: Packaging is meant to protect the commodity during storage, transportation, distribution and deterioration (physical, chemical or biological). It is generally done at point of production, processing or at distribution centre and helps to maintain the freshness of perishables and extend their shelf-life. So it is last link in the chain of production and first in chain in storage marketing and distribution. The following are among the more important general requirements and functions of perishables packaging materials/containers (Simson and Straus, 2010).

- It serves as an efficient handling unit for customers and dealers.
- It serves as convenient ware house or storage unit.
- It should protect fruit form mechanical injury.
- It should reduce the wastage of moisture loss.
- It should keep fruits in clean and hygienic manner.
- It should be convenient for transportation and service.
- It should be attractive to customers in different modes.

A well-designed package needs to be adapted to the conditions or specific treatments required for the product. If hydro-cooling or ice-cooling is required, the package must tolerate wetting without losing strength. For a product with a high respiratory rate, the packaging should have sufficiently large openings to allow good gas exchange. When produce dehydrates easily, the packaging should be designed to provide a good barrier against water loss, etc. Semipermeable materials make it possible to generate special atmospheres inside packages. This helps in maintaining produce freshness.

Use of Cushioning Materials: Cushioning materials are used in many stages during postharvest handling operations. But there are three main stages, where it becomes compulsory in order to maintain postharvest quality. The first stage is putting harvested produce into plastic crates or any rigid container. All crates have hard surfaces and while keeping produce inside, there is a chance of dropping off from little height, causing impact bruising popularly called touching marks. The second stage is transportation from field to pack house. Based on the condition of roads, there would be impact and vibration bruising; these bruising may not be visible immediately, but after few days, browning or blackening symptoms develop and finally produce starts rotting. Cushioning materials if used in plastic crates reduce these bruising and touching marks drastically. The third stage is transportation of packed produce from pack house to destination markets. Loading, unloading, and transportation jerks causes bruising. Therefore, it is recommended to use cushioning material to preserve postharvest quality of fresh produce. There may be many types of cushioning materials such as newspaper sheets, newspaper cuttings, rice straw, bubble sheet, specially designed foam nets, moulded trays, gunny bags, leaves, khaskhas and other locally available material.

I) Storage: Almost all fruits are seasonal in nature. Every year, harvesting season falls during a fixed period, say 2–3 months. This period may differ from state to state for the same fruit and for different fruits also. For example, in India, apple harvesting season falls from July to October in Himachal Pradesh and from August to November in Jammu and Kashmir every year. This may be little early or late due to prevailing weather conditions during growing periods. Demand for many fruits and vegetables are round the year. The demand of any fruit or vegetable beyond the harvesting season is called off-season demand. This demand can be fulfilled only if fruits are stored in the harvesting season and sold during off-season. The management of temperature, ventilation and relative humidity are the three most important factors that affect postharvest quality and storage life of horticultural produce. There may be many objectives of storage but the main objectives are:

- To minimize glut and distress sale in the market, that assuring good price to the farmers.
- To insure availability of food in off-season.
- Save horticultural produce from being spoiled.
- Storage in season when cost of produce is relatively low and marketing in off- season at a better price. This gives higher returns to growers and traders.
- To regulate the price of the commodity during season and also in off-season.

Lowering the temperature to the lowest safe level is of paramount importance for enhancing the shelf life, reducing the losses, and maintaining fresh quality of fresh produce. Storage of compatible groups of fruits and vegetables together (requires same temperature and RH) is advisable and necessary. Otherwise, quality of one produce affects the quality of other produce. Some basic requirements during storage are:

Temperature and relative humidity: Since fruits, vegetables, and flowers are alive after harvest, all physiological processes continue after harvest such as respiration and transpiration. Respiration results in produce deterioration, including loss of nutritional value, changes in texture and flavor, and loss of weight by transpiration. These processes cannot be stopped, but they can be reduced significantly by careful management of temperature and relative humidity during storage and transportation. Growth and multiplication of microorganism responsible for rotting and spoilage are also associated with low temperature. At sufficiently low temperature, many disease-causing microbes stop growth and multiplication. Respiration rates vary tremendously for different products. Lower the temperature, the slower will be its respiration rate and the growth of decay organisms. According to Van't Hoff Quotation (Q_{10}), the rate of deteriorative reactions doubles for each 10 °C rise in temperature. Generally, the higher the respiration rates of a fruit or vegetable, the greater the need for postharvest cooling.

Water is the main component found in fruits and vegetables. An important factor in maintaining postharvest quality is to ensure that there is adequate relative humidity (RH) inside the storage area. Water loss or dehydration means a loss in weight, which in turn affects the appearance, texture and in some cases the flavour also. Water loss also affects crispiness and firmness. Consumers tend to associate these qualities as poor with recently harvested fresh produce. For most fresh produce, relative humidity of about 90–95% is recommended for storage and transportation. Since transportation period is only few hours to days, maintenance of RH is not of much importance except for leafy vegetables. But in storage, maintenance of RH is compulsory. In modern cold stores, humidifiers are used for humidity generation.

Types of storage: Post-harvest deterioration can be controlled by reducing the storage temperature. Another factor that can be controlled to minimize the quality deterioration in horticultural products is the respiration rate. This can be achieved by modification of the gas atmosphere surrounding the product. Two types of technologies have evolved in order to provide this benefit: controlled atmosphere (CA) storage and modified atmosphere packaging (MAP).

Controlled atmosphere (CA): Innovation lies at the heart of developments in the horticultural sector, and it is principally adopted for fresh fruits and vegetables. CA is used to extend the utility and useful marketing period of fresh Fruits and vegetables during storage, transportation and distribution, so as to maintain quality and market value of the produce, in comparison to that achieved via the use of refrigeration or cold storage alone. CA works by altering O_2 and CO_2 concentrations during storage (i.e., reducing O_2 and increasing CO_2). Different CA storage regimes such as low O_2 (LO), ultra low O_2 (ULO) and more recently dynamic controlled atmosphere (DCA) or system (DCS) (Prange *et al.*, 2005) have been developed as a result of more accurate control of low levels of O_2 facilitated by automated measuring and control systems (Hoehn *et al.*, 2009). Control of O_2 and CO_2 determines indirectly the nitrogen (N_2) concentration, which is by far the most abundant component, followed by CO_2 and O_2, and trace amounts of inert gases. Optimal temperature and relative humidity (RH) must be adhered to in conjunction with optimal levels of O_2 and CO_2 to assure successful storage outcomes. The storage conditions are narrowly defined depending on the type of fruits and vegetables and may include setting the lowest temperature (to prevent freezing or chilling), the lowest level of O_2 (to attenuate the rate of respiration and delay the development of senescence), ensuring the highest and safest level of CO_2 (to minimize the rate of respiration and slow the senescence and ripening process), the lowest level of ethylene (to attenuate ripening and senescence) and maintenance of high levels of humidity (to reduce moisture loss from the stored product). CA has its most beneficial effects on climacteric fruits and vegetables at the pre-climacteric stage by prolonging this stage. The effects are less marked in climacteric fruits and vegetables at its ripening stage and in non-climacteric fruits at any stage. Climacteric fruits such as apples and pears are by far the leading crops for which CA technology has been adopted (Yahia, 2006), and to a lesser extent to cabbages, sweet onions, kiwifruit, avocados, persimmons and pomegranates (Kader, 2005). Major advantages of CA storage are:

- Decreases not only production of ethylene, but also the rate of response of the tissues to ethylene.
- Alleviates certain physiological disorders such as chilling injury of various commodities.
- Affects post-harvest pathogens directly or indirectly and consequently retards decay incidence and severity.
- Useful tool for insect control in some commodities.
- Increases the availability of fruits and vegetables even during off season.

Modified atmosphere packaging (MAP) MAP of fresh produce relies on modifying the atmosphere inside the package, achieved by the natural interplay between two processes: the respiration of the product and the transfer of gases through the packaging, which leads to an atmosphere richer in CO_2 and poorer in O_2. To design a proper MA package for a particular product is a complicated task, due to the complexity of the system and several variables involved. When fresh-cut fruits and vegetables is packed in MAP, it is exposed to high CO_2 and/or low O_2, and the sensitivity of it to modified atmospheres may be quite different from the whole fruits and vegetables (Lakakul *et al.*, 1999). If the packages are exposed to higher temperatures during distribution, there is a risk of anaerobic metabolic accumulation with concomitant risk of off-flavours. The characterization of the respiration rate process is a central point in the design of MAP systems. Mathematical models as a function of temperature, O_2/CO_2 concentrations describing O_2 consumption and CO_2 production are useful in packaging design and optimal MAP for extending shelf life of fruits and vegetables. The `pack-and-pray' approach with in-house trial-and-error experiments is commonly employed for choosing a suitable packaging material for MAP.

J)Transportation: Transportation may be a connecting link between producers and consumers. It holds key factor in postharvest quality maintenance of all fresh produce. Most fresh produce in India and other countries of the world is transported from farmer's field to nearby market or wholesale market and from wholesale market to terminal market up to final retailers' shops in open and non-refrigerated vehicles.Only few reputed firms use refrigerated vehicles for transportation and distribution in summer months only, starting from March to May/June in India. It is mainly due to the increased cost of transportation by reefer van. In open truck vehicles (Non-reefer), produce is always susceptible to a loss of quality. Ambient temperature alone spoils the produce. Other means of transport include rail transport (A/C and non A/C), air and ship. All imported fruits are transported in A/C containers by ships only. In every country, a dedicated port is assigned for receiving and dispatch of fresh produce containers. In India, a large number of fresh produce containers are received at Mumbai and Chennai port. After harvest, a number of vehicles (trucks, tractors, trains, boats, ships, utility vehicles, etc.) are used to transport the product from field to either packing houses or whole sale or retail markets. These vehicles are not equipped with refrigeration units and thus the produce decays faster, compared to that in refrigerated vehicles. If the produce is treated with edible wax or chemicals or additives after harvest, it can withstand little longer distances in open vehicles (non-reefer), without much damage. Refrigerated vehicles (trucks, trains, ships, airplanes, etc.) contain installed refrigeration units with sufficiently low temperatures to maintain freshness in fresh produce. These types of vehicles are sealed with insulation material inside the walls of the container, which maintains the inside container temperature at desired level and thus preserves maximum quality. This eliminates the possibility of product damage (chilling and freezing injury) when cooling at low temperatures during transport. Refrigeration temperatures can vary from 0 °C (32 °F) to 13 °C (55.4 °F) and RH from 70 to 95 %. It is therefore not advisable to transport mixed lots for long distance. However, for short distance there may not be any problem. Thus, the packaging must be designed to provide a partial barrier against movement of water vapour from the product.

Road Condition and Duration of Transportation: Both road condition and duration of transportation affect quality of fresh produce. In hilly tracks and rough road surface, more touching and bruising take place as compared to smooth surface. Longer duration during transportation also affects quality. Reefer van should not be hold unnecessary. It not only increases the cost of produce, but also affects quality.

Pattern of Loading: Pattern of loading also plays crucial role in maintaining quality of fresh produce. Here pattern of loading means number of packed boxes in one layer (stacking height). In case of fresh produce, stacking height depends on extent of perishable nature of packed commodities and strength of packing materials. If produce are more perishable or box strength is weak, stacking height is kept low and vice-versa.

K) Cold chain: A cold chain is a temperature-controlled supply chain. It is a logistic system that provides a series of facilities for maintaining ideal storage conditions for perishables from the point of origin to the point of consumption in the food supply chain. A well organised cool chain reduces spoilage, retains the quality of the harvested products and quality of the harvested products and guarantees a cost efficient delivery to the consumer.

L) Preparation for the Fresh Market: After harvest, fruits and vegetables need to be prepared for sale. This can be undertaken on the farm or at the level of retail, wholesale or supermarket chain. Any working arrangement that reduces handling will lead to lower costs and will assist in reducing quality losses. Market preparation is therefore preferably carried out in the field. However, this is only really possible with tender or perishable products or small volumes for nearby markets. Products need to be transported to a packinghouse or packing shed for large operations, for distant or demanding markets or for special operations like washing, brushing, waxing, controlled ripening, refrigeration, storage or any specific type of treatment or packaging. These two systems (field vs. packinghouse preparation) are not mutually exclusive. In many cases partial field preparation is completed later in the packing shed. Because it is a waste of time and money to handle unmarketable units, primary selection of fruits and vegetables is always carried out in the field where products with severe defects, injuries or diseases are removed. For distant markets, boxes prepared in the field are delivered to packinghouses for palletizing, pre- cooling and sometimes cold storage before shipping. Mobile packing sheds provide an alternative for handling large volumes in limited time. Harvest crews feed a mobile grading and packing line. On completion of loading, the consignment is shipped to the destination market. In mechanized harvesting, the product is transported to the packinghouse where it is prepared for the market. In many cases, harvest crews make use of an inspection line for primary selection in the field (Simson and Straus, 2010). Preparation and packing operations should be designed to minimize the time between harvest and delivery of the packaged product. Delays frequently occur in the reception area therefore the produce should be protected from the sun as much as possible. Produce is normally weighed or counted before entering the plant and in some cases, samples for quality analysis are taken. Records should be kept, particularly when providing a service to other producers. Water rinsing allows produce to maintain cleanliness and be free of soil, pesticides, plant debris and rotting parts. However, in some cases this is not possible because of insufficient water. If recycled water is used, it needs to be filtered and the settled dirt removed. It could be concluded that, fruits and vegetables need to be prepared for sale after harvest and this can be undertaken on the farm or at the level of retail, wholesale or supermarket chain.

M) Export Promotion: Export of horticultural produce in India is yet to make a dent in International market except few commodities like cashew, spices, mango and onion etc. Even in these commodities our export, market in not regular. In spite of largest producing country, our export potential is negligible.

For making export a sustained activity substantial investment are essential to establish auction houses specially for flowers, separate terminal under contoured environment and cool chain facility at airport and seaports. Although little attempt has been made in this direction by APEDA by setting perishables cargo centre at international airport in New Delhi for fresh fruits vegetables and flowers. Such facilities are needed at many airports in India.

Conclusion

Postharvest handling includes crop production immediately following harvest, including primary processing, cleaning, sorting, grading, cooling, packaging, transportation and storage. These losses at different steps vary differently in different perishables depending upon a season of production and area of production. However, these losses are to be reduced and shelf-life of fruits and vegetables is to be extended at different steps. For this purpose major objective is use to reduce the metabolic processes by various technology such as grading, sorting, post harvest treatments, transportation, storage etc. At every step scientific study to reduce these losses is required, in order to enhance the economy of farmers and country and to earn the foreign exchequer.

References

Atanda, S. A., Pessu, P. O., Agoda S., Isong I. U. and Ikotun. 2011. The concepts and problems of post-harvest food losses in perishable crops. *African Journal of Food Science*, 5 (11): 603-613.

Bastakoti, P. and Gautam, D. M. 2007. Effect of Maturity Stages and Postharvest Treatments on Shelf Life and Quality of Mandarin Oranges in Modified Cellar Store. *Journal of Institute of Agriculture and Animal Science*, 28: 65-74.

Buescher, R., Howard. L. and Dexter, P. 1999. Post harvest enhancement of fruits and vegetables for improved human health. *Hort Scienc.*, 34: 1167-1170.

Chauhan, S., Gupta, K. C. and Agrawal, M. 2014. Development of *A. vera* gel to control postharvest decay and longer shelf life of Grapes. *Int. J. Curr. Microbio. App. Sc.*, 3(3): 632-642.

Dhatt, A. S. and Mahajan, B.V. C. 2007. Horticulture post harvest technology harvesting, handling and storage of horticultural crops. Punjab Horticultural Postharvest Technology Centre, Punjab Agricultural University Campus, Ludhiana.

Douglas, M., Heys, J. and Smallfield, B. 2005. Herb spice and essential oil: post-harvest operation in developing country, 2nd Edition, pp. 45-55.

FAO 2004. Manual for the preparation and sale of fruits and vegetables: from field to market, FAO agricultural services bulletin no. 151. Food and Agriculture Organization of the United Nations, Rome.

Ferreira, M. D., Brecht, J. K., Sargent, S. A. and Aracena, J. J. 1994. Physiological responses of strawberry to film wrapping and pre-cooling methods. *Proceedings of the Florida State Horticultural Society*, 107: 265-269.

Goldman, I. L.., Kader, A. A. and Heintz, C. 1999. Influence of production, handling and storage on phytonutrient content of foods. *Nutrition Reviews*. 57 (9): 546-552.

Hodges, R. J., Buzby, J. C. and Bennett, B. 2011. Postharvest losses and waste in developed and less developed countries: opportunities to improve resource use. *Journal of Agricultural Science*, 149:37-45

Hoehn, E., Prange, R. K. and Vigneault, C. 2009. Storage technology and applications. In: EM Yahia (ed.), Modified and Controlled Atmospheres for Storage, Transportation, and Packaging of Horticultural Commodities, Woodhead Publishing Ltd, Cambridge, pp. 17-50.

Ingle, H. V., Rathod, N. G. and Patil, D. R. 2001. Effect of growth regulators and mulching on yield and quality of Nagpur mandarin. *Annals J. Plant Phys*, 15(1):85-88.

Kader, A. A. 1988. Influences of preharvest and postharvest environment on nutritional composition of fruits and vegetables, 18-32 In: B. Quebedeaux and F. A. Bliss (eds.). Horticulture and human health, contributions of fruits and vegetables. Prentice-Hall, Englewood Cliffs, N.]

Kader, A. A. 1995. Maturity, ripening, and quality relationships of fruit-vegetables. ISHS *Acta Horticult*, 434: 249–256.

Kader, A. A. 2005. Controlled atmosphere, Department of Pomology, University of California, Davis, CA, http://una:usda.gov/hb66/013ca.pdf.

Khalid, S., Malik, A. U., Khan, A. S. and Jamil, A. 2012. Influence of exogenous applications of plant growth regulators on fruit quality of young 'Kinnow' Mandarin (*Citrus nobilis* ×*C. deliciosa*) Trees. *Int. J. Agric. Biol*, 14(2): 229–234.

Lakakul, R., Beaudry, R. M. and Hernandez, R. J. 1999. Modelling respiration of apple slices in modified atmosphere packages. *J. Food Sci*, 64: 105-110.

Lin, D. and Zhao, Y. 2007. Innovation the development and application of edible coating for fresh and minimally processed fruits and vegetables. *Comprehensive Rev. Food Sc. and Food Safety*, 6: 60-75.

Mohamed, A.Y. E., Aboul, A. H. E. and Hassan, A. M. 2013. Utilisation of edible coating in extending the shelf life of minimally processed prickly Pea. *J. of App. Sc. Research*, 9(2): 1202-1208.

Parfitt, J., Barthel, M. and Macnaughton, S. 2010. Food waste within food supply chains: quantification and potential for change to 2050.

Park, H. J. 1999. Development of advanced edible coating of fruits. *Trends of Food Sci. and Technol.*, 10: 250-260.

Patel, A. B. and Katrodia, J. S. 1998. Effect of GA on shelf-life of sapota fruits after transportation. *Indian Journal of Horticulture*, 55(2): 127-129.

Pavlath, A. E. and Orts, W. 2009. Edible films and coating: Why, What and How? In: edible coating and film for food application, Edited by Embuscado, M. E., Hurber K. C., Springer Sc. Business Media, LIC. 1-24.

Prange, R. K., Delong, J. M., Daniels-Lake, B. J. and Harrison, P. A. 2005. Innovation in controlled atmosphere technology, Stewart Postharvest Rev, 3(9): 1-11.

Reina, L. D., Fleming, H. P. and Humphries, E. G. 1995. Microbial control of cucumber hydro-cooling water with Chlorine Dioxide. *Journal of Food Protection*, 58(5): 541-546.

Rojas-Grau, M. A., Montero-Calderon, M. and Martin- Belloso, O. 2008. Effect of packaging condition on quality and shelf life of fresh-cut pineapple *(Ananascomoves)*. *Postharvest Bio. & Techno*, 50: 182-189.

Shahi, N. C., Lohani, U. C., Chand, K. and Singh, A. 2012. Effect of pre-cooling treatment on shelf life of tomato in ambient condition. *International Journal of Food, Agriculture and Veterinary Sciences*, 2: 50-56

Simson, S. P. and Straus, M. C. 2010. Post-harvest technology of horticultural crops. Oxford Book Company/Mehra Offset Press, Delhi.

Singh, A. and Singh, A. K. 2011. Fruits and vegetables with longer shelf-life for extended availability in the markets. In: *Horti-business*. Published by Delhi Agri-Horticultural Society, India. p. 131-135.

Surenderanathan, K. K. 2005. Post-harvest biotechnology of fruits with special references to banana- Perspective and scope. *Indian Journal of Biotechnology*, 4: 39-46.

Tharantharn, R. N. 2003. Biodegradable films and composite casting: past, present and future, Trends in Food Sc. and Technology, 14(13): 71-78.

Tonini, G., Bertoloni, P. and Cimino, A. 1979. Controlled atmosphere storage of hydro-cooled Williams BC pears. Proceedings of the XVth International Congress of Refrigeration Venezia Italy 3: 640-646.

Yahia, E. M. 2006. Modified and controlled atmosphere for tropical fruits, Stewart Postharvest Rev, 5(6): 1-10.

11

Role of Microbes in Agriculture

Mukesh R. Jangra[1*], Sumit Jangra[2] and KS Nehra[1]

[1]Govt PG College, Hisar - 125 001, India

[2]Department of Molecular Biology, Biotechnology & Bioinformatics, CCS Haryana Agricultural University, Hisar - 125 001, India

*Corresponding author: sjangra.07@gmail.com

Abstract

Microorganisms are the unobserved major part of soil and make a huge part of life's genetic diversity. Many studies tried to investigate a variety of roles that soil microbes play in plant efficiency, productivity and diversity. Microorganisms are responsible for the maintenance of soil structures and sustainability of soil quality for efficient plant growth. Soil microbes play a significant role in regulating plant productivity, Plant and soil health, especially in nutrient-poor environment where plant symbionts are responsible for the acquirement of limiting nutrients. Microbes like Mycorrhiza fungi and some nitrogen-fixing bacteria are responsible for an average of between 5–20%, and up to 80% of all nitrogen and up to 75% of phosphorus, that is acquired by plants per annum. These microorganisms help by stimulating biological activity in the soil and plant. Apart from it, free-living microbes also strappingly regulate plant productivity, through mineralisation of, and competition for, nutrients that sustain plant productivity. Soil microbes including microbial pathogens are regulators of plant community, diversity and abundance. It has been figured out that nearly 20,000 plant species solely dependent on microbial symbionts growth and survival. Overall, soil microbes must be considered as important drivers of plant diversity, yield, and fertilisers enhancement.

Keywords: Agriculture, Microbes, Plant Symbionts, Mycorriza, Soil microbes

Introduction

Agriculture is a multifarious network of connections of plants with microorganisms. There is an emergent demand for ecologically compatible, atmosphere friendly technique in agriculture that might be able to make available plenty supply of nutrients for the increasing human population through improved quality and quantity of agricultural products and services. Under the changing environmental circumstances of comprehensive fluxes of key biogenic greenhouse gases (CO_2, CH_4, and nitrous oxide) and some other environmental problems, the application of useful microorganisms in the agriculture would give out an important alternative opportunity to some of the traditional agricultural practices. There are several microorganisms which are of agricultural significance and represent a key ecological approach for integrated management practices such as nutrient management, disease management and pest management in order to trim down the use of chemicals in agriculture as well to improve cultivar performance. Microbes present in our ecosystem are mostly microscopic small creatures. These are positioned in different groups such bacteria, fungi, protozoa, micro-algae and viruses. These organisms are omnipresent and everywhere like in soil, water, food, animal intestines and other harsh environments. A variety of microbial habitats imitates an enormous diversity of classical, biochemical and metabolic traits that have arisen by genetic variation, natural selection and mutation in microbial populations. Human beings use some of the microbial diversity in the production of fermented foods such as bread, yoghurt, wine and cheese. Some soil microorganisms release nitrogen that is used by plants for growth and give out gases that help to maintain the vital components of the Earth's atmosphere. Many other microbes reduce the food supply by infecting plants and cause yield-reducing diseases in food-producing plants. As the microbial world is the major unexplored pool of biodiversity on Earth, it is an important front line in biology under demanding investigation. Interest in the exploration of microbial diversity has become so important because of the fact that microbes are able to perform several functions crucial for the environment that include nutrient cycling and environmental detoxification. The immeasurable array of microbial activities and their importance to the biosphere and plants provide a strong basis for understanding their function, diversity, conservation, utilisation and exploitation for social benefit. The environmental "super challenges" of the twenty-first century have become fairly clear in the last several years. Continuous threats of pandemics, for example, the Asian flu, mad cow disease, biotic and abiotic stresses, dwindling agricultural productivity, and environmental contamination are some of the important issues which need attention. One of the solutions has come out in the form of microbes. Microbial resources present in our environment will help to address these challenges. These diverse communities make up "a metagenome of knowledge." This metagenome also extends to the microbial communities both within and surrounding of our system. Thus, selection of specific unexplored microbial habitats, natural or modified, may be of great significance in terms of benefits for the environment, agriculture and society. The present chapter is projected to focus on the surfacing of agriculturally important microorganisms, develop and expand an ideal agricultural system through efficient utilisation of nutrients and recycling of energy and in that way to preserve the natural ecosystem resources. The advancement to date in using the beneficial microbes in a variety of applications associated with agriculture together with key mechanism of action is also discussed in this chapter.

History

A novel life process called chemosynthesis was investigated by Sergei Nikolaevich Vinogradskii (Winogradsky) in 1890. His findings that some microbes could live exclusively on inorganic matter came into light during his physiological research in 1880s on sulphur, iron and nitrogen bacteria. During his nitrification research, Vinogradskii first suggested that microbiology could have many solutions for agricultural problems. His evaluation of agricultural chemists and Kochian-style bacteriologists spread this message to the broader agricultural community which boosted interest in biological, rather than chemical methods to explore soil related process and problems. His work in (1891 to 1910), the microbiological laboratory at the Imperial Institute of Experimental Medicine in St. Petersburg, Russia, expanded his chemosynthesis research to a broad investigation of the manifold importance of autotrophic organisms in soil related process. This work and his students attracted the serious interest of agricultural chemists and soil scientists in Russia and abroad, shifting effectively the way they understood and investigated the role of microbes in the soil. Vinogradskii's activities in the late 19th century imitate the changes occurring more largely in soil science.

Methods for Studying Soil Microbial Populations

Soil is considered a stockroom of microbes and their activity. These Living microfauna are estimated to consist of less than 5% of the total space occupied. Therefore, major microbial activity is restricted to "hot spots" such as aggregates with accumulated organic matter, rhizosphere, detritusphere and biosphere (Pinton *et al.*, 2001). Bacterial and fungal diversity associated with soil is often difficult to characterise, mainly because of their immense phenotypic and genotypic diversity, heterogeneity, cryptcity and due to lack of taxonomic knowledge. Bacterial populations in top layers of the soil profile can produce over 109 cells/g soil (Torsvik and Ovreas, 2002). Most of these cells are not culturable in lab conditions. The portions of the cells which make soil microbial biomass and diversity that have been cultured and studied in detail are less than 5% of the total microbial population. But still, soil can be studied for microbiological, biochemical and functional diversity using different approaches (Paul, 2007). Methods of studying microbial diversity can be generally divided into two categories: (1) culture-dependent and (2) culture-independent methods. Both methods have their unique limitations and advantages (Garbeva *et al.*, 2004).

1. Culture-dependent Methods

Variety of methods has been investigated so far to characterise microbial communities. It includes conventional plating like pour plate, spread plate, streak plate and drop plate methods. Pour plate method is usually the method of choice for counting the number of colony-forming bacteria present in a soil sample. In this method, fixed amount of inoculum (generally 1 ml) from a soil sample/ broth is placed in the centre of sterile petri plates by using a sterile pipette. Then, molten and cooled agar (15mL approx) is poured into the petri dish containing the inoculums and mixed well. After the solidification of the agar, the petri plate is inverted and incubated at 37°C for 24-48 hrs. Microbes will grow both on the surface and within the mediumin the form of small colonies. Each colony represents a "colony forming unit" (CFU). In spread plate method, the soil sample evenly distributed on media plate. It is the technique of isolation and enumeration of microorganisms in a mixed culture. This is the easier and rapid method to quantify soil bacteria in a solution. The streak plate method is a qualitative and quantitative isolation method which is used to isolate separate colonies by reducing the number in the inoculum by dilution. The consequential reduction in the population size of microbes ensures that individual cells will be satisfactorily far apart on the surface of the agar medium. This will help to identify and differentiate the different species of microbes present in the soil. Even though many types of streaking is performed, yet the quadrant streak is most commonly used. Another method is drop plate (DP) method. It is used to estimate the number of viable suspended bacteria in a known beaker volume. The drop plate method has some more advantages over the spread plate (SP) method. It is time-saving, effortless and faster method because less time and effort are required to dispense the drops onto an agar plate than to spread an equivalent total sample volume into the agar.

2. Culture-Independent Methods

Advancement in molecular technology and research studies has helped in better understanding of soil microbial communities. These molecular techniques include hybridization, DNA sequencing, restriction mapping, chromosome banding, amino acid sequencing, polymerase chain reaction (PCR) and RT-PCR, which use specific DNA or RNA in soil microbes. The highly conserved 16S or 18S ribosomal RNA (rRNA) or their genes (rDNA) represent useful markers for prokaryotes and eukaryotes, respectively and can also be used for phylogenetic studies. PCR products generated with primers based on conserved regions of the 16S or 18S rDNA using total DNA or RNA of the specific soil microbial community produce a mixture of DNA fragments representing all PCR handy species present in the soil sample. The mixed PCR products can be used for, (A) preparation of clone libraries and (B) an array of microbial community fingerprinting. Such clone libraries are useful for classification, identification and characterization of bacterial or fungal types in soil and thereby provide an image of microbial diversity (Garbeva *et al.,* 2004). Furthermore, a range of other techniques have been developed to fingerprint soil microbial communities, for example, denaturing or temperature gradient gel electrophoresis (DGGE/TGGE) (Muyzer and Smalla, 1998), amplified rDNA restriction analysis (ARDRA) (Massol-Deya *et al.,* 1995), terminal restriction fragment length polymorphism (T-RFLP) (Liu *et al.,* 1997), single-stranded conformational polymorphism (SSCP) (Schmalenberger and Tebbe, 2002), reverse sample gene probing (RSGP) and ribosomal intergenic spacer analysis (RISA)

(Ranjard and Richaume, 2001). Now a day's culture-independent methods have been used in combination with next-generation sequencing technologies to get insights into the structure and function of the soil and plant colonising microbial communities. Several techniques are commonly used e.g. marker gene analysis which targets amplification and taxonomically important informative marker gene from microbial populations. Another important technique is metagenomics that helps to retrieve genomic information for a microbial community by shotgun sequencing of arbitrarily sheared DNA from the community. In addition, metagenomics allows the characterization of the functional and metabolic capability of a microbial community by bioinformatic determination of encoded genes, functions, their taxonomic origin, and corresponding pathways from metagenomic sequences. Apart from these techniques, metatranscriptomics is also important to determine microbial diversity. It is the study of gene expression of a microbial community under certain conditions by reverse transcription and random shotgun sequencing by using isolated RNA from the microbial community. Such analyses can be further complemented with or metaproteogenomics. Metaproteomics determines expressed protein products of microbial populations under certain experimental conditions. As metaproteogenomics needs reference gene sequences for protein identification, synchronised metagenomic sequencing of a sample is used to amplify the number of identifiable proteins, which are not there in already sequenced isolate genomes. Functional diversity is the most important factor for identification, characterization and exploitation of microbial cultures. Likewise, functional genomics is considered a powerful tool for discovering new functions linked with microbial genome. Depending upon the use of the organism, they have been assign different names which specify their major functions in nature or under defined conditions. Nevertheless, it is prerequisite to obtain a novel class of compounds and functions, an intellectual design of test system and careful identification and selection of microbes for a successful screening strategy. Strain improvement methods (mutation, genetic exchange, recombination, protoplast fusion, and gene regulation) should be applied to a wild strain obtained from various reservoirs to increase the productivity and suitability of the culture in a specific location (Crueger and Crueger, 2003).

Agriculturally Important Microflora

AIMs have various applications in agriculture (fig 1), horticulture, and forestry. AIMs are often designated as a large group of habitually unknown or ill-defined microbes that interact positively with soil and plants to turn into beneficial effects which are sometimes complicated to predict. There are many microbes like *Azotobacter chroococcum, Azospirillum basilensis, Bacillus weihenstephanensis,* Bradyrhizobium sp., Paenibacillus sp., *Pseudomonas corrudeugate,* Rhizobium sp. etc. which have established their capability in plant growth promotion.

The microorganisms also play a major role in crop protection by increasing disease resistance capacity of plants against pathogens, exhibiting antagonistic activities or acting as biotic elicitors in opposition to different biotic and environmental factors. The term "Effective Microorganism" is generally used to denote a group of cultures of identified and beneficial microorganisms that are being used effectively as microbial inoculants (Higa and Wididana, 1991). These can be applied to increase the indigenous microbial diversity of soils and rhizosphere of the growing plants. Microorganisms are important in managing the invertebrates and vertebrate pests, plant diseases, weeds and other pests that usually damage the agricultural crops and forestry. Fungi act on upper parts of the plant and colonise plants, it provides its benefits to biotic and abiotic stresses (Singh *et al.*, 2011). Viruses can colonise the plant roots even at extreme situations (even at temperatures up to 115°F) e.g. In Yellowstone National Park, where virus build a symbiotic association with plants (Roossinck, 2011).

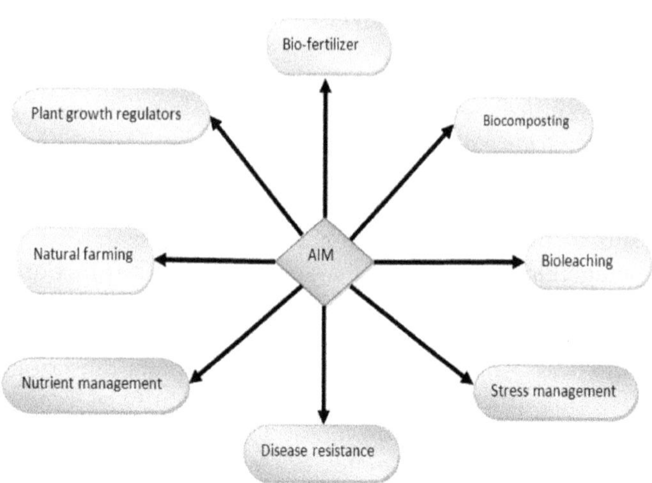

Fig. 1: Different applications of AIM

Table 1: AIMs and their contributions in target plant species

AIMs	Target plant	Beneficial effect	References
*Azospirillum*sp.,	*Cynara*	Known to increase radical and shoot length, shoot weight.	Jahanian *et al.*, (2012)
*Azotobacter*sp.	*Scolymus*	Have significant effect on germination	
Pseudomonas sp. strain PS1	*Vigna radiata*	It helps to increase plant dry weight, root nodule, total chlorophyll content, seed yield and seed protein	Ahemad and Khan, 2012a
*Bradyrhizobium*sp. strain MRM6	*Vigna radiata*	The bacterium can increase the growth parameters at tested concentrations of various herbicides	Ahemad and Khan, 2012b
Pseudomonas sp. strain A3R3	*Brassica juncea*	The bacterium has proved its ability in biomass increase	Ma *et al.*,(2011)

AIMs	Target plant	Beneficial effect	References
Rhizobium strain MRP1	*Pisum sativum*	This bacterium is known to increase the symbiotic properties of plants like nodulation and leghaemoglobin content, enhancement of nitrogen and phosphorous uptake, seed yield and seed protein	Ahemad and Khan, 2011a; Ahemad and Khan, 2010
Bacillus weihenstephanensis strain SM3	*Helianthus annuus*	The bacterium can increase plant biomass and accumulation of trace elements like Cu, Ni and Zn in the roots and shoot systems along with their mobilising potential	Rajkumar *et al.*,(2008)
Bacillus sp., *Paenibacillus*sp.	*Oryzae sativa*	The bacterium can promote the root and shoot growth	Beneduzi *et al.*,(2008)

Importance of Soil Microbes to Plants

The real deal is that without soil microbes, we all would die. The work these microbes do in our soil is unbelievably complicated but it all boils down to this: microbes eat, so we eat. Plants are unable to obtain the nutrients from soil without microbes working in the soil. Microbes are living and must have food to survive and that food comes from organic matter. As microbes take up food they assimilate various elements like nitrogen, carbon, oxygen, hydrogen, phosphorus, potassium and micronutrients for plants. It is the microbes that convert NPK and minerals into the form that can be utilised by the plant for their growth and development. Some important functions of soil microbes are given below.

Nutrient Management

Nutrient management is the science linked with optimal utilization of soil, weather, hydraulic factors and necessary NPK inputs (Miao *et al.*, 2011). Nutrient management includes the conservation strategies that directly or indirectly help in optimization of utilization in nutrient use efficiency so that plant quality, as well as soil health and environment, can be improved. (Turner *et al.*, 2013). Microbes present in soil and atmosphere plays a vital role in the nutrient management (Adhya *et al.*, 2015). Soil microbes, particularly bacteria and fungi, are found as essential in decomposing the soil organic matter and recycling of organic residues. There are many compounds which are released from different parts of the plant root system; it may create a special environment in the rhizosphere. These compounds are collectively termed as root exudates that actually belong to three main classes such as low-molecular weight, high-molecular weight and volatile organic compounds (VOCs) (Ortiz- Castro *et al.*, 2009). Low-molecular-weight compounds stand for the major portion of root exudates and consist of sugars, amino acids, organic acids, phenolics, vitamins and various secondary metabolites. High-molecular-weight compounds are mucilage and proteins, whereas CO_2, certain secondary metabolites, alcohols and aldehydes constitute volatiles organic compounds. Several environmental factors e.g. temperature, light, age and soil type can directly or indirectly affect the nature and timing of root exudation (Badri and Vivanco, 2009). In plants grown under low phosphate accessibility or in the presence of toxic concentration of aluminium, the exudation of organic acids like oxalic acid, malic acid, and citric acid is predominantly increased. These compounds might act as signals for microbial attraction or it can be used as carbon sources for microbial nutrition (Nihorimbere *et al.,* 2011). Potentiality of microbes such as *Aspergillus niger, A. chroococcum, Azospirillumbrasilense, Bacillus subtilis, Pseudomonas corrugata,* Rhizobium sp. and *Streptomyces nojiriensis* in enhancing plant growth as well aspest and disease suppression has been reported (Bhattacharyya and Jha, 2012). Antagonistic actinomycetes, native to soil habitat are also effective in controlling certain plant pathogens (Sarmah *et al.,* 2005) through proper nutrient supply. In rhizosphere, different bacteria compete more for nutrients and ecological space (Bhattacharyya *et al.,* 2015) and have developed various offensive tools for this intra and interspecies competition e.g. antibiotic substances, bacteriolytic enzymes and bacteriocins (Sood *et al.,* 2007). Beneficial microbes are able to make associations with plant roots that finally help in supplying important nutrients such as N, P and Po. Arbuscular mycorrhizal fungi (AMF) are a group of helpful microbiota identified for their symbiotic associations with the roots of higher plants (Salvioli *et al.,* 2016). They are obligate or sometimes saprophytic in nature and need a living host for survival. Due to this symbiotic association with plant roots, AMF helps in the absorption of minerals such as P, water and other important macro and microelements and make them obtainable for the growing plants. Symbiotic AMF (endo-mycorrhizae) may be applied in crops such as cereals, pulses, oilseeds and fruit crops to increase their nutritional requirement (Jeffries and Barea, 2001). Algal genera such as Anabaena, Aphanocapra, Chrococcus, Oscillatoria and Phormidiumhave the capacity to fix atmospheric N, especially in the paddy fields (Hasan, 2013). There are also certain blue-green algae that possess the ability of symbiotic association with a number of other beneficial microorganisms such as fungi, mosses, liverworts and aquatic ferns (Azolla) and thus contribute significantly to the nutrient management process.

Efficient Utilisation and Recycling of Energy

Photosynthesis which is a vital process in plant life depends on plant's ability to utilise solar energy in fixing atmospheric CO_2 into carbohydrate. The energy is further utilised for the biosynthesis of essential components such as amino acids and proteins. However, this process gets extremely low mainly due to the low utilisation of solar energy by green plants. Therefore, an integrated and efficient energy utilisation process is needed to be explored to increase the level of solar energy utilisation efficiency by green plants, so that maximum amount of atmospheric CO_2 can be converted into nutrients.

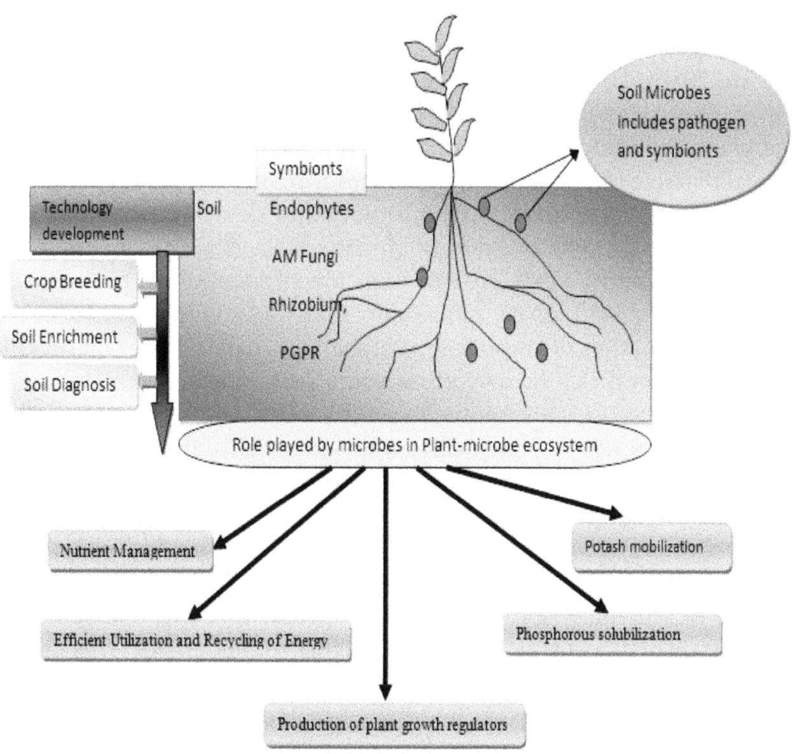

Fig. 2: Plant-microbe interaction, role of microbes and technology development

So, microorganisms are an important tool for efficient utilisation of solar energy, biogeochemical cycling of soil nutrients and recycling of organic molecules (Aislabie and Deslippe, 2013). The turnover and supply of nutrients is crucial for crop growth through the interconversion of different forms of nitrogen, sulphur and phosphorous molecules, interlinked with the carbon cycle. Microbial activity in soil is also essential to execute carbon losses to the atmosphere through respiration and methanogenesis, micro-remediation, through degradation of organic pollutions and immobilisation of heavy metals and thereby improving soil quality (Prosser, 2007). However, it is not an easy task to develop an ideal agricultural system until and unless the efficiency of utilising the solar energy by green plants is increased and the energy contained in existing organic molecules such as amino acids, peptides and carbohydrates is utilised by the plant. New technologies that can improve the economic viability of farming systems with little or no use of chemical fertilisers/pesticides are needed to develop better agricultural practices.

Production of Plant Growth Regulators

Plant growth regulators are synthetic substances like natural plant hormones. They are important in agriculture because growth of plants is regulated by them.

Microorganisms living in rhizosphere of growing plants are likely to synthesize and release a plant growth regulator called auxin, as secondary metabolites (Kapoor *et al.,* 2012). The ratios of plant hormones usually produced by roots as well as rhizosphere bacteria effect the plant morphogenic characters. Various soil microorganisms including bacteria, fungi and algae are responsible for producing physiologically active compounds such as various plants growth regulators which may exert physiological effects on plant growth and development (Ahemad and Kibret, 2014). Plant growth promoting rhizobacteria (PGPR) can modify root architecture and promote plant growth with the production of phytohormones such as indole acetic acid (IAA), gibberellic acid (GA), cytokinins (Kloepper *et al.,* 2007) and production of important metabolites such as siderophores, HCN and antibiotics. Several PGPRs as well as some pathogenic, symbiotic and free-living rhizobacterial species are known to produce IAA and gibberellic acid in the rhizosphere. These hormones have enormous potentiality in enhancing the root surface area and number of root tips in plants (Han *et al.,* 2005). Reports are also available explaining the potentiality of fungi in plant growth promotion (Murali *et al.,* 2012). Beneficial fungi destroy the destruction of fungal pathogens, production of antibiotics and elicitation of defence responses, etc. In addition, many beneficial fungi are able to parasitize spores, sclerotia or hyphae of pathogenic fungi and thereby contribute in biocontrol (Mejia *et al.,* 2008). Production of a large number of degradative enzymes, including chitinases, proteases and glucanases, is involved in this biocontrol process. Many Trichoderma strains are reported to colonise with diverse plant roots and thereby significantly assist in increasing plant growth and development (Saba *et al.,* 2012). In Arabidopsis, normal auxin perception is a prerequisite for growth enhancement when inoculated with Trichoderma (Contreras-Cornejo *et al.,* 2009). Root colonisation by Trichoderma sp. is frequently associated with the induction of both local and systemic resistance that depends on the production of an elicitor protein called Sm1 (small protein 1) (Djonovic *et al.,* 2006). Sm1 lacks toxic activity to plants and microbes. Native and purified form of Sm1 can trigger the production of reactive oxygen species (ROS) in rice and cotton seedlings and thereby induces the expression of defense-related genes both locally and systemically. The beneficial effects of fungi on plant growth and development might be associated with mechanisms such as mycoparasitism (Jeffries, 1995) followed by direct growth toward it, recognition of host-pathogen and its damage.

Phosphorus Solubilization

Plants obtain phosphorus from soil as phosphate anion. It is the least mobile element in plant in comparison to other macronutrients. Phosphorus solubilizing microorganisms (PSMs) play an important role in phosphorus uptake by increasing its accessibility to plants (Sharma *et al.,* 2013). Lowering soil pH by production of organic acids and mineralisation of organic phosphorus by microbes is an essential mechanism for phosphorus solubilization. A general module of functioning of PSMs in agriculture is represented in fig. 3.

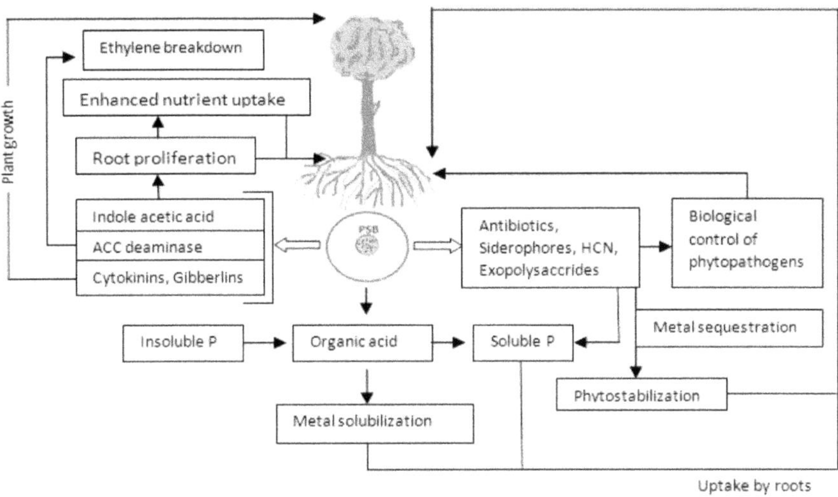

Fig. 3: Role of Phosphate solubilizing bacteria in agriculture

Phosphate solubilizing bacteria (PSB) work with maximum efficiency when they are co-inoculated with other beneficial microbes like bacteria or mycorrhizal fungi (Mohammadi, 2012). But bacteria are found more effective than fungi in phosphorous solubilization (Sharma *et al.,* 2013). *Penicillium bilaii* is a beneficial micro beand it helps in releasing phosphate from native soil. It may produce an organic acid that helps to dissolves the phosphate in soil and utilisation of phosphate in plant roots become very easy. Among the soil bacterial communities, some ectorhizospheric strains of Pseudomonas and Bacilli, Rhizobium, Enterobacter and endosymbiotic rhizobia are found as efficient strains for phosphate solubilization (Khan *et al.,* 2009). Recently it was suggested that phosphate solubilizing bacteria constitute about 1 to 50% in normal soil while phosphate solubilizing fungi have only 0.1 to 0.5% populations (Panhwar *et al.,* 2011). Still, some fungal genera such as Aspergillus and Penicillium are the most influential phosphorus solubilizers (Saxena *et al.,* 2013). Bacterial sp. i.e. *Bacillus megaterium, Bacillus circulans, Bacillus subtilis, Bacillus polymyxa, Bacillus sircalmous* and *Pseudomonas striata*are also considered as potent strains of phosphate solubilizers (Rodriguez and Fraga, 1999). A nemato fungus, *Arthrobotrys oligo spora* have the ability to solubilize the phosphate rocks. The diversity and dominance of beneficial microbial populations such as phosphate solubilizers depend on biotic and abiotic factors prevailing in a particular ecological niche.

Potash Mobilization

As we know, Potassium (K) is one of the vital components of plant growth. It performs multiple important biological functions so that the quality plant growth is maintained. K is generally plentiful in soils top soil ranges from 3000 to 1, 00,000 kg/ha (Bertsch and Thomas, 1985). It is found in four different forms in soil, first is water-soluble (solution K), second exchangeable, third non-exchangeable and fourth one is structural or mineral (Sparks and Huang, 1985). Soil parameters i.e. pH, moisture content, texture, oxygen level, soil tilling, temperature, topographical and biogeochemical characters directly affect the release of phosphate (Basak and Biswas, 2009). As microbes have the capability for K mobilisation so, microbes are used for mobilising K from rock phosphate or muriate of potash where unavailable K is converted into plant available form. Mobilisation of insoluble native soil K-source into plant available nutrient pool can be performed by microbes i.e. *Acidithiobacillus ferrooxidans*, Arthrobacter sp., Azotobacter sp., *Bacillus mucilaginosus, Bacillus edaphicus,* Frateuria sp., Klebsiella sp., Paenibacillus sp., Pseudomonas sp., Rhizobium sp. etc. (Liu *et al.,* 2012). Potassium mobilising bacteria are useful biofertilizer and if KMB is used in combination with low level of K fertilisers, it will, in turn, provide high crop yield with low cost and hold up eco-friendly crop production. Many important technologies such as biomineralization, bioremediation and biohydrometallurgy can be developed by utilising the knowledge on bio-dissolution by potash mobilizers (fig 4).

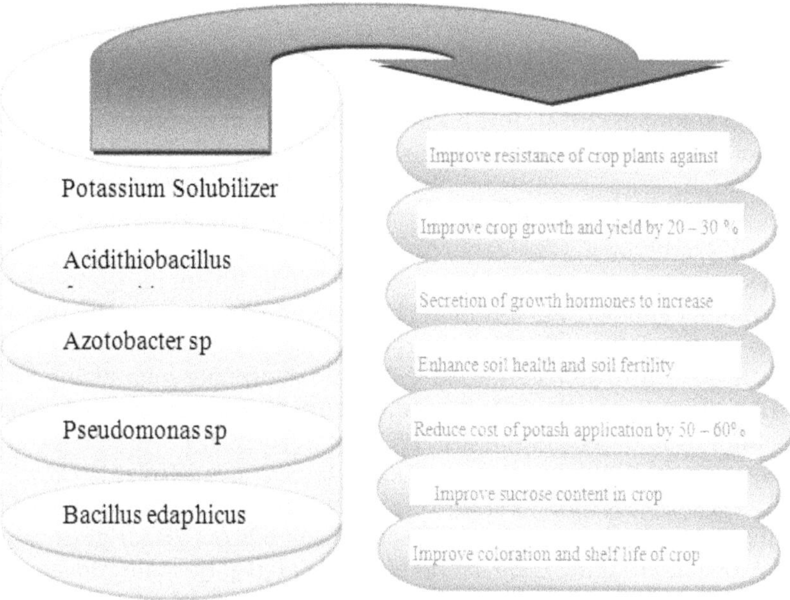

Fig. 4: Role of potassium solubilizers in agriculture

Microorganisms as Biofertilizers and Biopesticide

Most important tools for modern agriculture include the biofertilizers and biopesticides (Bhardwaj *et al.,* 2014). Microbial Biofertilizers consist of live microorganisms when these biofertilizers are applied on the seed, surface of plant or soil, it colonises the rhizosphere and help to promote plant growth due to increased supply of primary nutrients for the host (Bhattacharyya and Jha, 2012; Vessey, 2003). While microbial biopesticides are the microbes, that encourage plant growth by reducing phytopathogenic agents. Biopesticides use different mechanisms i.e. antibiotics production, siderophores production, HCN production, production of hydrolytic enzymes and acquired and induced systemic resistance (Chandler *et al.*, 2008). Many active microbes are generally used to make biofertilizer and biopesticide so that plant growth promotion as well pest and disease can be controlled. One of such bacterium is Rhizobium, that is mostly used in agriculture as an efficient biofertilizer. Rhizobia live in asymbiotic association (Shridhar, 2012) with leguminous plants (fig. 5). Being tolerant to a wide range of temperature, Rhizobium generally enters the root hairs, undergoes division there and forms root nodules where it can fix nitrogen and convert it to a plant usable form. Scientists have reported that rhizobium inoculants at different locations and soil types drastically increase the grain yield of Bengalgram, lentil, pea, alfalfa, sugar beet, berseem, groundnut and soybean. Rhizobium isolates obtained from wild rice have been found to supply nitrogen to the rice plant to promote growth and development. *Sinorhizobium meliloti* 1021, one of the species of Rhizobium, infects plants other than leguminous plants. It promotes growth by enhancing the endogenous level of plant hormone and photosynthesis performance to confer plant tolerance to stress.

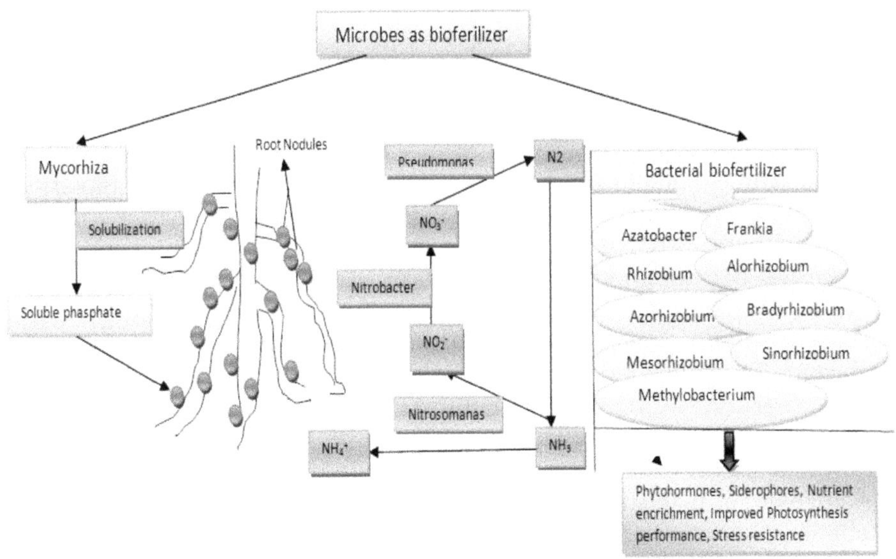

Fig. 5: Role of microbes as bioferilizer

Rhizobial symbiosis with plants provides defense to plants against pathogens and herbivores, such as, Mexican bean beetle. Now a day's, efficient Rhizobium strains are added in many crops because they help to improve soil fertility and encourage plant growth by improving nutrient availability. Rhizobial biofertilizers used in legumes were studied and it was found that when they are applied along with the chemical, they have the ability to replace chemical nitrogen fertilisers up to 30-35% in the field (Mia *et al.*, 2010). Likewise, Acetobacter, Alorhizobium, Aspergillus, Azorhizobium, Azospirillum, Azotobacter, Bacillus, Bradyrhizobium, Mesorhizobium, Penicillium, Pseudomonas, Rhizobium etc., have also satisfied the criteria of potent plant growth promoters (Vessey, 2003). So, scientific research, preparation and application of microbial formulation as biofertilizers are important in developing a sustainable agriculture. Azotobacter and its species *A. chroococcum, A. vinelandii, A. beijerinckii, A. nigricans, A. armeniacus* and *A. paspali* are used as biofertilizer and plays a crucial role in nitrogen cycle, produce vitamins such as thiamine and riboflavin and plant hormones viz., indole acetic acid (IAA), gibberellins (GA) and cytokinins (CK). *A. chroococcum* improves plant growth by increasing seed germination and advancing the root architecture by inhibiting pathogenic microorganisms around the root systems of crop plants viz., wheat, oat, barley mustard, seasum, rice, linseeds, sunflower, castor, maize, sorghum, cotton, jute, sugar beets, tobacco, tea, coffee, rubber and coconuts (Tchan and New, 1989). Another important bacterium is Azospirillum. It is free-living, motile, gram variable and aerobic bacterium that can flourish in flooded conditions. Different species of the genus Azospirillum including *A. lipoferum, A. brasilense, A. amazonense, A. halopraeferens* and *A. irakense* have been studied and found to improve productivity of a range of crops, alters the root morphology due the production of plant growth hormones, increases the number of lateral roots and root hairs formation, help to provide more root surface area to absorb sufficient nutrients and siderophore production. It was found that co-inoculation of *Azospirillium brasilense* and *Rhizobium meliloti* in combination with 2,4-D give encouraging results on grain yield and N, P, K content of *Triticum aestivum.* Other microbes such as fungus also act as very good biofertilizers. Mycorrhizae, for example, form mutualistic symbiotic associations with plant roots of more than 80% of land plants including many important crops and forest tree species (Gentili and Jumpponen, 2006; Rinaldi *et al.*, 2008). The two main types of mycorrhizae are ectomycorrhizae (ECM) and arbuscular mycorrhizae (AM) which can improve water uptake, nutrient acquisition, give protection from pathogens, but only a few families of plants are able to make useful associations with both AM and ECM fungi (Siddiqui and Pichtel, 2008). However, AM fungi are most commonly present in the rhizosphere roots of a wide variety of herbaceous and woody plants (Das *et al.*, 2007; Rinaldi *et al.*, 2008). They help in growth and development of plants because of roots colonisation with ectomycorrhiza which is able to absorb and accumulate N, P and C very quickly than nonmycorrhizal roots. ECM fungi also decompose complex substances in the soil and transfer these nutrients to the tree. Apart from providing protection from pathogens, ECM fungi also play an important role to increase the tolerance of trees to biotic and abiotic stresses. So, it is the mutual cooperation between plant roots and fungus. The plant roots provide food substances to the fungi and in return, the fungi give multiple advantages like transfer of nutrients and water and protection to the plant roots (Chen, 2006).

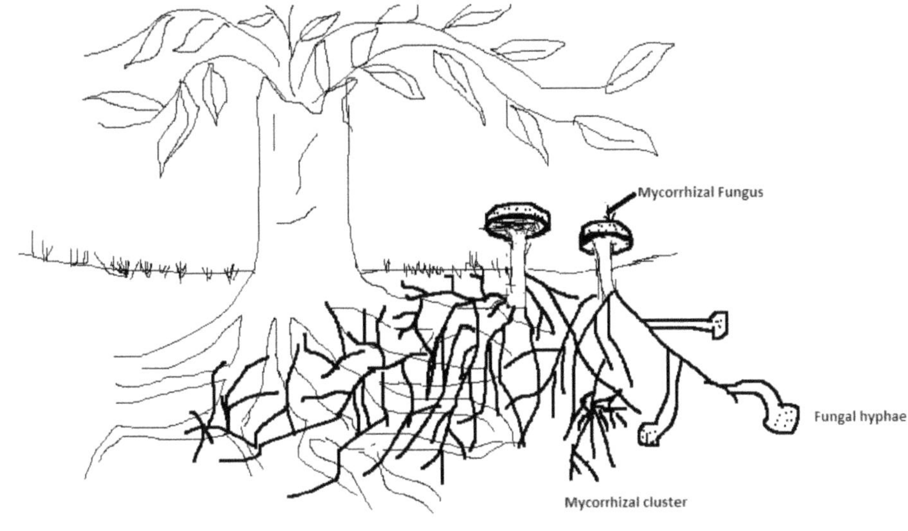

Fig. 6: Mutualistic symbiotic relationship of Mycorrhiza and plant roots

Another dominant mycorrhiza is Endomycorrhizal fungi. These are intercellular and infiltrate into the root cortical cells and form structures called vesicular-arbuscular mycorrhiza (VAM). AM fungi basically help plants by assimilating less available minerals such as copper, molybdenum, phosphorus and zinc (Yeasmin *et al.,* 2007). Like ECM, they also increase plant tolerance to biotic and abiotic stresses like drought, high temperatures, heavy metal toxicity and extreme pH ranges (Gupta *et al.,* 2000; Kannaiyan, 2002). Another indirect function performed by AM includes improvement of the soil structure, protection against pathogens and altered water relationships (Smith and Zhu, 2001). This results because of competition for colonisation sites or nutrients in the same root tissues and production of fungistatic compounds (Johansson *et al.,* 2004; Marin, 2006).

Production of VOCs

VOCs generally represent an orchestrated response to a wide range of stimuli in plants and microbes (Ortiz-Castro *et al.,* 2009). VOCs significantly vaporise and enter the atmosphere as they have higher vapour under normal conditions. Molecules with low molecular weight (<300 g/mol) like alcohols, aldehydes, ketones and hydrocarbons belong to this category. It was observed that VOCs are actively produced and used in plants for communication with other groups of organisms such as microbes (Choudhary *et al.,* 2008). Steeghs *et al.,* (2004) studied the Arabidopsis rhizosphere for VOC emission and observed that the induction of VOC is due to biotic stresses. Many volatiles such as alcohols, aldehydes, acids, ketones, esters and terpenes are induced specifically as a result of different symbiotic and non-symbiotic associations with microbes such as *B. subtilis, Bacillus amyloliquefaciens* and *Enterobacter cloacae* which promote the growth of *Arabidopsis thaliana* (Choong *et al.*, 2004). It was also studied that the rhizobacterial VOCs can act as signalling molecules which help to mediate plant–microbe interactions. It also helps to elicit the plant response in the direction of beneficial microflora colonisation. But still many aspects of VOCs are undiscovered. However, to know the exact mechanisms how VOCs functions in signalling plants against pathogens, efficient research on nature and accumulation of volatile components in plant-rhizobacteria system need to be explored.

Microbes as Biotic Elicitors

Elicitors are molecules which generally play role in plant defence mechanism (Thakur and Sohal, 2013). In most cases, elicitors are basically derived from microbes. Signalling molecules like salicylic acid, methyl jasmonate and NO help to accumulate a wide range of secondary metabolites such as indole glucosinolates, phytoalexins and alkamides. These metabolites have strong capability of communication with beneficial microbial communities (Ortiz-Castro *et al.*, 2009). As microbial responses in stress agriculture is a worldwide menace to agricultural productivity (Gornall *et al.*, 2010), so, it is an urgent need to purify and characterise specific elicitor binding so that the molecular basis of signal exchange process and activation of host defences can be understood. During drought, plants alter a series of physiological responses such as the stomatal closure due to an increase of abscisic acid content, the accumulation of compatible solutes, the increased expression of aquaporins and vacuolar H-pyrophosphatases for maintaining the cell turgidity through osmotic adjustments. When ethylene concentrations increase, it blocks plant growth (Burg, 1973) and in that way increase root to shoot ratio. An extensively larger root system stimulated to increase the water absorption area. Rhizosphere and endosphere are the two important regions where optimum colonisation of microbes is reported (Berg et al., 2014). *In vitro* as well as pot experiments established the potentiality of rhizosphere and endosphere microbes in improving the plant growth and stress tolerance. The growth of microbe's inoculated plants increased up to forty percent suggesting the potentiality of PGP microbes in agriculture (Perez-Montano *et al.*, 2014). The role of plant-associated microbiomes in plant adaptation toward drought is, thus, emerging. Microbiome composition at a particular environmental condition may differ significantly (Marasco *et al.*, 2012) as it strictly depends on the taxonomic variability of the associated plant species.

Molecular Manipulation of Agriculturally Important Microbes

Certain bacteria isolated from soil possess properties that allow them to exert beneficial effects on plants either by enhancing crop nutrition or by reducing damages caused by pathogens or pests. Some microbes such as rhizobia, azospirliia and agrobacterium have been traditionally brought out as seed inoculants in field. These microbes often led to gain in yield of important crop plants while others like pseudomonas have failed to do so. Genetically modified bacteria with improved traits have been constructed and released with plants in numerous experiments. These studies led to the monitoring of effects of modified bacteria on the environment and to assess their impact on surrounding microflora. Transgenic/ mutant microbes have been applied in agriculture for yield improvement, disease & pesticide resistant, phyto & bioremediations; weed control (Babalola and Glick, 2012). *Pseudomonas* sp. are typical examples of bacteria with specific plant-microbe interrelationships where a single strain can act as a biological control agent for several fungi. Many plant pathogens are not sufficiently pathogenic for weed control when used alone, but GEMs altered for hypervirulence, have a high potential for bioherbicidal activity (Babalola, 2009). Some examples of transgenic microbes and the reasons for their uses have been demonstrated (Table 2, 3).

Table 2: List of genetically modified microbes in agriculture

Transgenic microbes	Activity	References
P. putida R20	Enhanced microbial fertiliser and pesticidal activity	Staley and Brauer, 2006
Pseudomonas putida Biovar B. Strain *HS-2*	Phytoremediation and plant growth promotion	Rodriguez *et al.*, 2008
Fusarium arthrosporioides	Improved mycoherbicidal activity	Babalola, 2010
Alternaria cassiae	Mutants with improved mycoherbicidal activity	Babalola, 2009
Trichoderma koningi	Improved insecticidal activity against Asian corn borer (Ostriniafurnacalis)	Li *et al.*, 2012
Pseudomonad and *Rhizobium* Strain	Improved survival and threshold populations in soil environment with improved biocontrol, plant growthpromotion, bioremediation, and nodulation activity	Staley and Brauer, 2006

Some microbes are engineered for strain improvement and mutation-selection (Babalola, 2009). Others are engineered for enhanced microbial fertilizer and pesticidal activity, for example, *P. putida* R20 (Staley and Brauer, 2006). The interrelationship between plant root and the rhizosphere bacteria is appreciably more thorough than that found in the rhizoplane. It should be noted that the aim of biological weed control is not to eradicate weeds but to reduce them to a non-economic level. For this purpose, the microbial agent needs to be specific in action and directed to a targeted weed (Gentry, 2004). Transgenic microorganisms may also find use in bioremediation although it is important to avoid horizontal gene flow to wild or cultivated relatives.

Table 3: Examples of the use of transgenic microbes with potential for use in agriculture

Transgenic Microbes	Crop inoculated	Reason	Reference
Pseudomonas putida Biovar B. Strain HS-2	Non-transformed and transgenic canola plants	Phytoremediation and plant growth promotion	Rodriguez *et al.*, (2008)
Pseudomonas fluoresences Z30-97	Wheat	To monitor possible alterations of bacterial community structure	Bankhead et al., (2004)
Pseudomonas asplenii AC	Common reed (*Phragmites australis*)	Significant increases in shoot and root size and partial protection of the plant from inhibition by either copper or creosote	Reed *et al.*, (2005)
P. asplenii AC	Canola (*Brassica napus*)	Effective growth promotion and partial protection of the plant from inhibition by either copper or creosote	Reed and Glick (2005)

Manipulation of *Rhizobium* sp.

Rhizobium genomes ranging from 5.5 to 9 Mb have been sequenced recently and comparison of sequence data has led to a better understanding of genes involved in the symbiotic process. The sequence information of whole rhizobial genomes or symbiotic islands (Göttfert *et al.*, 2001) has provided an opportunity to improve nitrogen fixation by genetic engineering. A recombinant construct of *S. meliloti strains* that contain different copy numbers of a symbiotic gene region covering a regulatory gene (*nodD1),* the common nodulation genes *(nodABC)* and a gene essential for nitrogen fixation (*nifN).* Strains with a moderate increase in copy number of the symbiotic gene region showed significantly improved nodulation, nitrogenase activity, plant N content and growth (Castillo *et al.*, 1999). Several attempts have been undertaken to improve the N_2-fixation efficiency of rhizobial strains by genetic engineering (table 4). Nevertheless, some recombinant rhizobial strains have been commercialised, such as *Sinorhizobium meliloti* strain RMBPC-2, which has been approved by the US Environmental Protection Agency in 1997. This genetically engineered bacterium contains additional copies of *nifA* and *dctABD* to increase nitrogen fixation and thus yield of alfalfa (Bosworth *et al.*, 1994). The recombinant strain was tested at four field sites in Wisconsin, where yield was significantly increased by 12% compared with the biomass obtained with the wild-type strain at a site where soil nitrogen and organic matter content were low.

Table 4: Approaches to the Manipulation of Rhizobial Competition (Sessitsch *et al.*, 2002)

Characteristics	References
Selection of competent soil saprophytes, strains with enhanced motility, or more efficient attachment properties	Chetal *et al.*, 1968; Smit *et al.*, 1992
Altering the soil chemical or physical environment through management practices to advantage desirable strains	Almendras and Bottomley, 1987
Inoculation technology that provides a positional superiority to the desired strains	Duquenne *et al*, 1999; Danso and Bowen, 1989
Selecting legume hosts capable of expressing a preference for their micro-symbiont	Mpepereki *et al.*, 1996
Selecting strains with genotype-specific catabolite properties that confer an advantage to the desired strain under nutrient limitations	Phillips *et al.*, 1996 Rynne *et al.*, 1994
Selection of strains tolerant to edaphic stresses such as acidity, high temperature, salinity	Howieson *et al.*, 2000
Selecting or engineering strains producing anti-rhizobial peptides or bacteriocins	Triplett *et al.*, 1987; Oresnik *et al.*, 1999
Manipulating rhizobia for selective substrate utilisation to deliver a nodulation advantage	Van Dillewijn *et al.*, 2001

Manipulation of *Trichoderma* sp.

The main antagonist used in disease control in Agriculture is the fungus *Trichoderma harzianum* Rifai, while the choice of active *Trichoderma* strains is important in designing effective and safe biocontrol strategies. This antagonistic potential is the basis for the effective control of a wide set of phytopathogenic fungi (Papavizas, 1985) and the biodegradative capacity is a source of useful enzymes in different industrial sectors. Hundreds of separate genes and gene products in *Trichoderma* are involved in the processes of mycoparasitism, antibiosis, competition for nutrients or space, tolerance to stress through enhanced root and plant development, solubilization and sequestration of inorganic nutrients, induced resistance and inactivation of enzymes produced by pathogens (Monte 2001). Construction of transgenic *Trichoderma koningi* with *chit42* of *Metarhizium anisopliae* and analysis of its activity against the Asian corn borer has been studied by Li *et al.* (2012). The *chit42* gene from *Metarhizium anisopliae* CY1 was inserted into *Trichoderma koningii* T30 genome by protoplast transformation and mortality of the Asian corn borer (*Ostrinia furnacalis*) was assessed for efficacy of culture filtrates and conidial suspensions of transgenic *Trichoderma* strains against the insect. The inhibition of the expression of three genes associated with development and anti-stress response in the mid-gut of the Asian corn borer larvae were noticed. The results depicted significant lethal effect of transgenic Trichoderma strain harbouring *chit42* gene from *Metarhizium anisopliae* CY1 on Asian corn borer larvae. Similarly, understanding drug resistance phenomenon and drug efficacy testing have been done with transgenic Trichoderma carrying gene from *T. longibrachiatum*.

Manipulation of *Fusarium arthrosporioides*

Mycoherbicidal activity of *Fusarium arthrosporioides* was studied alone and in combination with cellulase enzyme by Babalola. He found an improved mycoherbicidal activity of this fungus with combination of enzymein controlling infection of *Orobancheae gyptiaca* on tomato. *F. arthrosporioides* plus *cellulase* was evaluated on tomato root systems to ascertain whether cellulase, a cell wall degrading enzyme, could accelerate fungus infection of *O. aegyptiaca* tubercles. Cellulase treatment alone was ineffective but its presence aggravates fungal infection but in combination with *Fusarium arthrosporioides* showed increased virulence to *O. aegyptiaca.* The findings add to the commercial value of *F. arthrosporioides as a* potential mycoherbicide when sufficiently virulent.

Manipulation of *Alternaria cassiae*

Alternaria cassiae is a mycoherbicide used for controlling Sicklepod (*Cassia senna L.),* coffee senna (*Cassia occidentalis*), cowpea, soybean, pea nut and cotton. But it is reported that*A. cassiae* not only control the target weeds but also have ahigh rate of spread to non-target plants and creating problem. The prevention of sporulation is an important factor in precluding the spread of transgenic biocontrol agents beyond desired limits, and asporogenic deletion mutants could prevent such spread so, asporogenic mutants of *Alternaria cassiae*have been generated by x-ray irradiation with reduced level of spread to the non-target plants (Babalola, 2009). The mutant *NEP1* fungus was more virulent to the target weed with reduced level of spread to the non-target plants

Manipulation of *Pseudomonad and Rhizobium*

Maintaining threshold populations of inoculum microorganisms in the soil environment is important for such practical applications as biocontrol, plant growthpromotion, bioremediation, and nodulation. Few studies on survival kinetics, particularly in relation to subtle alterations in soil acidity-related factors have been reported particularly for survival of a genetically modified root-colonizing *Pseudomonad* and *Rhizobium* strain (Staley and Brauer, 2006). A genetically modified strain *Pseudomonas putida* R20/*lacZY*or *Rhizobium leguminosarum*bv. *trifolii*162S7a/ *gusA* was introduced into conditioned, nonsterile loamy soil limed at four low levels (pH 4.71, 4.81, 4.92, and 4.99) or derivative soil solutions. In both soil and soil solution experiments, survival was negatively correlated ($R^2 \geq -0.914$) with pH and basic cation (Ca and Mg) and positively correlated ($R^2 \geq 0.933$) with Al, concentrations. This relationship of viability to soil solution chemistry was broadly confirmed for both bacterial strains by use of fluorescent probes, suggesting increased cell membrane damage at lower pH. These results demonstrate not only the alternative utility of using soil solutions, rather than nonsterile soils, for bacterial viability assessments but also the positive effect of low-level liming (~ 0.28 pH unit increase) on survival of beneficial root-colonizing bacteria in acidic soils.

Future prospects and challenges

Microbes and their metabolic activities are the most important players for the sustainability of a healthy environment. They recycle the elements on which main productivity depends on. Conservation of microbial diversity is very crucial because it helps in the maintenance of higher organism's species diversity. Microorganisms help directly in the management of strategies i.e. plant disease and nutrient management. Microbial communities produce different signals with different mechanisms for the establishment of altered communities or to elicit different plant responses. UK and USA are currently using bioremediation technologies to augment the fertility of soil and removing the soil contaminations. But in India, research on bioremediation has been initiated and consequently developed in applying microorganisms for restoration of impure and polluted soil. Some potent and transgenic fungal species i.e. Trichoderma formulations with strain mixtures have been prepared and applied. It promotes plant growth and performs better than individual strains for the management of pest and diseases of crop plants. So, to increase the productivity from the agriculture, it is urgent need to fasten the research programmes on microbial remediation. Remediation technologies need to be extended so that the decontaminated soil and water can be utilised for healthy agricultural purposes. The enduring microbial genomic studies will help to provide knowledge and architecture of microbial communities. Exciting developments in the use of microorganisms for the recycling of metal waste, with the formation of novel biomaterials with unique properties,still need to be exploited. Research should be concentrated on biotechnological production of microbes. Attention should be given to transgenic microbes to obtain superior quality of microbes and agriculture products. Genetic engineering and molecular approaches should be used to increase the shelf life of the microbial formulation by mutating wild strains to superior strains. But use of transgenic microbes also faces challenges. Farmers are not easily got ready to adopt new technologies. If they use these technologies, they would like to know how predictable the results will be season after season and what about the financial implications of using such technologies. Farmers do not want to take the risk of increased cost without getting any additional benefits. Unless farming is heavily subsidised by the government, still farmers hesitate to adopt a new technology whose benefits come in the long term. Recently, Microbial-bioinformatics has come into the light. Sequencing of microbial genome provides a better way for studying the biology of microorganisms. Genomic-based approaches will help to elucidate mechanisms on bacterial pathogenicity in the direction of agriculture development. But with the collection of genomic sequences, the challenge remains how to combine and construct biological picture of complex microbial world that can accurately reflecta proper microbial community. So, climate-smart farming technique can help the farmers struggling to adapt the worst climatic variations with limited land and water resources.

It would further give novel opportunities that can help to conserve water soil fertility and provide sustainable agriculture.

Conclusion

The excessive use of chemical fertilizers is creating challenges for agriculture because long term uses of these fertilisers wipe out a large proportion of soil's nutrients and reduce indigenous soil microflora. Without microbes, soil structure gets declined and slowly it loses its ability to keep hold of water, air and nutrients. Plants which are grown in such washed-out soil are extremely susceptible to biotic and abiotic stresses like diseases, insects and drought. So, the uses of constructive and advantageous microbes are the only way to defeat such type of situations. Microbe rich soil activates the natural immune system of the plants, restricts the population of phytopathogens, resists parasitic insects and creates idealenvironment for agriculture. Preferred crop productivity can be achieved by adopting integrated nutrient management strategy so that the soil fertility and optimum plant nutrient supply can be maintained. This can be achieved by draining all benefits from all possible sources of plant nutrients in an integrated manner. All the processes used should be green, recyclable and commercial that would serve as a plus point of sustainable agricultural practices. Climatic predictions are crucial because it directly affects the diversity distribution and activity of plant related soil microbes. Thus, this chapter focused that microorganisms play aunique role in agriculture. These microbes have unpredictable nature and their complex biosynthetic machinery make them adaptable to specific environmental and cultural conditions. This adaptation of microbes helps us to solve a variety of problems related with crop improvement, stress tolerance and disease control. Now a day's, microbial biotechnology and molecular biology's application in sustainable growth of agriculture are getting better attention. The purpose of the chapter is to further prioritise the importance of soil microorganisms and related technology in the agricultural sector for the maximum benefit of the farmers, scientific community and stakeholders.

References

Adhya, T.K., Kumar, N., Reddy, G., Podile, A.R., Bee, H. and Bindiya, S. 2015. Microbial mobilization of soil phosphorus and sustainable P management in agricultural soils. *Curr. Sci,* 108: 1280-1287.

Ahemad, M. and Khan, M.S. 2009. Toxicity assessment of herbicides quizalafop-p-ethyl and clodinafop towards Rhizobium pea symbiosis. *Bull Environ Contam Toxicol,* 82: 761-766.

Ahemad, M. and Khan, M.S. 2010. Comparative toxicity of selected insecticides to pea plants and growth promotion in response to insecticide-tolerant and plant growth promoting *Rhizobium leguminosarum*. *Crop Prot,* 29: 325-329.

Ahemad, M. and Khan, M.S. 2011b. Pseudomonas aeruginosa strain PS1 enhances growth parameters of green gram [*Vigna radiata* (L.) Wilczek] in insecticide-stressed soils. *J Pest Sci,* 84: 123-131.

Ahemad, M. and Khan, M.S. 2012. Alleviation of fungicide-induced phytotoxicity in green gram [*Vigna radiata* (L.) Wilczek] using fungicide-tolerant and plant growth promoting *Pseudomonas* strain. *Saudi J Biol Sci,* 19: 451-459.

Ahemad, M. and Khan, M.S. 2012b. Productivity of green gram in tebuconazole-stressed soil, by using a tolerant and plant growth promoting *Bradyrhizobium* sp. MRM6 strain. *Acta Physiol Plant,* 34: 245-254.

Ahemad, M. and Khan, M.S. 2012b. Pseudomonas Ahemad M, Khan MS. Productivity of green gram in tebuconazole-stressed soil, by using a tolerant and plant growth promoting Bradyrhizobium sp. MRM6 strain. *Acta Physiol Plant,* 34: 245-254.

Ahemad, M. and Kibret, M. 2014. Mechanisms and applications of plant growth promoting rhizobacteria: Current perspective. *J King Saud Univ Sci,* 26: 1-20.

Ahemad, M., and Khan, M.S. 2011a. Effect of tebuconazole-tolerant and plant growth promoting *Rhizobium* isolate MRP1 on pea- *Rhizobium* symbiosis. *Sci Hortic,* 129: 266-72.

Aislabie, J. and Deslippe, J.R. 2013. Soil microbes and their contribution to soil services. In: Dymond, JR, editor. Ecosystem Services in New Zealand - Conditions and Trends. Lincoln, New Zealand: Manaaki Whenua Press.

Almendras, A.S. and Bottomley, P.J. 1987. Influence of lime and phosphate on nodulation of soil-grown *Trifolium subterraneum* L. by indigenous *Rhizobium trifolii Appl Environ Microbiol,* 53: 2090–2097.

Babalola, O.O. 2009. Asporogenic mutants of Alternaria cassia generated by X-ray irradiation. *J. Culture Collect,* 6: 85-96.

Babalola, O.O. 2010a. Beneficial bacteria of agricultural importance. *Biotechnol Lett,* 32(11): 1559-1570.

Babalola, O.O. and Bernard, R.G. 2012. Indigenous African agriculture and plant associated microbes: Current practice and future transgenic prospects. *Sci Res Essay,* 7(28): 2431-2439.

Badri, D.V. and Vivanco, J.M. 2009. Regulation and function of root exudates. *Plant Cell Environ,* 32: 666-81.

Bankhead, S.B., Landa, B.B., Lutton, E., Weller, D.M. and Gardener, B.B.M. 2004. Minimal changes in rhizobacterial population structure following root colonization by wild type and transgenic biocontrol strains. *FEMS Microbiol Ecol,* 49(2): 307-318.

Basak, B.B. and Biswas, D.R. 2009. Influence of potassium solubilizing microorganism (*Bacillus mucilaginosus*) and waste mica on potassium uptake dynamics by sudan grass (*Sorghum vulgare* Pers.) grown under two Alfisols. *Plant Soil,* 317: 235-255.

Beneduzi, A., Peres, D., Vargas, L.K., Bodanese-Zanettini, M.H. and Passaglia, L.M. 2008. Evaluation of genetic diversity and plant growth promoting activities of nitrogen-fixing Bacilli isolated from rice fields in South Brazil. *Appl Soil Ecol,* 39: 311-20.

Berg, G., Grube, M., Schloter, M. and Smalla, K. 2014. Unraveling the plant microbiome: Looking back and future perspectives. *Front Microbiol,* 5: 1-7.

Bertsch, P.M. and Thomas, G.W. 1985. Potassium status of temperate region soils. In: Munson, RD, editor. Potassium in Agriculture. Madison, WI: American Society of Agronomy, Crop Science Society of America, and Soil Science Society of America, 131-62.

Bhardwaj, D., Ansari, M.W., Sahoo, R.K. and Tuteja, N. 2014. Biofertilizers function as key player in sustainable agriculture by improving soil fertility, plant tolerance and crop productivity. *Microb Cell Fact,* 13: 1-10.

Bhattacharyya, P.N. 2012. Diversity of Microorganisms in the Surface and Subsurface Soil of the Jia Bharali River Catchment Area of Brahmaputra Plains. PhD Thesis. Guwahati: Gauhati University.

Bhattacharyya, P.N. and Jha, D.K. 2012. Plant growth-promoting rhizobacteria (PGPR): Emergence in agriculture. *World J Microbiol Biotechnol,* 28: 1327-1350.

Bhattacharyya, P.N., Sarmah, S.R., Dutta, P. and Tanti, A.J. 2015. Emergence in mapping microbial diversity in tea (*Camellia sinensis*(L.) O. Kuntze) soil of Assam, North-East India: A novel approach. *Eur J Biotechnol Biosci,* 3: 20-25.

Bosworth, A.H., Williams, M.K., Albrecht, K.A., Kwiatkowski, R., Beynon, J., Hankinson, R.R., Ronson, C.W., Cannon, F., Wacek, R.J. and Triplett, E.W. 1994. Alfalfa yield response to inoculation with recombinant strains of *Rhizobium meliloti* with an extra copy of dctABD and/or modified nifA expression. *App. Env. Micbiol,* 60: 3815–3832.

Burg, S.P. 1973. Ethylene in plant growth. *Proc Nat Acad Sci* USA. 70: 591-597.

Castillo, M., Flores, M., Mavingui, P., Martínez-Romero, E., Palacios, R., and Hernández, G. 1999. Increase in alfalfa nodulation, nitrogen fixation and, plant growth by specific DNA amplification in *Sinorhizobium meliloti. Appl. Environ. Microbiol,* 65: 2716–2722.

Chandler, D., Davidson, G., Grant, W.P., Greaves, J. and Tatchell, G.M. 2008. Microbial biopesticides for integrated crop management: An assessment of environmental and regulatory sustainability. *Trends Food Sci Tech,* 19: 275-83.

Chatel, D.L., Greenwood, R.M, and Parker C.A. 1968. Saprophytic competence as an important character in the selection of *Rhizobium* for inoculation. In: *Proceedings of the Ninth International Congress of Soil Science Adelaide.* 65-73, V2.

Chen, X., Schauder, S., Potier, N., Van, Dorsselaer, A., Pelczer, I., Bassler, B.L. 2002. Structural identification of a bacterial quorum-sensing signal containing boron. *Nature,* 415: 545-549.

Choong-Min, R., Farag, M.A., Chia-Hui, H., Reddy, M.S., Kloepper, J.W. and Pare, P.W. 2004. Bacterial volatiles induce systemic resistance in Arabidopsis. *Plant Physiol*, 134: 1017-26.

Choudhary, D.K., Johri, B.N. and Prakash, A. 2008. Volatiles as priming agents that initiate plant growth and defence responses. *Curr Sci*, 94(5): 595-604.

Contreras-Cornejo, H.A., Macías-Rodríguez, L., Cortes-Penagos, C. and Lopez-Bucio, J. 2009. *Trichoderma virens*, a plant beneficial fungus, enhances biomass production and promotes lateral root growth through an auxin-dependent mechanism in Arabidopsis. *Plant Physiol*, 149: 1579-92.

Crueger, W. and Crueger, A. 2003. Biotechnology: A textbook of industrial microbiology, 2nd ED, Panima publishing corporation, New delhi.

Danso, S.K.A. and Bowen, G.D. 1989. Methods of inoculation and how they influence nodulation pattern and notrogen fixation using two contrasting strains of *Bradyrhizobium japonicum*. *Soil Biol. Biochem*, 8: 1053–1058.

Das, A., Prasad, R., Srivastava, A., Giang, H.P., Bhatnagar, K. and Varma, A. 2007. Fungal siderophores: structure, functions and regulation. In: Soil Biology Volume 12 Microbial Siderophores (eds. A. Varma and S.B. Chincholkar). Springer-Verlag Berlin Heidelberg: 1-42.

Djonovic, S., Pozo, M.J., Dangott, L.J., Howell, C.R. and Kenerley, C.M. 2006. Sm1, a proteinaceous elicitor secreted by the biocontrol fungus *Trichoderma virens*induces plant defense responses and systemic resistance. *Mol Plant Microbe Interact*, 19: 838-853.

Duquenne, P., Chenu, C., Richard, G., and Catroux, G. 1999. Effect of carbon source supply and its location on competition between inoculated and established bacterial strains in sterile soil microcosm. *FEMS Microbiol, Ecol*. 29: 331-339.

Gentili, F. and Jumpponen, A. 2006. Potential and possible uses of bacterial and fungal biofertilizers. In: Handbook of Microbial Biofertilizers (ed. M.K. Rai). Food products press: 1-28.

Gentry, T.J., Rensing, C., and Pepper, I.L. 2004. New approaches for bioaugmentation as a remediation technology. *Critical Rev. Environ. Sci. Technol*, 34(5):447-494.

Gornall, J., Betts, R., Burke, E., Clark, R., Camp, J. and Willett, K. 2010. Implications of climate change for agricultural productivity in the early twenty-first century. *Philos Trans R SocLond B Biol Sci*, 365: 2973-89.

Gottfert, M., Rothlisberger, S., Kundig, C., Beck, C., Marty, R. and Hennecke, H. 2001. Potential symbiosis-specific genes uncovered by sequencing a 410-kilobase DNA region of the *Bradyrhizobium japonicum* chromosome. *J Bacteriol*, 183: 1405–1412.

Göttfert, M., Rothlisberger, S., Kundig, C., Beck, C., Marty, R., and Hennecke, H. 2001. Potential symbiosis-specific genes uncovered by sequencing a 410–kilobase DNA region of the *Bradyrhizobium japonicum* chromosome. *J Bacteriol*, 183: 1405-12.

Gupta, V., Satyanarayana, T. and Garg, S. 2000. General aspects of mycorrhiza. In: Mycorrhizal Biology (eds. K.G. Mukerji, J. Singh and B.P. Chamola). Kluwer Academic/Plenum Plublishers, New York: 27-44.

Han, J., Sun, L., Dong, X., Cai, Z., Sun, X. and Yang, H. 2005. Characterization of a novel plant growth-promoting bacteria strain *Delftiatsuruhatensis* HR4 both as a diazotroph and a potential biocontrol agent against various plant pathogens. *Syst Appl Microbiol,* 28: 66-76.

Hasan, M.A. 2013. Investigation on the nitrogen fixing cyanobacteria (BGA) in rice fields of North-West region of Bangladesh. III. Filamentous (heterocystous). *J Environ Sci Nat Resour.* 6: 253-259.

Higa, T. and Wididana, G.N. 1991. The concept and theories of effective microorganisms. In: Higa, T, Parr, JF, editors. Beneficial and Effective Microorganisms for a Sustainable Agriculture and Environment. Atami, Japan: International Nature Farming. Research Centre.

Howieson J.G., Malden, J., Yates, R.J., and O'Hara, G.W. 2000a. Techniques for the selection and development of elite inoculant rhizobial strains in southern Australia. *Symbiosis,* 28: 33-48.

Jahanian, A., Chaichi, M.R., Rezaei, K., Rezayazdi, K. andKhavazi, K. 2012. The effect of plant growth promoting rhizobacteria (PGPR) on germination and primary growth of artichoke (*Cynarascolymus*). *Int J Agric Crop Sci,* 4: 923-9.

Jansson, J.K., Trevors, J.T., editors. Modern Soil Microbiology. New York, NY: CRC Press; p. 237-261.

Jeffries, P.1995. Biology and ecology of mycoparasitism. *Can J Bot,* 73: 1284-90.

Jeffries, P. and Barea, J.M. (2001). Arbuscular mycorrhiza: A key component of sustainable plant-soil ecosystems. In: Hock B, editor. The Mycota: Fungal Associations. Vol. IX. Berlin, Heidelberg, New York: Springer.

Johansson, F.J., Paul, R.L. and Finlay D.R. 2004. Microbial interactions in the mycorrhizosphere and their significance for sustainable agriculture. *FEMS Microbiology Ecology,* 48:1-13.

Kannaiyan, S. 2002. Biofertilizers for sustainable crop production. In: Biotechnology of Biofertilizers: (ed. S. Kannaiyan). Kluwer Academic Publishers: 9-50.

Kapoor, R., Kumar, A., Kumar, A., Patil, S. and Kaur, A.P.M. 2012. Indole acetic acid production by fluorescent Pseudomonas isolated from the rhizospheric soils of Malus and Pyrus. *Recent Res Sci Technol,* 4:6-9.

Khan, A., Jilani, G., Akhtar, M.S., Naqvi, S.M. and Rasheed, M. 2009. Phosphorus solubilizing bacteria: Occurrence, mechanisms and their role in crop production. *J Agric Biol Sci,* 1:48-58.

Kloepper, J.W., Gutierrez-Estrada, A. and Mclnroy, J.A. 2007. Photoperiod regulates elicitation of growth promotion but not induced resistance by plant growth-promoting rhizobacteria. *Can J Microbiol,* 53:159-67.

Li, Y., Fu, K., Gao, S., Wu, Q.,Fan, L., Li,Y. and Chen, J. 2013. Increased virulence of transgenic *Trichoderma koningi* **strains to the Asian corn borer larvae by** overexpressing heterologous *chit42* **gene with chitin-binding domains.** *J. envt. Sci, health,* 48(5): 376-383.

Li, Y.Y. 2012. Construction of Transgenic Trichoderma Koningi With Chit42 of Metarhizium Anisopliae and Analysis of Its Activity Against the Asian Corn Borer. *J Environ Sci Health B,* 47(7): 622-630.

Liu, D., Lian, B. and Dong, H. 2012. Isolation of Paenibacillus sp. and assessment of its potential for enhancing mineral weathering. *Geomicrobiology J,* 29: 413-21.

Liu, W.T., Marsh, T.L., Cheng, H. and Forney, L.J. 1997. Characterization of microbial diversity by terminal restriction length polymorphisms of genes encoding 16S rRNA. *Appl Environ Microbiol,* 63: 4516–4522.

Marasco, R., Rolli, E., Ettoumi, B., Vigani, G., Mapelli, F. and Borin, S. 2012. A drought resistance-promoting microbiome is selected by root system under desert farming. *PLoS One,* 7(10): 1-14.

Marin, M. 2006. Arbuscular mycorrhizal inoculation in nursery practice. In: Handbook of Microbial Biofertilizers (ed. M.K. Rai), Food products press: 289-324.

Mejia, L.C., Rojas, E.I., Maynard, Z., Bael, S.V., Arnold, A.E. and Hebbar, P. 2008. Endophytic fungi as biocontrol agents of Theobroma cacao pathogens. *Biol Control,* 46: 4-14.

Mia, M.A., Shamsuddin, Z.H., Wahab, Z. and Marziah, M. 2010. Effect of plant growth promoting rhizobacterial (PGPR) inoculation on growth and nitrogen incorporation of tissue-cultured Musa plantlets under nitrogen-free hydroponics condition. *Aust J Crop Sci,* 4: 85-90.

Miao, B., Stewart, B.A. and Zhang, F. 2011. Long-term experiments for sustainable nutrient management in China. A review. *Agron Sustain Dev,* 31: 397-414.

Mohammadi, K. 2012. Phosphorus solubilizing bacteria: Occurrence, mechanisms and their role in crop production. *Resour Environ,* 2: 80-85.

Monte, E. 2001. Editorial paper: Understanding Trichoderma: Between Agricultural Biotechnology and Microbiology Ecology. *Int Microbio,* 4:14.

Mpepereki S., Javaheri F., Davis P., and Giller K.E. 2000. Soybeans and sustainable agriculture. Promiscuous soyabeans in southern Africa. *Field Crops Res,* 65: 137–50.

Murali, M., Amruthesh, K.N., Sudisha, J., Niranjana, S.R. and Shetty, H.S. 2012. Screening for plant growth promoting fungi and their ability for growth promotion and induction of resistance in pearl millet against downy mildew disease. *J Phytology,* 4: 30-36.

Muyzer, G. and Smalla, K. 1998. Application of denaturing gradient gel electrophoresis (DGGE) and temperature gradient gel electrophoresis (TGGE) in microbial ecology. *Antonie Van Leeuwenhoek,* 73:127-141.

Nihorimbere, V., Ongena, M., Smargiassi, M. and Thonart P. 2011. Beneficial effect of the rhizosphere microbial community for plant growth and health. *Biotechnol Agron Soc Environ*, 15: 327-37.

Oresnik, I.J., Twelker, S., and Hynes, M.F. 1999. Cloning and characterization of a *Rhizobium leguminosarum* gene encoding a bacteriocin with similarities to RTX toxins. *Appl. Environ. Microbiol*, 65: 2833–2940.

Ortiz-Castro, R., Contreras-Cornejo, H.A., Macías-Rodríguez, L. and Lopez-Bucio, J. 2009. The role of microbial signals in plant growth and development. *Plant Signal Behav*, 4: 701-712.

Panhwar, Q.A., Radziah, O., Zaharah, A.R., Sariah, M. and Razi, I.M. 2011. Role of phosphate solubilizing bacteria on rock phosphate solubility and growth of aerobic rice. *J Environ Biol*, 32: 607-612.

Papavizas, 1985. Trichoderma and Gliocladium: Biology, Ecology, and Potential for biocontrol. *Annual Review of Phytopathology*, 23: 23-54.

Paul, E.A. 2007. Soil microbiology, ecology, and biochemistry, 3rd ed. Academic Press: San

Phillips, D.A., Streit, W.R., and Volpin, H. 1996. Applying rhizobia to the future. In: *Recent Progress in Research on Symbiotic Nitrogen Fixation*, Pate, J.S., Ed., CLIMA Occasional Publication 14.

Pinton, R., Varanini, Z. and Nannipieri, P. 2001. The rhizosphere as a site of biochemical interactions among soil component, plants and microorganisms. In: The rhizosphere: biochemistry and organic substances at the soil plant interface, bacteria in soil. *Res Microbiol*, 152: 707–716.

Reed, M.L.E. and Glick, B.R. 2005. Growth of canola (Brassica napus) in the presence of plant growth-promoting bacteria and either copper or polycyclic aromatic hydrocarbons. *Can. J. Microbiol*, 51(12): 1061-1069.

Reed, M.L.E., Warner, B.G. and Glick, B.R. 2005. Plant growth-promoting bacteria facilitate the growth of the common reed *Phragmites australis* in the presence of copper or polycyclic aromatic hydrocarbons. *Curr. Microbiol*, 51(6): 425-429.

Rinaldi, A.C., Comandini, O. and Kuyper, T.W. 2008. Ectomycorrhizal fungal diversity: separating the wheat from the chaff. *Fungal Diversity*, 33: 1-45.

Rodriguez, H. and Fraga, R. 1999. Phosphate solubilizing bacteria and their role in plant growth promotion. *Biotechnol Adv*, 17: 319-39.

Rodriguez, H., Vessely, S., Shah, S. and Glick, B.R. 2008. Effect of a nickel-tolerant ACC deaminase-producing Pseudomonas strain on growth of nontransformed and transgenic canola plants. *Curr Microbiol*, 57(2): 170-174.

Roossinck, M.J. 2011. The good viruses: Viral mutualistic symbioses. *Nat Rev Microbiol*, 9:99-108.

Rynne, F.G., Glenn, A.R., and Dilworth, M.J. 1994. Effect of mutations in aromatic catabolism on the persistence and competitiveness of *Rhizobium leguminosarum* bv. *trifolii*. *Soil Biol Biochem,* 26: 703–710.

Saba, H., Vibhash, D., Manisha, M., Prashant, K.S., Farhan, H. and Tauseef A. 2012. *Trichoderma* - A promising plant growth stimulator and biocontrol agent. *Mycosphere*. 3:524-531.

Salvioli, A., Ghignone, S., Novero, M., Navazio, L., Venice, F., Bagnaresi, P. 2016. Symbiosis with an endobacterium increases the fitness of a mycorrhizal fungus, raising its bioenergetic potential. *The ISME J,* 10(1): 130-144.

Sarmah, S.R., Dutta, P., Begum, R., Tanti, A.J., Phukan, I. and Debnath, S. 2005. Microbial bioagents for controlling diseases of tea. Proceeding International Symposium on Innovation in Tea Science and Sustainable Development in tea Industry. *China Tea Sci Soc Unilever Hangzhou China,* 58: 767-776.

Saxena, J., Basu, P., Jaligam, V. and Chandra, S. 2013. Phosphate solubilization by a few fungal strains belonging to the genera Aspergillus and Penicillium. *Afr J Microbiol Res,* 7:4862-4869.

Sessitsch, A., Howieson, J. G., Perret, X., Antoun, H., and Martínez-Romero, E. 2002. Advances in Rhizobium research. *Crit Rev Plant Sci,* 21: 323-78.

Sharma, S.B., Sayyed, R.Z., Trivedi, M.H. and Gobi TA. 2013. Phosphate solubilizing microbes: Sustainable approach for managing phosphorus deficiency in agricultural soils. *Springer Plus,* 2(587): 1-14.

Shridhar, B.S. 2012. Review: Nitrogen fixing microorganisms. *Int J Microbiol Res,* 3:46-52.

Siddiqui, Z.A. and Pichtel, J. 2008. Mycorrhixae: an overview. In: Mycorrhizae: Sustainable Agriculture and Forestry (eds. Z.A. Siddiqui, M.S. Akhtar and K. Futai). *Springer,* 1-36.

Singh, R., Singh, P. and Sharma, R. 2014. Microorganism as a tool of bioremediation technology for cleaning environment: A review. *Proc Int Acad Ecol Environ Sci,* 4:1-6.

Smit, G., Swart, S., Lugtenberg, B.J.J., and Kijne, J.W. 1992. Molecular mechanisms of attachment of *Rhizobium* bacteria to plant roots. *Mol. Microbio,* 6: 2897-2903.

Smith, S.E. and Zhu, Y.G. 2001. Application of arbuscular mycorrhizal fungi: Potentials and challenges. In: Bio-Exploitation of Filamentous Fungi (eds. B.P. Stephen and K.D. Hyde) Fungal Diversity Research Series, 6: 291-308.

Sood, A., Sharma, S., Kumar, V. and Thakur, RL 2007. Antagonism of dominant bacteria in tea rhizosphere of Indian Himalayan regions. *J Appl Sci Environ Manage,* 11: 63-66.

Sparks, D.L., Huang, P.M. 1985. Physical chemistry of soil potassium. In: Munson RD, editor. Potassium in Agriculture. Madison, WI: American Society of Agronomy. 201-76.

Staley, T.E. and Brauer, D.K. (2006). Survival of a genetically modified root-colonizing pseudomonad and rhizobium strain in an acidic soil. *Soil Sci. Soc. Am. J*, 70(6): 1906-1913.

Steeghs, M., Bais, H.P., Gouw, D.J., Goldan, P., Kuster, W. and Northway, M. 2004. Proton-transfer-reaction mass spectrometry as a new tool for real time analysis of root secreted volatile organic compounds in Arabidopsis. *Plant Physio*, 135: 47-58.

Tchan, Y.T. and New, P.B. 1989. *Azotobacteraceae.* In: Holt JG, Williams & Wilkins (Eds.) Bergey's Manual of Systematic Bacteriology Volume 1. Baltimore, USA, pp. 220-229.

Thakur, M. and Sohal, B.S. 2013. Role of elicitors in inducing resistance in plants against pathogen infection: A review. *ISRN Biochem*, 2013: 1-10.

Torsvik, V. and Ovreas, L. 2002. Microbial diversity and function in soil: from genes to ecosystems. *CurrOpinMicrobiol*, 5: 240–245.

Triplett, E.W. and Barta, T.M. 1987. Trifolitoxin production and nodulation are necessary for the expression of superior nodulation competitiveness by *Rhizobium leguminosarum* bv. trifolii T24. *Plant Physiol*, 85: 335– 342.

Turner, T.R., James, E.K. and Poole, P.S. 2013. The plant microbiome. *Genome Biol*, 14(209): 1-10.

Van Dillewijn, P., Soto, M.J., Villadas, P.J., and Toro, N. 2001. Construction and environmental release of a *Sinorhizobium meliloti* strain genetically modified to be more competitive for alfalfa nodulation. *Appl Environ Microbiol*, 67: 3860–3865.

Vessey, J.K. 2003. Plant growth promoting rhizobacteria as biofertilizers. *Plant Soil*, 255: 571-86.

Yeasmin, T., Zaman, P., Rahman, A., Absar, N. and Khanum, N.S. (2007). Arbuscular mycorrhizal fungus inoculum production in rice plants. *African J Agri Res*, 2: 463- 467.

12

Classical Biological Control: A Sustainable Approach for Insect Control

Prakash Chand Yadav[1*], K. C. Sharma[2] and A. K. Yadav[3]

[1]Department of Entomology, MPUAT, Udaipur - Rajasthan

[2]National Institute of Biotic Stress Management, Raipur - Chhattisgarh

[3]ICAR-Central Soil Salinity Research Institute, Karnal - Haryana

*Corresponding author: jadamprakash@gmail.com

Abstract

The use of beneficial organisms is a very important tactic in integrated pest management strategies of insect pests over the world. This chapter describes the approaches to using biological control and a historical perspective of each. The intent of biological control is not to eradicate pests, but to keep them at tolerable levels at which they cause no appreciable harm. In fact, because natural enemies require their prey or hosts for survival, biological control works best when there is always a small population of pests to sustain their natural enemies. This chapter describes these developments and the different types of approaches that have been used to implement biological control as a useful tool in IPM. It also describes how biological control interacts with other IPM tactics, and the potential for better integration into IPM programs.

Keywords: Beneficial organisms, Importation biocontrol History, Augmentation, Conservation biocontrol, Predators

Introduction

Biological control has been a valuable tool in pest management programs over the world for many years, but has undergone a resurgence in recent decades that parallels the development of IPM as an accepted practice for pest management. This chapter is not intended to be an exhaustive review of research involving biological control. Instead, it will try to focus on implementation of biological control practices in insect pest management programs. It will begin with an overview of the general concepts and challenges facing the use of beneficial organisms within each of the general approaches to biological control. A detail historical perspective of biological control follows. Next, the interaction of biological control with the different tactis of integrated pest management programs is considered. Existing implementation, as well as effective uses of biological control in pest management program are also considered.

Approaches to Biological Control

Natural enemies have been utilized in the pest management for centuries. However, this last 100 years has seen a dramatic increase in their use as well as our understanding of how they can better be manipulated as part of effective, safe, pest management systems. Recent advances in molecular systematics are shedding new light on classification of groups of beneficial insects such as the Hymenoptera (e.g. Sharkey, 2007), and delivery of this information on the internet makes it quickly and widely available (e.g. The Tree of Life Web Project at http://tolweb.org). Recent advances in the study of beneficial organism behavior (e.g. parasitoid foraging: van Nouhuys and Kaartinen, 2008) and reproductive biology (e.g. symbionts in parasitoids: Clark, 2007) are revealing surprising complexities in the life histories of these organisms. Understanding this complexity should lead to potential new methods for their uses. Despite the long history of utilizing natural enemies, it wasn't until 1919 that the term biological control was apparently used for the first time by the late Harry Smith of the University of California (Smith, 1919). There has been debate regarding the scope and definition of biological control brought about by technological advances in the tools available for pest management. In this chapter we will follow the definition presented by DeBach (1964) as the "study, importation, augmentation, and conservation of beneficial organisms to regulate population densities of other organisms". Biological control efforts conducted with predators and parasitoids still can be organized under three general approaches: importation, augmentation and conservation of natural enemies (Debach, 1964; Bellows and Fisher, 1999). Each of these approaches has been used to varying degrees in integrated pest management.

Importation Biological Control: Classical biological control (also called **importation of naturalenemies**) involves the importation, screening, and release of natural enemies to permanently establish effective natural enemies in new areas. Classical biological control usually targets introduced (**non-native**) pests, most of which arrive here without the natural enemies that control their populationsin their native lands.

It usually involves the re-uniting of natural enemies with pests that have escaped them into a new geographical area. Although the practice of introducing biocontrol agents from a related host species for the control of native arthropod pests has been used, in some cases effectively, this approach has been strongly criticized for its potential non-target effects.

Augmentation Biological Control: Augmentation biological control includes activities in which natural enemy populations are increased through mass culture, periodic release (either inoculative or inundative) and colonization, for suppression of native or non-native pests.

i) **Inoculative release:** Control expected from the progeny and subsequent generations only.

ii) **Inundative release:** Natural enemies mass cultured and released to suppress pest directly e.g. Trichogramma sp. egg parasitoid, Chrysoperlacarniapredato

Non-Target Impacts Augmentation biological control often utilizes exotic natural enemy species that have broad host ranges, and undoubtedly have some effect on populations of nontarget insects. However, augmentation does not face the same scrutiny as importation biocontrol over these potential non-target impacts. This is at least in part due to the temporary, non-persistent activity of released natural enemies (Lynch and Thomas, 2000; van Lenteren*et al.*, 2006).

Conservation Biological Control Conservation biological control seeks to understand human influences on resident natural enemies in a system, then manipulate those influences to enhance the ability of natural enemies to suppress pests. DeBach (1964) considered conservation biological control to be environmental modification to protect and enhance natural enemies.

Classical biological control involves numerous logistical and financial challenges related to exploring foreign lands for natural enemies, getting candidate natural enemies to a quarantine facility alive, and successfully rearing them in quarantine during the evaluation process.As a result, biological control workers often have a limited selection of candidate natural enemies with which to work.However, to the extent possible, natural enemies chosen for use in a classical biological control program should have these attributes:

- have a good ability to seek out and aggregate near the target pest
- have a life cycle well synchronized with that of the target pest
- be well adapted to local climate, including the ability to survive the winter
- have a short development time relative to that of the target pest
- engage in host-feeding, if a parasitic wasp
- disperse well
- attack a limited range of hosts or prey

Classical biological control achieved in India

- **1795-** Cochineal insect, *Dactylopiusceylonicus* was introduced from Brazil against carmine dye producing insect, *D. coccus.*

- 1983-1984- Exotic weevil, *C. Salviniae* from Australia against water fern, *Salviniamolesta* in a lily pond in Bangalore.
- 1982- Three exotic natural enemies were introduced viz., hydrophilic weevils –*Neochetinabruchi* (Ex. Argentina) and *N. eichhorniae* (Ex. Argentina) and galumnid mite *Orthogalumnaterebrantis* (Ex. Sout h America) against water hyacinth.
- 1926- The coccinellid beetle, *Rodoliacardinalis* against cottony cushion scale, *I.Purchasi*
- 1983- The encyrtid parasitoid Leptomastixdactylopii against *Planococcuscitri* and *P. lilacinus* from Trinidad, West Indies
- 1983-A chrysomelid beetle *Zygogrammabicolorata* against *parthenium* from Mexico
- 1988- The coccinellid predator, *Curinuscoeruleus* against *H. cubana* from Thailand
- 1921- the agromyzidseedfly, *Ophiomyialantanae* against *Lantana camara* from Hawaii (origin: Mexico) and released in south India
- 1941- Tingid lace bug, *Teleonemiascrupulosa,* against *L. camara* from Australia
- 1951- *C. Montrouzieri* against mealybugs
- 1963-The gallfly, *Procecidocharesutilis* against Crofton Weed, *Ageratinaadenophora* from New Zealand to Nilgiris (Tamil Nadu), Darjeeling and Kalimpong areas (West Bengal)
- 2010-Three exotic encyrtid parasitoids viz., *Acerophaguspapayae, Anagyrusloecki* and *Pseudleptomastixmexicana,* against papaya mealybug, *Paracoccusmarginatus*

Development of biological control and classical examples of biological Control

1. Early History

A. 200AD to 1200 AD :BC agents were used in augmentation

900 AD- First use of red ant, *Oecophyllas maragidna* to control leaf chewing insects on mandarin trees

1200AD-Ants were used for control of date palm pests in Yemen (south of SaudiaArabia).

Usefulness of ladybird beetles recognized in control of aphids and scales

B. 1300A.D. to 1799 A.D. : BC was just beginning to be recognized

1602 - Aldrovandi noted the hymenopteran parasite, *Apantelesglomeratus* laying eggs in the pupae of the cabbage butterfly, *Pierisbrassicae*

1726- The first insect pathogen was recognized by de Reaumur. It was a *Cordyceps* fungus on a noctuid

1762 - Indian mynah bird, *Graculareligiosa* exported from India to Mauritius to control red locust, *Nomadacrisseptemfasciata*

1776- Control of the bedbug, Cimexlectularius, was successfully accomplished by release of the predatory pentatomid ,*Picromerusbidens* in Europe

C. 1800 A.D. to 1849 A.D. : During this period advances were made in Europe which were both applied and basic

II. The Intermediate Period: 1888 to 1955

A. 1888- Vadalia beetle, *Rodoliacardinalis* was brought from Australia and introduced into California (Control) cottony cushion scale, *Iceryapurchasi* on citrus

It's a first well planned and successful classical biological control attempt made

The Cottony Cushion Scale Project

- In 1868 Cottony cushion scale, IceryapurchasiMaskell, was introduced into Californiain ca. around the Menlo Park (CA) area (near San Francisco) by 1887 it spread tosouthern California.
- C. V. Riley (Chief of the Division of Entomology, USDA) employed Albert Koebele and D. W. Coquillett in research on control of the cottony cushion scale and found nomethods to control.
- In1888Koebele was sent to Australia to collect natural enemies of the scale,He sent ca.12,000 individuals of *Cryptochaetumiceryae* and 129 individuals of *Rodoliacardinalis*(the vedalia beetle)
- 1898- Austalian*Cryptlaemusmontrouzieri* in India on *Coccusviridis*

B. 1900 to 1930: New faces and more Biocontrol projects

1902- The Lantana Weed Project in Hawaii (1902) First published work on BC ofweeds.

1911- Berliner described *Bacillus thuringiensis* as causative agent of bacterialdisease of the Mediterranean flour moth

1919- USDA laboratory for biological control established in France .

1927- The Imperial Bureau of Entomology created the Farnham House Laboratoryfor BC work in England .

C. 1930 to 1955: Expansion and decline of Biocontrol

- 1930 to 1940- Peak in BC activity in the world (57 different natural enemies established)
- WorldWar II caused a sharp drop in BC activity with switch to pesticide research
- 1920 - A parasitoid *Aphelinusmali* introduced from England into India to control Woolly aphid on Apple, *Eriosomalanigerum.*
- 1929-31 - *Rodoliacardinalis* imported into India (from USA) to control cottony cushion scale*Iceryapurchasi* on Wattle trees.
- 1947- The Commonwealth Bureau of Biological Control (CBBC) was established
- 1951- CBBC renamed as Commonwealth Institute for Biological Control (CIBC). Headquarters are currently in Trinidad, West Indies.
- 1955- The Commission Internationale de Lutte Biologiquecontre les Enemis des Cultures(CILB) was established.
- 1962- The CILB changed its name to the Organisation Internationale de Lutte Biologiquecontre les Animauxet les Plants Nuisibles.Also known as the International Organization for Biological Control (IOBC).

III. The Modern Period: 1957 to Present

1958-60 - Parasitoid *Prospatellaperniciosus* imported from China

1960 - Parasitoid *Aphytisdiaspidis* imported from USA

Both parasitoids used to control Apple Sanjose scale *Quadraspidiotusperniciosus*

1964 - Egg parasitoid Telenomus sp. imported from New Guinea to controlCastor semilooper*Achaea janata*

1964- Paul DeBach and Evert I. Schliner (Division of Biological Control,University of California, Riverside) published an edited volume titled "Biological Control of Insect Pests and Weeds"

1965 - Predator *Platymerislaevicollis* introduced from Zanzibar to control coconut Rhinoceros beetle, *Oryctes rhinoceros*

Biological Control agents:

A. **Predators:** A predator is catches and devours smaller or more helpless creatures by killing them in single meal

Types of predators

- 1. **Entirely predatory:** Eg: Ant lions (lace wings), tiger beetles, lady bird beetles except brinjal ladybird beetles.
- 2. **Mainly predatory, but occasionally harmful:**Eg. Odonata (Dragon flies) and Mantids, occasionally attack honeybees
- 3. **Mainly harmful but partly predatory:** Eg: Cockroach feeds on termites, Adult blister beetles feeds on flowers, but grubs predate on hopper eggs.
- 4. **Mainly scavenging and partly predatory:** Eg: Earwigs feed on decaying organic matter and also predate on fly maggots, both ways, they are helpful.
- 5. **Variable feeding habits of predator :** Tettigonidae (Omnovorous and Cornovorous), but damage crops by laying eggs.
- 6. **Stinging predators:** Nests are constructed and stocked with prey, which have been stinging and paralyzing the mother insect on which eggs are laid and then sealed up.. Eg: Spider Wasps and Wasps.

Qualities of insect predators:

1. **Narrow host range:** Generalized predators may be good natural enemies but they don't kill enough pests when other types of prey are also available
2. **Climatic adaptability:** Natural enemies must be able to survive the extremes of temperature and humidity that they will encounter in the new habitat.
3. **Synchrony with host (prey) life cycle:** The predator should be present when the pestfirst emerges or appears.
4. **High reproductive potential:** Good biocontrol agents produce large numbers of offspring.
5. **Efficient search ability:** In order to survive, effective natural enemies must be able tolocate their host or prey even when it is scarce. In general, better search ability resultsin lower pest population densities.
6. **Short handling time:** Natural enemies that consume prey rapidly or lay eggs quicklyhave more time to locate and attack other members of the pest population.
7. **Survival at low host (prey) density:** If a natural enemy is too efficient, it may eliminateits own food supply and then starve to death.
8. **Efficient eating ability:** They should kill or consume much prey.
9. **All should be predatory:** Males, females, immatures, and adults may be predator
10. **Should prey on all host stages:** They attack immature and adult prey

Important predators:

Order	Family	Scientific name	Prey
Coleoptera	**Coccinellidae**	*Coccinellaseptumpunctata*	Aphids
		Coccinellarependa	Aphids
		Scymnuscoccivora	*Pseudococcus sp.* Mealy bugs
		Rodoliacardinalis	Iceryapurchasi Cottony cushiony scale
		Harmoniaaxyridis (multi-colored Asian lady beetle)	Aphids, scales, and Psyllids
	Carabidae	*Parentalacticincta*	Coconut black headed caterpillar
		Ophiomea spp.	Rich BPH and Leaf folder
	Cicindellidae	*Cicindela spp.*	Soil dwelling insects
Neuroptera	Chrysopidae	Chrysoperlacarnea (Green lace wing)	Aphids, Scales, Bollworms, Mealy bugs
	Myrmeliontidae (Antlions)	*Myrmelion spp.*	Ants and soil dwelling insects
Hemiptera	Reduviidae (Assasin Bugs)	*Rhinocorisfuscipes*	*Helicoverpaarmigera*
		Platymerislaevicollis	Coconut rhinoceros beetle (*Oryctes rhinoceros)*
	Miridae (Mirid Bugs)	*Cyrtorhinuslividipennis*	Rice BPH (*Nilaparvathalugens*)

Other predators: Dragon flies, Damsel flies, Preying mantids, Giant water bug, Robber flies, Hover flies (Syrphids), Wasps and ants.

Parasitoids

A parasitoid is an organism that spends a significant portion of its life history attachedto or within a single host organism, which it ultimately kills (often consumes) inthe process.

Kinds of Parasitism:

Simple parasitism: The parasite attacks the host only once. *Apantelestaragame*on the larvae of *Opisinaarenosella; Gonizusnephatidis*on the larvae of *Opisinaarenosella.*

Super parasitism:A host has multiple attacks from parasites of the same species. Many individuals of the same species of parasite attack a single host. *Apantelesglomeratus*on *Pierisbrassicae, Trichospiluspupivora*on *Opisinaarenosella.*

Multiple parasitism: When more than one species of the parasite attckes the same hostsimultaneously. Eggs of *Scirpophagaincertulas*are attacked by *Trichogramma spp.Telenomus Spp. and Tetrastichus spp.*

Hyper parasitism: When the parasite itself is attacked by another parasite. *Gonozusnephantidis*is parasitised by *Tetrastichusisraeli.* Most of the braconids and Bethylidsare hyper-parasites.

Adelphoparasitism: Female develops as a primary parasitoid, but the male is a secondary parasitoid through females of its own species.

Cleptoparasitism:A parasitoid attacking a host, already parasitized by another species of parasitoid.

Parasitoids of agricultural importance

Parasitoids of agricultural importance mostly belongs to hymenoptera (90%) and Diptera (10%).

Successful examples of parasitoids

1. Apple wooly aphid (*Eriosomalangerum*) in Cooner area in TN was successfully controlled by *Aphelinusmali*(adult endoparasitoid) during 1940
2. Coconut black headed caterpillar *Opisinaarenosella*was successfully controlled usingpotential parasitoids (TABGET) parasitizing various stated of the pest. This is still inuse in certain parts of Godavari districts, and other parts of TN and Kerala.Egg parasitoid *Trichogrammaaustralicum*Early larval parasitoid *Apantelestaragame*Mid Larval parasitoid *Braconhebetor*Late Larval Parasitoid *Goniozusnephantidis*Pre Pupal Parasitoid *Elasmusnephantidis*Pupal Parasitoid *Trichospiluspupivora*
3. Control of shoot borers of sugarcane, stem borers of paddy and sorghum with eggparasitoid *Trichogrammaaustralicum*@ 50,000/acre/week for 4-5 weeks from onemoth after transplanting.
4. Cotton pink bollworm (*Pectinophoragossypiella*) can be successfully controlled by*Trichogrammachilonis*@20,000/acre/week from flowering to boll formation.
5. Against *Helicoverpaarmigera*on cotton, release of *Trichogrammachilonis*@50,000/acre/week for 4 weeks from 30-40 DAS gave good results.
6. Sugarcane top shoot borer (*Triporyzanivella*) can be controlled successfully releasinglarval parasitoid *Isotomajavensis.*
7. Castor semilooper *Achaea janata*naturally controlled by larval parasitoid *Microplitisophiusae.*

Conclusions

Since the late 1940s, insect control has relied heavily on synthetic chemical insecticides. Insecticides are relatively easy to use and have generally provided safe and effective pest control.They will certainly continue to be a component of most pest management programs.Many newer pesticides made available in the past decade or so are more selective and less hazardous than most of the older compounds. Nevertheless,most insecticides have at least some of these undesirable attributes: they usually present some degree of hazard to the applicator and other people who may come in contact with them; they can leave residues that some find unacceptable; they can contaminate soil and water and affect wildlife, aquatic life, and other nontarget organisms; they can interfere with beneficial organisms, such as pollinating insects and the natural enemies of pests; and insects can develop resistance to insecticides, effectively eliminating those materials as pest management options. In addition, organic standards prevent the small but rapidly growing number of organic growers and processors from using synthetic chemicals. For these reasons, many farmers and gardeners are exploringand adopting methods that reduce pesticide use.

References

Bellows, T.S. and Fisher, T.W. 1999. *Handbook of Biological Control*. Academic Press, San

DeBach, P. 1974. *Biological Control by Natural Enemies*. Cambridge University Press, London.

Lynch, L.D. and Thomas, M.B. 2000. Nontarget effects in the biocontrol of insects with insects,

Management 49(3):205–214.

Sharkey, M.J. 2007.Phylogeny and Classification of Hymenoptera. In: Zhang, Z.Q. and Shear, W.A. (eds), *Linnaeus Tercentenary: Progress in Invertebrate Taxonomy. Zootaxa,*1668: 521–548.

Smith, H.S. 1919. On some phases of insect control by the biological method. *Journal of Economic Entomology,* 12:288–292.

vanLenteren, J.C. (ed). 2003a. *Quality Control and Production of Biological Control Agents:*

Theory and Testing Procedures. CABI Publishing, Wallingford.

Van Nouhuys, S. and Kaartinen, R. 2008. A parasitoid wasp uses landmarks while monitoring

potential resources. *Proceedings of the Royal Society, B* 275:377–385.

13

Plant Growth Regulators: A Timely Approach for Sustainable Agriculture

Priyanka Pal, Kuldeep Yadav and Narender Singh*

Department of Botany, Kurukshetra University, Kurukshetra, Haryana 136119, India

**Corresponding author: nsheorankuk@yahoo.com*

Abstract

Growing population projections have indicated that approximate increase in world population could be 10 billion by 2050, as par increment of 1.2 per cent annually which in turn upsurge the food consumption. Sustainable crop production increase by understanding physiological and biochemical regulation of plant growth and development in regards to crop yield boosting agronomic techniques likewise plant growth regulators, nutrients and disease resistant hybrid plants that help in increasing the economic yield per unit area. The higher-value staple crops contributed in displacing the traditional crops including wild leafy vegetables and legumes that were essential sources of critical micronutrients likewise vitamin A, iron and zinc etc. Intake of vegetables in the daily diet hold overall sound health, improvement in vision and cut down hazard of various deadly diseases. Gibberellic acid(GA_3) regarded as environmental friendly PGR as it affect stem elongation, increased number of leaves, leaf area, photosynthesis, nitrogen utilization, mineral uptake, enzyme activity, sugar synthesis and shelf life. Gibberellic acid has great potentialities to promote sustainable agriculture production by improving plant growth, yield and quality attributes.

Keywords: Sustainable, Crop, Vegetable, Gibberellic Acid, Yield.

Introduction

As the world population increases exponentially, much more effort and innovation will be mandatory in order to achieve sustainably increase agricultural production, upgrade the global supply chain to eradicate hunger within next generations. India being an agriculture country; the agriculture progress will be most proficient social assurance against hunger and poverty. There is need to employ new prospective concerning agriculture productivity which is not only profitable but also socially and environmentally sustainable. Green revolution program introduce for improving crop yield in rural and agricultural countries like India (Swaminathan, 2007). Green revolution prevented the world from immense starvation by increasing the staple food crops (wheat, rice and maize) production. Downside of green revolution has also lead to lesser intake of vital micronutrients by displaced traditional crop including wild leafy vegetables and legumes which are essential sources of critical micronutrients likewise vitamin A, iron and zinc (Webb, 2009).

Vegetables are important component of agriculture and nutritional security owing to their short duration, nutritional richness, high yield, economic vitality and capacity to procreate on-farm and off-farm employment. Vegetables supplemented several vitamins (C, A, B1, B6, B9, E), plant fibers, mineral elements, carbohydrates and proteins in human diet (Dias and Ryder, 2011). It holds a vast variety of phytochemicals like flavonoids, alkaloids; tannins and steroids promulgate health benefits beyond basic nutrition. Various phytochemicals acting as strong antioxidants against free-radical damage by modification in metabolic activity and process altering the course of tumour cells conquer numerous death dealing diseases (Herrera *et al.*, 2009). A World Health Report (2007) revealed that unbalanced diet with stumpy vegetable intake and less consumption of dietary fibre are estimated to account 2.7 million deaths per year, so contributing in top 10 risk factors leads to age related diseases and mortality. In human diet, vegetables are essential as impart overall sound health, improvement in vision and cut down hazard for some forms of cancer, heart disease, diabetes, stroke, anaemia, gastric ulcer, rheumatoid arthritis and other cardiovascular diseases (Mullie and Clarys, 2011).

India is second highest producer of world As per National Horticulture Database published by National Horticulture Board, 169.478 million metric tonnes of vegetables were produced in India under the area 9.542 million hectares during 2014-15. China produces half of worldwide vegetables, ranks first in vegetable production. Major vegetable producing countries of the world are China (51.6% world production), India (14% world production), USA (3.1% world production), Turkey (2.2% world production) and Egypt (2% world production).Total vegetable exports from India was accounted for 4702.78 crores rupees in 2014-15, sharing 2.25% of total agricultural exports and 0.23 % of total national exports during the period. Nonetheless India shares only 0.1% in the global market.

Agriculture is regarded as the engine of national economic growth and poverty alleviation by government. Moreover, agriculture occupies a prominent position in nation's economic policies for its highest contribution in India's gross domestic product (GDP) and share 17.4% of total GDP during 2015-16 (Indian economic survey, 2016). Improvement in agriculture produce (cereals, vegetables and fruits etc.) will not only raise country's economic growth at faster rate but also proved to be beneficial for about 65% population depend on agriculture as their principal means of livelihood (Tyagi, 2012).Crop production increase by understanding physiological and biochemical regulation of plant growth and development in regards to crop yield boosting agronomic techniques that help in increasing the economic yield per unit area.

These days, researchers preferred the use of plant growth regulator (PGRs) that holds great prospect for meeting the demand of yield and crop quality. PGRs help in maintain sustainable agriculture as they are natural and synthetic compound, controlling the plant growth and development by dramatically change the level of endogenous hormones and active in minute amount. Endogenous plant growth regulators are known to control vital physiological and biochemical processes of plants (Sharma *et al.*, 2013). Growth and development in plants is regulated in a coordinated fashion by the activity of several plant growth hormones likewise indole acetic acid (IAA), gibberellins (GA), cytokinins, abscisic acid (ABA) and ethylene which mediate numerous vital physiological process.

History of Gibberellic Acid

Existence of gibberellic acid (GA_3) first came to the attention of Japanese scientist in 1926, which focused their studies on fungal disease named "bakane" (foolish seedling) in Japanese on rice plants grow by local farmers. The bakane disease induced many symptoms likewise stunting, seedling blight, root and crown rot and the classic sign of etiolation and abnormal elongation which oppose their own weight and die (Nicholson *et al.*, 1998). A couple of plant pathologist revealed that particular symptoms in rice crop possessed by distinct chemical secretion of Gibberella fujikuroi (Fusariuam moniliforme) the causative fungus of the 'Bakanae' disease. In 1930s Japanese scientist Yabuta and Hyashi, collect tainted crystal of compound retaining plant growth enhancing activity by culturing and assaying the culture filtrate of fungus Gibberella in laboratory, was named gibberellin A. The Tokyo university researcher in 1950s isolated and outline 3 distinct gibberellins from gibberellin A sample and recognised them gibberellin A_1, gibberellin A_2 and gibberellin A_3. Later on with the onset of World War II, studies on gibberellins cease in Japan but these discoveries highly acknowledged in world and fascinate international scientific interest. Imperial Chemical Industries in Britain and US Department of Agriculture (USDA) in Illinois started a new phase of research by illustrate the chemical compound that purified from Gibberella culture filtrate which contains physiological properties of gibberellin A isolated by the Japanese scientist, Yabuta and samuki (1938) and named "gibberellic acid". This compound had molecular formula $C_{19}H_{22}O_6$. Gibberellin A_3 produced by ICI enlighten as gibberellic acid (Takahashi, 1986). Other discoveries would confirm that higher plant also synthesized this hormone (Mitchell *et al.*, 1951). Isolation of the gibberellic acid from a plant extract involved in this response was achieved in the 1958 by discovery of GA_1 from immature seeds of Phaseolus cocineus. Presently more than a hundred Gibberellins identified from plant that are designate by their chemical structure instead of their biological activity (Davies, 2002). Only a scanty of them are consider to function as biologically active gibberellins such as GA_1, GA_3, GA_4 and GA_7 and numerous non biologically active gibberellins remain in plants as precursors of the bioactive forms (Yamaguchi, 2008). Subsequently gibberellic acid frequently produced commercial for agronomic, horticultural and other scientific purposes.

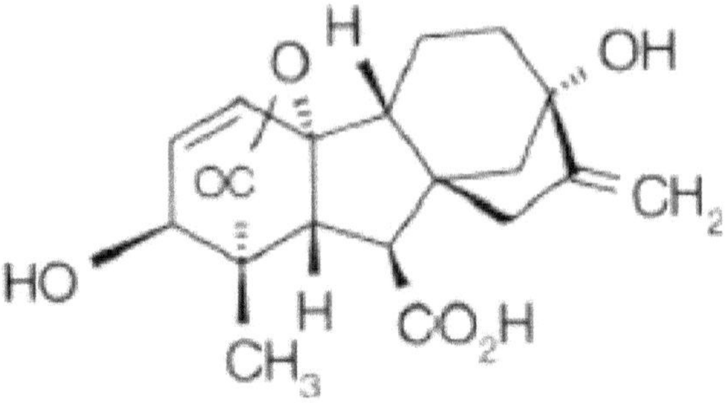

Fig. 1: Chemical structure of gibberellic acid

Biosynthesis of GA_3:

Gibberellins are tetracyclic endogenous phytohormones, comprise a huge family of diterpenoid acids. In 1983, Hanson revealed that the gibberellic acid biosynthesized as a secondary metabolite. Its biosynthetic pathway has been well documented by genetic and cell biology studies and using biochemical techniques. In plant, stem is the site of gibberellic acid synthesis (Vysot-Skaya *et al.*, 2007). All the diterpenes are synthesized from geranylgeranyl diphosphate (GGDP), a common C_{20} precursor prevalent throughout the plant kingdom and derivedfrom isopentenyl pyrophosphate (IPP) by two enzymes, IPP-isomerase and GGPP-synthase in plastids of higher plants. Gibberellins are derived from GGDP in plants via three distinct classes of enzymes: terpene synthases (TPSs), cyctochrome P450 monooxygenases (P450s) and 2-oxyglutarate- dependent dioxygenases (2ODD). First, the conversion of Geranyl geranyl pyrophosphate into ent-copalyl pyrophosphate (CPP) catalyzed by enzyme ent-copalyl diphosphate synthase (CPS) and cyclized to tetracyclic hydrocarbon intermediate ent-Kaurene by another enzyme, ent- Kaurene synthase bounded in plastid (Helliwell *et al.*, 2001). Further the cyclization of ent-Kaurene into ent-kaurenoic acid (KA) involving two P450s enzymes. The oxidation of the C-19 methyl (CH_3) group of Kaurene catalyzed by ent Kaurene oxidase (KO) localized in the outer membrane of plastids (Cove, 2006). P450s enzyme ent Kaurene oxidase (KO) is encoded by GA_3 in Arabidopsis and produce firmly dwarfed plant by invariably mutation in this gene (Shinjiro, 2008). GA_{12} derived from ent-kaurenoic acid by other P450s, ent-kaurenoic oxidase confined in endoplasmic reticulum. In final step, $GA_{4,}$ a bioactive form of gibberellins formed from GA_{12} oxidationonC-20 and C-3 by GA 20-oxidase and GA 3-oxidase respective which are soluble 2-oxoglutarate - dependent dioxygenases.

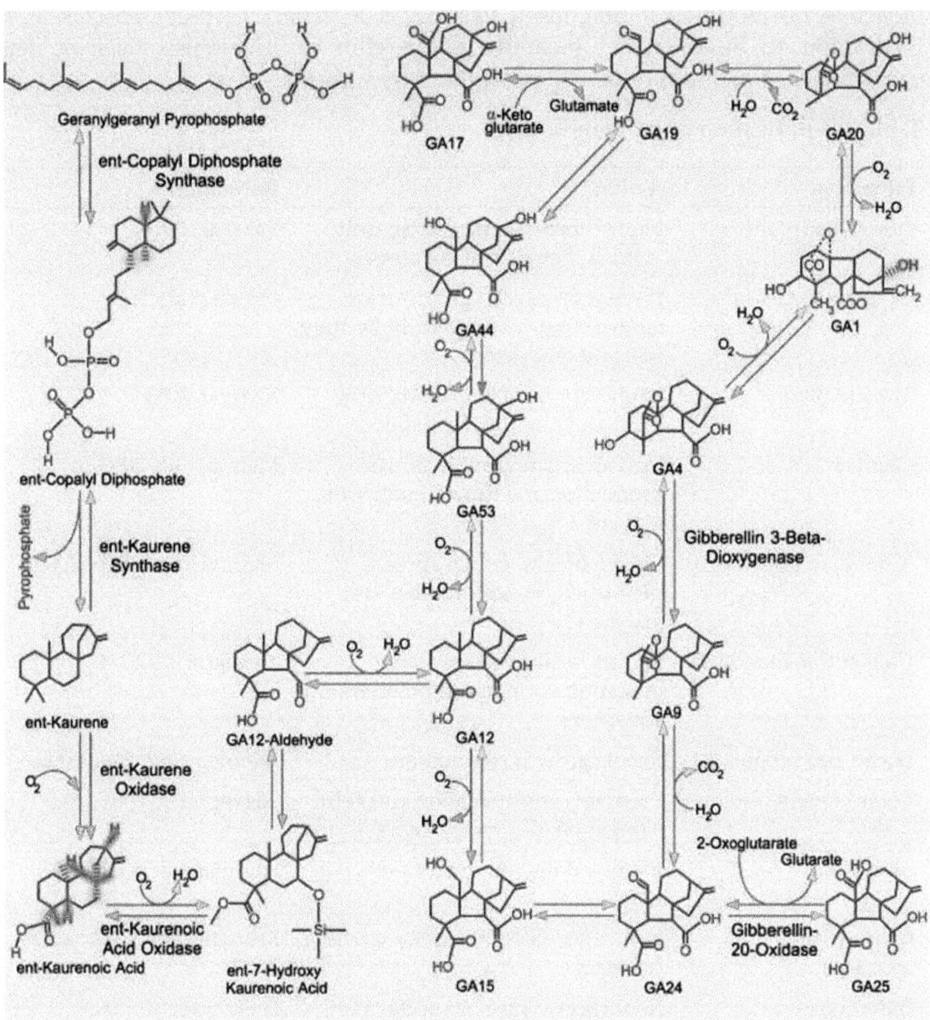

Fig. 2: Scheme of gibberellins biosynthesis in Arabidopsis (Ogawa et al., 2003).

Properties and Application of Gibberellic Acid:

Gibberellins (GAs) constitute a group of tetracyclic diterpenes. Hooley (1994) classified responses of plant cell and tissue with gibberellins in three categories: cell growth and development in vegetative tissues, aleurone layer of seed reserve endosperm for mobilization, development of flower and fruits. Many bioactive forms of gibberellins includes GA_1, GA_3, GA_4 and GA_7 which are formed from a basic diterpenoid carboxylic acid skeleton and generally have a hydroxyl group on C-3β, a carboxyl group on C-6 and a lactone between C-4 and C-10 (Yamaguchi, 2008). In GA_5 and GA_6, 3β-hydroxyl group can be displaced by other functional group at C-2 and C-3 to perform as bioactive forms. GA_1 consider as widespread bioactive hormone as it diagnosed usually in a variety of plant species. GA_4 is thought to be the chief bioactive Gibberellin in *Arabidopsis thaliana,* few Cucurbitaceae members and also prevail in many species.

Table 1. Application of gibberellic acid

Objective	Action	Reference
Cucumber plant	Higher increase in growth, fruit yield and quality of cucumber	Pal *et al.*, 2016
Opuntia seeds	Promotion of seed germination (above 90%), increase carbohydrate and amolytic activity	Ochoa *et al.*, 2015
Tomato plant	Higher increase in growth, fruit yield and quality of tomato.	Kazemi, 2014
Gerbera	increase plant growth, flower production and flower quality of gerbera	Mehraj *et al.*, 2013
'Chandler' strawberry	Crown height, crown spread, petiole length, leaf number and leaf area	Sharma and Singh, 2009
Clementine mandarin	Affects fertilization by acting in ovulation under cross-pollination conditions	Mesejo *et al.*, 2008
Dwarf pea seeds	Shoots growth stimulation	Baumgartner *et al.*, 2008
Potato tubers	Promotes multiplication and cell elongation, dormancy breakage	Alexopoulos *et al.*, 2007
okra plant	Increased the number of pods per plant	Abduljabbar *et al.*, 2007
Grape (flame seedless)	induce seedlessness in the seedless cultivars	Srivastava and Srivastava, 2007
Bittergourd	Reduction in number of days to fruit set	Dostogir *et al.*, 2006
litchi fruit	Increment in fruit weight	Chang and Lin, 2006

It has been reported that gibberellic acid play significant role in enhancing growth and biochemical parameters including sprouting, stem elongation, increased number of leaves, leaf area, photosynthesis, nitrogen utilization, mineral uptake, enzyme activity, sugar synthesis and shelf life (Chauhan *et al.*, 2010). Gibberellic acid delays senescence affecting plant morphology improves growth and development of chloroplasts and intensifies photosynthetic efficiency which could lead to increased yield. It is also used widely in other horticultural crops for stimulating fruit set and yield. Forthwith, gibberellic acid is attaining enormous attention all over the world by considering its effective use in agriculture, nurseries, tissue culture, horticulture etc. (Shukla *et al.*, 2005).

GA_3 role in growth

Gibberellic acid promotes plant growth by its effect on fast growth that occurs as a result of both the superlative number of cells formed and elongation of individual cells. Numerous study have shown that the application of gibberellic acid resulted in maximum plant height, number of branches, leaf per plant, leaf area, fresh and dry weight of plant. Mechanism of gibberellins action proposes that DELLA protein restrain the plant growth whereas the gibberellins signal promotes growth by overcoming Della-mediated growth restraint (Achard and Genschik, 2009). Cell elongation is the result of higher amylase production induced by gibberellic acid which might enhance the sugar contents, thereby increasing the osmotic pressure of the cell sap leading to entry of water into the cell causing cell wall stretching. It was found that gibberellic acid applied at root soaking stage is more effective in increasing the plant height, number of flower and yield than vegetative stage whereas similar results obtained with gibberellic acid application at flowering stage (Rahman *et al.*, 2015). Increase in leaf area by applied gibberellic acid and potassium which may automatically intensify the solar energy harvesting efficiency causing a significant enhancement in photosynthates production, translocation and ultimately total dry matter production of plant. Potassium and gibberellic acid promote DNA, RNA, protein synthesis, ribose and polyribosome multiplication contribute toward biomass production in vegetative parts. They enhance enzyme activity acceleration resulting in biomass accumulation in plants (Marschner and Marschner, 2012). Consequently, the certain hormones and sugar control flowering emergence time and uniformity. The additional growth that occurs with applied gibberellin also reflects in stimulate flowering. Menesy *et al.* (1991) studied that gibberellic acid application significantly reduced the flowering time and improved the flowering characters of marigold. Gibberellic acid shortened the period of fruit setting, ripening and accelerated the harvesting time (Manju and Rawat, 2015).

GA_3rolein Water use Efficiency

GA_3 has been shown to cause synthesis of α-amylase in aleurone cells. This enzyme converts starch to reducing sugars as a results solute potential (ψs) of the cells increases. Exogenous supply of gibberellic acid was found to induce high relative water content and membrane stability in maize leaves and provide strength to the plant in eliminating the inhibitory effect of water stress (Kaya *et al.*, 2006). Spraying foliage with gibberellic acid increases the leaf's water potential and then leaf stomatal conductance but only at appropriate dose. Increment in relative water content percentage in plants treated with gibberellic acid due to decrease in loss of water from leaves which might be ascribed to the reduction in transpiration depending the osmotic potential of mesophyll cells and controlled by the opening and closing of stomata to a larger extent. Foliar spray of gibberellic acid has been found to protect the plant membranes from dehydration and heat stress injury. Moreover, this protection is presumably due to role of GA_3 in foliar spray protecting the plant lipid from peroxidation through reducing the amount of malonyldialdehyde (MDA) content (Sayed and Gadallah, 2013).

GA_3 role in Gaseous Exchange

Stomata occupy a central position in the pathway for the CO_2 and O_2 transport as well as indirectly involve in passive absorption of water, ascent of sap and absorption of mineral salts by regulating the transpiration. Stomatal conductance is known to be the highest in young actively growing plant parts and it declined as soon as these plant parts mature (Siddique *et al.*, 1999). Supplementation of gibberellic acid resulted in increase the Rubisco content and its activity in broad bean and soybean. It also mediates Rubisco protein biosynthesis at translational rather than transcriptional level. Eventually, Rubisco is closely associated with stomatal conductance because of that gibberellic acid enhances the stomatal conductance (Yuan and Xu, 2001). Application of GA_3 to *Vicia faba* and *Fritillaria imperialis* persuade a transient opening of stomata in the dark, however a substantial accumulation of potassium in guard cell was not found (Goring *et al.*, 1990). Gibberellic acid is known to initiate a process that leads to increase in stomatal activity by inducing an influx of Ca^{2+} into the endoplasmicreticulum of guard cells and thus enhance stomatal conductance (Assmann *et al.*, 1999).

GA_3 role in Photosynthetic Pigments

Photosynthesis and carbohydrate synthesis increased due to chlorophyll content in response to growth regulators treatment (Lamrani *et al.*, 1996). Proficiency of photosynthetic process mainly depends on the quantity and maintenance of the numerous types of chlorophyll. Plant growth regulators stimulate the transformation of etioplasts into chloroplasts which show positive effect on increase of chlorophyll content by ceasing senescence and ultimately result into increased photosynthesis (Ouzounidou and Illias, 2005). Numerous studies have shown that gibberellic acid treatment leads to increment in greenness index. These increments occur due to an interaction with phytochrome, thereby, chlorophyll is enhanced by phytochrome. Inadequate application of GA_3 considerably decrease the photosynthetic rates, changes in the reaction centers of Photosystem II and suppress the enzymatic processes of the calvin system. Additional plant growth occurs with gibberellic acid treatment due to its effect on ceasing chlorophyll degradation and senescence which increase chlorophyll level (Jordi *et al.*, 1994). Pigments content as well as the rate of photosynthesis increased with GA_3 application. Carotenoid and photosynthetic pigment content depends generally on the dosage of hormones supplied because hormones primarily affect the biosynthesis of terpenoids through modulating the metabolic route that controlled by porphyrins and isoprenoids (Hakimeh *et al.*, 2011). Alteration in carotenoid content mainly associated with activity of Rubisco. Tuna *et al.* (2010) also observed an increase in carotenoid content in salinity stressed maize plants after foliar application of GA_3.

GA_3 role in Storage Products

Gibberellins may function aspromising phytohormone in the regulation of sucrose phosphate synthase (SPS) activity. A gradual increase in the activity of sucrose phosphate synthase enzyme also resulted in increase the photosynthetic sucrose formation in soybean and spinach (Kozlowska *et al.*, 2007). Glucose is synthesized from hydrolysis of starch granules by β-amylase that is exported from chloroplast, as substrate for transglucosylation reaction, generating glucosylated acceptor molecule and thus glucose (Smith *et al.*, 2005). Sugar provides not only energy source but also molecules controlling metabolism, development and gene expression in prokaryotes and eukaryotes (Kumar *et al.*, 2008). The amount and translocation of carbohydrate is mainly responsible for development of flower and overall growth. The most pronounced effect of gibberellic acid and potassium on soluble sugar synthesis via starch hydrolysis obtained at appropriate concentration as compared to higher and lower concentration in cucumber (Pal *et al.*, 2016). A gradual increase in the carbohydrate accumulation in plant is mainly responsible for the increase in the amount of sugar content. An application of gibberellic acid persuades physiological changes in plant to draw more carbohydrates by increased endogenous auxin level directly or indirectly may be ascribed to rapid metabolic transformation in to soluble compound (Singh *et al.*, 1993).

Progressive increases in NAR reflected in higher level of protein synthesis and intensify the total soluble protein. It also expressed the utilization of nitrogenous compounds in the plants for many metabolic activities. In addition, most fraction of total soluble protein consists of Rubisco enzyme, which is a carboxylation/ oxygenation enzyme that incorporates CO_2 into plants during photosynthesis. So, the photosynthetic efficiency increased with increase of total soluble protein in plant (Kuttimani *et al.*, 2013). Pal *et al.* (2016) while working on cucumber noticed highest level of total soluble proteins (TSP) at vegetative stage. At later stages of growth, the total soluble proteins decreased. It has been underlined that gibberellic acid controls various enzymes by affecting the endogenous level of gibberellins that alter the basic processes of translation/transcription involve in protein synthesis (Huttly and Phillips, 1995). NAR and its activity are very much unsteady and depend on presence of hormones and also important for the initiation of nitrate assimilation which ultimately manifests in the form of protein synthesis. Enhancement in protein synthesis with increase in NRA level brought about by pivot role of gibberellic acid (Shah, 2008).

GA_3 role in Mineral Elements

Gibberellic acid was also found to play an important role in enhancing the N, P and K content in leaves (Soad, 2005). Nitrogen is a necessary element for plant growth as it performed as an active participant of chlorophyll and protein. It shows pivotal role in plant nutrition and also known as a factor limiting plant growth. Nitrogen element takes up by plant in the form of nitrate. Nitrate reductase enzyme is effective enzymes involved in accumulation of nitrogen in plants by revive nitrite ion in the conversion of ammonium to nitrate in plant. Maximum content of leaves K, P, Ca were observed with gibberellic acid foliar spray (Mosa *et al.* 2015). Shah *et al.* (2006) in their study on *Nigella sativa* reported that GA_3 treatment enhanced plant growth parameters through intensifies assimilation and absorption potential of mineral nutrient all along vegetative growth stage. Additionally, Eid *et al.* (2006) noted that GA_3 application raised the absorption and accumulation of macro and micro nutrients. Increment of minerals was attributed to early growth and higher crop yield. Gibberellic acid increases the Ca^{2+} activity affiliated with endoplasmic reticulum and activates Ca^{2+} flux across the plasma membrane (Bush *et al.*, 1993).

GA_3 role in Production

Plant growth hormones influences leaf light interception, CO_2 fixation, mineral uptake, delay senescence, reduction in competition for nutrients and photo assimilates within plant organ and also well-developed chloroplast enhance photosynthetic efficiency. This could enhance in productivity of mother plant results in boosting yield of plant (El-otmani *et al.*, 2000). In fruit growing, GA_3 is frequently applied as it affects the shape of fruits and increase number of fruit and weight per plant (Davies and Zalman, 2006). Biemelt *et al.* (2004) suggested that high concentration of gibberellic acid showed negative impact on plant production and mainly economic yield. A great influence of gibberellic acidon the production of crop as it delays senescence improves the growth and development of chloroplasts and intensifies photosynthetic efficiency which could lead to increased yield but only at appropriate GA_3 concentration. Biemelt *et al.* (2004) also suggested that high concentration of gibberellic acid showed negative impact on plant production and mainly economic yield. Economic yield of the crop strongly influence by photosynthetic area, photosynthetic time, photosynthetic capacity, photosynthates consumption and distribution in plant. Photosynthesis is the fundamental way of increasing crop yield because it accounts for more dry matter in plant. On the other side, insufficient leaf potassium levels lead to reduced photosynthesis per unit leaf area and cause overall reduction in plant economic yield. Gibberellic acid play effective role in maintaining higher economic return and benefit cost ratio as reported in bittergourd (Biradar, 2008) and brinjal (Al-Imran, 2013).

Future Scope and Conclusion:

Benefits to farmers:

Increase in crop yield and production.

Boost soil productivity

Benefits to Environment:

Enhanced in soil quality

Reduction in soil erosion

Minimize pollution with reduced utilization of chemical fertilizers.

Benefits to National Economy

Sustainable agriculture productivity

Lessen reliance on fossil fuel for fertilizer production

Today with prosperity on the upsurge and dignifying surging, farmers also need a good standard of living which attained with increasing yield and quality produce as it mediate increment in farmer's income? Nevertheless, Indian crop yield are lowest in comparison to world which continued to be passive or desist bygone time. Forthwith, daily intake of vegetable rich diet is vital for health as provided various phytoneutricals much needed by human to get rid of death dealing human diseases. Now is the time, we should consider what type of methodological advances could be propagandized in sustainable agriculture and gibberellic acid is one of the yield stimulants or environmentally sound plant growth regulator. Therefore, a comprehensive strategy required for achieving new prospective to attain global food production, as the government requisite to work towards enhancing yields significantly. Crop growth, sustainable productivity and quality mainly depend on its genetic potential and its interaction to fertigation and exogenous supplementation of growth substances in addition to its response with the environmental conditions.

References

Abduljabbar, I.M., Abduljabbar and Shukri H.S. 2007. Effect of sowing date, topping and some growthregulators on growth, pod and seeds yield of okra (*Abelmoschus esculentus* L.M.). *African Crop Science Conference Proceedings, 8:* 473-478.

Achard, P. and Genschik, P. 2009. Releasing the brakes of plant growth: how GAs shutdown DELLA proteins. *J. Exp. Bot*, 60: 1085–1092.

Alexopoulos, A.A., Akoumianakis, K.A., Vemmos, S.N. and Passam, H.C. 2007. The effect of postharvest application of gibberellic acid and benzyladenine on the duration of dormancy of potatoes produced by plants grown from TPS. *Postharvest Biology Technol.*, 46: 54–62.

Al-Imran, M.D. 2013. Morphological and yield attributes of brinjal in response to gibberellic acid. MSc Thesis, Department of Horticulture, Sher-e-banglaAgricultural University, Dhaka, Bangladesh, pp. 20-50.

Assmann, S.M. and Armstrong, F. 1999. Hormonal regulation of ion transporters: the guard cell system. In Hooykaas *et al.* (eds) Biochemistry and MolecularBiology of Plant Hormones Elsevier, Amsterdam, pp. 337-361.

Baumgartner, S., Shah, D., Schaller, J., Kämpfer, U., Thurneysen, A. and Heusser, P. 2008. Reproducibility of dwarf pea shoot growth stimulation by homeopathic potencies of gibberellic acid. *Complement Ther Med.*, 16: 183–91.

Biemelt, S., Tschiersch, H. and Sonnewald, U. 2004. Impact of Altered Gibberellins Metabolism on Biomass Accumulation, Lignin Biosynthesis, and Photosynthesis in Transgenic Tobacco Plants. *Plant Physiol*, 135:254-265.

Biradar, G.S. 2008. Effect of plant growth regulators on physiology and quality in bittergourd (*Momordica charantia*). M.Sc thesis. University of agricultural sciences, Dharwad.

Bush, D.S., Biswas, A.K. and Jones, R.L. 1993. Hormonal regulation of Ca^{2+} transport in the endomembrane system of the barley aleurone. *Planta,*189: 507–515.

Chang, J.C. and Lin, T.S. 2006. GA_3 increases fruit weight in `Yu Her Pau` litchi. Sci. Hort., 108: 442–43.

Chauhan, J.S., Tomar, Y.K., Badoni, A., Indrakumar, N., Seema, S. and Debarati, A. 2010. Morphology, germination and early seedling growth in *Phaseolus mungo* L. with reference to the influence of various plant growth substances. *J. Am. Sci,* 6: 34-41.

Davis, P.J. 2002. Gibberellins: Regulators in plant height (L Taiz and Zeigler E (eds) *Plant Physiology* (3rd Edn), *Sinauer Assocn. Inc,* USA. pp. 461- 492.

Davies, F.S. and Zalman, G. 2006. Gibberellic acid, fruit freezing, and postfreeze quality of Hamlin oranges. *Hort. Tech,*16: 301-305.

Dias, J. S. and Ryder, E. 2011. World Vegetable Industry: Production, Breeding, Trends. *Hort. Review,* 38: 299-356.

Dostogir, H., Karim, A., Habibur, M., Rahman Pramanik, M. and Syedur Rahman, A.A.M. 2006. Effect of gibberellic acid (GA_3) on flowering and fruit development of bittergourd (*Momordica charantia* L.). *Intl. J. Bot.,* 2(3): 329-332.

Eid, R.A. and Abou-leila, B.H. 2006. Response of croton plants to gibberellic acid, benzyl adenine and ascorbic acid application. *World J. Agric. Sci,* 2: 174–176.

El-otmani, M., Coggins, C.W., Agusti, M. and Lovatt, C.J. 2000. Plant growth regulators in citriculture. World current issues. *Crit. Rev. Plant Sci,* 19: 395-447.

Goring, H., Koshuchowa, S. and Deckert, C. 1990. Influence of Gibberellic acid on Stomatal movement. *Biochem. Physiol. Pflenzen,* 186: 367-374.

Hakimeh, M., Zahra, A. and Ryszard, A. 2011. The response of terpenoids to exogenous gibberellic acid in Cannabis sativa L. at vegetative stage. *Acta Physiol. Plant,* 33: 1085-1091.

Helliwell, C.A., Sullivan, J.A., Mould, R.M., Gray, J.C., Peacock, W.J. and Dennis, E.S. 2001. A plastid envelope location of Arabidopsis entkaurene oxidase links the plastid and endoplasmic reticulum steps of gibberellin biosynthesis pathway. *The Plant Journal,* 28: 201-208.

Herrera, E., Jimenez, R., Aruoma, O., Hercberg, I. and Sanchez-Garcia, Fraga, C. 2009. Aspects of Antioxidant Foods and Supplements in Health and Disease. *Nutrition Reviews,* 67: 140-144.

Hooley, R. 1994. Gibberellins: perception, transduction and responses. *Plant Mol. Biol.,* 26: 1529- 55 .

Huttly, A.K. and Phillips, A.L. 1995. Gibberellin regulated plant genes. *Physiol. Plant,* 9: 310–317.

Jordi, W., Pot, C.S., Stoopen, G.M. and Schapendonk, A.H.C.M. 1994. Effect of light and gibberellic acid on photosynthesis during leaf senescence of Alstroemeria cut flowers. *Physiol. Plant*, 90: 293-298.

Kaya, C., Levent Tuna, A., Alfredo, A. and Alves, C. 2006. Gibberellic acid improves water deficit tolerance in maize plants. *Acta physiol. Plant*,28: 331-337.

Kazemi, M. 2014. Effect of gibberellic acid and potassium nitrate spray on vegetative growth and reproductive characteristics of Tomato. *J. Biol. Env. Sci*, 8: 1-9.

Kozlowska, M., Rybus-Zaiac, M., Stachowiak, J, and Janowska, B. 2007. Changes in carbohydrate contents of *Zantedeschia* leaves under gibberellin-stimulated flowering. *Acta Physiol. Plant*, 29: 27-32.

Kumar, S. 2008. Oxidative stress and antioxidant system during ripening and storage of ber (*Ziziphus mauritiana* Lamk.) fruits. PhD Thesis, Department of Biochemistry, CCS Haryana Agricultural University, Hisar, India.

Kuttimani, R., Velayudham, K. and Somasundaram, E. 2013. Growth and yield parameters and nutrient uptake of banana as influence enhanced by integrated nutrient management practices. *Inter. J. Recent Scientific Res*, 4: 680-686.

Lamrani, Z., Belakbir, A., Ruiz, J.M., Ragala, L. and Lopez- Cantarero, I. 1996. Romero. Influence of nitrogen, phosphorus, and potassium on pigment concentration in cucumber leaves. *Commun. Soil Sci. Plant Anal*, 27: 1001-1012.

Mehraj, H., Taufique, T., Ona, A.F., Roni, M.Z.K. and Jamal Uddin, A.F.M. 2013. Effect of spraying frequency of gibberellic acid on growth and flowering in gerbera. *J. Expt. Biosci.*, 4(2): 7-10.

Manju and Rawat, S.S. 2015. Effect of bioregulators on fruit growth and development of local malta (*Citrus sinesis* Osbeck) under valley conditions of Garhwal Himalaya. Int. *J. Pl. An and Env. Sci*, 5: 105-108.

Marschner, H. and Marschner, P. 2012. Marschner's Mineral Nutrition of Higher Plants.Elsevier, london, UK.

Mesejo, C., Martínez-Fuentes, A., Reig, C. and Agustı, M. 2008. Gibberellic acid impairs fertilization in *Clementine mandarin* under crosspollination conditions. *Plant Science*, 175:267–71.

Menesy, F.A., Nofal, E.M.S. and El-Mabrouk, E.M. 1991. Effect of some growth regulators on *Calendula officinalis* L. *Egyptian J. Appl. Sci*, 6:1-15.

Mitchell, J.W., Skaggs, D.P. and Anderson, W.P. 1951. Plant growth stimulating hormones in immature seeds. *Science*, 114: 159.

Mosa, W. F. Abd El-G., Nagwa, A. Abd El-M., Aly, M.A.M. and Paszt, L.S. 2015. The Influence of NAA, GA3 and Calcium Nitrate on Growth, Yield and Fruit Quality of "Le Conte" Pear Trees. *Am. J. Exp. Agr*, 9: 1-9.

Mullie, P. and Clarys, P. 2011. Association between Cardio- vascular Disease Risk Factor Knowledge and Lifestyle. *Food Nutr. Sci*, 2: 1048-1053.

Nicholson, P., Simpson, D.R., Weston, G., Rezanoor, HN., Lees, AK., Parry, D.W. and Joyce, D. 1998. Detection and quantification of *Fusariumculmorum* and *Fusarium graminearum* in cereals using PCR assay. *Plant Pathology*, 53: 17–37.

Ochoa, M.J., González-Flores, L.M., Cruz-Rubio, J.M., Portillo, L. and Gómez-Leyva, J.F. 2015. Effect of substrate and gibberellic acid (GA_3) on seed germination in ten cultivars of Opuntia sps. *J. Prof. Assoc. Cactus*,17: 50-60 51.

Ogawa, M., Hanada, A., Yamauchi, Y., Kuwahara, A., Kamiya, Y. and Yamaguchi, S. 2003.Gibberellin biosynthesis and response during Arabidopsis seed germination. Plant Cell, 15(7): 1591-604.

Ouzounidou, G. and Ilias, I. 2005. Hormone-induced protection of sunflower photosynthetic apparatus against copper toxicity. *Biol. Plant*, 49: 223–228.

Pal, P., Yadav, K., Kumar, K., Singh, N. 2016. Effect of gibberellic acid and potassium foliar sprays on productivity and physiological and biochemical parameters of parthenocarpic cucumber cv. seven star f1. *Journal of Horticultural Research*, 24(1): 93-100

Rahman, M.S., Haque, M.A. and Mostafa, M.G. 2015. Effect of GA_3 on biochemical attributes and yield of summer tomato. *J. Biosci. Agric. Res*, 3: 73-78.

Sayed, S.A. and Gadallah, M.M.A. 2013. Gibberellic acid ameliorates the adverse effects of acid mist and improved antioxidant defense, water status and growth of acid misted sunflower plants. *J. Biol. Earth Sci*, 3: B275-B285

Shah, S.H., Ahmad, I. and Samiullah. 2006. Effect of gibberellic acid spray on growth, nutrient uptake and yield attributes during various growth stages of black cumin (*Nigella sativa*). *Asian J Plant Sci*, 5: 881 -884.

Shah, S.H. 2008. Effects of nitrogen fertilization on nitrate reductase activity, protein, and oil yields of *nigellasativa* L. as affected by foliar GA_3 application. *Turk J. Bot*, 32: 165-170.

Sharma, R.R. and Singh, R. 2009. Gibberellic acid influences the production of malformed and button berries, and fruit yield and quality in strawberry (Fragaria x ananassa Duch.). *Scientia Hortic.*, 119: 430–433.

Sharma, I., Chin, I., Saini, S., Bhardwaj, R. and Pati, P.K. 2013. Exogenous application of brassinosteroid offers tolerance to salinity by altering stress responses in rice variety Pusa Basmati-1. *Plant Physiol. Biochem*, 69: 17–26.

Shinjiro, Y 2008. Gibberellin Metaboilism and its regulation, *Anne. Rev. plant Biol.*, 59: 225-251.

Shukla, R., Chand, S. and Srivastava, A.K. 2005. Batch kinetics and modelling of gibberellic acid production by *Gibberella fujikuroi*. *Enzyme Microb. Technol*, 36: 492–497.

Siddique, M.R.B., Hamia, A. and Islam, M.S. 1999. Drought stress effects on photosynthetic rate and leaf CO2 exchange of wheat. *Bot. Bull. Acad. Sin*, 40: 141-145.

Singh, S., Singh, I.S. and Sungh, D.N. 1993. Physico-chemical changes during development of seedless grapes (*Vitis vinifera*, L.). *Orissa J. Hort*, 21: 43-46.

Smith, A.M., Zeeman, S.C. and Smith, S.M. 2005. Starch degradation. *Ann. Rev. Plant Biol*, 56: 73-98.

Soad, M.M.I. 2005. Response of vegetative growth and chemical composition of jojoba seedlings to some agriculture treatments. Ph.D. thesis, Faculty of Agriculture Minia University, Egypt.

Srivastava, N.K. and Srivastava, A.K. 2007. Influence of gibberellic acid on 14 CO_2 metabolism, growth, and production of alkaloids in Catharanthus roseus. *Photosynthetica*, 45: 156-60

Swaminathan, M.S. 2007. India's tryst with destiny in agriculture. *Yojana*, 51: 61-65.

Tyagi, V. 2012. India's agriculture: challenges for growth and development in present scenario. *Int. J. Phy. Soc. Sci.*. 2: 116-128.

Tuna, A.L., Kaya, C. and Ashraf, M. 2010. Potassium sulfate improves water deficit tolerance in melon plants grown under glasshouse conditions. *J. Plant Nutr*, 33: 1276-1286.

Vysotskay, L.B., Timergalina, L.N., Veselov, Yu.S. and Kudoyarova, G.R. 2007. Effect of the Nitrogen-Containing Salt on the Content of Cytokinins in Detached Wheat Leaves. *Russian J. Plant Physiol.*, 54(2): 191–195.

Webb, P.J.R. 2009. Fiat Panis: For a World Without Hunger, ed Eiselen H (Hampp Media/Balance Publications, Stuttgart), pp. 410–434.

Yabuta, T and Sumiki, Y. 1938. Biochemical studies on "Bakanae" fungus. Crystals with plant growth promoting activity. *J. Agr. Ch. Soc Japan*, 14:1526.

Yamaguchi, S. 2008. Gibberellin metabolism and its regulation. *Annu. Rev. Plant Biol.*, 59: 225-25.

Yuan, L. and Xu, D.Q. 2001. Stimulation effect of gibberellic acid short-term treatment on leaf photosynthesis related to the increase in RuBisCO content in broad bean and soybean. *Photosynthesis Res*,68: 39–47.

14

Ecologically Based Management Techniques of Rodent Pests

Ravikant[1] * and Tejpal Dahiya[2]

[1] Department of Zoology, CCSHAU, Hisar 125004 Haryana, India

[2]Department of Zoology GCW, Tohsam, Tosham-127040 (Bhiwani) Haryana, India

*Corresponding author: rkantazad@gmail.com

Abstract

Rodents are the most abundant and diversified order of living mammals in the world. Rodents have been identified as the most important mammalian agricultural pests at the global level. Many rodents are common in fields with cereal crops, which constitute the largest part of the total area used for agriculture in the world. The management of rodents has focused on conventional methods, mainly the use of rodenticides as a symptomatic treatment approach. These methods are supported by government, especially to contain outbreaks. However, conventional control methods have remained largely ineffective. An ecological approach for management of rodent outbreaks is not widely practiced for lack of basic experimental data to substantiate its efficacy. Measures that are practiced on a limited scale but have a wide scope for future management of rodents include various techniques of environmental manipulation that specifically focus on altering the suitable habitats for rodents to reduce their carrying capacity. For the future, a more pragmatic approach is required for ecologically focused rodent management techniques. The development and implementation of control activities, may assist in alleviating the damage and losses due to rodents in the future.

Keywords: Management strategies, Ecological, techniques, Rodents, Pest,

Introduction

Rodents form the largest and most successful group of mammal species (Parshad 1999). Rodents comprise about 43% of mammalian species, surpassing all other mammalian orders in abundance of individuals as well as in number of genera and species. Rodents are nearly cosmopolitan in distribution and occur in a diverse array of habitats. The range of ecological settings that they are able to exploit has contributed to an equally impressive array of social systems, some of which are among the most complex known for any mammalian species (Lacey *et al.*, 2003).

Rodents of the order Rodentia, any of more than 2,050 living species of mammals characterized by upper and lower pairs of ever-growing rootless incisor teeth. Rodents are the constituting almost half the class Mammalian's approximately 4,660 species. They are indigenous to every land area except Antarctica, New Zealand, a few Arctic and other oceanic islands, although some species have been introduced even to those places through their association with humans. Order Rodentia comprising 27 different families, including not only the rats and mice but such diverse groups as porcupines, beavers, squirrels and chinchillas. Within Rodentia, a division can be made between field rodent species and commensal rodent species (in Europe: *Rattus norvegicus, Rattus rattus* and *Mus musculus*).

Worldwide, rodents are the most important group of mammals in terms of the problems they create in agriculture, horticulture, forestry and public health. They show a wide range of adaptation, enabling them to successfully colonise and inhabit almost any type of habitat. Rodents play an important role as reservoirs and carriers of zoonotic diseases for which some epidemics have afflicted mankind for centuries. In various Afro-Asian countries some rodent-borne diseases constitute a serious burden on the human population in endemic areas.

Characteristics of rodents

- They are omnivorous *i.e.* they are able to eat anything and everything.
- They cannot vomit and it is because of this character that they always try to sample the various diets before actually consuming them.
- They can be diurnal, nocturnal, arboreal or terrestrial.
- They show a bimodal feeding activity and show one peak in the morning and the other during the evening.
- They become sexually mature when they are 6-8 months old.
- The gestation period is 18-21 days. Gestation period is the time interval between the conception and the delivery of young ones.
- The litter size (the number of young ones at the time of delivery) varies from 1 to 22.
- They have ever growing incisors. If the growth of these ever growing incisors is not checked, they may grow so much that they may pierce the forehead of the rodent and it may die. In order to avoid this, the rodents always nibble something or the other because of which they keep a check upon the growth of incisors.

- They do not have canines. The space between the incisors and the premolars is called diastema.
- They are fossorial i.e. they are adapted for digging and they make their burrows inside the field.
- They have a life span of 1-2 years.

Economic importance

Agriculture plays an important role in the life of an economy. It is the backbone of Indian economy and contributes to the overall economic growth of the country. It also determines the standard of life for more than 50% of the Indian population. Though it contributes only about 14% to the overall GDP but its impact is observed in the manufacturing and the service sector as the rural populace has become a major consumer of goods and services in the last couple of decades. Agriculture not only provides food and raw material but also employment opportunities to a very large proportion of the population. However, the damages inflicted to the crops by rodents, especially rats and squirrels, lead to huge productivity losses along with crop contamination. In Asia, rats cause an average of 5–10% loss in rice yield every year. Also, farmers lose an estimated average of 37% of their rice crop to pests and diseases every year.

Singleton and Petch (1994) observed that rodents have three major impacts, the first is the substantial damage they can cause at any stage of the growing crop, second is the losses they cause during post-harvest to storage and third, often overlooked, impact is on the health of farmers, as the rodents are carriers of at least 20 severely debilitating human diseases (Meerburg *et al.*, 2009). According to Tripathi (2007) in Andhra Pradesh the tiller damage to *Kharif* rice ranged from 12-32 per cent and up to 40 per cent to *Rabi* rice.

Rodents are by far the greatest vertebrate pest problem in the world. They are responsible for substantial damage to food and cash crops, industrial and domestic property. More than 25 species of rodents have been recorded as pests in agriculture, causing a wide range of damage and losses in cereals, legumes, vegetables, root crops, cotton and sugarcane. Pest species occupy a diversity of habitats, including cultivated fields, urban environments and domestic areas. Other than being instrumental in crop damage, they are also reservoirs and carriers of zoonotic diseases.

Rodents have played havoc with man's economy as they damage each and every food item in fields, godowns, storage houses, poultry farms and residential premises (Kocher and Kour 2013). Pre-harvest and post-harvest losses due to rodents of grain were 25% and 25-30% respectively (Hart 2001). The rat is now the number one pre-harvest pest for rice crops in Indonesia (Singleton and Petch 1994). In Asia, claims of annual pre-harvest losses in rice production by rodents range from 5% in Malaysia to 17% in Indonesia. In India, it is estimated 25-30% of post-harvest grains are lost each year due to rodents. In terms of the nation's stored food and seed grain market, more than $5 billion (in U.S. money) losses caused by the rodents. In terms of pre-harvest rice lost due to rodents, the nation-by-nation percentages is Malaysia 4-5 percent, Philippines 3-5 percent (alternate sources claim 30-50 percent), Thailand 6-7 percent and Vietnam 10-35 percent (Singleton 2003).

Role as Vector

Whilst rodents are often associated with infrastructural damage and eating or spoiling of stored feed and products, their zoonotic risks are frequently underestimated. Huq *et al.*, (1985) reported that their predatory and depredatory habits have a pronounced impact on human economies and potentially more seriously, they are major vectors of human and domestic animal diseases worldwide. Hiett *et al.*, (2002) concluded that wild rodents can be reservoirs of a number of agents that cause zoonotic disease in food, animals and humans (e.g. *Leptospira* spp., *Salmonella* spp., *Campylobacter* spp., *Toxoplasma* spp.). Zoonosis are the diseases that are transmissible between animals and humans (WHO 2013). This group of diseases is generally classify according to zoonotic agents itself, the degree of humans-to- human transmissibility or through the route of transmission of the diseases from animal to humans (Lloyd-smith *et al.*, 2009; Karesh *et al.*, 2012). The classification according to the route of transmission group zoonoses into diseases mainly transmitted through food or food borne zoonosis, such as Salmonellosis (EFSA 2013).

Rodent Management

The management of rodents has focused on conventional methods, which are supported by government includes the use of rodenticides as a symptomatic treatment approach. However, conventional control methods have remained largely ineffective.

An ecological approach for the management of rodent outbreaks is not widely practiced. Measures that are practiced on a limited scale but have a wide scope for future management of rodents in East Africa include various techniques of environmental manipulation that specifically focus on altering the suitable habitats for rodents to reduce their carrying capacity. Strategies for management of rodent populations in urban areas, in post-harvest crop systems and in response to disease outbreaks are not well developed. For the future, a more pragmatic approach is required, involving, better planning of urban housing schemes, sanitation and hygienic measures; improved storage structures and practices; and ecologically focused rodent management techniques.

Rodents are well recognized pests and are considered the number one enemy of man. These animals, though considered primitive, have been so successful and abundant worldwide that they alone account for more than half of the known mammalian fauna. They can subsist on a variety of food and even without water for a long time. Due to their prowess of adaptability, they have successfully adapted to varied ecological and physiological niches, and therefore are inhabitants of practically each and every latitude and altitude. They multiply at super fast speed because they become sexually mature when they are just 6-8 months old. They pose a challenge to man in all respects. According to Prakash and Mathur (1988) in India there are 135 species and about 300 sub-species of rodents belonging to 46 genera. At least 15 of these rodent species are known to be the serious pest of public health hazard. Rodents not only damage the crops in the fields, where the infestation occurs at all stages of food production, processing, storage and distribution, they are also responsible for a number of diseases in the human beings. Rodents attack the-crops shortly after sowing, during the vegetative growth and during the ripening of the seed heads. They damage the various crops of the *Kharif* season (jowar, maize, ground nut and cotton etc.) as well as the crops of the *Rabi* season (wheat and barley etc). They damage other crops also which include paddy, sugarcane, soyabean and Bengal gram etc. Hence some well planned rodent management is the need of the time and before going for the planning of any control method in a given situation it is essential to have an estimation of the population and the species involved (Steven 2007).

Ecologically-based rodent management (EBRM)

Experience has shown that armed with the right knowledge and tools it is possible to sustainably reduce pest rodent populations in a cost-beneficial way. In recent years, applied research on 'Ecologically-based rodent management' has taken place in many countries throughout Asia and Africa, involving a number of research and extension institutions working together in collaboration with farming communities to develop effective, sustainable and cost-beneficial rodent management strategies (Herwig *et al.*,2014).

Current rodent management strategies

Use and choice of rodenticides

The use of rodenticides to control rodent outbreaks is not widely practiced on an individual farm basis. In Tanzania, the government has organised control campaigns since the mid 1970s, but in areas where major outbreaks do not occur, farmers do not feel the need to control rodents. Success in the use of rodenticides whether acute poisons or anticoagulants has been influenced by three factors, as outlined below:

i) Availability of the required rodenticides

ii) Acceptability of bait formulations to rodents

iii) Timing of bait application

Rodenticides have commonly been used for symptomatic treatment to reduce damage when rodent populations are already high. This necessitates the application of large amounts of rodenticides. If knowledge of rodent population dynamics is available, this could be used to suggest appropriate timing of prophylactic treatment to alleviate the damage caused in rodent outbreaks.

- ◈ Availability of the required rodenticides, often influenced by available funds for their purchase.
- ◈ Acceptability of bait formulations to rodents, often influenced by palatability under field conditions. The availability of other food resources for rodents in the field may determine the level of bait consumption. In Tanzania, a highly acceptable bait formulation of bromadiolone, targeted against *Mastomys natalensis,* was developed in 1988.
- ◈ The timing of bait application. This is critical for alleviating damage. One hypothesis that has not been widely tested under different agro-climatic and agro ecological conditions is that rodents should be effectively controlled during the season when the population is low and before animals start breeding. However, with small farm holdings, which can be easily re- colonised, prophylactic treatment, may produce less than expected results, especially for those species which breed prolifically (Myllymaki 1987).
- ◈ Zinc phosphide has been most commonly used for controlling rodent outbreaks throughout the world. The choice of zinc phosphide by farmers with little disposable income is based on low cost and a reasonably quick effect relative to anticoagulants, but even this rodenticide is not easily available to most farmers.

Physical Measures

Physical measures are used widely to control rodents in East Africa. The measures commonly practiced by farmers include trapping, digging and flooding burrows (*Tachyoryctes* spp., *Tatera* spp.), exclusion and hunting (*Hystrix* spp., *Cricetomys gambianus* and *Thryonomys* spp.). There are many different types of traps used to capture rodents in different areas within East Africa, but basically they are of two main designs: kill and live-traps. Traps are widely used to control rodents within houses, storage structures and in crop fields. Trapping methods are generally popular among peasant farmers who lack other resources for rodent pest management and in few places are used to capture rats to supplement the diet. To make rodent control highly successful the traps should be combined with other control methods such as environmental sanitation, proper storage of food, and when necessary, application of rodenticides. When rodenticides are available, trapping is a less favoured method because it is labour intensive and is less effective in controlling outbreaks. However, pitfall especially those which also combine drowning in tins or buckets half filled with water, have been claimed by farmers to be effective during times of rodent outbreaks.

Exclusion by rat proofing of the storage house or structure is recognised as an effective method to reduce post-harvest losses in rural communities (Hall 1970).

The practice of storing grain in the ceiling or stores built within traditional houses, usually with walls constructed with mud, is common in East Africa. The structural nature of the houses makes it difficult to protect the grain from rodent damage because an effective barrier between the commodity and rodents cannot be created.

Improving storage structures, therefore, will be the most appropriate long-term strategy to reduce rodent damage to crops during storage. This should entail, first, raising the basement of the structure to one metre above ground level and, secondly, incorporating a sleeve or a band of sheet metal that makes the surface too slippery for rodents to traverse when fitted closely to the slit.

Further, rodents can be kept away by the removal of vegetation around the vicinity of the storage structure, maintaining the stores in a good state of repair and ensuring the surrounds have minimal food residues and other rubbish on which rodents feed. Although rodent proofing of outdoor storage structures is very effective in excluding synanthropic rodents, it has not been adopted by many rural communities.

Exclusion using rodent proof containers, such as steel drums and clay pots, is common in rural areas. These are used for storage of seed and smaller quantities of grain. The containers provide adequate protection against Mus *musculus* and *Rattus rattus* for stored grain in the household.

Environmental Manipulation

Habitat manipulation has been encouraged in a few in East Africa. It assumes that shelter and food are the main factors affecting rodent numbers in any given habitat. This approach also focuses on the fact that rodents are extremely dependent on shelter for survival. For numbers to increase conditions must be favourable for breeding and survival of the young to reproductive stage. Disruptions of the environment, caused by harvesting and ploughing, lead to a decrease in the shelter available for rodents and possibly expose them to predators and reduce their population density. Taylor and Green (1976) noted that arable fields were unstable habitats for rodents. Leirs *et al.*, 1997 observed a rapid decrease in rodent abundance immediately after ploughing and planting but densities increased again after a few days. It has been suggested that in some parts of East Africa, the destruction of the natural environment has displaced many predators of rodents, while the new conditions created are favourable for high rodent population density. Environmental manipulation as a rodent control strategy in East Africa has not been very successful because it has not been extensive enough or incorporated as a component of the small holder farming system.

In order to benefit many farmers it must not be confined to a few individual fields. Where environmental manipulation has been carried out. It must includ one or more of the following practices:

Grazing, Regular Bush Clearing and Grass Cutting

Areas that are regularly cleared of bushes or that support grazing usually have a lower carrying capacity for rodent populations. However, pasture land that is not grazed regularly can support high populations of rodents, especially granivorous species *(A. niloticus,* M. *natalensis and Otomys angoniensis*). Green and Taylor (1975) found that cover was an important population regulating factor for these species in Kenya. But when cover was removed it resulted in depletion of rodent populations. The population density of *A. niloticus* is markedly affected by regular cutting of vegetation, which reduces suitable habitats for this species

Observations in the plague outbreak villages of Lushoto District Northeast Tanzania, showed that clearance of bushes, especially of the perennial *Rumex usambarensis,* removed pockets of *A. niloticus* populations near houses. It also has been suggested that the population of *A. niloticus* in Uganda, dropped considerably because the municipal council was continuously cutting grass around the city (De Graaf 1981).

Grazing in the fields immediately after harvest and in the fallow land between farms might help to destroy vegetation cover and remove food sources for rodents. This practice is common in many areas in East Africa, although the common purpose is not to control rodents, but to make use of the stubble and crop remains as animal feed during the dry season. However, there is a delicate balance between vegetation cover and soil erosion. The risks of soil erosion due to overgrazing must always be considered in using this approach.

Regular weeding has been reported to affect rodent population density in cultivated fields. Mwanjabe (1993) reported that clean, weeded farms were less severely attacked than unweeded farms throughout the year. A potential widespread management strategy could be the application of herbicides to control weeds over a large area but the economic and environmental implications do not allow this method to be implemented on a wide scale.

Agricultural Practices and Land Management Strategies

In Asia, the rains are seasonal with one or two rain seasons in a year. This also determines the cropping patterns, with intensive agricultural practices being found where the rainfall is well distributed over the year. Over a vast area the farming system is composed of small farms that are 0.5-2 ha, forming a mosaic of fallow land interspersed with cultivated areas, which is ideal for maintaining large rodent numbers that invade crop fields. In these areas, rodent management strategies require changes in land practices, producing less fallow patches that are a refuge for rodents and from where invasion of crops occurs.

To reduce the damage to crops by rodents, cropping needs to be synchronised over a large area. In general there is no common approach to land management that is aimed at reducing rodent damage to crops in Asia. Encouragement of large block farms and clearing of headlands reduces the density of rodents considerably. Taylor (1968) noted that large, mechanically cultivated monocultures were often not highly infested with rodents, but these are not common at the small holder farmer level.

Other practices like efficient harvesting of cereals and cotton are recommended in most rodent outbreak areas to reduce the available food resources for rodents. Many farmers burn their fields in the aftermath of the harvest or immediately before planting. This probably changes the habitat for a short duration, but most likely it has no detrimental effect on the future population size of rodents because burnt areas soon have new vegetation and are reinvaded rapidly by species from other areas.

In many parts of East Africa, the harvesting time coincides with the beginning of the dry season. It is common for farmers to leave the crop in the fields, a form of temporary storage, for extended periods to allow further drying before threshing. This practice is common for cereals, especially maize, sorghum and millet. A crop left in the field for extended periods is predisposed to severe attack by rodents. Among the practices that are encouraged are early harvesting and storage of the crops in improved, rodent-proofed cribs constructed for the dual purpose of storage and in-storage drying to reduce both rodent and insect damage of crops.

Ecologically-based rodent management in practice

There was much anecdotal evidence of rodent pest problems in field crops, but there was a need to show the actual impact of rats on people's livelihoods. Research activities showed that 5 to 10 percent of stored paddy rice was lost to rodents over each 3 month storage period, with each farming household losing an approximate of 200 kg per year. In common with most of developing countries in Asia, most farmers stated they plant about 2 rows of rice for the rats for every 8 rows sown. Our assessments showed pre-harvest losses from rats ranged from 5 - 17 percent in irrigated and rain-fed rice fields. Farmer damage assessments highlighted some of the more overlooked impacts of rodents, namely physical damage to houses, personal possessions, roads and fields.

Through surveys and questionnaires with farmers and community members, we were able to assess the effectiveness of existing rodent management actions carried out by farmers and households. In common with most developing countries, farmers had access to some rodent control tools and methods. However, because they were not used properly, or were not well adapted to local situations, they were often not very effective. This led to apathy and widespread acceptance of rodent pests in the environment.

Rodenticides are frequently used to control rodents. Misuse of these poisons is unfortunately common. More importantly, when a rodenticide is not used correctly, it may not significantly reduce the rodent population. Other rodent management methods involving trapping and environmental management can be more appropriate for the rural and peri-urban situations found in developing countries. Adopting an ecologically-based rodent management strategy is increasingly seen as more sustainable, both economically and environmentally, than the traditional use of acute poisons.

Step I: Know Your Enemy

As with any IPM strategy, the main principle is to "know your enemy". Not all rodent species are the same; each species has different breeding rates, habitats and species-specific behaviours. These factors will affect their pest status and the methods of control. For example, some rats like to live up high in trees or the roofs of people's houses, while others like to burrow in the ground or the walls of mud-brick houses. Knowing where rats live is important when targeting control actions.

Rodents are also highly adaptable, and the same species may exploit different foods or habitats when found in different environments. Once armed with the basic knowledge about the rodents, where and when they cause damage and the types and extent of damage caused to different crops, stored food and health, it becomes possible to address all the problems rats cause in an integrated way. This information improves peoples' understanding of the costs of doing nothing about rats on their livelihoods and allows an assessment of potential cost-benefits when developing a management strategy.

Step II: Know Your end User

In addition to understanding the local rodent biology and ecology, Ecologically-based rodent management must also consider the knowledge, attitudes and practices of the people affected. Effective rodent control practice must be based on the financial and time constraints of the people suffering from rodent pest problems. Rodent-human interactions can be complex, with rats seen as food, pests, and even involved in witchcraft or religious beliefs. Understanding existing practices and knowledge helps in the design of a strategy that will be locally acceptable and sustainable.

For example, few small-scale farmers understand the difference between acute and chronic rodent poisons, and will often choose acute poisons as they see dead bodies in the morning, which they rarely see when using chronic poisons. However, chronic poisons can work well and effectively reduce pest populations, but the effects are not so clearly seen as the poisoned rodents die in their burrows.

Step III: Know Your Technology

The use of rodenticides which work by interfering with blood clotting remains a powerful tool, particularly in urban environments and for large-scale agriculture. However, their financial and environmental sustainability is questionable for the majority of situations found in rural and peri-urban communities engaged in small-scale agriculture.

Fig. I, II, III: Showing live burrows and nibbled fruits. Source: Author

Not all pest species are the same. Knowing your enemy is the first step of a successful pest management approach. Because rats are mobile, moving over large distances in their daily foraging, the main principle of ecologically-based management is that farming communities must act together. Individuals acting on their own in their house or crop field will have little impact on the overall rodent population, with rats quickly migrating back into areas from where they have been removed. This implies that communities must coordinate and communicate effectively over a large scale, and it is important to encourage high levels of community cohesion for EBRM to be successful. This can be a challenge, particularly in more peri-urban situations. The cost-benefits of working together for rodent management means that individual investment costs are low, as the overall effort is shared by many. EBRM must, therefore, be a community-based effort.

Monitoring the Costs and Benefits of EBRM

The initial stages of implementing EBRM are often challenged with a lack of interest and doubt in local farming communities. This is because small-scale farmers who have tried to control rodent pests usually see very little benefit, often because their actions are ad-hoc, one-off, and unco-ordinated. And as is generally the case with any pest management, such actions are too little, and come too late. A final challenge in implementing EBRM is encouraging communities to assess success by looking at the changes in their lives, and not only at the number of dead rodents have they collected. These challenges favour education and extension programmes that strongly focus on demonstration and community participation.

Our work with EBRM in Asian countries showed a reduction in the impact of rodents by 60 – 80 percent for different measurable indicators. This was established through comparing intervention villages with non-intervention villages. Similarly, farmer assessments showed that these strategies cost about the same (in terms of money and time) as the former practices, but with a much higher benefit. As a result, the 3-step approach is now being extended widely through southern Africa *via* the *'Ecorat project'*. Once basic information is collected about the rats, end users and management tools, EBRM can be developed for a variety of local agro-ecological contexts. Once a few communities see the difference this type of management makes to their lives, up-scaling and dissemination to other nearby communities can occur through traditional extension channels. Rodent pests have been a largely neglected problem in the developing world, but an ecologically based approach can triumph where poisons alone have failed, particularly when communities work together to overcome the multiple impacts of rodents on their lives.

Constraints on Developing and Iimplementing EBRM

i. Basic biological knowledge

The complex biology and behaviour of rodents is probably the main reason why integrated pest management in rodent control has not evolved as fast as that for insect or weed control. For many rodent species, even for the common pest species, knowledge of their population ecology is far from adequate. In some instances even the taxonomy of the rodent pests is not well defined. Although a farmer may not care which species is destroying his crop, management strategies based on ecological characteristics require knowledge on which species are present, their habitat use, breeding patterns and population dynamics. Sometimes, taxonomic differences are of paramount importance to a rodent problem. For example, within the African genus *Mastomys,* there are two morphologically similar species that can be recognised only by the number of chromosomes. One of them, *Mastomys natalensis,* is resistant to, and a reservoir for, the plague *(Yersinia pestis),* while the other, *Mastomys coucha,* is very susceptible to the plague (Isaacson *et al.,* 1981). Knowing which species are present in an urban or semi-urban environment has major implications for management strategies, especially when resources are limited, as is the case in most eastern African rural communities. Krebs claims that it is now time to move on from descriptive studies of rodent pests to experimental work. Unfortunately this is true for a few pest species in Afroasia where sufficient knowledge is available. In most instances, however, more basic research is required. Information at community or landscape level is even more scarce.

The development of specialised techniques like immune-contraception requires a detailed knowledge of the reproductive physiology of the target species and a thorough understanding of its population ecology. In developing countries, such a broad knowledge base is available for just a few species of rodent pests. This shortage of biological knowledge often reflects a limited understanding of its importance by funding organisations, the low number of interested and adequately trained scientists, and the poor economic situation of many of the countries that we discuss here. Added to this is the reality that most of the research underpinning EBRM in developing countries has to be done *in situ,* where scientists are often working in isolation from international colleagues because of language problems, poor communication infrastructure and or international political issues.

ii. Ownership of rodent problems

In many countries, national rodent control programs or services have been established, often for good reasons. Yet, some of these programs contribute to a perception that farmers should participate in control actions directed by government technicians. Moreover, rodent problems are often considered to be a fact of life and a common view held by farmers is that nothing can be done about them. As a result, there is often not a strong commitment by farmers and other end-users in applying strategies that may require a long-term involvement.

Few governments in developing or developed countries are involved in early tactical management of rodent pests. Instead, governments generally involved when problems suddenly reach a level at which emergency actions are required. After this 'crisis management', rodent pests again fall back to the bottom of the political agenda. If farmers take a lead from how governments handle rodent pests, then it makes it harder to convince farmers that they need to adopt early tactical control rather than crisis management.

An ecologically-based approach also requires an acceptance that activities in one place may have a dampening effect on rodent populations in another place. For example, EBRM may recommend changes in land management or in irrigation schemes where the individual who applies the actions is not always the one (or the only one) who will benefit. Socioeconomic, cultural or political aspects may then dominate the development of EBRM strategies (Singleton and Petch 1994).

iii. Access to Information

EBRM is, by necessity, targeted to specific agro-ecosystems and pest species. Solutions are rarely generic; they cannot simply be transplanted between places. Therefore, large information campaigns cannot be set up easily. On top of that, the rather complex message of EBRM has to compete with straightforward techniques of pesticide application. Promotion of the latter is nearly always driven by commercial objectives, and therefore often is better organised and has more financial support. Where national authorities have a large influence on rodent control they will search for national solutions. These will not always take into account the specific local conditions that are important in EBRM.

iv. Investment capital

Most EBRM strategies are, aiming for long term results. In countries such as Indonesia, Tanzania and Vietnam, most farmers are risk-aversive, low-capital entrepreneurs. Thus, many will not invest in solutions which immediately are more expensive, even though they may payoff in the long run. Therefore, EBRM should be marginally more expensive to implement than existing practices, even if these practices are less effective and more expensive than EBRM in the long term.

Maintaining the momentum for ecologically-based rodent management

Training the next generation of scientists

Although we have discussed a number of factors which are likely to constrain the development and implementation of EBRM, the most basic and important of them is the lack of biological knowledge. Fortunately, there is a growing interest among zoologists and wildlife managers to spend the time and energy on understanding the population biology of rodent pests. In order to secure this development, particularly in developing countries, there is an urgent need for more Young population ecologists and wildlife managers levels (MSc., Ph.D.) with a strong emphasis on pure science, but in the context of applied, strategic research. Far too often there has been a wide gap between basic and applied research, leading to technologists providing incremental improvements on methods that are part of a classical but unimaginative template, while pure scientists conduct their research and gather new insights outside agricultural systems. Today, we have a need for scientists and managers to be adept in both fields.

Once young scientists have been trained formally, they require the opportunity to conduct strategic research. In an EBRM context this would consist of basic biological research under the field conditions where rodents cause problems. This kind of research will not quickly provide new techniques and solutions meaning. Also, the results of strategic research need to be transferred to practitioners. This technical transfer should focus on new EBRM methods that have been tested under field conditions in replicated, experimental studies conducted on an appropriate scale. It is very important that at this stage the science can be integrated with particular socio-economic systems.

Short-term Thinking for Long-term Problems

EBRM requires a long-term approach, both in development and in application. Much effort will be needed to convince policy makers. On the other hand, the impacts of rodent pests will not abate until we have developed appropriate EBRM techniques. Therefore, there will be continued pressure for urgent, short-term solutions, particularly in developing countries. Consequently, EBRM proponents must be prepared to develop interim management practices, even though they are unlikely to be sustainable.

Maintaining Credibility with Farmers and Other End-users

Ecologically-based rodent management will rarely provide spectacular quantities of dead rodents. Indeed, the concept of EBRM is to maintain rodent densities below levels that cause significant economic losses to agricultural produce or significant health problems. On the other hand, as discussed above, EBRM will require investments in material or labour. For these reasons, farmers may be sceptical about applying such methods. Therefore, a major challenge is to convince farmers about the benefits of EBRM.

An equally large challenge is to ensure that new methods are only promoted when they have proven their value. There is a risk that enthusiastic scientists, impatient funding organisations or politicians may promote strategies that are not yet well tested but that are popular. If these strategies are less effective than announced, the loss of credibility would make it difficult to later convince end-users to adopt an improved version. A typical example in Southeast Asia is the trap-barrier system plus trap crop. This method has good potential to be the cornerstone for developing EBRM in rice ecosystems, but a number of possible weaknesses have been identified. This management system is being strongly promoted by some government agencies in the region, yet its performance has not been assessed at the village level (Singleton *et al.*, 1998).

Incorporating and Integrating Different Strategies

As a holistic approach, EBRM will rarely focus on one single element in the rodent's biology. Management strategies will therefore implicate different techniques and methods. Some may be slight modifications of existing technologies, others may be more innovative. The challenge will be to find a balance, ensuring that focusing too much on one technique does not compromise the benefits of an EBRM approach.

The re-emergence of the importance of population ecology and an emphasis on management raises hopes that rodent pest management will begin to match the progress made by entomologists and botanists in controlling insect and weed pests. We are confident that the next decade will see rapid advances in ecologically-based rodent pest management. Therefore, by necessity, we will have to develop more environmentally benign and sustainable methods for rodent reducing our sole reliance on rodenticides as killing agents.

References

Advani, R. and Mathur, R.P. 1982. Experimental reduction of rodent damage of vegetable crops in Indian villages. *Agro-Ecosystems*, 8: 39-45.

De Graaf, G. 1981. The rodents of southern Africa. Pretoria, Buttersworth, 267p.

European food safety authority (EFSA) 2013. http://www.efsa.europ.eu/en/topics/topic/*salmonella*.htm.accessed 14 aug 2013.

Green, M. and Tay tor, KD. 1975. Preliminary experiments in habitat alteration as a means of controlling field rodents in Kenya. In: Hansson, L. and Nilsson, 8th ed., Biocontrol of rodents. Ecological Bulletin No. 19. Stockholm, Swedish Natural Science Research Council,175-181.

Hall, D.W. 1970. Handling and storage of food grains in tropical and sub-tropical areas. Rome, FAO (Food and Agricultural Organization of the United Nations), 350p.

Hart, K. 2001. Postharvest losses. In: Pimentel D, ed. Encyclopedia of pest management. New York: Marcel Dekker. Electronic web access at http://dekker.com/ servlet/product/DOI/10.1081-E-EPM-100200058/ main.

Herwig L., Grant R. S. and Hinds, Lyn A. 2014. Ecologically-based Rodent Management in developing countries.http://aciar.gov.au/files/node/323

Hiett, K.L., Stern, N.J., Fedorka-Cray P., Cox, N.A., Musgrove, M.T. and Ladely, S. 2002. Molecular subtype analyses of *Campylobacter* spp. from Arkansas and California poultry operations. *App. Environ. Microbiol*, 68: 6220-6236.

Huq, M.M., Karim, M.J. and Sheikh, H. 1985. Helminth parasites of rats, house mice and moles in Bangladesh. *Pakistan J. Vet, Sci,* 5 (3): 143-144.

Isaacson, M., Arntzen, L. and Taylor, P. 1981. Susceptibility of members of the *Mastomys natalensis* species complex to experimental infection with *Yersinia pestis. J. Infec. Dis,* 144: 80.

Karesh,W.B., Dobson, A., Lloyd- smith, J.O., Lubroth, J., Dixon, M., Bennet, M., Aidrich, S., Harrington, T., Formenty, P., Loh, E.H., Machalaba, C.C., Thomas, M.J. and Heymann, D.L. 2012. Ecology of zoonosis: natural and unnatural histories. *Lancet,* 380 (9857):1936-1945.

Kocher, D.K. and Kaur, N. 2013. Synergistic effect of bromadiolone and cholecalciferol (vitamin D3) against house rat, *Rattus rattus. Int. J. Res. in Bio. Sci,*2(1): 73-82.

Lacey, E.A. and Solomon, N.G. 2003. Social biology of rodents: trends, challenges, and future directions. *J. of Mammalogy,* 84(4): 1135-1140.

Leirs, H., Verhagen, R, Sabuni, c.A., Mwanjabe, P. and Verheyen, W.N. 1997. Spatial dynamics of *livastorms natalensis* in a field-fallow mosaic in Tanzania. *Belgian J. of Zoology,* 127:29-38.

Lloyd-smith, J.O., George, D., Pepin, K.M., Pitzer, V.E., Pulliam, J.R.C., Dobson, A.P., Hudon, P.J. and Grenfell, B.T. 2009. Epidemic dynamics at the human-animal interface. Science (New York), 326(5958):1362-1367.

Meerburg, G.M., Singleton, G.R. and Kijlstra, A. 2009. Rodent-borne diseases and their risks for public health. *Critical Rev.Microbiol,* 35:221-270.

Mwanjabe, P.S. 1993. The role of weeds on population dynamics of *Mastomys natalensis* in Chunya (Lake Rukwa) valley. In: Machang'u, RS., ed., Economic importance and control of rodents in Tanzania. Workshop Proceedings, 6-8 July 1992. Morogoro, Sokoine University of Agriculture, 34-42.

Myllymaki, A. 1987. Control of rodent problems by use of rodenticides: rationale and constraints. In: Richards, e. G.J. and Ku, T.Y., ed., Control of mammal pests. London, Taylor and Francis, 83-111.

Parshad, V. P. 1999. Rodents control in India. *Integrated Pest Manage. Rev,* 4:97-126.

Prakash, I., and Mathur, R. M. 1988. Rodents problems in Asia. pages 67-84 in I. Prakash, ed.1988. Rodent pest management CRC press. Boca Raton, FL USA 480 p.

Singleton, G.R. and Petch, D.A. 1994. A review of the biology and management of rodent pests in Southeast Asia.ACIAR Technical Report No. 30.*Canberra (Australia): Australian Centre for Int. Agri. Res,* p 65.

Singleton, G.R. 2003. Impacts of Rodents on Rice Production in Asia. *International Rice Research Institute (IRRI) Discuss paper Series,* 45: 1.

Singleton, G.R. and Petch, D.A. 1994. A review of the biology and management of rodent pests in Southeast Asia. Canberra, ACIAR Tech. Reports, 30, 65p.

Singleton, G.R., Sudarmaji and Suriapermana, S. 1998. An experimental field study to evaluate a trap-barrier system and fumigation for controlling the rice field rat, *Rattus,* in rice crops in West Java. *Crop Protection,* 17:55-64.

Steven, R. B. 2007. Rats: an ecologically-based approach for managing a global problem In: LEISA Magazine 23.4

Taylor, KD. 1968. An outbreak of rats in agricultural areas of Kenya in 1962. *East African Agric. and Forestry J.* , 34: 66-77.

Taylor, KO. and Green,M.G. 1976. The influence of rainfall on diet and reproduction in four African rodent species. *J. of Zoology,* 175:453-471.

Tripathi, R.S. 2007. Recent trends in Coordinated Research on Rodent Control. *Rodent Newsletter,* **31**: 1-4.

World Health organization WHO 2013. Zoonosis: anthrax. http://www.who.int/vaccine_ reseach/ Sdiseases/zoonosis/en/index1 .html.accessed 19 aug 2013.

15

Vermicompost to Metagenomics for Agriculture Sustainability

Tejpal Dahiya[1*] and Ravikant[2]

[1]Department of Zoology, Government College for Women, Tosham-127040 (Bhiwani) Haryana

[2]Department of Zoology, CCSHAU Hisar-125004 Haryana India

*Corresponding author: dahiyatejpal@yahoo.co.in

Abstract

Vermicomposting is an environmental friendly, low-technology process used to treat organic waste and the resulting Vermicompost, has positive impacts on plant growth and health in agriculture. This can be used to improve soil structure and fertility which hence plant growth, development and yield. It also minimise the risk of spreading human or animal or plant pathogens by recycling organic matter back into the food production chain. In metagenomics studies, microbial composition of vermicast is characterised. It is an efficient technology and contribute in better understanding of the process of vermicomposting also changes the microbial community in it. As the organic matter passes through the gizzard of the earthworm it is grounded into a fine powder after which the digestive enzymes, micro-organisms and other fermenting substances act on them further aiding their breakdown within the gut, and finally passes out in the form of "casts" which are later acted upon by earthworm gut associated microbes converting them into mature product, the "vermicomposts". Earthworms influence soil physical, chemical and biological properties due to their role, these are considered as 'Soil Engineers' and indicators of soil quality.

Keywords: vermicompost, techniques, metagenomics, agriculture, sustainability

Introduction

The production of organic waste from municipalities, industry, domestic and agricultural farm, sources have led to a problem of landfill space, its proper disposal and pose serious threat to the environment. The compostable materials *viz.* paper, food waste, dung, fallen leaves, dead animals and their body organs and grass clippings; create a need to process and recycle nutrients, which intern reduces the negative environmental impacts of insufficient organic waste management. The release of unprocessed organic waste such as animal manure, crop residues and household waste can pose a potential risk of spreading human; animal and plant pathogens. For this composting is carried out by the farmers without the use of earthworm. It is a controlled aerobic process that degrades organic waste to stable material, with resident microbial community mediating the biodegradation and conversion process. In this procedure organic waste material is kept in ditches for longer period to act upon by microbes.

Earthworms

Earthworms are capable of transforming garbage into 'gold'. Charles Darwin described earthworms as the 'unheralded soldiers of mankind'; similarly Aristotle called them as the 'intestine of earth', due their capability to digest a wide variety of organic materials (Darwin and Seward 1903; Martin 1976). The soil volume, microflora and fauna influenced by earthworms is termed as "drilosphere" and the soil volume includes the external structures produced by earthworms such as surface and below ground casts, burrows, middens, diapauses chambers as well as the earthworms body surface and internal gut associated structures in contact with the soil (Brown *et al.,* 2000).

Earthworms play an essential role in carbon turnover, soil formation, participates in cellulose degradation and humus accumulation. Their activity profoundly affects the physical, chemical and biological properties of soil. These are voracious feeders of organic wastes and these utilize only a small portion of these wastes for their growth and excrete a large proportion of wastes consumed in a half digested form (Jambhekar 1992). Its intestine contains a wide range of micro-organisms, enzymes and hormones help in rapid decomposition of half-digested material transforming them into vermicompost in a short time of nearly 4–8 weeks (Nagavallemma *et al.,* 2004) in comparison to traditional composting process which takes the advantage of microbes alone and thereby requires a prolonged period (nearly 20 weeks) for compost production (Sanchez-Monedero *et al.,* 2001). As the organic matter passes through the gizzard of the earthworm it is grounded into a fine powder after which the digestive enzymes, micro-organisms and other fermenting substances act on them further aiding their breakdown within the gut, and finally passes out in the form of "casts" which are later acted upon by earthworm gut associated microbes converting them into mature product, the "vermicomposts" (Dominguez and Edwards 2004).

Earthworms are classified into epigeic, anecic and endogeic species based on definite ecological and trophic functions (Bhatnagar and Palta 1996).

a). Epigeic Eearthworms

These are smaller in size, having uniformly pigmented body, with short life cycle, high reproduction rate and regeneration. These dwell in superficial soil surface within litters, feeds on the surface litter and mineralize them. These are phytophagous and rarely ingest soil. These contain an active gizzard and aids in rapid conversion of organic matter into vermicomposts. In addition to this, these are efficient bio-degraders and nutrient releasers, tolerant to disturbances, aids in litter comminution and early decomposition and hence can be efficiently used for vermicomposting. These include *Eisenia foetida, Lumbricus rubellus, L. castaneus, L. festivus, Eiseniella tetraedra, Bimastus minusculus, B. eiseni, Dendrodrilus rubidus, Dendrobaena veneta, D. octaedra.*

b). Endogeics Earthworms

These are smaller to larger sized worms, with slightly pigmented body, have life cycle of medium duration, moderately tolerant to disturbance, forms extensive horizontal burrows and they are geophagous i.e. feeding on particulate organic matter and soil. These bring about pronounced changes in soil physical structure and can efficiently utilize energy from poor soils, hence can be used for soil improvements. These include *Aporrectodea caliginosa, A. trapezoides, A. rosea, Millsonia anomala, Octolasion cyaneum, O. lacteum, Pontoscolex corethrurus, Allolobophora chlorotica* and *Aminthas* sp. These are further classified into three type i.e. polyhumic endogeic earthworms with small sized, rich soil feeding earthworms, dwelling in top soil (A1); mesohumic endogeic earthworms with medium sized worms, dwelling in A and B horizon, feeding on bulk (A1) soil; and oligohumic endogeic earthworms: these are very large worms, dwelling in B and C horizons, feeding on poor, deep soil.

c). Aneceics

These are larger, dorsally pigmented worms, with low reproductive rate, sensitive to disturbance, nocturnal, phytogeophagous, bury the surface litter, forms middens and extensive, deep, permanent vertical burrows, and live in them. Formation of vertical burrows affects air water relationship and movement from deep layers to surface helps in efficient mixing of nutrients. *Lumbricus terrestris, L. polyphemus* and *Aporrectodea longa* are examples of aneceics earthworms (Kooch and Jalilvand 2008).

Epigeics and aneceics are used largely for vermicomposting (Asha *et al.,* 2008). Epigeics namely *Eisenia foetida, Eudrilus eugeniae, Perionyx excavatus* (Suthar and Singh 2008) and *Eisenia anderi* (Munnoli *et al.,* 2010) have been used in converting organic wastes into vermicompost. These surface dwellers capable of working on litter layers converting them into manure are of no significant value in modifying the soil structure. In contrast, aneceics such as *Lampito mauritii* are efficient creators of an effective drilosphere as well as excellent compost producers. Earthworms thus act as natural bio-reactors, altering the nature of the organic waste by fragmenting them.

Vermicomposting

It is a non-thermophilic biological oxidation process in which organic material are converted into vermicompost which is a peat like material, exhibiting high porosity, aeration, drainage, water holding capacity and rich microbial activities (Arancon *et al.*, 2004), through the interactions between earthworms and associated microbes. Vermiculture is a cost-effective tool for waste management (Asha *et al.*, 2008). Earthworms are the crucial drivers of the process, as they aerate condition and fragment the substrate and thereby drastically alter the microbial activity and their biodegradation potential (Lazcano *et al.*, 2008). Several enzymes, intestinal mucus and antibiotics in earthworm's intestinal tract play an important role in the breakdown of organic macromolecules. Biodegradable organic wastes such as crop residues, municipal, hospital and industrial wastes pose major problems in disposal and treatment. Using unprocessed animal manures into agricultural fields contaminates ground water causing public health risk, so, vermicomposting is the best alternative to conventional composting and differs from it in several ways; it fastens the decomposition process by 2–5 times, thereby quickens the conversion of wastes into valuable biofertilizer and produces much more homogenous materials compared to thermophilic composting (Atiyeh *et al.*, 2000).

Distinct differences exist between the microbial communities found in vermicomposts and composts and hence the nature of the microbial processes is quite different in vermicomposting and composting. The active phase of composting is the thermophilic stage characterized by thermophilic bacterial community where intensive decomposition takes place followed by a mesophilic maturation phase (Vivas *et al.*, 2009). Vermicomposting is a mesophilic process characterized by mesophilic bacteria and fungi.

Vermicomposting comprises of an active stage during which earthworms and associated microbes jointly process the substrate and the maturation phase that involves the action of associated microbes and occurs once the worm's moves to the fresher layers of undigested waste or when the product is removed from the vermireacter. The duration of the active phase depends on the species and density of the earthworms involved (Aira *et al.*, 2011). A wide range of organic wastes *viz.*, horticultural residues from processed potatoes, mushroom wastes, horse wastes, pig wastes, brewery wastes, sericulture wastes, municipal sewage sludge, agricultural residues, weeds, cattle dung, industrial refuse such as paper wastes, sludge from paper mills and dairy plants; domestic kitchen wastes, urban residues and animal wastes can be vermicomposted (Sharma *et al.*, 2005).

In composting there are three successional phases driving chemical and microbial changes through time. These phases are determined primarily by the changes in temperature. The first phase is mesophilic phase, in which moderate temperature rises to 45 degree Celsius. The second phase is thermophilic, in which the temperature of composting material is rise to 70 degree Celsius and the third is Curing phase, in which the composting material is cooled to ambient temperature (Ryckeboer *et al.*, 2003 and Ashraf *et al.*, 2007). The high temperature of thermophic phase is useful in removal of plant and animal pathogens. It has higher electrical conductivity that's why compost is less preferred in horticulture (Tilman *et al.*,

2002). The primary types of commercial composting methods are windrow, aerated static pile (ASP) and vermicomposting.

Windrow

Windrow involves placing a mixture of organic waste materials into long, narrow piles on a composting pad which are turned frequently (Ashraf *et al.*, 2007). Piles were mixed with a bucket loader and were capped with manure/bedding and were managed to maintain a temperature between 55–77°C for a minimum of 15 days and turned with a bucket loader a minimum of five times to ensure all materials have been subjected to the minimum temperature requirements.

Aerated Static Pile (ASP)

ASP systems force air throughout the pile and does not require turning once the pile has been formed, thus allowing for larger piles to be produced. Piles were mixed with a bucket loader, placed in a three-sided ASP bay, and capped with manure/bedding. Each ASP pile was built to a height of 2 to 2.5 m after settling. Initial piles were 57–76 m^3 when placed in ASP bays. Piles were aerated in place with an in-floor air delivery system that uses a 20 cm layer of wood chips between the duct-work and pile to evenly distribute air. Blower fans were managed with speed control and timers, to meet 'Process to Further Reduce Pathogens' (PFRP) requirements as dictated by Vermont and NOS regulations. Briefly, piles were aerated in the ASP system for 3–6 weeks so the pile attained temperatures of 55°C for a minimum of three days. Piles were then re-stacked on composting pads, and turned regularly to continue composting (Ashraf *et al.*, 2007).

Vermicompost Method

Vermicompost is a mesophilic process that employs earthworms to stabilize organic residues. Material entering the Continuous-Flow Worm Reactor was taken from piles on the composting pad, after they have been through the ASP procedure (outlined above), and re-stacked on the pads. Material had already met PFRP, and was generally four to six weeks old. Fresh material (0.76 m^3) was spread out weekly in a 3.8 cm layer on top of the bed, where it was allowed to continue to decompose, and be consumed by earthworms (*Eisenia fetida*). The bed was 1.52 m wide, 12.19 m long and 0.6 m deep. The worm bed was housed in an indoor, heated room, and the compost temperature was 21–27°C. The compost remained in the worm bin for 60 to 90 days, the time it took for the fresh compost to be decomposed and move downward and out of the bottom of the bed. For the better conversion of organic material to usable form vermicomposting is being carried out and it doesn't have thermophilic phase (Ahmad *et al.*, 2007; and Anastasi *et al.*, 2005).

In vermicomposting, the same feedstock is passed across the alimentary canal of earthworm. It may be called as bio-oxidation of organic matter facilitated by worms. The most commonly used earthworms in vermicomposting are *Eisenia foetida* and

Eudrilus eugeniae. It is produced under mesophilic conditions, and although micro-organisms degrade the organic matter biochemically but earthworms are the crucial drivers of the process. Earthworms has very important role in this process.

Metagenomic Analysis

For the metagenomical analysis of vermicompost produced from the organic matter, a standard commercial recipe of HCC is composted through windrow piles, aerated static pile, and vermicompost. Then the samples are collected from the various ages throughout the thermophilic and curing phases of the compost process for windrow, aerated static pile, and vermicomposting. Here, it is advised to the researcher that the samples must be analysed in duplicate at each time point for each composting method for getting more information with accuracy.

DNA extraction

Genomic DNA may be extracted by using many DNA extraction and isolation kits (Lauber *et al.,* 2009), for example the MoBio PowerSoil™ kit (MoBio, Carlsbad, CA, USA).

PCR (Polymerase Chain Reaction) Amplification

PCR amplification of the 16S rRNA gene (for bacteria and archaea) or the internal transcribed spacer region (ITS1) of the nuclear ribosomal RNA gene (for fungi) followed the standard approach. For the amplification each sample must amplified in triplicate, and the amplicons must be composited together in equimolar concentrations prior to sequencing. PCR reactions may be carried out by using PCR-grade water, 5 Prime Hot Master Mix, each of the forward and reverse primers and genomic DNA (diluted 1◎10 with PCR-grade water). Reactions should be carried out at 94°C for 3 min to denature the DNA, with amplification proceeding for 35 cycles at 94°C for 45 s, 50°C for 60 s, and 72°C for 90 s; a final extension of 10 min at 72°C was added to ensure complete amplification (Fierer *et al.,* 2012; Caporaso *et al.,* 2012).

Sequencing

For sequencing and analysing the bacterial and archaeal microbials, the PCR primers targeted the V4 region of the 16S rRNA gene amplified from bacterial samples and for the fungal analyses, PCR primers (ITS1-F/ITS2) may be used to amplify the ITS1 spacer. These both primer pairs containing a particular number of base pairs barcodes which are unique to each sample and the appropriate adapters to permit sequencing on the Illumina MiSeq platform (Fierer *et al.,* 2012; Caporaso *et al.,* 2012; Gardes and Bruns 1993).

Data Analysis

Quality filtering, assignment of sequences to samples based on their barcodes, and clustering of sequences into operational taxonomic units (OTUs) should be done following the standard QIIME pipeline with sequence data quality-filtered. OTUs may be determined by using an open reference-based approach that implements reference-based clustering followed by *de novo* clustering using the UCLUST algorithm (Caporaso *et al.*, 2010). Clustering may be conducted at the 97% similarity level using pre-clustered versions of the October 2012 Green genes database (for 16S rRNA) and November 2012 UNITE database (fungal ITS gene) for the sequence reference set (McGuire *et al.*, 2013). Sequences may be assigned to taxonomic groups using the RDP classifier. To keep sequencing depth consistent across all samples, the sequence data may be rarefied by randomly sub-sampling 2,000 and 100 reads per sample before downstream analyses of the 16S and fungal ITS rRNA datasets, respectively (Wang *et al.*, 2007; Edgar R. C. 2010; and McDonald *et al.*, 2011). Amplicon sequences deposited in the public EMBL-EBI database (http://www.ebi.ac.uk/) and may be accessed using their respective accession numbers.

Statistical Analysis

It may be done by using the PRIMER v.6 software package (PRIMER-E, Plymouth, WA, USA) for the calculation of pair-wise differences in community composition (Bray-Curtis distances) and the subsequent analyses of the pair-wise dissimilarity matrices *via* principal coordinate analysis (Clarke and Gorley 2001; and R Core Team 2012) and permutational multivariate analysis of variance (PERMANOVA). PERMANOVA is used to assess the effects of compost type and process type on the composition of the bacterial and fungal community compositions. For compost type, the farm identity may be included as a random factor in account for variation between farms. The differences in the relative abundance of specific taxa across recipe and methods may be determined using multiple Kruskal-Wallis tests in *R* and applying false discovery rate corrections to *p*-values to account for the multiple comparisons (Deborah *et al.*, 2013). Tests should be only performed for the more abundant taxa (those with median relative abundances greater than 1.0% in any of the recipes or processes).

Role of Earthworm in Vermicomposting

i) **Aeration**: Earthworms aerate the organic matter by moving in between and replacing top material with the lower layer. It eats the organic matter and breaks it into small particles which adsorbed more molecules of air gases *viz.* oxygen and nitrogen.

ii) **Conditioning and fragmentation of substrate**: When the organic waste moves across the alimentary canal of earthworms; the bigger sized particles are broken down into smaller particles; this process of breaking down bigger organic matter particles into smaller particles is called as fragmentation. During the process of fragmentation, a large amount of mucus substance is mixed with organic matter and also the soil particles eaten by the earthworms along with organic matter; mucus makes it soft and peat like substance, which become favourable for the growth of plants in it. The conditioning and fragmentation further reduces the C/N ration which results into more availability of Nitrogen for the plants.

iii) Surface Area: Earthworms increases the surface area of feedstock by fragmenting it into smaller particles, due to which more micro-organisms can act upon it and make it more suitable for agricultural use.

iv) The passing of organic matter across the alimentary canal of earthworms make more favourable for micro-organisms to act it upon.

v) Earthworms after passing organic matter from their alimentary canal convert it into more uniform sized and finally give it earthy appearance with heterogeneous appearance.

vi) Earthworm activity engineers the soil by forming extensive burrows which loosen the soil and makes it porous. These pores improve aeration, water absorption, drainage and easy root penetration. Soil aggregates formed by earthworms and associated microbes, in the casts and burrow walls play an indispensable role in soil air ecosystem. These aggregates are mineral granules bonded in a way to resist erosion and to avoid soil compaction both in wet and dry condition. Earthworms speed up soil reclamation and make them productive by restoring beneficial microflora. Thus degraded unproductive soils and land degraded by mining could be engineered physically, chemically and biologically and made productive by earthworms. Hence earthworms are termed as ecosystem engineers (Munnoli *et al.*, 2010).

Sustainable Agriculture

It can be defined as a set of practices that conserve resources and the environment without compromising human needs by use of organic fertilizers such as animal manure. Sustainable agriculture leads to soil conservation, environmental safety and good food quality (Tilman *et al.*, 2002).

The results of several long-term studies have shown that the addition of compost improves soil physico-chemical and biological properties:

Soil Physico-chemical Properties

a) **Bulk density**: Vermicompost decreases the bulk density of the soil after a long-term use of it.

b) **Soil water holding capacity**: It increases the soil water holding capacity, due to which soil hold more water and do not allow the irrigated or rain fed water to seep deep up to water table. The soil requires less amount of water in comparison to chemical fertilizers used soil.

c) **Soil organic carbon**: Use of Vermicompost increases organic carbon in the soil, Vermicomposting of paper mill and dairy sludge resulted in 1.2–1.7 fold loss of organic carbon as CO_2. In contrast to the parent material used, vermicomposts contain higher humic acid substances (Albanell *et al.*, 1988).

d) **Nutrients**: It increases the amount of some nutrients into the soil. These are required by the plants and lead to better crop quality and productivity. Thus vermicomposting increases the ash content and accelerates the rate of mineralization which is essential to make nutrients available to plants. The observed increase of total phosphorous (TP) in vermicompost is due to mineralization and mobilization of phosphorus resulting from the enhanced phosphatase activity by micro-organisms in the gut epithelium of the earthworms (Garg *et al.*, 2006). Vermicomposts showed a significant increase in exchangeable Ca^{2+}, Mg^{2+} and K^{+} compared to fresh sludge indicating the conversion of nutrients to plant-available forms during passage through the earthworm gut (Yasir *et al.*, 2009).

e) **Humic/Fulvic acid ratio**: Vermicompost increases H/F ratio, which lead to better soil health. Humic acid substances occur naturally in mature animal manure, sewage sludge or paper-mill sludge, but vermicomposting drastically increases the rate of production and their amount from 40–60 percent compared to traditional composting. The enhancement in humification processes is by fragmentation and size reduction of organic matter, increased microbial activity within earthworm intestine and soil aeration by earthworm feeding and movement (Dominguez and Edwards, 2004).

f) **Cation exchange capacity**: It increases cation exchange capacity of soil which results into base saturation. Which intern stabilise the soil and pH remains constant.

g) **C/N ratio**: use of Vermicompost reduces C/N ratio increases more availability of nitrogen to the growing plants. C:N ratio is an indicator of the degree of decomposition. During the process of bio-oxidation, CO_2 and N are lost and loss of N takes place at a comparatively lower rate. Comparison of compost and vermicompost showed that vermicompost had significantly less C:N ratios as they underwent intense decomposition (Lazcano *et al.*, 2008).

h) **Electrical conductivity (EC):** EC indicates the salinity of the organic amendment and minor production of soluble metabolites such as ammonium and precipitation of dissolved salts during vermicomposting lead to lower EC values. As compared to the parent material used, vermicomposts contain less soluble salts and greater cation exchange capacity.

i) **pH**: Earthworm activity reduced pH and C:N ratio in vermicompost. The Chemical analysis of vermicompost showed, a lower pH, EC, organic carbon (OC), C:N ratio, nitrogen and potassium and higher amounts of total phosphorous and micronutrients in comparison to the parent materi*al.*, Slightly decreased pH values of vermicompost compared to traditional compost might be attributed due to mineralization of N and P, microbial decomposition of organic materials into intermediate organic acids, fulvic acids, humic acids (Lazcano *et al.,* 2008) and concomitant production of CO_2 (Garg *et al.,* 2006).

j) **Mineralization:** Mineralization is the process in which the chemical compounds in the organic matter decompose or oxidise into forms that could be easily assimilated by the plants. Increase in ash content increases the rate of mineralization and during vermicomposting there is a significant increase in ash concentration suggests that vermicomposting accelerates the rate of mineralization (Albanell *et al.,* 1988). Vermicomposting of cow manure using earthworm species *E. andrei* (Atiyeh *et al.,* 2000) and *E. foetida* (Hand *et al.,* 1988) favoured nitrification, resulting in the rapid conversion of ammonium-nitrogen to nitrate-nitrogen; and increased the concentration of nitrate-nitrogen to 28 fold after 17 weeks, while in conventional compost there was only 3-fold increase (Atiyeh *et al.,* 2000).

k) **Controlling Pollution:** Ash is an alkaline substance which hinders the formation of H_2S as well as improves the availability of O_2 and thereby renders composts odourless. Vermicomposts contain higher nutrient concentrations, but less likely to produce salinity, than composts. Additionally, vermicomposts possess outstanding biological properties and have microbial populations significantly larger and more diverse compared to conventional composts. Soil supplemented with vermicompost showed better plant growth compared to soil treated with inorganic fertilizers or cattle manure.

Biological properties of soil

i) **Microbial biomass**: It increases microbial biomass, which includes bacteria, fungi, viruses and protozoans. These are mostly useful microbes and lead to better soil health and more productivity of agricultural crops.

ii) **Microbial community**: Use of vermicomposting in the agricultural fields is resulting in the establishment of microbial communities (bacteria, fungal) in the rhizosphere of the root of the plants

iii) **Microbial activity**: long-term use of vermicompost lead to a great increase microbial diversity and further microbial activity in the soil. Which lead to Nitrogen fixation and changing it to nitrates.

iv) **Soil enzymes**: vermicompost is produced from the organic waste by passing it from the alimentary canal of earthworms; during the passing vermicompost receives a large number of useful enzymes which further enhance the soil quality and lead to easily availability.

Quality of Vermicompost and Agriculture

The effects of vermicompost on agricultural and horticultural crops are depending upon many factors. These are given below:

a) **Cultivation system:** The positive effect of vermicompost on crop plants depends upon the cultivation system i.e. spacing in-between the plants and either theses are planted in row or broadcasted by the hand; and line to line spacing also matter a lot.

b) **Properties of vermicompost:** The physico-chemical and biological properties of vermicompost play a very important role in the productivity of the crops (discussed earlier).

c) **Feedstock quality:** The quality of original feedstock used by the earthworms also related to the quality of vermicompost. It is also depends upon the earthworm preferences towards the feedstock.

d) **Species of earthworms:** It also depends upon the species of earthworms used to produce vermicompost. In vermicomposting mostly *Essinia feotida* is used.

e) **Production methods:** The process and methods used to produce vermicompost also intern affects the quality of vermicompost, which further directly lead to agricultural and horticultural crop productivity.

f) **Age of vermicompost:** Lastly, the age of vermicompost

Effects of Vermicompost on Plant Growth

i) Vermicompost significantly stimulates the growth of plants species (tomato, pepper, garlic, aubergine, strawberry, sweet corn, green gram and some aromatic and medicinal plants), cereals, rice, fruits (banana, papaya) ornamental plants such as geranium, marigold, petunia and chrysanthemum; when applied in right proportions.

ii) In forestry plant species such as acacia, eucalyptus and pine tree, when applied positive effects on growth and productivity has been observed when vermicompost is applied long-term.

iii) It may be used as partial or total substrate as a mineral supplier to the large number of plant species.

iv) It stimulate seed germination in several plant species *viz.* green gram, tomato plants, petunia and pine tree; when applied during the soil preparations.

v) Vermicompost has positive effects on vegetative growth of plants. It stimulates shoot and root development. It alters the seed morphology, increased leaf area and causes extensive root branching in the agricultural plants.

vi) It stimulates plant to better flowering. Use of vermicompost in the agriculture lead to increase in the number and biomass of the flowers produced. In tern, it leads to increase in fruit yield or productivity of the crops.

vii) Use of vermicompost to the vegetable crops is found to very fruitful because it increases the nutritional quality of some vegetable crops *viz.* tomatoes, chineese cabbage, spinach, strawberries, lettuce and sweet corn.

viii) It accelerates the maturation process of plants in many species. Which lead to harvesting of crops well in time and further farmers will gain good price of their crops.

ix) Using vermicompost in many species of plants *viz.* maritime, pine and others; lead to increased nutrient uptake capacity and nutrient use efficiency by modifying root exudation patterns.

x) After application of vermicompost in sweet corn, a significant microbial community is observed in rhizosphere, which is beneficial to the crops.

xi) In the light of above point, it is clear that the vermicompost constitutes a promising alternative to inorganic fertilizers in promoting the plant growth.

Plant Growth Mechanisms

Vermicompost is a microbiologically-active, nutrient-rich, peat-like substance, which when added to plant growing media may influence plant growth directly or directly through different chemical, physical and biological mechanisms. In both kinds of mechanisms, the proposed mechanisms are competition, antibiosis and parasitism (Lazcano and Dominguez, 2011).

Direct Mechanisms

Vermicompost has direct effects on the plant growth by slow-release of macro- and micronutrients. It enhances the plant growth by the following mechanisms:

a) The feedstock or organic matter is taken up by the earthworms which are changed into vermicompost. This vermicompost has humic substances in it. This humic substance has similar compound to Indole-3-acetic acid (IAA) which induces lateral root initiation and growth by promoting the activity of H^+-ATPase enzyme activity across the cell membrane.

b) These humic substances present in vermicompost are absorbed by the plant and further these activate the transcription of Auxin responsive genes.

c) It is further confirmed by analysing the extracts obtained from *Aporrectoda caliginosa, Lumbriscus rubellus, L. terrestris* and *Eisinia fetida* have Indole compounds in it.

Indirect Mechanisms

Vermicompost has a wide range of indirect effects on plant growth, development and maturation

a) It mitigates or suppresses the plant diseases (microbial, insect pests and plant parasitic diseases caused by nematodes)

b) Antibiosis: There are many biological suppressive agents present in vermicompost mitigate many pathogens *viz. Pythium, Rhizoctonia,*

Verticellium and *Plectosporium* and fungal pathogens (*Phytophthora cryptogea, P. nicotianae, Botrytis cinerea, Sclerotinea sclerotiorum, Corticium rolfsii, Rhizoctonia solani, Fusarium oxysporium* and *F. lycopersici.*

c) Vermicompost which when added to the crops, significantly reduces the incidents of insect pests, mites (*Heteropsylla cubana*), the sucking insect (*Aproaerema modicella*), jassids, aphids, beetles and spider. Vermicompost reduces plant infestation as well as population of plant parasitic insects in the soil.

Metagenomics of Vermicompost

Agricultural wastes, dung of domestic animal (cow, buffaloes and others) and wastes from households are used as a starting material for vermicomposting. It has a large number of bacteria, viruses and fungi in it; after passing across the earthworms' alimentary canal.

Viruses

The vermicompost has both types of DNA and RNA viruses. It contains around 90% and 10% DNA and RNA viruses respectively. It contains Phage particles of DNA viruses. These are Microviridae, Siphoviridae, Podoviridae and Myoviridae families of viruses. The viruses of these virus families are mainly found to be associated with a wide variety of bacterial species. Vermicompost also has insect infecting virus families namely Iridoviridae; which present in greater proportions i.e. around 20% of total number of species. It has only Picornaviridae family of viruses, which is mainly associated with the intestinal diseases and disorders in animals and human being (Anne-Lie Blomstrom *et al.,* 2016).

Vermicompost is also has a large number of RNA viruses, which belongs to Discitroviridae, Partitiviridae and Leviviridae families. It also has plant viruses belonging to Sobemoviridae (Anne-Lie Blomstrom *et al.,* 2016).

Bacteria

Around the world a large number of scientists has isolated and identified a large number of bacterial species from the vermicompost. It is observed that the vermicompost has a greater number of bacterial species in it as compared to the starting feedstock materi*al.,* A large number of bacteria are isolated and identified mainly from the three portions / layers *viz.* top (*Aeromonas* spp., *Bacteriodes* spp., *Mesorhizobium* spp., *Rhizobium* spp. and *Clostridium* spp.), middle (*Aeromonas* spp., *Salmonella* spp., *Mesorhizobium* spp., *Rhizobium* spp. and *Clostridium* spp.) and bottom (*Methylocystis* spp., *Bradyrhizobium* spp., *Mesorhizobium* spp., *Rhizobium* spp. and *Gemmata* spp.) of the vermibed/vermicast (Anne-Lie Blomstrom *et al.,* 2016). These are discussed below:

Phylum-Proteobacteria

This phylum has a large number of bacterial species which are associated with N_2-fixation in the agricultural soil. This Phylum-Proteobacteria is further divided in to six classes but out of these six classes, the three classes namely- Zetaproteobacteria, Alphaproteobacteria and Gammaproteobacteria are found in vermicompost. These classes of bacteria included useful species of bacteria but the species belonging to orders Rhizobiales, Aeromonades, Burkholderiales has potential pathogenic bacteria namely *Brucella* sp., *Yersinia* sp., *Salmonella* sp. (an enetrobacteria) and *Escherichia coli* are present in soil, water and in vermicompost (Anne-Lie Blomstrom *et al.*, 2016)

Phylum-Actinobacteria

It contains *Microbacterium* sp., *Mycobacterium* sp., *Corynebacterium* sp., *Streptomyces* sp. and *Cellulomonas* sp. (Anne-Lie Blomstrom *et al.*, 2016)

Phylum-Firmicutes

This phylum has *Clostridia* species of bacteria. The CFU (cell forming unit) of these bacterial species is found to be higher in comparison to starting material (substrate). There is significant improvement in the removal of pathogenic bacteria i.e. *Salmonella* sp., faecal coliforms, enteric viruses and helminthes ova in comparison to the system without earthworms. Vermicomposting reduces the load of pathogenic microbes. In metagenomics analysis and investigation, *E. coli*, *Enterobacter* spp., *Enterococcus* spp. and *Clostridium* spp. were detected and; *Clostridium* spp. was detected in a higher abundance. This *Clostridium* spp. is anaerobic sulphite-reducing bacteria commonly found in the environment and, in compost and vermicompost, it decomposes the cellulose. It is also observed that handling of vermicompost in comparison to organic material is easy and safe due to reduced temperature because of reduction in ammonia and carbon lowers the heat production in the vermicompost (Anne-Lie Blomstrom *et al.*, 2016).

Fungi in Vermicomposting

The fungi function in vermicomposting both as a source of enzymes to help to facilitate decomposition as well as a preferred food source for the earthworms (Pizl and Novakova 2003), correspondingly, earthworms affect fungal populations through grazing and dispers*al.*, Fungi have been found to be favoured food source for earthworms, which demonstrate preferential feeding of specific fungi (Cooke 1983). Earthworms have been found to impact fungal community composition in vermicompost by influencing spore germination and creating microsites that may or may not favour the development of fungi (Teunov and Scheu 2000), although there is some contradictory evidence of this. It has been demonstrated that *Eisenia andrei* can survive on fungi as its sole food source. Although earthworms consume fungi along with organic matter, fungal biomass and diversity in the finished vermicompost is generally equal to or higher than, the fresh substrate (Pramanik 2007). Fungal biomass in vermicompost is also higher when earthworms are present and lower when earthworms leave the substrate (Aira *et al.*, 2007). The species of fungi acive in vermicomposting can be variable by the site, substrate utilized and batch. One study using a culture-depending next generation sequencing approach found members of the Arthrobotrys (Nematode –Trapper fungi), Pezzales, Microascaecae, Zopfiecella, Agaricomycetes and Mortierella to the highest in abundance in vermicompost (Neher *el al.*, 2013). This is in contrast to the dominant fungi, observed in Windrow composting, while included unclassified Sordariomycetes, Acremoium amd unclassified Basidiomycota group and those found in aerated static pile composting, which included an unclassified family of Pezzales, various unclassified Sordariomycetes and unclassified Sordariales.

Another study using traditional culture-based methods found *Aspergillus fumigatus, A. flavus, Geotrichum candidumi, Penicillium expansum, P. roquefortii, Fusarium ventricosum* and *Rhizopus stolonifer* to be the dominant fungal taxa in vermicompost, with *A. fumigatus, G. candidumi, P. expansum, Mucor circinelloides* and *Scopulariopsis brevicaulis* dominating the intestine of the earthworms (Pizl and Novakova 2003). These fungal taxas are are predominant in saprotrophs of food waste. Vermicompost is generally found to harbour a greater diversity of fungi than the starting substrate. Vermicomposting also tends to result in a greater diversity of fungi than Windrow or aerated state pile composting, possibly due to the fact that the digestive tract of earthworms houses taxonomically distinct communities of fungi that add to the overall fungal diversity of vermicompost (Neher *et al.*, 2013). However, many of the fungi found in vermiculture substrate and earthworm intestine are typical of those normally involved in soil organic matter decomposition. Earthworm-processed substrate tends to have increased mycelial biomass in comparison to starting substrates. Earthworms have been found to activate fungal growth, triggering increased cellulose decomposition during vermicomposting.

Vermicompost Employed as a Suppressive Agent towards Phytopathogens

Prepared vermicompost has been shown to be suppressive towards plant pathogens such as *Fusarium oxysporum, Phytophthora nicotianae, P. infestans* and *Plasmodiophora brassicae* (Szczech 1999). A vermicompost water extract was also shown to be effective in controlling *Blumeria graminis horde*, a powdery mildew of barley. On study characterized the mechanisms of phytopathogens, suppression by comparing the effects of heat-sterilized versus unsterilized vermicompost on the suppression of *Fusarium oxysporium lycopersici* (Szczech 1999). It is observed that vermicompost lost its suppressive activity after heating and that sterilized water extracts stimulated rather than hindered that the growth of *F. oxysporium* in vitro including that it was the biological and not chemical component of vermicompost that explained its suppressive action.

Role of Vermicompost in Plant Disease Management (Plant Pathogen control)

Soils with low organic matter and microbial activity are more prone to plant root diseases and addition of organic amendments can effectively suppress plant disease. Several investigators reported the disease suppressive properties of thermophilic compost on a wide range of phytopathogens viz., *Rhizoctonia, Phytopthora, Plasmidiophora brassicae* and *Gaeumannomyces* and *Fusarium*.

Microbial antagonism might be one of the possible reasons for disease suppression as organic amendments enhances the microbial population and diversity. Traditional thermophilic composts promote only selected microbes while non-thermophilic vermicomposts are rich sources of microbial diversity and activity and harbour a wide variety of antagonistic bacteria thus acts as effective biocontrol agents aiding in suppression of diseases caused by soil-borne phytopathogenic fungi (Singh *et al.*, 2008).

Earthworm feeding reduces the survival of plant pathogens such as *Fusarium* sp. and *Verticillium dahliae* and increases the densities of antagonistic fluorescent pseudomonads and filamentous actinomycetes while population densities of *Bacilli* and *Trichoderma* spp. remain unaltered. Earthworm activities reduce root diseases of cereals caused by *Rhizoctonia*. It has been proved that earthworms decreased the incidence of field diseases of clover, grains, and grapes incited by *Rhizoctonia* spp. and *Gaeumannomyces* spp. Earthworms *Aporrectodea trapezoides* and *Aporrectodea rosea* act as vectors of *Pseudomonas corrugata,* a biocontrol agent for wheat take-all caused by *G. graminis* var. *tritd*. Greenhouse studies on augmentation of pathogen infested soils with *L. terrestris* showed a significant reduction of disease caused by *Fusarium oxysporum* f. sp. *asparagi* and *F. proliferatum* on susceptible cultivars of asparagus (*Asparagus officinalis*), *Verticillium dahliae* on eggplant (*Solanum melongena*) and *F. oxysporum* f. sp. *lycopersici* race 1 on tomato. Plant weights increased by 60-80% and disease severity reduced by 50-70% when soils were augmented with earthworms. Incorporation of soil with vermicompost effectively suppressed *R. solani* in wheat, *Phytophthora nicotianae;* and *Fusarium* in tomatoes, *Plasmodiophora brassicae* in tomatoes and cabbage, *Pythium* and *Rhizoctonia* (root rot) in cucumber and radish, *Botrytis cineria* and *Verticillium* in strawberry and *Sphaerotheca fulginae* in grapes.

Vermicompost application drastically reduced the incidence of 'Powdery Mildew', 'Color Rot' and 'Yellow Vein Mosaic' in Lady's finger (*Abelmoschus esculentus*). Substitution of vermicompost in the growth media reduced the fungal diseases caused by *R. solani, P. drechsleri and F. oxysporum* in gerbera. Amendment of vermicompost at low rates (10-30%) in horticulture bedding media resulted in significant suppression of *Pythium* and *Rhizoctonia* under green house conditions. Research findings proved that vermicompost when added to container media significantly reduced the infection of tomato plants by *P. nicotianae* var. nicotianae and *F. oxysporum* sp. *lycopersici* (Szczech 1999). Club-rot of cabbage caused by *P. brassicae* was inhibited by dipping cabbage roots into a mixture of clay and vermicompost. Potato plants treated with vermicompost were less susceptible to *P. infestans* than plants treated with inorganic fertilizers. Aqueous extracts of vermicompost inhibited mycelial growth of *B. cineria, Sclerotinia sclerotiorum, Corticium rolfsii, R. solani* and *F. oxysporum,* effectively controlled powdery mildew of barley and affected the development of powdery mildews on balsam (*Impatiens balsamina*) and pea (*Pisum sativum*) caused by *Erysiphe cichoracearum* and *E. pisi,* respectively in field conditions (Singh *et al.,* 2003).

Role of Vermicompost in Arthropod Pest Control

Addition of organic amendments helped in suppression of various insect pests such as European corn borer, other corn insect pests, aphids and scale insects (Huelsman *et al.,* 2000) and brinjal shoot and fruit borer. Several reports also evidenced that vermicompost addition decreased the incidence of *Spodoptera litura, Helicoverpa armigera,* leaf miner (*Apoaerema modicella),* jassids (*Empoasca kerri*), aphids (*Aphis craccivora)* and spider mites on groundnuts (Rao 2003) and psyllids (*Heteropsylla cubana*) on a tropical leguminous tree (*Leucaena leucocephala*). Vermicompost amendment decreased the incidence of sucking pests under field conditions and suppressed the damage caused by of two-spotted spider mite (*Tetranychus* spp.), aphid (*Myzus persicae*) and mealy bug (*Pseudococcus* spp.) under green house conditions (Arancon *et al.,* 2007). Vermicompost substitution to soil less plant growth medium MetroMix 360 (MM360) at a rate less then 50% reduced the damage caused by infestation of pepper seedlings by *M. persicae* and *Pseudococcus* spp. and tomato seedlings by *Pseudococcus* spp., cabbage seedlings by *M. persicae* and cabbage white caterpillars (*Pieris brassicae* L.).

Greenhouse cage experiments conducted on tomatoes and cucumber seedlings infested with *M. persicae,* citrus mealybug (*Planococcus citri*), two spotted spider mite (*Tetranychus urticae*); striped cucumber beetles (*Acalymna vittatum*) attacking cucumbers and tobacco hornworms (*Manduca sexta*) attacking tomatoes proved that treatment of infested plants with aqueous extracts of vermicompost suppressed pest establishment, and their rates of reproduction. Vermicompost teas at higher dose also brought about pest mortality (Edwards *et al.,* 2010b). Suppression of aphid population gains importance since they are key vectors in transmission of plant viruses. Addition of solid vermicompost reduced damage by *A. vittatum* and spotted cucumber beetles (*Diabotrica undecimpunctata*) on cucumbers and larval hornworms (*Manduca quinquemaculata)* on tomatoes in both greenhouse and field experiments. Combined application of vermicompost and vermiwash spray to chilli (*Capiscum annum*) significantly reduced the incidence of 'Thrips' (*Scirtothrips dorsalis*) and 'Mites' (*Polyphagotarsonemus latus*).

Mechanisms That Mediate Pest Control

Plants grown in inorganic fertilizers are more prone to pest attack than those grown on organic fertilizers. Inorganic nitrogen fertilization improves the nutritional quality and palatability of the host plants, inhibits the raise of secondary metabolite concentrations, enhances the fecundity of insects dieting on them, attracts more individuals for oviposition and increases the population growth rates of insects. Though organic fertilizer has an enhanced nutritional composition they release nutrients at a slower rate hence plants grown with organic fertilizers possess decreased N levels and have higher phenol content resulting in resistance of these plants to pest attack. Similarly vermicomposts exhibit a slow, balanced nutritional release pattern, particularly in release of plant available N, soluble K, exchangeable Ca, Mg and P. Vermicomposts are rich in humic acid and phenolic compounds. Phenolic compounds act as feeding deterrents and hence significantly affect pest attacks. Soil containing earthworms contained polychlorinated phenols and their metabolites. An endogenous phenoloxidase present in *L. rubellus* bioactivate compounds to form toxic phenols viz., *p*-nitrophenol. Monomeric phenols could be absorbed by humic acids in the gut of earthworms. Uptake of soluble phenolic compounds from vermicompost, by the plant tissues makes them unpalatable thereby affecting pest rates of reproduction and survival (Edwards *et al.*, 2010).

Role of Vermicompost in Nematode Control

It has been well documented that addition of organic amendments decreases the populations of plant parasitic nematodes (Akhtar and Malik 2000). Vermicompost amendments appreciably suppress plant parasitic nematodes under field conditions. It also suppressed the attack of *Meloidogyne incognita* on tobacco, pepper, strawberry and tomato (Edwards *et al.*, 2007) and decreased the numbers of galls and egg masses of *Meloidogyne javanica*.

Mechanisms that Mediate Nnematode Control

There are several mechanisms that suppress plant parasitic nematodes by vermicompost application and involving biotic and abiotic factors. Organic matter when added to the soil stimulates the population of bacterial and fungal antagonists of nematodes e.g. *Pasteuria penetrans*, *Pseudomonas* spp. and chitinolytic bacteria, *Trichoderma* spp., and other typical nematode predators including nematophagous mites viz., *Hypoaspis calcuttaensis*, Collembola and other arthropods which selectively feeds on plant parasitic nematodes. It promotes fungi capable of trapping nematode and destroying nematode cysts and increased the population of plant growth-promoting rhizobacteria which produce enzymes toxic to plant parasitic nematodes. When added to soils planted with tomatoes, peppers, strawberry and grapes has a significant reduction of plant parasitic nematodes and increased the population of fungivorous and bacterivorous nematodes as compared to inorganic fertilizer (Arancon *et al.*, 2002). In addition, few abiotic factors *viz.*, nematicidal compounds such as hydrogen sulphide, ammonia, nitrates, and organic acids released during vermicomposting, as well as low C/N ratios of the compost cause direct adverse effects while changes in soil physiochemical characteristics *viz.*, bulk density, porosity, water holding capacity, pH, EC, CEC and nutrition posses indirect adverse effects on plant parasitic nematodes (Thoden *et al.*, 2011).

Enzymes in Vermicompost

Vermicompost has also been documented with various enzymes activities including cellulose, amylase, invertase, protease, peroxidise, urease, phosphatise and dehydrogenase in which the maximum enzyme activity is contributed by gut microbes. Though vermicomposts have a wide range of enzyme activities, fluctuations are there during the composting period that the maximum enzyme activities were observed during 21-35 days in vermicomposting, whereas in conventional composting it was noticed on 42-49 days (Devi *et al.*, 2009). Earthworms influence soil physical, chemical and biological properties due to their role, these are considered as 'Soil Engineers' and indicators of soil quality.

References

Ahmad, R., Jilani,G., Arshad, M., Zahir, Z. A. and Khalid, A. 2007. Bio-conversion of organic wastes for their recycling in agriculture: an overview of perspectives and prospects. *Ann. Microbiol*, 57: 471-479.

Aira, M., Gomez-Brandon, M., Gonzalez-Porto, P. and Dominguez, J. 2011. Selective reduction of the pathogenic load of cow manure in an industrial-scale continuous-feeding vermireactor. *Bioresour. Technol*, 102: 9633-9637.

Aira, M., Monroy, F. and Dominguez, J. 2007. Earthworms strongly modify microbial biomass and activity triggering enzymatic activities during vermicomposting independently of the application rates of pig slurry. *Sci. Total Environ*, 385: 252-261.

Akhtar, M. and Malik, A. 2000. Role of organic amendments and soil organisms in the biological control of plant parasitic nematodes: a review. *Bioresour. Technol*, 74: 35-47.

Albanell, E., Plaixats, J. and Cabrero, T. 1988. Chemical changes during vermicomposting (*Eisenia fetida*) of sheep manure mixed with cotton industrial wastes. *Biol. Fertil. Soils*, 6: 266-269.

Anastasi, A., Varese, G. C. and Marchisio, V. F. 2005. Isolation and identification of fungal communities in compost and vermicompost. *Mycologia*, 97: 33-44.

Anne-Lie Blomstrom, Lalander, C., Komakech, A. J., Vinneras, B. and Boqviat, S. 2016. A metagenomic analysis displays the diverse microbial community of vermicomposting system in Uganda. *Infection ecology and Epidemiology*, 6: 324-53.

Arancon, N. Q., Edwards, C. A. and Lee, S. 2002. Proceedings Brighton Crop Protection Conference- Pests and Diseases. Management of plant parasitic nematode populations by use of vermicomposts, pp. 705-716.

Arancon, N. Q., Edwards, C. A., Atiyeh, R. and Metzger, J. D. 2004. Effects of vermicomposts produced from food waste on the growth and yields of greenhouse peppers. *Bioresour Technol*, 93:139-144.

Arancon, N. Q., Edwards, C. A., Yardim, E. N., Oliver, T. J., Byrne, R. J. and Keeney, G. 2007. Suppression of two-spotted spider mite (*Tetranychus urticae*), mealy bug (*Pseudococcus* sp.) and aphid (*Myzus persicae*) populations and damage by vermicomposts. *Crop Prot*, 26:29-39.

Asha, A., Tripathi, A. K. and Soni, P. 2008. Vermicomposting: A Better Option for Organic Solid Waste Management. *J. Hum. Ecol,* 24: 59-64.

Ashraf, R., Shahid, F. and Ali, T. A. 2007. Association of fungi, bacteria and actinomycetes with different composts. *Pak. J. Bot,* 39: 2141-2141.

Atiyeh, R. M., Dominguez, J., Subler S. and Edwards, C. A. 2000. Changes in biochemical properties of cow manure during processing by earthworms (*Eisenia andrei,* Bouche) and the effects on seedling growth. *Pedobiologia,* 44: 709-724.

Bhatnagar, R. K. and Palta, R. K. 1996. Earthworm-Vermiculture and Vermicomposting. Kalyani Publishers, New Delhi.

Brown, G. G., Barois, I. and Lavelle, P. 2000. Regulation of soil organic matter dynamics and microbial activity in the drilosphere and the role of interactions with other edaphic functional domains. *Eur. J. Soil Biol,* 36:177-198.

Caporaso, J. G., Kuczynski, J., Stombaugh, J., Bittinger, K., Bushman, F. D., 2010. QIIME allows integration and analysis of high-throughput community sequencing data. *Nat. Methods,* 7:335-336.

Caporaso, J. G., Lauber, C. L., Walters, W. A., Berg-Lyons, D., Lozupone, C. A. 2012. Ultra-high-throughput microbial community analysis on the Illumina HiSeq and MiSeq platforms. The ISME J, 6: 1621-1624.

Clarke, K. R. and Gorley, R. N. 2001. Plymouth Routines in Multivariate Ecological Research (Primer) Version 5: User Manual/Tutori*al.,* Primer-E Ltd, Plymouth, UK.

Cooke, A. 1983. The effects of fungi on food selection by *Lumbricus terrestris* L. *In: Earthworm Ecology: From Darwin to Vermicultur.* Satchell, J.E. (ed.). London: Chapman and Hall, pp.365-373.

Darwin , F. and Seward, A. C. 1903. More letters of Charles Darwin. In: John M (ed) A record of his work in series of hitherto unpublished letters, vol 2, London, p 508.

Deborah, A. N., Thomas, R. W., Scott,, T. B., Jonathan W. L. and Noah, F. 2013. Changes in Bacterial and Funagl Communities across Compost Recipes, Preparation Methods, and Composting Times. *PLoS ONE,* 8:11.

Devi, S. H., Vijayalakshmi, K., Pavana Jyotsna, K., Shaheen, S. K., Jyoti, K. and Surekha Rani, M. 2009. Comparative assessment in enzyme activities and microbial populations during normal and vermicomposting. *J. of Environ. Bio,* 30(6): 1013-1017.

Dominguez, J. and Edwards, C. A. 2004. Vermicomposting organic wastes: A review. In: Shakir Hanna SH, Mikhail WZA (eds) Soil Zoology for sustainable Development in the 21st century, Cairo, pp 369–395.

Edgar, R. C. 2010. Search and clustering orders of magnitude faster than BLAST. *Bioinfo,* 26:2460-2461.

Edwards, C. A., Arancon, N. Q., Bennett, M. V., Askar, A., Keeney, G. and Little, B. 2010. Suppression of green peach aphid (*Myzus persicae*) (Sulz.), citrus mealy bug (*Planococcus citri*) (Risso), and two spotted spider mite (*Tetranychus urticae*) (Koch.) attacks on tomatoes and cucumbers by aqueous extracts from vermicomposts. *Crop Prot*, 29:80-93.

Edwards, C. A., Arancon, N. Q., Emerson, E., Pulliam, R. 2007. Supressing plant parasitic nematodes and arthropod pests with vermicompost teas; *Biocycle*, pp. 38-39.

Fierer, N., Leff, J. W., Adams, B. J., Nielsen, U. N., Bates, S. T.2012. Cross-biome metagenomic analysis of soil microbial communities and their functional attributes. *PNAS*, 5:245-56

Gardes, M., and Bruns, T. D. 1993. ITS primers with enhanced specificity for basidiomycetes-application to the identification of mycorrhizae and rusts. *Mol Ecol*, 2: 113-118.

Garg, P., Gupta, A. and Satya, S. 2006. Vermicomposting of different types of waste using *Eisenia foetida*: a comparative study. *Bioresour. Technol*, 97: 391-395.

Hand, P., Hayes, W. A., Frankland, J. C. and Satchell, J. E. 1988. Vermicomposting of cow slurry. *Pedobiologia*, 31:199-209.

Huelsman, M. F., Edwards, C. A., Lawrence, J. L. and Clarke-Harris, D. O. 2000. A study of the effect of soil nitrogen levels on the incidence of insect pests and predators in Jamaican sweet potato (*Ipomoea batatus*) and Callaloo (*Amaranthus*) Proc Brighton Pest Control Conference: Pests and Diseases. 8D-13:895-900.

Jambhekar, H. 1992. Use of earthworm as a potential source of decomposes organic wastes. *Proc. Nat. Sem. Org. Fmg.*, Coimbatore, pp 52-53.

Kooch, Y. and Jalilvand, H. 2008. Earthworm as ecosystem engineers and the most important detritivores in forst soils. *Pak. J. Boil. Sci*, 11:819-825.

Lauber, C. L., Hamady, M., Knight, R. and Fierer, N. 2009. Pyrosequencing-based assessment of soil pH as a predictor of soil bacterial community structure at the continental scale. *Appl. Environ. Microbiol*,75: 5111-5120.

Lazcano, C. and Dominguez, J. 2011. The use of vermicompost in sustainable agriculture: impacts on plant growth and soil fertility. Soil Nutrients, Nova Science Publishers, Inc. ISBN 978-1-61324-785-3.

Lazcano, C., Gomez-Brandon, M. and Dominguez, J. 2008. Comparison of the effectiveness of composting and vermicomposting for the biological stabilization of cattle manure. *Chemosphere* 72:1013-1019.

Martin, J. P. 1976. Darwin on earthworms: the formation of vegetable moulds. Bookworm Publishing, Ontario.

McDonald, D., Price, M. N., Goodrich, J., Nawrocki, E. P., DeSantis, T .Z. 2011. An improved Green genes taxonomy with explicit ranks for ecological and evolutionary analyses of bacteria and archaea. ISME J 6: 610–618.

McGuire, K. L., Payne, S. G., Palmer, M. I., Gillikin, C. M., Keefe, D. 2013. Digging the New York City skyline: Soil fungal communities in green roofs and city parks. *PLoS One* , 8:e58020.

Munnoli, P. M., Da Silva, J. A. T. and Saroj, B. 2010. Dynamics of the soil-earthworm-plant relationship: a review. *Dynamic soil, dynamic plant,* pp. 1-21.

Nagavallemma, K. P., Wani, S. P., Stephane, L., Padmaja, V. V., Vineela, C., Babu, Rao, M., and Sahrawat , K. L. 2004. Vermicomposting: Recycling wastes into valuable organic fertilizer. *Global Theme on Agrecosystems* Report no.8. Patancheru 502324. International Crops Research Institute for the Semi-Arid Tropics, Andhra Pradesh, p 20.

Neher, D. A., Weicht, T. R., Bates, S. T., Leff, J. W. and Fierer, N. 2013. Changes in bacterial and fungal communities across compost recipes, preparation methods, and composting times. *PLOS One* ,8:e79512.

Pizl, V. and Novokova, A. 1993. Interactions between microfungi and *Eisenia andrei* (Oligochaeta) during cattle manure vermicomposting. *Pedobiologia,* 47:895-899.

Pramanik, P., Ghosh, G. K., Ghos*al.*, P. K., and Banik, P. 2007. Changes in organic-C, N, P and K enzyme activities in vermicompost of biodegradable organic wastes under liming and microbial inoculants. *Bioresour. Technol,* 98:2485-2494.

R, Core, Team. 2012. A language and environment for statistical computing. R Foundation for Statistical Computing, Vienna, Austria. Available: http://www.R-project.org/. Accessed 7 January 2013.

Rao, K. R. 2003. Influence of host plant nutrition on the incidence of *Spodoptera litura* and *Helicoverpa armigera* on groundnuts. *Indian J. Entomol,* 65:386-392.

Ryckeboer, J., Mergaert, J., Vaes, K., Klammer, S., de Clercq D. 2003. A survey of bacteria and fungi occurring during composting and self-heating processes. *Ann. Microbiol,* 53:349-410.

Sanchez-Monedero, M. A., Roig,, A., Paredes C. and Bernal, M. P. 2001. Nitrogen transformation during organic waste composting by the Rutgers system and its effects on ph, EC and maturity of the composting mixtures. *Bioresour. Technol,*78:301-308.

Sharma, S., Pradhan, K., Satya, S. and Vasudevan, P. 2005. Potentiality of earthworms for waste management and in other uses-A Review. *The J. of American Sci,*1:4-16.

Singh, R., Sharma, R. R., Kumar, S., Gupta, R. K. and Patil, R. T. 2008. Vermicompost substitution influences growth, physiological disorders, fruit yield and quality of strawberry (*Fragaria* x *ananassa* Duch.) *Bioresour. Technol,* 99:8507-8511.

Singh, U. P., Maurya, S. and Singh, D. P. 2003. Antifungal activity and induced resistance in pea by aqueous extract of vermicompost and for control of powdery mildew of pea and balsam. J. *Plant. Dis. Protect,* 110:544-553.

Suthar, S., and Singh, S. 2008. Vermicomposting of domestic waste by using two epigeic earthworms (*Perionyx excavatus* and *Perionyx sansibaricus*). *Int. J. Evniron. Sci. and Technol*, 5:99-106.

Szczech, M. M. 1999. Suppressiveness of vermicomposts against *Fusarium* wilt of tomato. *J Phytopathology*, 147:155-161.

Thoden, T. C. and Korthals, G. W. (2011). Termorshuizen Organic amendments and their influences on plant-parasitic and free living nematodes: a promosing method for nematode management. *Nematology*, 3:133-153.

Tilman, D., Cassman, K.G., Matson, P.A., Naylor, R.and Polasky, S. 2002. Agricultural sustainability and intensive production practices. *Nature*, 418, 671-677.

Tiunov, A. V. and Scheu, S. 2000. Microfungal communities in soil litter and casts of *Lumbricus terrestris* (Lumbricidae): a laboratory experiment. *Appl. Soil Ecol*, 14:17-26.

Vivas, A., Moreno, B., Garcia-Rodriguez, S. and Benitez, E. 2009. Assessing the impact of composting and vermicomposting on bacterial community size and structure, and functional diversity of an olive-mill waste. Bioresour. *Technol*, 100:1319-1326.

Wang, Q., Garrity, G. M., Tiedje, J. M. and Cole, J. R. 2007. Naive Bayesian classifier for rapid assignment of rRNA sequences into the new bacterial taxonomy. *Appl. Environ. Microbiol*. 73:5261-5267.

Yasir, M., Aslam, Z., Kim, S. W., Lee, S. W., Jeon, C. O. and Chung, Y. R. 2009. Bacterial community composition and chitinase gene diversity of vermicompost with antifungal activity. *Bioresour. Technol*,100:4396-4403.

16

Microbes: An Eco-friendly Tool in Crop Improvement

Yogesh Yadav[1,2], Sunil Kumar[3] and Narender Singh[2*]

[1]Department of Botany, Govt. P.G. College, Narnaul- 123001, India

[2]Department of Botany, Kurukshetra University, Kurukshetra- 136119, India

[3]Department of Botany, CRM Jat PG College, Hisar - 125 001, India

*Corresponding author: nsheorankuk@yahoo.com

Abstract

Microbes are generally microscopic small organisms placed in different groups such bacteria, fungi, protozoa, micro-algae and viruses. These can be a better weapon for copping stress over plant in various conditions. Stress is the condition which include brutal environmental factors limiting the productivity of crop plants because most of the crop plants are sensitive to at least few of stresses caused by biotic and abiotic factors such as high concentrations of salts in the soil, and the area of which affected by it is increasing day by day, high temperature, water scarcity, low temperature, high humidity that favours fungal infection, bacterial infection and viral infection. For all important crops, average yield losses are somewhere between 20% and 50% of record yields; these losses are mostly due to drought and high soil salinity, environmental conditions which will worsen in many regions because of global climate change. A wide range of adaptations and mitigation strategies are required to manage with such conditions. Efficient resource management for evolving better varieties can help to overcome stressful conditions. However, such strategies being long drawn and cost intensive, there is a requirement to develop simple and low cost biological methods for stress management, which can be used on short term basis. Microorganisms could play a significant role in this respect, if we use their unique properties such as tolerance to stressful conditions, genetic diversity, synthesis of compatible solutes, enzymes, production of plant growth promoting hormones, bio-control potential, and their interaction with crop plants and thus crop productivity.

Keywords: Microbes, Eco-friendly, Bio-fertilizer, Climate, Bio-insecticide

Introduction

Changing climatic conditions and increasing population in current environment created pressure on agriculture system. In this changing environment particularly due to global warming, and other potential climate abnormalities associated with it, crops typically facing an increased number of abiotic and biotic stress combinations, which severely affect their growth and yield (Ramegowda and Kumar, 2015). Simultaneous occurrence of abiotic stresses such as drought and heat has been shown to be more destructive to crop production than these stresses occurring separately at different crop growth stages (Mittler, 2006; Prasad *et al.*, 2011). According to FAO, out of a US$1.3 trillion annual food production capacity worldwide, the biotic stresses caused by insects, diseases and weeds cause 31–42 percent loss (US$500 billion), with an additional 6–20 percent (US$120 billion) lost post harvest to insects and to fungal and bacterial rots. Crop losses due to pathogens are often more severe in developing countries (e.g. cereals, 22 percent) when compared to crop losses in developed countries (e.g. cereals, 6 percent). In the same way abiotic stress causes another 6–20 percent (US$120 billion) is estimated to be lost to abiotic causes (drought, flood, frosts, nutrient deficiencies, various soil and air toxicities). One of the most significant abiotic stress reducing crop yields is water stress, both water deficit stress (drought) and excess water stress (flooding, anoxia). So, there is a continuous need of resistant crops to avoid such conditions, but these are not available for all type of stress, that's why there is need of alternate methods, in these methods chemical control is the most common method but its caused hazardous effect on environment and human health. In these conditions microbes are the centre of interest in such conditions due the enormous potential of microbes favouring in such conditions sustainably. Micro-organisms are mostly microscopic small creatures are placed in different groups such bacteria, fungi, protozoa, micro-algae and viruses. These organisms live in soil, water, food, animal intestines and other different environments. Various microbial habitats reflect an enormous diversity of biochemical and metabolic traits that have arisen by genetic variation and natural selection in microbial populations. Men used some of microbial diversity in the production of fermented foods such as bread, yogurt, and cheese. Some soil microbes release nitrogen that plants need for growth and emit gases that maintain the critical composition of the Earth's atmosphere. Microbes interact with plants in agricultural system, their nutrition and ability to resist biotic and abiotic stress and have the potential to be manipulated such that their positive effects are enhanced Some of the benefits of microbial control over chemical control are as shown in table

S.No.	Characterstic	Biological control	Chemical control
	No.of ingredients tested(approx.)	2000	>3.5 million
	Success ratio	1: 10	1: 200000
	Time of production	1-10 years	5-15 years
	Cost of production	2 million US$	150 million US$
	Resistance risk	Small	Large

S.No.	Characterstic	Biological control	Chemical control
	specificity	Large	small
	Harmful side-effects	Nil/very less	many

Table showing effectiveness of biological control over chemical control

Various efficient forms of microbes

Although countless importance of microbes, some of the commonest are discussing here

1. As Biofertilizers

Phosphate and nitrogen are important for the growth of plants. These compounds exist naturally in the environment but plants have a limited ability to extract them. Phosphate plays an important role in crop stress tolerance, maturity, quality and directly or indirectly, in nitrogen fixation. A fungus, *Penicillium bilaii* helps to unlock phosphate from the soil. It makes an organic acid, which dissolves the phosphate in the soil so that the roots can use it. Biofertilizer made from this organism is applied by either coating seeds with the fungus as inoculation, or putting it directly into the ground. *Rhizobium* is a bacteria used to make biofertilizers, This bacterium lives in the plant's roots in cell collections called nodules. The nodules are biological factories that can take nitrogen out of the air and convert it into an organic form that the plant can use. This fertilization method has been designed by nature. With a large population of the friendly bacteria on its roots, the legume can use naturally-occurring nitrogen instead of the expensive traditional nitrogen fertilizer. Biofertilizers help plants use all of the food available in the soil and air, thus allowing farmers to reduce the amount of chemical fertilizers they use. This helps preserve the environment for the generations to come. Rhizobacterial application to plant increases growth via increased nutrient availability. This occurs in many different ways, for example, via N_2 fixation, P solubilisation and production of siderophores and increased Fe availability (Andrews *et al.*, 2003; Vessey, 2003). Rhizosphere and endophytic bacterial application on plant enhances growth by influencing plant hormone balance. Hormonal effects occur when rhizo- and endophytic bacteria either produce or metabolise chemical signalling compounds that directly impact plant growth and function (Dodd *et al.*, 2010). There are many reports on the bacterial production of plant hormones which stimulate root growth and alter root structure such that the nutrient uptake is increased (Andrews *et al.*, 2003; Vessey, 2003; Dodd *et al.*, 2010). There is also strong evidence that rhizo- and endophytic bacteria can improve plant performance via effects on root-to-shoot hormonal signalling which regulate leaf growth and gas exchange (Dodd *et al.*, 2010). Along with plant growth rhizosphere and endophytic bacterial interaction helps in cope abiotic stress ((temperature, water, salt, heavy metal) by production of a range of compounds including plant hormones and antioxidants (Yang *et al.*, 2009; Belimov *et al.*,2009). Microbes provides endophytic and symbiotic N_2 fixation-to legumes, which are especially important in agricultural system. Generally the plant symbiont is the crop but for some legumes and other N_2-fixing associations are also available such as p *Azolla–Anabaena* in rice crops, the plant with symbiont is used as green manure (Ran *et al.*, 2010).

2. As Bio-pesticides

Microorganisms found in the soil are all not so friendly to plants. These pathogens can cause disease or damage the plant. As scientists developed biological "tools," which use these disease-causing microbes to control weeds and pests naturally, certain microorganism suppresses of plant diseases, plant disease suppression occur via competitive exclusion of pathogens by production of antibiotics and the induction of systemic resistance that combats foliar disease (Lugtenberg & Kamilova, 2009). Protection against plant parasitic nematodes occurs by some saprophytic fungi which are used for the control of nematodes, for example, the facultative nematode parasite *Pochonia chlamydosporia* has been developed as a biological control agent of cyst and root-knot nematodes. This fungus, which is saprophytic in the soil, infects nematode cysts or egg masses. In the current time, intensive use of pesticides have caused various ecological and environmental problems, the reappearance of various insect and mites pests, contaminated food commodities, have adversely affected non-target organisms and have also progressively increased poisoning cases in developing countries. So, use of biopesticides open up new door of scientific advancement, which requires sustained efforts and will certainly reap long-term benefits. Bio-pesticides encompass a broad array of microbial pesticides, bio-chemicals derived from microorganisms and other natural sources, and processes involving the genetic incorporation of DNA into agricultural commodities that confer plant protection. Over 400 species of fungi and more than 90 species of bacteria, reported for use as biopesticide. Bacteria do their job as obligate pathogens, crystalliferous spore-formers, facultative pathogens and potential pathogens. These are effective only when eaten by insects with a specific (usually alkaline) gut pH and the specific gut membrane structures required to bind the toxin. So, when the insect binds eventually lead to gut paralysis of the infected insect. As a consequence, the insect stops feeding and dies from the combined effects of starvation and tissue damage.

Viruses another group of microbes used for this purposes, particularly a family of viruses called baculoviruses is the most popular choice for microbial control, which have been used regularly for pest control since the 1950s. The majority of baculoviruses used as biological control agents belong to genus *Nucleopolyhedrovirus*. Viruses invade an insect's body via the gut, they replicate in tissues and can disrupt components of an insect's physiology, interfering with feeding, egg laying, and movement, finally death takes place within 3-8 days. Examples are *Helicoverpa zea nuclear polyhedrosis virus* (HzSNPV), *Orgyia pseudotsugata* (Ot), and *Lymantria dispar* (Ld) MNPV.

Over 750 species of fungi kill insects. Entomopathogenic fungi are widely distributed throughout the fungal kingdom, although the majority occur in the Deuteromycotina and Zygomycotina families. Out of these less than 20 have received serious attention as control agents of insect pests (Copping 2004). These fungi invade their hosts using spores that grow through the cuticle, and hence they are particularly suited for control of pests with piercing mouthparts, such as aphids and whiteflies, which are unlikely to acquire pathogens through feeding. These also invade their hosts by direct penetration of the host exoskeleton or cuticle, infected hypha penetrates down through the host cuticle and eventually emerges into the haemocoel of the insects , followed by enzyme assisted degradation with mechanical pressure eventually leads to death which is caused by tissue destruction and, occasionally by toxins produced by the fungus. Various fungi protect against herbivore pest such as grasses from herbivores pest in temperate grasses and can occur in various ways, one of the method by production of toxic alkaloids that deters herbivores.

3. As Bioherbicides

Weeds are the problem for farmers. They not only compete with crops for water, nutrients, sunlight, and space but also harbor insect and disease pests, clog irrigation and drainage systems, undermine crop quality, and deposit weed seeds into crop harvests. Bio-herbicides are another way of controlling weeds without environmental hazards posed by synthetic herbicides. The microbes possess invasive genes that can attack the defence genes of the weeds, thereby killing it. The benefit of using bioherbicides is that it can survive in the environment long enough for the next growing season where there will be more weeds to infect. It is cheaper than synthetic pesticides thus could essentially reduce farming expenses if managed properly. Further, it is not harmful to the environment compared to conventional herbicides and will not affect non-target organisms.

4. As Bioinsecticides

Biotechnology can also help in developing alternative methods to synthetic insecticides to fight against insect pests. Microorganisms in the soil attack fungi, viruses or bacteria, which cause root diseases. These are used in the form of coatings on the seed (inoculants) which carry these beneficial organisms can be developed to protect the plant during the critical seedling stage. Bioinsecticides do not persist long in the environment and have shorter shelf lives; they are effective in small quantities, safer to humans and animals compared to synthetic insecticides; they are very specific, often affecting only a single species of insect and have a very specific mode of action; slow in action and the timing of their application is relatively critical. Bacteria protect against insect pests, A wide range of naturally occurring associative bacteria, fungi and viruses can protect plants to some extent from insect pests. The bacterium *Bacillus thuringiensis* (Bt) is the most widely used microorganism 'insecticide'. Bt produces insecticidal crystal proteins during its sporulation phase. These proteins differ depending on the Bt subspecies and as a result there is a range of Bt types that are toxic to different insect species. Bt can be both sprayed onto or genetically engineered into crops.

Fungi as Bioinsecticide- Fungi cause diseases in some 200 different insects and this disease producing traits of fungi is being used as bioinsecticides. Fermentation technology is used to mass production of fungi. Spores are harvested and packaged so these are applied to insect-ridden fields. When the spores are applied, they use enzymes to break through the outer surface of the insects' bodies. Once inside, they begin to grow and eventually cause death. Fungal agents are recommended by some researchers as having the best potential for long-term insect control. This is because these bioinsecticides attack in a variety of ways at once, making it very difficult for insects to develop resistance.

Viruses as Bioinsecticides- Bacteriophage protect against many diseases. The use of bacteriophages (viruses that infect bacteria) to control bacterial plant diseases has been trialled in a range of crops. Baculoviruses affect insect pests like corn borers, potato beetles, flea beetles and aphids. One particular strain is being used as a control agent for bertha army worms, which attack canola, flax, and vegetable crops. Traditional insecticides do not affect the worm until after it has reached this stage and by then much of the damage has been done.

5. In Soil Health

Soil is under demand for increased crop production to feed the ever-growing population. This has led to intensification of agriculture with extensive use of chemicals, exploitation of surface and ground water for irrigation. According to one of ecologist,

The continued capacity of soil to function as a vital living system, within ecosystem and land-use boundaries, to sustain biological productivity, promote the quality of air and water environments, and maintain plant, animal and human health (Doran et al., 1997)

Soil health assessed by biological indicators, which includes microbial biomass, C and N content, dehydrogenase activity, acid and alkaline phosphatase activity, urease activity, arylsulphatase activity, soil respiration and ergosterol concentration, microbial diversity, and functional groups of soil fauna, etc. Microbe mediated processes like decomposition, fixation, solubilization, filtration, remediation and suppression of pathogens make the soil healthy and fertile, which enhances crop yield. In healthy soil conditions, numerous nutrient cycles influenced by microbes operate simultaneously in soil. Soil thus contributes nutrients uninterruptedly to the nutrient pool of soil ensuring balanced plant nutrition. Moreover, healthy soil conserves nutrients and releases them more or less synchronously with the demand of growing crops. Thus, nutrient use efficiency and nutrient conversion ratio remain favorable for reasonable production. Some of important genera of microbes maintain soil health are *Rhizobium leguminoarum, R. rhizogenes, R. endophyticum, R. lusitanum, R. tibeticum, Arthrobacter viscosus, Neorhizobium vignae, N. alkalisoli, Sinorhizobium americanum, Azotobacter, Azospirillum, Acetobacter, Azoarcus, Burkholderia, Herbaspirillum, Pseudomonas fluorescens, P. putida, P. Gladioli, Clostridium spp., Corynebacterium spp., Erwinia spp., Nitrosomonas spp., Nitrobacter spp., Rhizoctonia, Fusarium, Trichoderma spp., Pythium spp., Armillaria spp., Helminthosporium spp., Ophiobolus., Phytophthora spp., Plasmodiophora spp., Pythium spp., Rhizoctonia spp., Sclerotium spp., Thielaviopsis spp., Verticillium spp.,*

Anabaena spp., Calothrix spp., Oscillatoria spp., Aulosira spp., Nostoc spp., Scytonema spp., Tolypothrix spp., Colpoda spp., Pleurotricha spp., Heteromita spp., Cercomonas spp., Oikomonas spp., Phalansterium spp. Microbes provide stabilisation of soil aggregates and structure, Bacteria provides this with the production of exopolysaccharide , mycorrhiza by hyphae enmeshing soil particles and production of glycoprotein particularaly glomalin (Alami *et al.*, 2000; Hinsinger *et al.*, 2009; Smith *et al.*, 2010; Rillig, 2004; Bedini *et al.*, 2009.)

Microorganisms sustain growth of plants in saline soil- Salinity is one of the important worldwide problem for agriculture, especially for crops that are grown under irrigation. This is because salt is inhibitory to the growth of a large number of different plants about 6% of the total global land mass, or about 20% of the world's cultivated area (Flowers., 2004) Saline solution inhibits plant growth and development with adverse effects such as osmotic stress, Na^+ and Cl^- toxicity, ethylene production, plasmolysis, nutrient imbalance, production of reactive oxygen species (ROS), and interference with photosynthesis, consequences of such changes leads to inhibition of seed germination, seedling growth and vigor, flowering and fruit set, hence crop yield. Bacteria helps by fixing atmospheric nitrogen and supply it to plants; synthesize and secrete siderophores which can solubilize and sequester iron from the soil and provide it to plant cells; synthesize different phytohormones, including auxins, cytokinins and gibberelins; solubilize minerals such as phosphorus which then become more readily available for plant growth; and they may synthesize an enzyme that can modulate plant ethylene levels (Patten and Glick 1996, 2002), through the action of the enzyme ACC deaminase expressing on bacteria, bind to the surface of either plant seeds or roots and, in response to exuded tryptophan (or other small molecules), the bacteria synthesize and secrete indole acetic acid (IAA) (Fallik *et al.* 1994; Patten and Glick 2002), some of which is taken up by the plant. Together with the endogenous pool of plant IAA, the IAA taken up by the plant can stimulate plant cell proliferation and elongation. ACC converted to ketobutyrate and ammonia by ACC deaminase, both of which are readily metabolized by the bacteria. So, ACC deaminase-expressing rhizospheral environment protect plants against the growth inhibition that might otherwise result following flooding, extremes of temperature, the presence of organic or inorganic toxicants, phytopathogens, drought, or high salt. Another microbe Arbuscular mycorrhizal (AM) fungi helps in alleviation of salt stress. The mechanism involves (1) improvement of mineral nutrition leading to plant growth promotion; (2) variation in the plant accumulation of Na and K as well as soluble sugars and electrolytes; (3) modification of some physiological processes and enzymatic activities involved in plant antioxidative reactions; and (4) alteration of the root architecture facilitating water uptake by the plant. Overall, differential gene expression/protein synthesis induced by the establishment of the symbiosis between the fungi and plant under salt stress is at the base of these strategies. Example are *Glomus geosporum and G.intraradices, G. mosseae, G. fasiculatum, G. macrocarpum Gigaspora rosea, Hebeloma crustuliniforme, Scleroderma bermudense, Paxillus involutus,*

Microorganisims enhances soil health by promotion of mineral availability by phosphate solubilization- An important nutrient phosphorus (P) required for growth and development of plants and promotes N_2 fixation. It is also involved in photosynthesis, energy transfer, signal transduction, macromolecular biosynthesis, and respiration. However, most of the soils throughout the world are phosphorus deficient, to balance these synthetic phosphorus fertilizers are applied. The repeated and injudicious applications of these fertilizers, however, lead to the loss of soil fertility, disturbance to microbial diversity and their associated metabolic activities, and reduced yield of agronomic crops. Moreover, after application, a considerable amount of P is rapidly transformed into less available forms by forming a complex with Al or Fe in acid soils or with Ca in calcareous soils. For avoiding such situation microbes can be used for the better productivity of crops. Phosphorus is present in soils both in organic and inorganic forms, Organic P compounds can be slowly mineralized as available inorganic P or they can be immobilized as part of the soil organic matter. The process of mineralization (solubilisation) affected by rhizosphere microbes, especially phosphate solubilizing microbes, serves as an alternative to chemical phosphatic fertilizers and provides the available forms of P to plants. The phosphorus solubilisation trait is correlated with the production of organic acids via the direct oxidation pathway that occurs on the outer face of the cytoplasmic membrane, with drop in pH value. The inverse correlation between pH of the culture and the release of P consolidate the hypothesis of organic acid involvement in the solubilisation of insoluble P.

The organic acids released by microbes, notably bacteria and fungi chelate mineral ions or drop the pH to bring P into solution. Organic acids produced by bacteria and fungi leads to acidification of microbial cells and their surroundings, consequently, the release of P-ions from the phosphorus mineral by H^+ substitution for Ca^{2+}. But insoluble P could be solubilized by mechanisms other than acidification process. In this process, the fungal-solubilizing activity attributed both to chelation and to reduction processes, which could also play a role in the bio-control of plant pathogens. Along with this process enzymes released from microbes also helps in solubilisation of phosphorus, among these impotant ones are phosphatases, phytases, phosphonatases and C–P lyases enzymes are generally used. Out of these phosphatases, acid phosphatases are commonly found in fungi. Examples *Bacillus, Pseudomonas, Aspergillus, Penicillium, Rhodococcus, Arthrobacter, Serratia, Chryseobacterium, Gordonia, Phyllobacterium, Arthrobacter, Delftia sp., Azotobacter, Xanthomonas, and Enterobacter, Pantoea, and Klebsiella* and some N_2 fixing bacteria *Rhizobium leguminosarum bv. trifolii* (Abril *et al.* 2007), *R. leguminosarum bv. viciae* (Alikhani *et al.* 2007) and *Rhizobium* species nodulating *Crotalaria* species (Sridevi *et al.* 2007) improved plant P-nutrition by mobilizing inorganic and organic P. Microbes also enhances uptake of water and Nutrients, for example mycorrhizal association occurs with around 90% of plants and supply the associated plant with additional P, N and water under certain conditions. Increased nutrient uptake with mycorrhizas is related to their smaller diameter, greater degree of branching and longevity in comparison with root hairs, greater secretion of low molecular weight metabolites and of enzymes making inorganic P and soluble low molecular mass organic N compounds available (Lambers *et al.*, 2008; Raven, 2010; Smith *et al.*, 2010.

Microbes as Efficient Tool in Combating Abiotic Stress

For growth, development and reproduction plant requires light, water, carbon and mineral nutrients. Extreme in any of these result stresses in many ways that lessen their nutrition. Prolonged water stress decreases leaf water potential and stomatal opening, reduces leaf size, suppresses root growth, reduces seed number, size, and viability, delays flowering and fruiting and limits plant growth and productivity (Osakabe *et al.*, 2014; Xu *et al.*, 2016). Under severe conditions, loss in plant productivity is observed (Li *et al.*, 2009). Both freezing injury and/or an increase in temperature are major cause of crop loss (Koini *et al.*, 2009; Pareek *et al.*, 2010). Various edaphic factors like acidity, salinity, and alkalinity of soils (Bromham *et al.*, 2013; Bui, 2013), pollutant contamination and anthropogenic perturbations (Emamverdian *et al.*, 2015) severely affect plant development and adversely influence crop production. Induced Systemic Tolerance (IST) is the term being used for microbe-mediated induction of abiotic stress responses. The role of microorganisms to alleviate abiotic stresses in plants has been the area of great concern in past few decades. Microbes with their potential intrinsic metaboloic and genetic capabilities, contribute to alleviate abiotic stresses in the plants (Gopalakrishnan et. al., 2015). The role of several rhizospheric occupants belonging to genera *Pseudomonas, Azotobacter* (Sahoo et. al., 2014a, b), *Rhizobium* (Remans et. al., 2008; Sorty et. al., 2016), *Pantoea* (Egamberdiyeva and Hoflich, 2003), *Bacillus* (Tiwari et. al., 2011; Vardharajula et. al., 2011; Sorty et. al., 2016), *Enterobacter* (Nadeem et. al., 2007), *Bradyrhizobium* (Panlada et. al., 2013), *Methylobacterium* (Meena et. al., 2012), *Burkholderia* (Oliveira et. al., 2009), *Trichoderma* (Ahmad *et al.*, 2015) and cyanobacteria (Singh *et al.*, 2011) in plant growth promotion and mitigation of multiple kinds of abiotic stresses has been documented. *Trichoderma* help in mitigating stress due to upregulation of aquaporin, dehydrin and malonialdehyde genes along with various other physiological parameters. Rhizospheric bacteria induce changes in the levels of phytohormones, defense-related proteins and enzymes, antioxidants and epoxypolysaccharide help in resilence of **drought stress.** A microbe *Burkholderia phytofirmans* strain PsJN mitigates drought stress in maize (Naveed *et al.*, 2014b) and wheat (Naveed *et al.*, 2014a), *Piriformospora indica* induces drought tolerance in Chinese cabbage (Sun *et al.*, 2010) by increasing the levels of antioxidants. It is reported that *Trichoderma* enhances oil content in NaCl affected Indian mustard (*Brassica juncea*) by improving the uptake of essential nutrients, enhanced accumulation of antioxidants and osmolytes and decreased Na^+ uptake, ameliorates **salinity stress** by producing ACC-deaminase (1-aminocyclopropane-1-carboxylate deaminase) to lower plant ethylene levels, often a result of various stresses, another microbe *Burkholderia phytofirmans* strain PsJN mitigates salt stress in *Arabidopsis* (Pinedo et. al. 2015), The root fungal endophyte *Piriformospora indica* induces salt tolerance in barley

(Baltruschat *et al.*, 2008). Plants under **extreme cold conditions** survive either through avoiding super cooling of tissue water or through freezing tolerance. Certain species of plants have developed an ability to tolerate super-cooling or freezing temperatures by increasing their anti-freezing response within a short photoperiod, a process called cold acclimation. Alleviation strategies in plants against **heat stress** involve activation of mechanisms that support maintenance of membrane stability and induction of mitogen-activated protein kinase (MAPK) and calcium-dependent protein kinase (CDPK) cascades. Besides, scavenging of ROS, accumulation of antioxidant metabolites and compatible solutes, chaperone signaling and transcriptional modulation are certain parallel activities that help cells to sustain heat stress, along with these reports many Soil-inhabiting microbes belonging to genera *Achromobacter, Azospirillum, Variovorax, Bacillus, Enterobacter, Azotobacter, Aeromonas, Klebsiella* and *Pseudomonas* have been shown to enhance plant growth even under unfavorable environmental conditions (Ortiz *et al.*, 2015; Kaushal and Wani, 2016) The selection, screening and application of stress-tolerant microorganisms, therefore, could be viable options to help overcome productivity limitations of crop plants in stress-prone area.

Conclusion

An ideal sustainable agricultural system is one which maintains and improves human health, benefits producers and consumers both economically and spiritually, protects the environment, and produces enough food for an increasing world population. One of the most important constraints to agricultural production in world is the stress conditions existing in the environment. Plant-associated microorganisms can play an important role in conferring resistance to such stresses. These organisms could include rhizoplane, rhizosphere and endophytic bacteria, cyanobacteria and symbiotic fungi and operate through a variety of mechanisms like triggering osmotic response, providing growth hormones and nutrients, acting as biocontrol agents and induction of certain genes in plants. The development of stress tolerant crop varieties through genetic engineering and plant breeding is essential but a long drawn and expensive process, whereas microbial inoculation to alleviate stresses in plants could be a more cost effective environmental friendly option which could be available in a shorter time frame. Taking the current leads available, determined future research is needed in this area, particularly on field evaluation and application of potential organisms as biofertilizers in stressed soil.

References

Abril, A., Zurdo-Pin˜eiro J. L., Peix A., Rivas, R. and Vela´zquez, E. 2007. Solubilization of phosphate by a strain of *Rhizobium leguminosarum* bv. trifolii isolated from *Phaseolus vulgaris* in El Chaco Arido soil (Argentina). In: Velazquez E, Rodriguez-Berrueco C(eds) Book Series: *Developments in Plant and Soil Sciences*. Springer, The Netherlands. 135–138.

Ahmad, P., Hashem, A., Abd-Allah, E. F., Alqarawi, A. A., John, R., Egamberdieva, D. 2015. Role of *Trichoderma harzianum* in mitigating NaCl stress in Indian mustard (*Brassica juncea* L) through antioxidative defense system. *Front. Plant Sci.* 6:868.

Alami, Y., Achouak, W., Marol, C. and Heulin, T. 2000. Rhizosphere soil aggregation and plant growth promotion of sunflowers by an exo-polysaccharide-producing *Rhizobium* sp. Strain isolated from sunflower roots. *Appl. and Environ. Microbiol.* 66:3393–3398.

Alikhani, H. A., Saleh-Rastin, N., Antoun, H. 2007. Phosphate solubilization activity of rhizobia native to Iranian soils. In: Velazquez E, Rodriguez-Berrueco C(eds) Book Series: *Developments in plant and soil sciences.* Springer, The Netherlands. 135–138.

Andrews, M., James, E.K., Cummings, S.P., Zavalin, A.A., Vinogradova, L.V. and McKenzie, B.A. 2003. Use of nitrogen fixing bacteria inoculants as a substitute for nitrogen fertiliser for dryland graminaceous crops: progress made, mechanisms of action and future potential. *Symbio.* 35:209–229.

Baltruschat, H., Fodor, J., Harrach, B. D., Niemczyk, E., Barna, B., Gullner, G. 2008. Salt tolerance of barley induced by the root endophyte Piriformospora indica is associated with a strong increase in antioxidants. *New Phytol.* 180:501–510.

Bedini, S., Pellegrino, E., Avio, L., Pellegrini S., Bazzoffi, P., Argese, E. and Giovannetti, M. 2009. Changes in soil aggregation and glomalin related soil protein content as affected by the arbuscular mycorrhizal fungal species *Glomus mosseae* and *Glomus intraradices. Soil Bio. and Biochem.* 41:1491–1496.

Belimov, A.A., Dodd, I.C., Hontzeas, N., Theobald, J.C., Safronova, V.I. and Davies, W.J. 2009. Rhizosphere bacteria containing ACC deaminase increase yield of plants grown in drying soil via both local and systemic hormone signalling. *New Phytol.* 181:413–423.

Bromham, L., Saslis-Lagoudakis, C. H., Bennett, T. H. and Flowers, T. J. 2013. Soil alkalinity and salt tolerance: adapting to multiple stresses. Biol. Lett. 9:20130642.

Bui, E. N. 2013. Soil salinity: a neglected factor in plant ecology and biogeography. *J. Arid Environ.* 92:14–25.

Copping, L.G. 2004. The manual of biocontrol agents. *British Crop Protection Council, Farnham, UK.* 752.

Dodd, I.C., Zinovkina, N.Y., Safronova, W.I. and Belimov, A.A. 2010. Rhizobacterial mediation of plant hormone status. *Ann. of Appl. Bio.*, 157:361–380.

Egamberdiyeva, D., and Höflich, G. 2003. Influence of growth-promoting on the growth of wheat in different soils and temperatures. *Soil Biol. Biochem.* 35:973–978.

Emamverdian, A., Ding, Y., Mokhberdoran, F. and Xie, Y. 2015. Heavy metal stress and some mechanisms of plant defense response. *Sci. World J.* 2015:756120.

Fallik, E., Sarig S., Okon, Y. 1994. Morphology and physiology of plant roots associated with Azospirillum. In: *Okon. Y. (ed) Azospirillum/Plant associations.* CRC, Boca Raton, FL, 77–85.

Flowers, T., 2004. Improving crop salt tolerance. *J. Exp. Bot.* 55:307–319.

Gopalakrishnan, S., Sathya, A., Vijayabharathi, R., Varshney, R. K., Gowda, C. L. and Krishnamurthy, L. 2015. Plant growth promoting rhizobia: challenges and opportunities. *Biotech.* 5:355–377.

Hinsinger, P., Bengough, A.G., Vetterlein, D. and Young, I.M. 2009. Rhizosphere: biophysics, biogeochemistry and ecological relevance. *Plant and Soil.* 321:117–152.

Jorge, T. F., Rodrigues, J. A., Caldana, C., Schmidt, R., van Dongen, J. T., Thomas-Oates, J. 2015. Mass-spectrometry-based plant metabolomics: metabolite responses to abiotic stress. *Mass Spectrom. Rev.* 35:620–649.

Kaushal, M. and Wani, S. P. 2016. Plant-growth-promoting rhizobacteria: Drought stress alleviators to ameliorate crop production in drylands. *Ann. Microbiol.* 66:35–42.

Koini, M. A., Alvey, L., Allen, T., Tilley, C. A.,Harberd, N. P. and Whitelam, G.C. 2009. High temperature-mediated adaptations in plant architecture require the bHLH transcription factor PIF4. *Curr. Biol.* 19:408–413.

Lambers H., Raven J.A., Shaver G.R., Smith S.E. 2008. Plant nutrient-acquisition strategies change with soil age. *Tren. in Eco. and Evol.*. 23:95–103.

Li, Z., Setsuko, W., Fischer, B. B. and Niyogi, K. K. 2009. Sensing and responding to excess light. *Annu. Rev. Plant Biol.* 60:239–260.

Lugtenberg, B., Kamilova F. 2009. Plant-growth-promoting rhizobacteria. *Ann. Rev. of Microbiol.*, 63:541–556.

Meena, K. K., Kumar, M., Kalyuzhnaya, M. G., Yandigeri, M. S., Singh, D. P., Saxena, A. K. 2012. Epiphytic pink-pigmented methylotrophic bacteria enhance germination and seedling growth of wheat (*Triticum aestivum*) by producing phytohormone. *Antonie Van Leeuwenhoek* 101:777–786.

Mittler, R. 2006. Abiotic stress, the field environment and stress combination. *Trends Plant Sci.* 1115–19.

Nadeem, S. M., Zahir, Z. A., Naveed, M., and Arshad, M. (2007). Preliminary investigations on inducing salt tolerance in maize through inoculation with rhizobacteria containing ACC deaminase activity. Can. *J. Microbiol.* 53:1141–1149.

Naveed, M., Hussain, M. B., Zahir, Z. A., Mitter, B., and Sessitsch, A. 2014a. Drought stress amelioration in wheat through inoculation with *Burkholderia phytofirmans* strain PsJN. *Plant Growth Regul.* 73:121–131.

Naveed, M., Mitter, B., Reichenauer, T. G.,Wieczorek, K., and Sessitsch, A. 2014b. Increased drought stress resilience of maize through endophytic colonization by *Burkholderia phytofirmans* PsJN and *Enterobacter sp* FD17. *Environ. Exp. Bot.* 97:30–39.

Oliveira, C. A., Alves, V. M. C., Marriel, I. E., Gomes, E. A., Scotti, M. R., Carneiro, N. P. 2009. Phosphate solubilizing microorganisms isolated from rhizosphere of maize cultivated in an oxisol of the Brazilian Cerrado biome. *Soil Biol. Biochem.* 41: 1782–1787.

Ortiz, N., Armadaa, E., Duque, E., Roldánc, A. and Azcóna, R. 2015. Contribution of arbuscular mycorrhizal fungi and/or bacteria to enhancing plant drought tolerance under natural soil conditions: effectiveness of autochthonous or allochthonous strains. *J. Plant Physiol.* 174:87–96.

Osakabe, Y., Osakabe, K., Shinozaki, K. and Tran, L.-S. P. 2014. Response of plants to water stress. *Front. Plant Sci.* 5:86.

Panlada, T., Pongdet, P., Aphakorn, L., Rujirek, N.-N., Nantakorn, B., and Neung, T. 2013. Alleviation of the effect of environmental stresses using co-inoculation of mungbean by Bradyrhizobium and rhizobacteria containing stress-induced ACC deaminase enzyme. *Soil Sci. Plant Nut.* 59:559–571.

Pareek, A., Sopory, S. K., Bohnert, H. K. and Govindjee. 2010. Abiotic Stress Adaptation in Plants: *Physiol., Mol. and Gen. Foundation*. Dordrecht: Springer, 526.

Patten, C. L. and Glick, B. R. 1996. Bacterial biosynthesis of indole-3-acetic acid. *Can. J. Microbiol.* 42:207–220.

Patten, C. L. and Glick, B.R. 2002. The role of bacterial indoleacetic acid in the development of the host plant root system. *Appl. Environ. Microbiol.* 68:3795–3801.

Pinedo, I., Ledger, T., Greve, M., and Poupin, M. J. 2015. *Burkholderia phytofirmans* PsJN induces long-term metabolic and transcriptional changes involved in Arabidopsis thaliana salt tolerance. *Front. Plant Sci.* 6:466.

Prasad, P. V. V., Pisipati S. R., Momcilovic I., Ristic Z. 2011. Independent and combined effects of high temperature and drought stress during grain filling on plant yield and chloroplast EF-Tu expression in spring wheat. *J. Agron. Crop Sci.* 197:430–441.

Ramegowda, V., Senthil, M. K. 2015. The interactive effects of simultaneous biotic and abiotic stresses on plants: mechanistic understanding from drought and pathogen combination. *J. Plant Physiol.* 176 47–54.

Ran, L., Larsson, J., Vigil-Stenman, T., Nylander, J.A.A., Ininberg, K., Zheng, W.W., Lapidus, A., Hasselkorn, R. and Bergman B. 2010. Genome erosion in a nitrogen-fixing vertically transmitted endosymbiotic multicellular cyanobacterium. *PLoS ONE*, 5: 11486.

Raven, J. A. 2010. Why are mycorrhizal fungi and symbiotic nitrogen-fixing bacteria not genetically integrated into plants? *Ann. of App. Biol.* 157:380–392.

Remans, R., Ramaekers, L., Shelkens, S., Hernandez, G., Garcia, A. and Reyes, G. L. 2008. Effect of *Rhizobium, Azospirillum* co-inoculation on nitrogen fixation and yield of two contrasting *Phaseolus vulgaris* L. genotypes cultivated across different environments in Cuba. *Plant Soil* 312:25–37.

Rillig, M. C. 2004. Arbuscular mycorrhizae, glomalin, and soil aggregation. *Can. J. of Soil Sci.* 84:355–363.

Sahoo, R. K., Ansari, M. W., Dangar, T. K., Mohanty, S. and Tuteja, N. 2014a. Phenotypic and molecular characterisation of efficient nitrogen-fixing *Azotobacter* strains from rice fields for crop improvement. *Protoplasma* 251:511–523.

Sahoo, R. K., Ansari, M. W., Pradhan, M., Dangar, T. K., Mohanty, S., and Tuteja, N. 2014b. A novel *Azotobacter vinellandii* (SRIAz3) functions in salinity stress tolerance in rice. *Plant Sig. Behav.* 9:e29377.

Singh, D. P., Prabha, R., Yandigeri, M. S., and Arora, D. K. 2011. Cyanobacteria mediated phenylpropanoids and phytohormones in rice (*Oryza sativa*) enhance plant growth and stress tolerance. *Antonie Van Leeuwenhoek* 100:557–568.

Smith, S.E., Facelli, E., Pope, S. and Smith, F.A. 2010. Plant performance in stressful environments: interpreting new and established knowledge of the roles of arbuscular mycorrhizas. *Plant and Soil.* 326:3–20.

Soil. 255:571–586.

Sorty, A. M., Meena, K. K., Choudhary, K., Bitla, U. M., Minhas, P. S. and Krishnani, K. K. 2016. Effect of plant growth promoting bacteria associated with halophytic weed (*Psoralea corylifolia* L.) on germination and seedling growth of wheat under saline conditions. *Appl. Biochem and Biotechnol.* 180:872–882.

Sridevi, M., Mallaiah K. V., Yadav, N.C.S., 2007. Phosphate solubilization by Rhizobium isolates from *Crotalaria* species. *J. Plant Sci.* 2:635–639.

Sun, C., Johnson, J., Cai, D., Sherameti, I., Oelmüeller, R., and Lou, B. 2010. *Piriformospora indica* confers drought tolerance in Chinese cabbage leaves by stimulating antioxidant enzymes, the expression of drought-related genes and the plastid-localized CAS protein. *J. Plant. Physiol.* 167:1009–1017.

Tiwari, S., Singh, P., Tiwari, R., Meena, K. K., Yandigeri, M., Singh, D. P. 2011. Salt-tolerant rhizobacteria-mediated induced tolerance in wheat (Triticum aestivum) and chemical diversity in rhizosphere enhance plant growth. *Biol. Fertil. Soils* 47:907–916.

Vardharajula, S., Ali, S. Z., Grover, M., Reddy, G., and Bandi, V. 2011. Droughttolerant plant growth promoting Bacillus spp.: effect on growth, osmolytes, and antioxidant status of maize under drought stress. *J. Plant Int.* 6:1–14.

Vessey, J.K. 2003. Plant-growth-promoting rhizobacteria as biofertilisers. *Plant and*

Wang, L. J. and Li, S. L. 2006. Salicylic acid-induced heat or cold tolerance in relation to Ca^{2+} homeostasis and antioxidant systems in young grape plants. *Plant Sci.* 170:685–694.

Xu, Z., Jiang, Y., Jia, B. and Zhou, G. 2016. Elevated CO_2 response of stomata and its dependence on environmental factors. *Front. Plant Sci.* 7:657.

Yang, J., Kloepper, J., W. and Ryu, C.M. 2009. Rhizosphere bacteria help plants tolerate abiotic stress. *Tren. in Pl. Sci.* 14:1–4.

17

Microbial Control of Insect Pests with Special Reference to Bacteria

Krishna rolania[1*], Tarun Verma[1] and Rakesh Sangwan[2]

[1]Department of Entomology [2]Department of Pathology CCS Haryana Agricultural University, Hisar 125004

*Corresponding author: krish81rolania@rediffmail.com

Abstract

Microbial pathogens of insect pests *viz.* viruses, bacteria, fungi, nematodes, protozoa and rickettsiae are intensively investigated to develop environmental friendly and sustainable pest management strategies in agriculture. The most successful example of insect pathogen is the bacterium, *Bacillus thuringiensis* (*Bt*) which is used extensively for management of certain lepidopteran insect-pests. Bacterial pathogens can be applied as dusts, sprays or baits either alone or integrated with predators, parasites and other pathogens, chemical insecticides, as well as most other control procedures such as chemosterilants, pheromones etc. Entomopathogenic fungi like *Metarhizium anisopliae, Beauveria bassiana, B. brongniartii, Lecanicillium spp., Hirsutella thompsonii, Nomuraea rileyi and Isaria fumosorosea* are generally used in the crop pest control in recent years due to the simple and cheaper mass production techniques. Baculoviruses comprising nuclear polyhedrosis virus (NPV) and granulosis virus (GV) formulations are mainly used for lepidopteran pests like *Helicoverpa armigera* (HaNPV) and *Spodoptera litura* (SlNPV) in India. *Trichoderma sp.* is the most commonly used fungi for preparation of microbial pesticides. The nematode/bacterial complex is highly virulent, killing its host within 48h through the action of the mutualistic bacteria. *Streptomyces* are the most important actinomycetes found abundantly in soil. Large numbers of secondary metabolites produced by these actinomycetes act as antibiotics. Avermectin produced by *Streptomyces avermitilis* is being commercially used as pesticide.

Key words: Microbial pesticides, Entomopathogens, *Bacillus thuringiensis*, Nuclear Polyhedrosis Virus, Granulosis Virus, Mycoinsecticides

Introduction

Chemical insecticides cause's health hazards, environmental pollution, pest resurgence, pest out break etc. so there is a need of some alternative methods of insect management which offer an adequate levels of pest control for long time and become environmental friendly. One such alternative is the use of microbial insecticides (insecticides that contain microorganisms or their by-products). Microbial insecticides are especially valuable because of their low toxicity to non-target organisms and safety to both pesticide user and consumers of treated crops. Microbial insecticides are also known as biological pathogens and biological control agents. The global market for biopesticides was valued at $2.78 Billion in 2016 (Annonymous, 2017). Over 3000 micro-organisms have been reported to cause diseases in insects. But scientists familiar with specific pathogens group agree that a very large number of insect pathogens remain undiscovered or unidentified. It is very likely that the insect pathogens outnumber the species of insects (Maddox, 1994). Worldwide about 700 products of different microbial are available. In India, 38 fungal formulations based on *Trichoderma, Metarhizium, Beauveria,* about 16 commercial preparations of *Bacillus thuringiensis* and about 45 baculovirus based formulations of *Helicoverpa* and *Spodoptera* are available. Microbial pesticides are expected to replace at least 20 per cent of the chemical pesticides. About 128 units in the country (80 private companies) are supplying biotic agents (Wahab, 2010).

Fungi (Deuteromycetes and Entomopthora) cause mycosis in insects. More than 800 fungal species comprising of 125 genera have been reported to infect insects. *Beauvaria bassiana* (the white muscardine fungus) is a potential bio-inoculant against 700 species of insect pests. Similarly, *Metarhizium anisopliae* (the green muscardine fungus) has been reported to effectively control a large number of target organisms. About 90 per cent mortality of pumpkin caterpillar, *Diaphania indica* larvae by fungus *Paecilomyces furinosus* have been reported by Kuruvilla and Jacob (1980). In soyabean, *Nomuraea rileyi* (Farlow) Samson is the only important factor regulating the populations of velvet caterpillar, *Anticarsia gemmatalis* from Brazil and cloverworm, *Plathpena scabra. Verticillium lecanii* in the form of a wettable powder gives upto 90 per cent reduction in glasshouse whitefly infestation (Revensberg *et al.* 1990).

Viruses have been isolated from more than 1000 species of insets from at least 13 different insect order (Flexner and Belnavis, 2000). Payne (1982) reported baculoviruses as promising biocontrol agent in IPM programme due to their virulence and specificity against many pests. Among six main families of insect viruses' nuclear polyhedrosis virus (NPV) and granulosis virus (GV) are the most suitable and effective pest control agents (Lacey *et al.*, 2001). About 523 species of insect from 52 families and 8 orders were recorded as hosts of NPVs. Of these 523 species, 455 species are in the order Lepidoptera and 107 species in the family Noctuidae, which contains many agricultural pests of economic importance. The order hymenoptera has 31 species listed as host of NPVs and 19 of which are in the family diprionidae and are pests of forest and shade trees. The order Diptera has 27 species as hosts of NPVs, 20 of them are in the family culicidae. Most NPVs are pathogenic to only a few closely related species of insects. However, a few NPV such as the *Autographa californica* (alfalfa looper) NPV, *Syngrapha falcifera* (celery looper) NPV and the *Mamestra brassicae* NPV were found to have broader spectrum (upto 30 species) of infesting many species of Lepidoptera in the family noctuidae and several species in other families of Lepidoptera. Most NPVs loose 50% or more of their original activity within one or two days when sprayed on plants in the field by UV rays and plant foliage factors. However, NPV can persist for years in the soil which acts as a reservoir of virus and provides inoculums to intimate new infections each year.

There are about 1200 species of entomopathogenic protozoa and microsporidia, causing diseases in insects. *Microsporidia* contains many species that have promise for biological control. Microsporidian are relatively slow acting organisms, taking days or weeks to debilitate their host, reduce host reproduction or feeding rather than killing the pest outright. These microsporidia must be eaten to infect an insect, but also natural transmission within a pest population, for example by predators and parasitoids. The pathogen enters the insect body via the gut wall, spreads to various tissues and organs, and multiplies, sometimes causing tissue breakdown and septicemia. Infected insects may be sluggish, smaller than normal, difficulty in molting, sometimes with reduced feeding and reproduction. Death may follow if the level of infection is high. One advantage of this type of infection is that the weakened insects are more likely to be susceptible to adverse weather and other mortality factor (Anon, 2011b). In America protozoa, *Perezia pyraustae* has been successfully utilized for the control of European corn borer, *Ostrinia nubilalis*. Similarly, *Nosema* sp. was found parasitizing 13.8 per cent of *Spodoptera litura* in Andhra Pradesh (Sridhar and Prasad, 1996) and upto 100 per cent in castor semilooper in Maharashtra (Phadke and Rao, 1978). The grasshopper pathogen, *Nosema locustae* Canning has been registered and used commercially (Henry and Oma, 1981). *Nosema locustae* is the only commercially available species of microsporidium, marketed under several labels for the control of grasshoppers and crickets. It is applied with insect-attractant bait. Because of its slow mode of action, this product is better suited to long-term management of rangeland pests than to the more intensive demands of commercial crop or even home garden production. Under greenhouse conditions the potential of *Nosema marucae* for the control of *C. partellus* on sorghum was studies by Odindo (1992). *Nosema* treatment reduced the hatching by 37 % of stem borer larvae as well as increased the larval mortality by 92.7%. *Vairimorpha necatrix* is microsporidium with a wide host range including corn earworm, European corn borer, various armyworms, fall webworm, and cabbage looper. It can be more virulent than other species and infected insects may die within six days of infection.

Presence of nematodes as international parasite on various beetles, grasshoppers, moths and other insects reported by Tanada in 1959 and he recommended that they may be used successfully for the control of these insects. The citrus root weevil, *Diaprepes abbreviatus* in citrus; the black vine weevil, *Otiorhynchus sulcatus* in nurseries and cranberries; the black cutworm, *Agrotis ipsilon* and mole crickets, *Scapteriscus* spp. in turfgrass and the peach borer moth, *Carposina niponensis,* Walsing-ham in apples have been successfully controlled by nematodes. They are similar to pathogens because of their association with mutualistic bacteria in the genera Xenorhabdus (for steinernematids) and Photorhabdus (for heterohabditids). The nematode/bacterial complex is highly virulent, killing its host within 48h through the action of the mutualistic bacteria. Aggarwal and Mahal (2010) reported highest mortality of 47.9, 53.3 and 39.9, 46.6 per cent after 7 and 10 days of treatment with *Steinernema feltiae* in case of yellow stem borer and leaf folder of rice, respectively. Nowadays DD 136 (*Neoaplectana carpocapsae*) is used against the control of various insect pests of Lepidoptera and Coleoptera. Similarly three sprays of DD 136 based on *S. feltiae* (Filipjev) checked the population of sawfly on radish (Narayanan and Gopalakrishnan, 1988). However, maximum mortality 95 and 77 per cent of wax moth larvae, *Galleria mellonella* and vine mealy bug after 24-48 h exposure to cell water suspension of *Xenorhabdus nematophila* and its metabolites from *Steinernema pakistanense* (strain HAM-10) was reported by Shahina *et al.* 2011. Similarly Ensigen *et al.* (2002) reported that the symbiotic bacteria from the third stage infective juvenile of entomopathogenic nematode secrete a wide variety of toxic metabolites which can be applied as insecticides. Once inside the host, nematodes invade the haemocoel and release the bacteria which multiply rapidly and cause septicemia, kill the host within 24-48 hours. Dudney (1970) reported control of fire ant, *Solenopsis invicta* with the cells suspension of *X. nematophila* and its toxic metabolites and larvae of beat army worm, *Spodoptera exigua* (Elawad *et al.,* 1999). However, studies conducted by Jan *et al.* (2008) proved that the toxin and the metabolites of bacterial symbionts were found very effective against different insect pupae. The other nematode species are Neotylenchus, Mermis, Rhabditis, Hercamermis, Protrellus, Sternienonema, Pseudonymus *etc.*

More than 100 species of bacteria pathogenic (Bacillaceae, Lactobacillaceae, Streptococcaceae, Micrococcacaceae, Pseudomonadaceae and Enterobacteriaceae) to insect pests (Dhaliwal and Koul, 2007), but hardly one per cent among these is used as bio control agents. These can be applied against more than 175 larval species of the most important economic pests and can be mixed with a number of commercial insecticidal formulations. Several species of bacteria for insect-pest control *viz. Aerobacter aerogenes, Escherichia coli, Bacillus popillae, B. sphaericus, B. moritai, B. lentimorbus* and *B. thuringiensis* are already used in commercial preparations. The above mentioned genus of bacteria attack mostly insects belonging to order Hemiptera, Diptera, Coleoptera and Lepidoptera. Between 1920 and 1960, the most extensive trials have been carried out with *B. thuringiensis,* which has provided effective control of many Lepidopteran insects of field crops and forests. There has been virtually no commercial development of rickettsia or protozoans and only limited development of fungi. Bacteria accounts for most of the commercial use of microbial agents. In this chapter the main emphasis will be given on the use of bacterial control agents, their field efficacy and development of resistance, if any.

Table 1: Estimated global market value of microbial pesticides by 2008 (US $ million)

Country	Bacteria	Virus	Fungi
Europe	15	10	25
NAFTA	80	15	45
Latin America	10	10	20
Africa	5	2	3
Asia	20	5	15
Oceania	30	10	20
Total	160	42	128

Wahab, 2010

Table 1.1 : Commercial Availability of Entomopathogens at Global Level

Category	Organism	No. of products	No. of countries
Bacteria	*Bacillus thuringiensis* subsb. *Aizawai*	36	45
	Bt subsp. *Isralensis*	50	79*
	Bt subsp. *Kurstaki*	220	113*
	Bt subsp. *Tenebrionis*	7	20
	Bt subsp. *Thuringiensis*	10	6
Fungi	*Beauveria bassiana*	94	44
	Metarhiaium anisopliae	73	63*
	Paecilomyces fumosoroseus	18	20
	Lecanicillium lecanii	43	24
Nematodes	*Heterorhabditis bacteriophora*	19	25
	H. megidis	9	21
	Steinernema carpocapsae	37	25
	S. feltiae	39	23
Viruses	*Cydia pomonella* GV	13	30
	Helicoverpa spp. NPV	18	9
	Spodoptera spp. NPV	16	14

**Includes upto 34 sub Saharan African nations; Source: CPL Business Consultants (2010)*

Microbial Control: This is an extension of biological control where microbes (pathogens) or their products (toxins) are employed for the control of insects, animals and plants in a particular area. The term microbial control was coined in 1949 by Steinhaus. **Microbial control** refers to the exploitation of disease causing organisms (viruses, bacteria, protozoa, fungi, mycoplasma, rickettsia and nematodes) to reduce the population of insect pests below the damaging levels. Some of these pathogens may be quite common and are frequently the cause of epizootics in natural insect populations while others are rarely observed.

Microbial insecticides: Insecticides that contain microorganisms or their by-products. These are also known as biological pathogens, and biological control agents. Microbial insecticides are especially valuable because their toxicity to non-target animals and humans is extremely low.

History

Insect diseases and their symptoms have been well known as far back as 2700 BC in China with the honeybee, *Apis mellifera* Linnaeus and the silk worm, *Bombyx mori* (Linnaeus). Aristotle was the first in Europe to mention that the bees suffered from disease. Agostino Bassi published his great work in 1835 on muscardine disease of silkworm, caused by *Beauveria bassiana* (Balsamo) Vuillemin. First systemic experiment for the control of injurious insects with microorganisms was conducted by a Russian Entomologist, Metschnikoff in 1879, by infecting grubs of the grain beetle, *Anisoplia austricaca* with fungus *Metarrhizium anisopliae* (Metchnikoff) Sorokin. Burges and Hussey (1971) also reported this fungus as effective against the sugarbeet curculio, *Cleonus punctiventris* (Germ.).

First successful establishment of an exotic virus disease was accomplished by Balch (1946), who introduced the polyhedrosis virus of the Europena spruce-saw fly (*Diprion heraynial*) into Newfound land. The first successful attempt to introduce a virus disease not present previously in the North American continent was made by Bird (1953) who applied a polyhedrosis virus of the European pine sawfly (*Neodiprion sertifer*) obtained from Swedon. The use of baculoviruses especially NPV in insect pest management was initiated sometime back in 19th century. One of the first reported attempts to use such a virus on an operational scale was the introduction of a NPV into the population of nun moth, *Lymantria monacha* during its outbreak in Germany.

The bacterium for the first time was discovered by a Japanese bacteriologist (S. Ishiwata) in the body fluid of ill silkworm in 1901, and designated it as *Bacillus sotto.* However, *B. thuriengiensis* name was given in 1915 by Berliner, who isolated it from the Mediterranean flour moth, *Ephestia kuehniella* Zeller in Thuringia, Germany in 1911. It was initially used as a pesticide in 1920s in the southeast Europe. Milky disease A and B in Japanese beetle, *Papilio japonica* was observed to be caused by bacteria *Bacillus popilliae* and *B. lentimorbus,* respectively in the beginning of 1940 in the United States. Infected larvae are crushed and mixed into the soil where they parasitize and kill the larvae. Hannay in 1953 described the parasporal inclusion in sporulating cultures of *Bt,* as these inclusions were proteinaceous and might have insecticidal entity (Hannay and Fitz-James, 1955). The insecticidal activity of these parasporal inclusions were demonstrated by Heimpel and Augus (1959). In 1961 *Bt* was registered as a pesticide by the Environmental Protection Agency (EPA), USA.

Definitions and concepts: Some of the important terms and concepts used in insect pathology and microbial control are briefly introduced in this section (Tanada and Kaya, 1993).

Pathogens

Pathogens are microorganisms generally invade and multiply in an insect and spread to infect other insects. Infectious microorganisms can be separated broadly into potential, facultative and obligate pathogens.

Potential pathogens are those microorganisms that are incapable of invading the host, either through the body wall or through the digestive tract, without assistance of external factors that lower the insect resistance or enhance the ability of the microorganisms to invade the insect.

Obligate pathogens require living insect hosts for survival and replication. They may occur outside of the insect in a dormant stage, such as spore, cyst, viral occlusion body etc. Viruses, microsporidia (protozoa), most nematodes and certain fungi and bacteria are obligate pathogens.

Facultative pathogens are those microorganisms that do not require an insect weakened by external factors to cause infection. Moreover, the survival of both potential and facultative pathogens is not dependent entirely on the insect host and they can live and multiply independently of the insect. These pathogens are readily reared on non-living media in the laboratory. Most of them are bacteria and fungi.

Infectivity is the ability of a microorganism to produce infection. Generally an infection results in detectable pathologic effects, such as injuries or dysfunctions. An infection may or may not be harmful. Mutualistic symbiont infections do not result in a diseased condition.

Virulence is the disease producing power of a microorganism. It is the ability of a microorganism to invade and cause injury to the host.

Pathogenicity refers to the disease producing power of a group of species of microorganism. Thus we may say that pathogenicity of *B. thuringiensis* is high for lepidopteran species, but its virulence may differ depending on conditions, such as methods of rearing, storage, formulation and environmental factors. Virulence is a variable property and can be intensified or reduced.

Microbial toxins: A disease can be brought about in a susceptible host by the pathogen through the effects of chemical or toxic substances, the mechanical destruction of cells and tissues, and the combination of these two actions. There are two general types of toxins produced by entomopathogenic organisms, catabolic and anabolic substances.

The **catabolic toxins** result from decomposition brought about by the activity of the pathogen. They may arise from the host substrate or from the decomposition of the pathogen itself. For example, the breakdown of proteins, carbohydrates and lipids by the pathogen may produce toxic alcohols, acids, mercaptans, alkaloids etc.

The **Anabolic toxins** are substances synthesized by the pathogens. These may be classified as exotoxins and endotoxins.

The **exotoxin (ectotoxins)** is excreted of passed out of the cell of the pathogen. Bacteria and fungi are known to produce exotoxins. The **endotoxins** produced by the pathogen, are confined to the cell and are liberated when the pathogen dies or degenerates. In *Bt* the delta endotoxin is released when the sporangial wall disintegrates.

Precaution to Be Taken in the Use of Entomopathogens:

- Correct identification of pest, selection of the correct variety or strain of microbial.
- As older caterpillars are more difficult to control, therefore, monitor plants frequently to make applications against young larvae.
- Some of these are sensitive to sunlight and has a short activity period, multiple applications may be necessary.
- Carefully read the label and follow all directions regarding storage, mixing, application and safety.

Bacteria: Bacteria are prokaryotic, unicellular organisms with rigid cell walls and may be spherical (cocci), rod shaped (bacilli) or spiral (spirilla, spirochaetes), whereas those without cell walls (mollicutes) are plemorphic. Insect pathogenic bacteria occur mostly in the families Bacillaceae, Pseudomonadaceae, Enterobacteriaceae, Streptococcaceae and Micrococcaceae; in the order Rickettsiales and class Mollicutes (Tanada and Kaya, 1993). **Entomogenous bacteria can be classified into**

i) **Spore forming bacteria:** *B. popillae, Clostridium brevifaciens, B. thuriengiensis, B. cerus*

ii) **Non-spore forming bacteria**: *Pseudomonas, Aerobacter, Serratia* spp.

Spore forming bacteria

Bacillaceae: It includes genus *Bacillus* (aerobic or a facultative aerobe) and *Clostridium* (anaerobic). These produce endospores and are gram positive motile or non motile rods. Vasiliki *et al.* (2004) under laboratory assay conditions found that *Bacillus* sp. (strain Vj02i- 18) and two strains (Vj02e-1, Vj02e-2, Vj02i-3) of Micrococcaceae family significantly decreased the time for 50% mortality of the honey bee mite, *Varroa destructor* population (up to 57%) and may be used as potential control agents.

Clostridium: This bacterium was found to cause brachytosis disease in tent caterpillar, *Malacosoma pluvial* which manifests desiccation and shrinking of the body (Bucher, 1961).

B. cereus: Closely related to *B. thuriengiensis* and *B. anthracis* (vertebrate pathogen) widely distributed and found in soil. Absence of a protoxin parasporal body (facultative acrystalliferous) in *B. cereus* differs it from *B. thuringiensis. B. cereus* has been isolated from over 30 insect species mainly in the orders Coleopteran, Hymenoptera and Lepidoptera. The pathogenic strain produces a toxin lecithinase (phospholipase C) which is active at a gut pH of 6.6 to 7.4. *B. cereus* produces a number of exotoxins known to affect vertebrate.

***B. scelestes* (Banks):** *Eupeodes corolla* (Fab.) along with *B. scelestes* (Banks) reduces the woolly apple aphid population by 23.8-34.8 per cent at 4 sites (Thakur *et al.* 1991).

***B. sphaericus* Neide:** The bacterium is ubiquitous, aerobic, rod shaped, gram variable, found in both soil and soil aquatic system, first described by Neide in 1904. It produces terminally or sub terminally swollen spherical spores. This bacterium has attracted attention due to excellent control over *Culex, Psorophora* and *Mansonia* mosquito larvae and its long persistency. Akpabey and Ocran (2010) reported that when *B. sphaericus* (Trade name: VectoLex CG (Corn Cob)) was used against *Culex* mosquito larvae in choked gutters at Labadi, Ashaiman and Nungua in the Greater Accra Region caused 99 to 100% mortality.

***B. laterosporus*:** Insecticidal effects of different bacterial isolates was studied by Erturk and Demirba (2006) on *Cydia pomonella.* Highest insecticidal effect was determined by *Bacillus laterosporus* (65% mortality) within eight days, as compared with *Proteus rettgeri, Eschericia coli, Pseudomonas stutzeri, Pseudomonas aeroginosa, Micrococcus sp., Proteus vulgaris and Deinococcus sp.* (22, 19, 25, 60, 20, 57 and 17 per cent, respectively).

***B. popilliae* and *B. lentimorbus* :** These obligate pathogens were reported to cause type A and type B milky disease in Japanese beetle, *P. japonica* Newman in 1933. The bacteria causing type A and Type B milky diseases were identified as *B. popilliae* and *B. lentimorbus,* respectively. *B. popilliae* host range within the scarabaeids is about 70 insect species. *B. popilliae* has been used as a biopesticide since 1937. In India, *B. popilliae* is known to occur naturally on whitegrubs, *Holotrichia consanguinea* Blanchard, *H. serrata* and *Leucopholis lepidophora.* The treatment is most effective in areas where winter temperature do not fall below 15^{0}C. Vyas *et al.* (1986) reported that *B. popilliae* Dutky caused upto 71.4 per cent mortality of *Holotrichia* in Gujarat in 1984 and was cross infective to *H. serrata* (Fabr.), *Anomala bengalensis* (Frey) and *Antoserica vathani* (Blanch). *H. consanguinea* in groundnut in the sub tropics was effectively managed by *B. popillae* var. *holotrichiae* @ 5 billion spores per square meter (Vyas *et al.* (1991)). About 10 biovars of *B. popilliae* are known throughout the world. Battu ad Arora (1997) reported that the *in vivo* production technique makes the product very expensive. *B. popilliae* spores enter the larval midgut after ingestion during feeding where these germinates. The vegetative cells proliferate and enter the haemocoel and continue to multiply. In larvae infected by *B. lentimorbus* for extended periods of time, there is a build up of blood clots, causing the haemolymph to look brownish in color instead of milky white. *B. popilliae* Dutky "Doom" and "Japidemic" (marketed by Fairfax Biological Laboratory Inc.) and "Milky spore" (by Reuter Laboratory Inc.) used in the USA, when used in Gujarat in late seventies against *Holotrichia* spp. gave 20-25 per cent reduction in population.

B. thuringiensis: *B. thuringiensis* is a ubiquitous microbe found commonly in soil and also on leaf surface. Steinhaus (1951) grew the *Bt* on nutrients agar, harvested spores again re-suspended in water and applied for the control of alfalfa caterpillar. After application the population was reduced within a few days. Angus (1956) observed that toxicity which led to paralysis and death was associated with the crystals and the crystal could be activated by the gut juice of silk worm. Till now a large number of crystals protein have been recognized. The crystal protein (Cry) gene incoding for these problems have been classified on the basis of crystal protein structure and host range. Based on amounts used in agricultural applications, molecules of Bt toxin are 80,000 times potent than organophosphates and 300 times more potent than pyrethroids. *Bt* sprays comprise one to two per cent of the global insecticide spray market, estimated to be U.S. $ 8 billion per annum. At one time, *Bt* sprays constituted $100 million in annual sales, but with the advent of transgenic plants engineered with ICP genes, sales have decreased to $40 million (Feitelson *et al.* 1992). So far, more than 40,000 species of *Bacillus thuringiensis* have been isolated and identified as belonging to 39 serotypes. These organisms are active against either Lepidoptera, or Diptera or Coleoptera (Anon. 2011).

Non Spore Forming Bacteria

Enterobacteriaceae: It includes many species which live in the digestive tracts of animals, including insects. They are gram negative facultative anaerobic, non spore forming, rod shaped bacteria with peritrichous flagella. Tanada and Kaya (1993) reported that the entomopathogenic forms usually lack invasive ability to penetrate the midgut wall but generally are high virulent when enter the insect's haemocoel.

Serratia: The bacteria live in digestive tracts of animals and many insects. These are gram negative facultative anaerobic rod shaped bacteria with peritrichous flagella. Tanada and Kaya (1993) reported that entomopathogenic forms are generally highly virulent when they enter the insect's haemocoel, but usually lack invasive ability to penetrate the midgut wall. O'Callaghan *et al.* (1996) reported that *Serratia marcescens* and *S. proteamaculans* have potential as control agents against Diptera. *S. marcescens* have been reported as a pathogen of several insect pests *viz. Athalia proxima* larvae (Bogawat *et al.* 1966), boll weevil, *Anthonomus grandis* (Ourth and Smalley, 1980), the tobacco hornworm, *Manduca sexta* (Dunn and Drake, 1983), housefly, *Musca domestica* (Benoit *et al.* 1990), apple maggot fly, *Rhagolestis pomonella* (Louzen *et al.* 2003) as well as beneficial insects *viz.* honey bee, *Apis mellifera* (El Sanousi *et al.* 1987). Bacteria found to infect insect's belonging to orthoptera, coleoptera, hymenoptera, lepidoptera and diptera. New Zealand grass grub, *Costelytra zealandica* (White) infected with *S. entomophila* stop feeding within a few days and dies in 4-6 weeks. The liquid formulation of this bacterium is marketed with product name Invade® (Jackson *et al.* 1992). *S. marcescens* was also found to kill substantial population of lemon butterfly, *Papilio* spp. (Singh, 1994) and mango leaf webber, *Orthaga euadrusalis* Walker (Srivastava, 2004).

***Xenorhabdus* spp.:** These bacteria are mutualistically associated with entomopathogenic nematodes. The nematode assists the bacteria in penetrating the insect haemocoel and in return the bacteria provide nutrients essential to the nematode.

Other Bacterial Pathogens

Actinomycetes: Gram positive bacteria with a mycelia growth habit. Most important among these are *Streptomyces* found abundantly in soil. Large numbers of secondary metabolites produced by these actinomycetes act as antibiotics. Avermectin produced by *Streptomyces avermitilis* is being commercially used as pesticide. Jasmine (2010) reported abamectin 1.9EC @ 22.5 and 18.5 g a.i./ha was found to be more effective against cotton bollworms. The order of relative efficacy was abamectin 1.9EC @ 22.5>18.5>spinosad 45 SC @ 75g a.i./ha>abamectin 1.9EC @ 14.5 g/ha>10.8. Citromycin produced by *S. hygroscopicus,* aureothin produced by *S. thiolutens,* piericidins by *S. mabaraensis* and spinosyns by *Saccharopolyspora spinosa* are other compounds with insecticidal activity.

Rickettsia: These are gram negative, obligate intracellular pathogens with typical bacterial cell walls and no flagella. Genus *Rickettsiella* have some entomopathogenic species. After ingestion infectious elementary forms penetrate the insects' midgut wall and replicate in tissues. The small rickettsiae within cytoplasmic vacuoles undergo transformation to bacteria like forms that multiply by binary fission. Tanada and Kaya (1993) reported 'blue disease' in Japanese beetle caused by *Rickettsiella popilliae,* while *R. melolonthae* found infecting *Melolontha melolontha* (Linnaeus) and other scarabaeid beetles.

Mollicutes: Wall less prokaryotic microorganisms surrounded by a unit membrane. Cells are small, highly pleomorphic, varying from coccoid to helical to filamentous and ranging in sizes from 50-150 nm. It includes *Mycoplasma* (cause lethargie in *M. melolontha*) and *Spiroplasma* (found infecting insects in orders hymenoptera, hemiptera, diptera and coleoptera predominantly in the gut lumen and /or haemolymph) (Tanada and Kaya, 1993).

Among several species of bacteria, *B. thuringiensis* (*B.t*) is true entomopathogenic bacteria having many subspecies with vast differences in spectrum of activity.

The classification of *B. thuringiensis* subspecies based on the serological analysis of the flagella (H) antigens was introduced in the early 1960s (de Barjac and Bonnefoi, 1962). Until 1977, all *Bt* subspecies (only 13) were toxic to lepidopteran larvae only. The discovery of other subspecies toxic to dipteral, coleoptera and apparently *Nematoda* (Narva *et al.,* 1991) enlarged the host range and markedly increased the number of subspecies. Based on flagellar H- serovars over 67 subspecies had been identified upto 1998.

Mode of Action

The toxic crystal *Bt* protein is only effective when eaten by insects with a specific (usually alkaline) gut pH and the specific gut membrane structures required to bind the toxin. Not only the insect have the correct physiology, a susceptible stage of development, but the bacterium must be eaten in sufficient quantity. When ingested by a susceptible insect, the protein toxin damages the gut lining, leading to gut paralysis. Affected insects stop feeding and die from the combined effects of starvation and tissue damage. *Bt* spores do not usually spread to other insects or cause disease outbreaks at their own as occurs with many pathogens.

Symptoms

Larvae affected by *Bt* become inactive, stop feeding, and may regurgitate or have watery excrement. The head capsule may appear to be overly large for the body size. The larva becomes flaccid and dies, usually within days or a week. The body contents turn brownish-black as they decompose. Other bacteria may turn the host body red or yellow.

Host range: Facultative *B. thuringiensis* has been found to be generally effective against a wide range of Lepidopterous larvae and used as a feed additive for cattle and poultry. Bacteria fed animals' dropping found inhibiting the development of larvae of the housefly, *Musca.*

Field efficacy of Bt

B. thuringiensis Berliner has found to be effective in controlling potato tuber moth in Peru, Tunisia and India by several workers (Amonkar *et al.* 1979; Lal 1987 and Chandla *et al.* 1993). In California ultra low volume (ULV) application of *Bt* Berliner as an alternative method to conventional insecticides against peach twig borer, *Anarsia lieatella* (Roltsch *et al.* 1995), monitoring with sex pheromones and post bloom spray with *Bt* showed promising results (Quarles, 1997). *B. thuringiensis* has been used in management of several insect-pests globally particularly Diamond Back Moth, *Plutella xylostella* larvae on cruciferous vegetables, *Helicoverpa* spp. and *Spodoptera litura.* In India this bacterial insecticide is used for the management of many insect-pest, *viz. P. xylostella* larvae on vegetables (Asokan *et al.* 1996), *S. litura* on cabbage (Datta and Sharma, 1997), fruit borer of okra (Sathpathy and Panda, 1997), *H. armigera* on pigeonpea (Putambekar *et al.* 1997) and tobacco (Sreedhar *et al.* 2010).

Significant and concentration dependent decrease in the food consumption and growth indices of surviving larvae at various sub lethal dosages have been reported by Singh *et al.* (2000). Whereas moderate to high level of control was reported (Battu *et al.* 1993, Arora *et al.* 2003) in cotton bollworms, brinjal fruit and shoot borer, cabbage semilooper, tobacco caterpillar and castor hairy caterpillar. In bioassays studies Chang *et al.* (1993) showed that 13 strains of *B. thuringiensis* isolated from soil increased mortality of the honey bee mites, *V. destructor* about

15%. Under laboratory conditions, two species of *Bt* against fourth instar larvae of *S. obliqua* gave 100 % mortality within 24 h (Chahal *et al.* 2001). Whereas, Majumder *et al.* (2010) reported 100 per cent mortality of 2 days old larvae of looper, *Hyposidra* spp. (@0.1% and 0.2% conc.) and 53.33 to 63.33 per cent mortality of 7 days old larvae (@0.2% conc.) after 72 hours of feeding with *B. thuringiensis* var. *kurstaki.* However, 25-30 per cent mortality and 20-25 per cent nematodes (juveniles) static effect on root knot nematode within 42-48 hours of exposure to *Bacillus* have also been reported (Chandla and Verma, 2005). Bait containing castor leaves and jiggery in 4:1 treated with *Bt* at 1.0 per cent provided 77.04 per cent cumulative mortality (Jethva and Kapadia, 2006). Erturk (2007) reported the insecticidal effects of different *Bt* toxins against *P. xylostella* were 100, 73.5, 45.5, 47.0 and 55.3 per cent using toxin HD-1 (isolated from the Harry Dumagae strain of *B. thuringiensis),* toxin BTS-1 (isolated from *tenebrionis* strain of *B.t*), *B. thuringiensis, B. thuringiensis israelensis* and *B. thuringiensis Kurstaki,* respectively. Grossi *et al.* (2007) reported the susceptibility of *A. grandis* (Cotton boll weevil) and *S. frugiperda* (fall armyworm) to a Cryi Ia-type toxin from a Brazilian *B. thuringiensis* strain. Brownbrindge and Onyango (1992) conducted bioassay studies at ICIPE, Kenya with four commercial *Bt* formulations (Thuricide, SAN 415, Certan and Dipel 2X) based on var. *kurstaki* against stem borer larvae by incorporation into artificial diet. Dipel 2X was found most toxic formulation followed by SAN 415 and Thuricide.

Fruit: Birds played important role in dispersal of the bacterium (Vyas *et al.* 1988). *Bt* sub sp. *Kurstaki* as Dipel (0.05%) was found quite effective against early instars of lemon butterfly (Sharma, 2004).

Okra: *Bt* (Biolep) @ 2 kg/ha (Satpathy and Panda, 1997) and 0.5 Kg/ha (Dipel) (Krishanaiah *et al.* 1981) reported to keep *Earias* spp. under check.

Sweet Corn: Control of *H. zea* has been reported with heavy doses of *Bt.* An encapsulated formulation and clay granular and emulsifiable formulation of *Bt* were effective against the European corn borer. The tobacco hornworm and budworm were successfully controlled.

Cole crops: Different formulations (Dipel, Delfin, Halt, Biobit, Bioasp etc.) of *Bt* have been evaluated by several workers and found effective in reducing the populations of Diamondback moth under field conditions (Asokan *et al.* 1996, Asokan and Mohan, 1999, Anon., 2002, Narayanan, 2004). *Bt* formulations have found recommendations against this pest on cabbage and cauliflower in many states of our country. DBM was found to be effectively controlled by two applications of Dipel 8L/Biolep/Biobit all at 0.75 l or kg/ha to cauliflower crop in Punjab, India. Arora *et al.* (2000) observed larval mortality five days after the second spray of *Bt,* was more than 80 per cent as compared to 53 per cent in fenvalerate (standard). Kol-chevskii *et al.* (1991) tested several concentration of *B. thuriengiensis* var. *Kurstaki* (Berliner) and found 61-100 per cent mortality of 3rd instar larvae of *Pieris brassicae.* Dhawan *et al.* (2010) reported 62.2 per cent mortality of *P. brassicae* when Dipel (1x 10ki power 9) LE was applied @ 0.2%.

Oilseeds: *Bt* cause mortality upto 52.3 per cent to castor semi looper, *A. janata* (Kulshreshtha *et al.* 1965). However, *B. thuringiensis* Berliner suspension @ 0.05-0.20 per cent gave 50 per cent and 100 per cent larval mortality within 3 and 4 days, respectively against mustard saw fly under field conditions (Saxena, 1979). *Bt* have been found to be effectively managing the population of groundnut leaf miner, *Aproaerema modicella* up to 36.4 per cent by Rajagopal *et al.* (1998).

Cucurbits : Sood *et al.* (2010) used washed and fermented bacterial preparations of symbionts (*Klebsiella oxytoca* and *Pantoea agglomerans*) living in the gut of fruit flies, *Bactrocera* spp. for their attractancy to these flies under laboratory and field conditions. As fermented bacterial preparation, *K. oxytoca* (48-h old culture) in combination with jaggery attracted maximum fruit flies (mean 9.33 adults/30 min.) when applied on potted plants. Under field conditions, *K. oxytoca* applied as foliar application and as bait in combination with insecticides resulted in significant reduction in fruit fly infestation (65.46%) over untreated control (79.56%).

Cotton: During evaluation of different *B.t.* formulations against bollworm complex of cotton spraying of thuricide/biotrol at 1.25 l/ha found effective against *H. armigera.* Application of Biobit @ 1.0-1.5 kg/ha was as effective as fenvalerate, cypermethrin and acephate against cotton bollworms in Karnataka (Dhawan, 1999). However, these formulations were not found effective against pink bollworm.

Pulses: Different bioagents *viz. Bt* and viral pathogens (NPV, cytoplasmic polyhedrosis virus) have been reported to be naturally occurring in the field and also isolated from Lepidopteran pests of pulses. Granular formation of *Bt* based on wheat meal and yeast extract found to protect it against degradation and reported more effective with their application during evening, proved more successful. Careful selection of these formulations is needed to select not only efficient natural enemies but also the species not detrimental to non target beneficial insects. Mahapatro and Gupta (1999) reported that pre feeding of *Helicoverpa* larvae with *Bt* var. *Kurstaki* strain HD-1 does not prevent *Microplistis coroceipes* from ovipositing the infected host larvae and also feeding *Bt* preparation with honey does not harm the wasps.

Commercialization: As early as 1960, *Bt* was produced and sold as an insecticide. Nowadays, it is registered and sold for use as an insecticide in over 50 countries for use against leaf-eating caterpillars. The first major commercial *Bt* product "Dipel" an isolate of sub sp. *Kurstaki* reported by Dulmage (1970) was produced by Abbot Laboratories. HD-1 strain is the most important *Bt* product registered for nearly 30 crops and against over 100 pest insect species worldwide. Prof. S.K. Majumdar and his team of research workers at the Central Food Technological Research Institute (CFTRI), Mysore has got credit for production of an indigenous *Bt* product on a pilot scale in 1968. *Bt* var. *Kenyae* isolated from *Ephestia cautella* (Walker) by scientists at Bhaba Atomic Research Centre (BARC), Trombay, has shown effectiveness against several economically important pest insect species. Battu and Arora (1997) reported that local strains of *Bt* are being investigated for development of bacterial insecticides in almost all the agricultural universities/ research institutes in India. *B. thuriengiensis* (Berliner) and its various sub species, strains and geographic isolates have been studied in detail for their efficacy in field and the laboratory against major Lepidopterous pests such as *H. armigera, S. litura, Leucinodes orbonalis* (Guenee) and *P. xylostella* (L.) etc.

In India several *Bt* based products effective against Lepidopteran pests are being marketed by various agrochemical industries. Some of the trade names of the products containing *B. thuringiensis* var. *thuringiensis* are Bakthane L-69®, Parasporin® and Thuricide® in the USA and Sporeine and Biospore2808 (superscript) in Europe. The other trade names are Dipel®, Javelin®, Worm Attack®, Caterpillar Killer®, Bactospeine®, and SOK-Bt®. Certan®, formulated from *Bacillus thuringiensis* var. *aizawai*, has very target specific used exclusively for the control of wax moth larvae in honeybee hives. Other examples of commercial products are Halt, Biolep, Bioasp and Biobit. All commercial formulation showed bacterial concentration in terms of billions of spores per gram powder. Both spray (wettable powder: 25-100 billion spores/g powder) and dust formulations (2.5 to 7.5 billion spores/g) of a bacterial insecticide could be used. There are 67 registered *Bt* products with more than 450 formulations. Formulations based on *Bt* account for nearly 90 per cent of the total biopesticide sales worldwide (Neale, 1997) with annual sales of nearly US$90 million (Lambert and Perferoen, 1992).

Bacterial pesticides registered in the United States

Species	**Pest controlled**
B. thuriengiensis Berliner subsp. *Kurstaki*	Lepioptera larvae
B. thuriengiensis subsp. *Israelensis*	Dipteran larvae
B. popillae Dutky	Japanese beetle larvae
B. lentimorbus Dutky	
B. thuriengiensis var. *Sandiego*	Coleopteran larvae
B. thuriengiensis subsp. Kurstaki strain *EG2371*	Lepidopteran larvae
Bacillus sphaericus Neide Dipteran larvae	
B. thuriengiensis subsp. *tenebrionis*	Coleopteran larvae
B. thuriengiensis subsp. Kurstaki strain *EG2348*	Lepidopteran larvae
B. thuriengiensis subsp. Kurstaki strain *EG2424*	Lepidopteran larvae
Coleopteran larvae	
B. thuriengiensis subsp. *aizawai*	Lepidopteran larvae
B. thuriengiensis subsp. *aizawai* strain *GC 91*	Lepidopteran larvae

Bacillus thuriengiensis formulations available in India

Product	Variety of B.t	Insecticide potency	Manufacturer/supplier
Bioasp	*Kurstaki*	3200 lu/mg	Bio Tech Internation
Agree	*Kurstaki*	NA	Hindustan Ciba Geigy Ltd.
Biolep, B.t.K-I, B.t.k.-II	*Kurstaki*	3000 lu/mg	Bio Tech Internation
Spicturin	*Galleriae*	1x10^{6} /ml (Sproe count)	Tuticorin Alkali Chemicals
Biobit	*Kurstaki*	3200 lu/mg	Rallies India
Dipel 8L	*Kurstaki*	17600 /ml	Lupin India/Cheminova India Ltd.
Delfin	*Kurstaki*	53000 su/mg	Syngenta India/Margo India
Halt	*Serovar Kurstaki* (Indigenous)	53000 lu/mg	Wockhardt Life Sciences
Lepidocide, B.t.t.	*Thuriengiensis*	1500 lu/mg	Bio Tech Internation
Biotoxibacillin	*Thuriengiensis*	3000 lu/mg	Bio Tech Internation
Dendrobacillin	*Dendrolimus*	3000 lu/mg	Bio Tech Internation

Subspecies	Product	Manufacturer
Kurstaki	Biobit, Dipel, Foray	Abbott Laboratories
	Bactee, Bernan	Bactec Corp.
	BMP 123	Becker Microbial Products,
	Condor, Crymax, Cutlass, Raven	Inc
	Forawabit	Ecogen Inc.
	Costar, Javelin, Thuricide, Vault	Forward International Ltd.
	Bactosid K	Thermo Trilogy
	Agrobac	Sanex Inc
	Troy BT	Tecomag
	MVP, MVP II, M-Peril	Troy Biosciences Inc.
	Biodart	Mycogen Corp.
	Caterpiller Attact	ICI (Zenece)
	Bactospeine	Ringer Coporation
	Toarow ct	Novo Nordisk
		Taogosei Chem
Israelensis	Bactimos, Gnatrol, VectoBac, Skeetal,	Abbott Laboratories
	Aquabac, Aquabac Primary	Becker Microbial Products
	Powder, Vectocide	Inc Sanex Inc
Aizawai	Xen Tari	Abbott Laboratories
	Agree Design	Thermo Trilogy
	Mattch	Mycogen Corp.
Tenebrionis	Novodor	Abbott Laboratories
	M-trak	Mycogen Corp.
Morrisoni	Bactec Bernam	Bactec Corp.

Compatibility of *Bt* : Compatibility of *Bt* viz., *Kurstaki, Aizawai* etc. with endosulfan have been studied by various workers against *Helicoverpa* (Lone *et al.* 2010). Prabhuraj *et al.* 2004 studied the effects of combination of local isolates of *Heterorhabditis indica* and *B. thuriengiensis* against *H. armigera* under laboratory conditions and reported that combination of *H. indica* and *B. thuriengiensis* @ 150 ljs + 0.264 lu/mg, recorded 100 per cent mortality in 3rd instar larva at 36 hour after inoculation. Similar results were also observed in case of 4th instar larva but level of mortality was less than to 3rd instar larva. This clearly indicates that 3rd instar larvae were more susceptible than 4th instar larva. Further *H. indica* + *B. thuriengiensis* at lethal doses resulted in quicker mortality as compared to alone after 48 hours.

Development of resistance : The commercial use of *Bt* var. *kurstaki* was considered reliable due to its unique mode of action, chances of development of resistance against it are remote (Kreig and Langenbruch, 1981). But evolution of resistance threatens their continued efficacy. The Indian meal moth was the first insect to develop resistance to *Bt Kurstaki* in laboratory experiments due to high selection pressure. McGaughey and Whalon (1992) also reported resistance development in tobacco budworm, the colarado potato beetle and other insect species under laboratory conditions. About 50 fold reduction in the affinity of the Cry1Ab toxin to the brush border membrane of the midgut in case of *M. unipunctata* and 200-fold in case of *P. xylostella* have also been reported by Tabashnik *et al.* (1996). Number of reports based on field trials had indicated reduced efficacy of *Bt* to *P. xylostella.* Substantial level of resistance to *Bt* by diamond backmoth was confirmed by Tabashnik *et al.* (1990) at Hawaii. Later on, Shelton and Wyman (1992) reported resistance in field population of *P. xylostella* from Phillipines, Thailand, Malaysia, Japan and USA. Resistance under laboratory conditions has also been observed in tobacco budworm, the Colorado potato beetle and other insect species. The gypsy moth also shows potential for developing *Bt* resistance. Some insects, such as the diamondback moth and the tobacco budworm, exhibit resistance to multiple *Bt* strains. Development of resistance occurs faster when larger amounts of a pesticide are used (Carrie, 1994). Since insect can develop resistance to *Bt* toxin this bacterial insecticide should be used judiciously (Gelenter, 1997). Under field conditions only two insects have developed resistance against *Bt* spray i.e. DBM in many countries (Dhaliwal and Koul, 2010) and *Trichoplusia ni* in vegetable green houses in British Columbia, Canada (Jammaat, 2007). There are several reports related to less effectiveness and ineffectiveness of *Bt* formulations particularly against noctuids. Srinivasan *et al.* (1994) found Delfin was ineffective in controlling *H. armigera* on chickpea. Similarly, Dertan formulation of *Bt* in bioassay test was found almost non-toxic to sorghum stem borer (Brownbrindge and Onyango, 1992).

Insects showing resistance to *Bt*

Insect Resistance to *Bt*	*Bt*
Plodia interpunctella	*Btk*
Plutella xylostella	*Btk*

Heliothis virescens	*Btk*
Spodoptera litura	*Btk*
Spodoptera exigua	*Btk*
Spodoptera littoralis	*Btk*
Cadra cautella	*Btk*
Leptinotarsa decemlineata	*Btn*
Aedes aegypti	*Bti*
Culex quinquefasciatus	*Bti*
Musca domestica	*Bti*
Drosophila melanogaster	*Bti*

Narayanan, 2004

Advantages of Microbial Insecticides

- Microbial insecticides are essentially nontoxic and nonpathogenic to wildlife, humans, and other organisms not closely related to the target pest. Most of them target the specific insect and in turn protect beneficial insects.
- No residual hazards to human and other animals make scope for their application in crop almost ready for harvest.
- Most inoculants are easily culturable within limited space and are inexpensive to produce larger quantities of inoculums.
- Development of resistance is slow in case of microbial pathogens.
- Where chemical control is not feasible (insect in cavities) microbial control proved effective.

Disadvantages or factors affecting microbial agents:

- **Environmental conditions**: Can greatly influence the effectiveness of microbial agents. Bacteria and fungi generally lose their virulence below 18 °C and many viruses do not replicate rapidly unless temperature are between 21 and 29 °C. Not very effective at low temperatures. Many viruses and bacteria are quickly killed when exposed to sunlight (UV light).
- **Soil conditions:** Composition of the soil can be critical; fungi seem to survive best in soil high in organic matter. Soil ph or ph of the foliage is important. Acid conditions are unfavourable for *B. papillae* spores (milky disease of the Japanese beetle), while alkaline conditions can destroy the polyhedral structure of polydedrosis viruses.

- **Timing of application and thorough coverage:** Most effective against early instars. Necessity for careful and correct time of application. Forest tent caterpillars, for example, should be treated when the caterpillars are 3/8 to 1/2 inch long or the tree leaves are one half of their full size. *Bt* should be applied to dry foliage and have a droplet size of 75-150 microns (Anon. 2011a).
- **Persistency :** Protein endotoxin is not very persistent (amount of toxin ingested dependent on the amount of food consumed by the larvae)
- **Specificity:** Due to host specificity complex of pest species is not controlled. As a result other types of pests present in the treated area will survive and may continue to cause damage.
- **Development of resistance:** Development of insecticide resistance. For eg. *P. xylostella* (Linn.) and *P. interpunctella* (Hubner) have been reported to develop resistance to *Bt.*
- Difficulty in producing some obligate and facultative pathogens on a large scale.

Environmental Fate

Bt generally persists only a short time (half life of crystal protein: 9 days) in soil. However, small amounts can be quite persistent. *Bt* spore numbers declined by one order of magnitude after 2 weeks, but then remained constant for 8 months following application. The effectiveness of *Bt* deposited on the upper side of leaves is less (1-2 days) as compared to underside of leaves (7-10 days). It is possible for it to be significantly more persistent. However, viable spores of *Bt Kurstaki* were recovered from white spruce foliage (one year after application) and in Japan, *Bt* persisted for two years in a citrus orchard. *Bt Kurstaki* has been found to drift over 3,000 meters downwind during an aerial application. *Bt* was measured in air for up to 17 days following an application.

The future: Microbial biopesticides can play a key role in controlling pests' undoubtedly and make a significant contribution to the reduction of chemical inputs. But commercialization of microbial products is difficult because they have to compete with cheaper conventional pesticides already on the market.

Conclusion

Entomopathogens appear to be promising as an important component of IPM as these proved ecologically safe, compatible with other components and sustainable in changing scenario of Indian agriculture (Sachithanandam *et al.* 1988, Pankaj *et al.* 2003, Prabhuraj and Virabtamath, 2004 and Prabhuraj *et al.* 2004). But there is need to ascertain the compatibility of these chemicals under field conditions. It is ascertain that not adequate resources in terms of money, research and time is directed towards naturally occurring or safe alternate methods of pest control due to mind tuned towards shortcuts and profits. The use of bio inoculants or their by products is under exploited. The dangers of chemical poisons are very well known right from water bodies to food chains. Due to the accumulation of toxic metabolites in the air as well as soil system, it will result in creating immunity to dangerous pests. Use of microbial agents can provide greater protection for both flora and fauna.

References

Aggarwal, Naveen and Mahal, M. S. 2010. Evaluation of two entomopathogenic nematodes,*Steinernema riobravae* and *Steinernema* feltiae against rice stem borer, *Scirpophaga* incertulas (Walker) and rice leaf folder, *Cnaphalocrosis* medinalis (Guenee). In : *3rd Conference Biopesticides : Emerging Trends (BET 2010) Biopesticides in Food and Environment Security Abstracts.* Organised by Society of Biopesticide Sciences India Jalandhar and Deptt. Of Entomology, CCS Haryana Agricultural University, Hisar, 175-176.

Akpabey, Felix Jerry and Ocran, Michael. 2010. An assessment of the use of formulations of the microbial agents *Bacillus thuringiensis* and *Bacillus sphaericus* in the control of larvae of mosquitoes that cause malaria and filarial diseases. *J. Ent. & Nemato,* 2(2): 18-24.

Amonkar, S. V., Pal, A. K., Vijaylakshmi, L. and Rao, A. S. 1979. Microbial control of potato tuber moth (*Phthorimaea operculella* Zell.). *Indian J. Exptl. Biol,* 17: 1127-1133.

Angus T. A. 1956. Association of toxicity with proteincrystalline inclusions of Bacillus sotto Ishiwata. *Can. J. Microbiol,* 2: 122-131.

Anonymous, 2002. *Package of Practices of Horticultural Crops,* CCS Haryana Agricultural University, Hisar.

Anonymous, 2011a. Bt MicrobialInsecticides.

http://www.dnr.state.mn.us/treecare/forest_health/ spraying/bt_insecticide.html

Anonymous, 2011b. http://www.eolss.net/ebooks/Sample%20Chapters/C17/E6-58-05-08.pdf.

Anonymous, 2017. Global Bio Pesticides Market, 2017-2022: Market is estimated to reach $6.55 billion-Research and Markets. News provided by research and market, DUBLIN, May,17 2017/PRNewswire

Arora, R., Battu, G. S. and Dhaliwal, G. S. 2003. Microbial bipesticides for management of crops pests in Punjab: Progress and potential. In : O. Koul, G. S. Dhaliwal, S. S. Marwaha and J. K. Arora (eds.) *Biopesticides and Pest Management.* Campus Books Internatinal, New Delhi, 1: 40-55.

Arora, R., Battu, G. S. and Bath, D. S. 2000. Management of insect pests of cauliflower with biopesticides. *Indian J. Ecol,* 27(2): 156-162.

Asokan, R. and Mohan, K.S. 1999. Use of *Bacillus thuringiensis* Berliner for the management of fruit and vegetable crop pests- an appraisal. *Agril. Rev.,* 41: 238-248.

Asokan, R., Mohan, K. S. and Gopalakrishnan, C. 1996. Effect of commercial formulations of *Bacillus thuringiensis* Berliner on yield of cabbage. *Insect Environ*, 2: 58-59.

Balch, R. E. 1946. A disease of the European spruce sawfly. Canad. Dept. Agr., Forest Insect Invest. *Bi-mon. Prog. Rept*, 2(5): 1.

Battu, G. S. and Arora, R. 1997. Insect pest management through microorganisms. In : K.R. Dadarwal (ed.) *Biotechnological Approaches in Soil Microorganisms for Sustainable Crop Production.* Scientific Publishers, Jodhpur, 223-246.

Battu, G. S., Ramakrishnan, N. and Dhaliwal, G. S. 1993. Microbial pesticides in developing countries : Current status and future trends. In: G.S. Dhaliwal and B. Singh (eds). *Pesticides : Their Ecological Impact in Developing Countries.* Commonwealth Publishers, New Delhi, India, 270-334.

Benoit, T. G., Wilson, G. R., Pryor, N. and Bull, D. L. 1990. Isolation and pathogenicity of *Serratia marcescens* from adult house flies infected with *Entomophthora muscae*. *J. Invert. Pathol*, 55: 142-144.

Berliner, E. 1915. Uber die Schlaffsucht der Mehlmottenraupe (Ephestia kühniella, Zell.) und ihren Erreger Bacillus thuringiensis, n. sp. *Zeitschr. Angew. Ent*, 2: 29-56.

Bird, F. T. 1953. The use of a virus disease in the biological control of the European pine sawfly, Neodiprion sertifer (Geoffr.). Canad. Ent, 85: 437-46.

Bogawat, J. K., Rangarajan, M., and Chakravarthi, B. P. 1966. Bacterial mortality of the grubs of mustard sawfly, *Athalia proxima* Klug (Hymenoptera : Tenthredinidae). *Indian J. Ent*, 28: 417-421.

Brownbrindge, M. and Onyango, T. 1992. Laboratory evaluation of four commercial preparations of *Bacillus thuringiensis* (Berliner) against the spotted stem borer, *Chilo partellus* (Swinhoe) (Lep., Pyralidae). *J. Appl. Ent*, 113: 159-167.

Bucher G. E. 1961. Artificial culture of *Clostridium brevifaciens* N. Sp. and *C. malacosomae* N. Sp., the causes of brachytosis in tent caterpillars. *Can. J. Microbiol*, 7: 641-655.

Burges, H. D. and Hussey, N. W. 1971. In Burges H.D. and Hussey N.W. (eds.) *Microbial Contrrol of Insects and Mites*, Academic Press, New York.

Carrie, Swadener 1994. *Bacillus thuringiensis* (*B.t.*). Northwest Coalition for Alternatives to Pesticides, Eugene, *J. Pesticide Reform*, 14(3).

Chahal, V.P.S, Singh Gurdip and Chahal P.P.K. 2001. In : *Biopesticides Emerging Trends BET 2001*, Chandigarh, India, 136.

Chandla, V. K. and Verma, K. D. 2005. Pests of Potato. In L.R. Verma, A.K. Verma, D.C. Gautam (eds.), *Pest Management in Horticultural Crops, Principles and Practices.* 387-409.

Chandla, V. K., Trivedi, T. P., Misra, S. S., Sharma, D. C. and Kashyap, N. P. 1993. Potato tubermoth menace in Himachal Pradesh and its management (Abstract). *J. Indian Potato Assoc*, 20 (1): 60-61.

Chang, Y. D., Woo, K. S. and Lee, H. R. 1993. Taxonomy, ecology and control of honey bee mites in Korea. *Korean J. Apic*, 8: 35–47.

Datta, P. K. and Sharma, D. 1997. Effect of *Bacillus thuringiensis* Berliner in combination with chemical insecticides against *Spodoptera litura* Fabricius on cabbage. *Crop Sci.*, 14: 481-486.

De Barjac, H. and Bonnefoi, A. 1962. Essai de classification biochimique et serologique de 24 souches de *Bacillus* de type *thuriengiensis. Biocontrol*, 7: 5-31.

Dhaliwal, G. S. and Koul, O. 2007. Microbial Control. In : G. S. Dhaliwal and Opender Koul (eds.) *Biopesticides and pest management Conventional and Biotechnological* Approaches, 72-132

Dhaliwal, G. S. and Koul, O. 2010. *Quest for Pest Management : From Green Revolution to Gene Revolution.* Kalyani Publisher, New Delhi.

Dhawan, A. K. 1999. Major insect pests of cotton and their integrated management. In : R.K. Upadhyay, K.G. Mukherji and O.P. Dubey (eds.). *IPM System in Agriculture* Vol. 6 : *Cash Crops.* Adiya Books Pvt. Ltd., New Delhi, 165-225.

Dhawan, A. K., Kaur, Amandeep, Singh, Rajinder, Mondal, Shaonpius and Kaur, Gurpreet. 2010. Relative toxicity of novel insecticides against cabbage butterfly, *Pieris brassicae* Linnaeus. *J. Insect Sci*, 23(3) : 318-321.

Dudney, R. A. (1970). Use of Xenorhabdus nematophilus Im/1 and 1960/1 for fire ant control. US patent, No.5616318.

Dulmage, H. T. 1970. Insecticidal activity of HD-1, a new isolate of *Bacillus thuringiensis* var. *alesti. J. Invertebr. Pathol*, 15: 232-239.

Dunn, P. E. and Drake, D. R. 1983. Fate of bacteria injected into maive and immunized larvae of tobacco horn worm, *Manduca sexta. J. Invert. Pathol*, 41: 77-85.

EL Sanousi, S. M., EL Sarag, M. S. A. and Mohamed, S. E. 1987. Properties of *Serratia marcescens* isolated from diseased honey bees (*Apis mellifera*) larvae. *J. Gen. Microbiol*, 133: 215-219.

Elawad, S. A., Gowen, S. R. & Hague, N. G. M. 1999. Efficacy of bacteria symbionts from entomopathogenic nematodes against the beet army worm (Spodoptera exigua). Tests of Agrochemical and Cultivars No. 20. (Supplement). *Ann. Appl. Biol*, 134: 66-67.

Ensign, J. C., Bowen, D. J., Tenor, J. L., Ciche, T. A., Petell, J. K., Strickland, J. A., Orr, G. L., Fatig, R. O., Bintrim, S. B. & Ffrench-Constant, R. (2002). Proteins from the genus Xenorhabdus are toxic to insects upon oral exposure. US Patent Application no. 0147148 A1.

Erturk, O. 2007. Insecticidal effects of selected biological control agents on the larvae of *Plutella xylostella* (Lepidoptera : Plutellidae). *Entomol. Res*, 37: 122-124.

Erturk, O. and Demirba, Zihni 2006. Studies on bacterial flora and biological control agent of *Cydia pomonella* L. (Lepidoptera: Tortricidae). *African J. Biotech,* 5(22): 2081-2085.

Feitelson, J. S., Payne, J. and Kim, L. 1992. *Bacillus thuringiensis;* Insect and beyond. *Biotechnology,* 10: 271-275.

Flexner, J. L. and Belnavis, D. L. 2000. Microbial insecticides. In : J.E. Rechcigl and N.A. Rechicigl (eds.) *Biological and Biotechnological Control of Insect Pests.* Lewis Publishers, Boca Raton, Florida, 35-61.

Gelernter, W. D. 1997. Resistance to microbial insecticides : the scale of the problem and how to manage it. In H.F. Evans (ed.) *Microbial Insecticides Novelty or Necessity? BCPC Sym. Proc. No.* 68: 201-214.

Grossi, M. F., Magalhaes, M. Q., Dias, S. C., Nakasu, E. Y., Brunetta, P. S. F. and Oliveira, G. R. 2007. Susceptibility of *Anthonomus* grandis (cotton boll weevil) and *Spodoptera* frugiperda (fall armyworm) to a CryIIa-type toxin from a Brazilian *Bacillus* thuringiensis strain. *J. Biochem. Mol. Biol.,* 40: 773-782.

Hannay, C. L. and Fitz-James, P. C. 1955. The protein crystals of *Bacillus thuringiensis* Berliner. *Can. J. Microbiol,* 1: 694-710.

Heimpel, A. M. and Augus, T. A. 1959. The site of action of crystalliferous bacteria in Lepidoptera larvae. *J. Insect Pathol,* 1: 152-170.

Henry, J. E. and Oma, E. A. 1981. Pest control by *Nosema locustae,* a pathogen of grasshoppers and crickets. In Burges, H.D. (ed.). *Microbial Control of Pests and Plant Diseases 1970-1980.* London, UK; Academic Press, 573-586.

Jackson, T. A.; Pearson, J.F.; O'Callaghan, M.; Mahanty, H.K.; Willocks, M. 1992: Pathogen to product - development of *Serratia entomophila* (Enterobacteriaceae) as a commercial biological control agent for the New Zealand grass grub (*Costelytra zealandica*). In: T.A. Jackson; T.R. Glare (eds.) *Use of Pathogens in Scarab Pest Management.* Intercept, Andover, UK, 191-198.

Jammaat, A. F. 2007. Development of resistance to the biopesticide *Bacillus thuringiensis kurstaki.* In: C. Vincent, M.S. Goettel and G. Lazarovits (eds.), *Biological Control : A Global Perspective.* CAB International, Wallingford, UK, 179-184.

Jan, N. D., Mahar, G. M., Mahar, A. N., Hullio, M. H., Lanjar, A. G. & Gowen, S. R. 2008. Susceptibility of different insect pupae to the bacterial symbiont, Xenorhabdus nematophila, isolated from the entomopathogenic nematode, Steinernema carpocapsae. *Pak. J. Nematol,* 26: 59-67.

Jasmine, R. Sheeba. 2010. Field evaluation of abamectin 1.9 EC against cotton bollworms. *Res. J. Agril. Sci,* 1(4): 396-403.

Jethva, D. M. and Kapadia, M. N. 2006. Evaluation of *Bacillus thuriengiensis* var. *kurstaki* based bait against fourth instar larvae of *Spodoptera litura* (Fabricius). *Indian J. Entomol,* 68: 384-386.

Kol-chevskii A. G., Rybina L. M. and Kolomiets V. Y. 1991. Influence of residual quantities of bacterial preparations on the cabbage butterfly. *Sel' skokhozyaistvennaya* Biologiya. % : 124-129.

Kreig, A. and Langenbruch, G. A. 1981. Susceptibility of arthropod species to *Bacillus thuringiensis*. In H.D. Burges (ed.) Microbial control of pests and plant diseases 1970-1980. Academic Press, 837-840.

Krishanaiah, K., Jaganmohan, N. and Prasad, V.G. 1981. 93. Efficacy of *Bacillus thuringiensis* Berliner for the control of lepidopterous pest of vegetable crops. *Entomon*, 6 : 87-93.

Kulshreshtha, J. P., Sanghi, P. K. and Ravindranath, V. 1965. Microbial control of castor semi-looper *Achaca janata.. Indian J. Ent*, 27: 353-354.

Kuruvilla, Suma and Jacob, Abraham. 1980. Pathogenicity of the entomogenous fungus *Paecilomyces furinosus* (Dickson Ex Fries) to several insect pests. *Entomon*, 5(3): 175-176.

Lacey, L. A., Frutos, R., Kaya, H. K. and Vail, P. 2001. *Biol. Contr*, 21: 230-248.

Lal, L. 1987. Studies on natural repellents against potato tuber moth (*Phthorimna operculella* Zeller) in country stores. *Potato Res*, 30: 329-334.

Lambert, B. and Peferoen, M. 1992. Insecticidal promise of *Bacillus thuriengiensis. Bioscience*, 42: 112-121.

Lauzon, R. C., Bussert, G., Teresa, Sjogren, E. Robert and Prokopy, J. Ronald. 2003. *Serratia marcescens* as a bacterial pathogen of *Rhagolestis pomonella* flies (Diptera : Tephritidae). *Eur. J. Entomol*, 100: 87-92.

Lone, G. M.; Siddiqui, A. U. and Parihar, Aruna. 2010. Enthomopathogens and their compatibility with other components for insect pest management in IPM. In : Bio Pest Management Eds. H.C.L. Gupta, A.U. Siddiqui and Aruna Parihar, 420-455.

Maddox, J.V. 1994. Insect pathogens as biological control agents. In : R.L. Metcalf and W.H. Luckmann (eds.) *Introduction to Insect Pest Management*. John Wiley & Sons, New York, USA, 199-244.

Mahapatro, G. K. and Gupta, G. P. 1999. Evenings suitable for spraying Bt formulations. *Insect Environ*, 5: 126-127.

Majumder, A. B., Pathak, S. K., Talukdar, T. and Banerjee, D. K. 2010. Bioefficacy of *Bacillus thuringiensis* var. *kurstaki* against two species of *Hyposidra* infesting tea. In : *3rd Conference Biopesticides : Emerging Trends (BET 2010) Biopesticides in Food and Environment Security Abstracts*. Organised by Society of Biopesticide Sciences India Jalandhar and Deptt. Of Entomology, CCS Haryana Agricultural University, Hisar, 139-1405.

McGaughey, W. H. and Whalon, M. E. 1992. Managing insect resistance to *Bacillus thuriengiensis* toxins. *Science*, 258: 1451-1455.

Narayanan, K. 2004. Insect defence: its impact on microbial control of insect pests. *Current Sci*, 86(6): 800-814.

Narayanan, K. and Gopalakrishnan, C. 1988. Effect of DD-136 (*Steinernema feltiae*) for the control of mustard sawfly, *Athalia proxima* on radish and its persistence in soil. Abstr. 10th International Soil Zoology Colloquium, 7-13 August, Bangalore, India, 15.

Narva, K. E., Payne, J. M., Schwab, G. E., Hickle, L. A., Galasan, T. and Sick, A. J. 1991. Novel *Bacillus thuringiensis* microbes active against nematodes, and genes encoding novel nematode-active toxins cloned from *Bacillus thuringiensis* isolates: *European Patent Application* EP0462 721A2, Munich, Germany.

Neale, M. C. 1997. Bio-pesticides : Harmonization of registration requirement with EU directive 91-414. An industry view. *Bulletin OEPP*, 27: 89-93.

O'Callaghan, M., Garnham, M. T., Nelson, T. L., Baird, D. and Jackson, T. A., 1996. The pathogenicity of *Serratia* strains to *Lucilia sericata* (Diptera: Calliphoridae). *J. Invert. Path, 68*: 22-27.

Odindo, M. O. 1992. Screenhouse trials on the potential of *Nosema marucae* for the control of *Chilo partellus. Entomol. Exp. Appl*, 65: 283-289.

Ourth, D. D. and Smalley, D. L. 1980. Phagocytic and humoral immunity of the adult cotton boll weevil, *Anthonomus grandis* (Coleoptera : Curculionidae) to *Serratia marcescens. J. Invert. Pathol*, 36: 104-112.

Pankaj, K.; Mishra and Tandon, S. M. 2003. Compatibility of entomopathogenic *Bacillus sphaericus* strain R3 with chemical insecticides. *Indian J. Microbiology* 43(4): 265-266.

Payne, C. C. 1982. Insect viruses as control agents. *Parasitology*, 84: 35-77.

Phadke, C. H. and Rao, V. G. 1978. Studies On The Entomogenous Fungus Nomuraea Rileyi (Farlow) Samson. *Current Sci*, 47: 511-512.

Prabhuraj, A. and Viraktamath, C. A. 2004. Compatibility of *Steinerma glaseri* with two inorganic and one neem based insecticides. *Indian J. Ent*, 66(4): 377-379.

Prabhuraj, A., Shivaleela, Girish, K. S. and Patil, B. V. 2004. Effect of combination of *Heterorhabditis indica* (Rhabditida : Heterorhabditidae) and BT against *Helicoverpa armigera* (Hubner). *Indian J. Ent*, 66(4): 369-372.

Putambekar, U. S., Mukherjee, S. N. and Ranjekar, P. K. 1997. Laboratory screening of different *Bacillus thuringiensis* strains against certain lepidopteran pests and subsequent field evaluation on the pod boring pest complex of pigeonpea (*Cajanus cajan*). *Antonie Van Lenwenhoek*, 71: 319-323.

Quarles, W. 1997. Mating disruption for the peach twig borer. *IPM-Practitioner*, 19(5-6): 1-8.

Rajagopal, D., Malikarjunappa, S. and Gowda, J. 1998. Occurrence of natural enemies of the groundnut leaf miner, *Aproaerema modicella* Deventer (Lepidoptera : Gelechiidae). *J. Biol. Contr*, 2(1): 129-130.

Revensberg, W.J., Malais, M., and Van-der-Schaaf, D.A. 1990. *Verticillum lacanii* as a microbial insecticide against glasshouse whitefly. *Proceedings of the Brighton Crop Protection Conference, Pests and Diseases,* Brighton, England, 265-268.

Roltsch, W. J., Zalom, F. G., Barry, J. W., Kirfman, G. W. and Edstrom, J. P. 1995. Ultra low volume aerial applications of *Bacillus thuringeinsis* variety *Kurstaki* for control of peach twig borer in almond trees. *Applied Engineering in Agriculture,* 11(1): 25-30.

Sachithanandam, S., Ranbindra, R. J. and Jayaraj, S. 1988. Compatibility of NPV of *Spodoptera* litura (Fabricius) with certain fungicides. *J. Biol. Control,* 2(2): 137-138.

Sathpathy, C. R. and Panda, S. K. 1997. Effect of commercial *B.t.* formulation against fruit borers of Okra. *Insect Environ,* 3: 54.

Saxena, R. C. 1979. Efficiency of *Bacillus thuriengiensis* Berliner for the control of *Athalia proxima* Klug. *Indian J. Ecology,* 6: 33-35.

Shahina, F., Kazmi, A. R., Tabassum, K. A., Salma, J. and Mahreen, G. 2011. Pathogenicity of *Xenorhabdus nematophila* (Strain HAM-10) against wax moth larvae and vine mealy bug in Pakistan. *Pak. J. Nematol,* 29(1): 15-23.

Sharma, D. R. 2004. Advances in integrated pest management of citrus. In B.S. Chhillar, V.K. Kalra, S.S. Sharma and Ram Singh (eds.) *Advances in the Integrated Pest Management of Horticultural, Spices and plantation Crops,* 26-34.

Shelton, A. M. and Wyman, J. A. 1992. In: *Proc. Second International Workshop* (Talekar,N.S. ed.), *10-14 December 1990.*Tainan, Taiwan, AVRSC, Taipei, 447-454.

Sridhar, V. and V.D. Prasad 1996. Life table studies on natural population of *Spodoptera litura* on groundnut. *Ann. Pl. Protec. Sci,* 4: 142-147.

Singh, A. P., Arora, R. and Battu, G.S. 2000. Laboratory evaluation of three *Bacillus thuringiensis* Berliner based biopesticides against the diamondback moth, *Plutella xylostella (Linnaeus). Pestic. Res. J,* 12(1): 54-62.

Singh, S. P. 1994. Biological suppression of fruit pests. In : *Fifteen Years of AICRP on Biological Control.* Project Directorate of Biological Control, Bangalore, 139-171.

Sood, P., Prabhakar, C. S. and Mehta, P. K. 2010. Ecofriendly management of fruit flies through their gut bacteria. *J. Insect Sci,* 23(3): 275-283.

Sreedhar, U., Gunneswararao, S. and Krishnamurthy, V. 2010. Biopesticides in tobacco pest management : Prospects and problems. In : *3rd Conference Biopesticides : Emerging Trends (BET 2010) Biopesticides in Food and Environment Security Abstracts.* Organised by Society of Biopesticide Sciences India Jalandhar and Deptt. Of Entomology, CCS Haryana Agricultural University, Hisar, 74.

Srinivasan, G., Sundarababu, P. C., Sathiah, N. and Balasubramanian, G. 1994. Field efficacy of HaNPV alone and in combination with Bt for control of *Helicoverpa armigera* on chickpea *Pest Mgmt. Econ. Zool,* 2: 45-48.

Srivastava, R. P. 2005. Pests of mango. In L. R. Verma, A. K. Verma, D. C. Gautam (eds.), *Pest Management in Horticultural Crops, Principles and Practices*, 320.

Steinhaus, E.A. 1951. Possible use of *Bacillus thuriengiensis* Berliner as an aid in the biological control of the alfalfa caterpillar. *Hilgardia*, 20: 359-381.

Tabashnik, B. E., Cushing, N. L., Finson, N. and Johnson, M.W. 1990. Field development of resistance to *Baedlus lhuringiensis* in diamondback moth (Lepidoptcl'a: Plutellidae). *J. Econ. Ent*, 83 : 1671-1676.

Tabashnik, B. E., Groetus, F. R., Finson, N., Liu, Y. B., Johnson, M. W., Heckel, D. G., Luo, K. and Adang, M. J. 1996. Resistance to *Bacillus thuriengiensis* in *Plutella xylostella*. In : T. M. Brown (ed.) *Molecular Genetics and Evolution of Pesticide resistance*. American Chemical Society, Washington DC, 130-140.

Tanada, Y. and Kaya, H. K. 1993. *Insect Pathology*. Academic Press Inc., Harcourt Brace Jovanovich Publishers, San Diego.

Tanada, Y. 1959. Microbial control of insect pests. *Annu. Rev. Ent*, 4: 277-302.

Thakur, J. N., Pawar, A. D. and Rawat, U. S. 1991. Apple woolly aphid, Erisoma lanigerum Hausman(Hemiptera : Aphidae) and post release impact of its natural enemies in Kullu valley(H.P.). *Plant Prot. Bull. (Faridabad)*, 44(3): 18-20.

Vasiliki, T., Alexandra, L., Dimitrios, L., Nikolaos, E. and George 2004. Newly isolated bacterial strains belonging to Bacillaceae (*Bacillus* sp.) and Micrococcaceae accelerate death of the honey bee mite, *Varroa destructor* (*V. jacobsoni*), in laboratory assays. *Biotechnology Letters*, 26: 529–532.

Vyas, H. G., Joshi, D. P., Patel, K. C., Yadav, D. N. and Dodia, J. F. 1986. Natural incidence of milky disease of white grubs in Gujarat. *Indian J. Agril. Sci.*, 56: 213-214.

Vyas, R. V., Parasharya, B. M. and Yadav, D. N. 1988. Dispersal of milky disease organism *Bacillus popilliae* var. *holotrichiae* of white grub (*Holotrichia consanguinea*) through birds. *Indian J. Agril. Sci.*, 58: 243-244.

Vyas, R. V., Yadav, D. N. and Patel, R. J. 1991. Efficiency of *Bacillus popillae* var. *holotrichiae* against white grub (*Holotrichia consanguinea*). *Indian J. Agric. Sci.*, 61(1): 80-81.

Wahab Seema. 2010. Biotechnological approaches in the management of plant pests, diseases and weeds for sustainable agriculture. In Opender Koul, G.S. Dhaliwal, Sucheta Khokhar and Ram Singh (eds.), *Souvenir 3rd Conference Biopesticides: Emerging Trends (Bet 2010) Biopesticides in Food and Environment Security*, 15-40.

18

Pros and Cons of GM Crops

Ram C. Yadav[1*], Sumit Jangra[2] and Neelam R. Yadav[2]

[1]Centre for Plant Biotechnology, CCS Haryana Agricultural University, Hisar, India

[2]Department of Molecular Biology, Biotechnology & Bioinformatics, CCS Haryana Agricultural University, Hisar, India

*Corresponding author: rcyadavbiotech@yahoo.com

Abstract

In the present world genetically modified crops have become an integral part of our agricultural system. The area under GM crops is continuously increasing, since their introduction. Developing countries are taking a keen interest in the GM crops and for the last consecutive five years developing countries are planting more GM crops than the developed countries. The introduction of GM crops has resulted in reduced use of nitrogen fertilizers and chemical pesticides, which are a great threat to the environment. The introduction of abiotic stress tolerant varieties like maize will accelerate the crop production. Recently, the interest of how the GM crops are affecting the environment is increasing. Several questions have been raised that the introduction of GM crops will lead to increased weediness, formation of superweeds, reduce biodiversity and will target non-target organisms. Despite of all the potential of GM crops to increase the crop production cannot be ignored and is the only way at present to attain food security.

Keywords: GM Crops, Gene Flow, Transgenes, Scenario of GM Crops

Introduction

Crop productivity is decreasing day by day and several biotic and abiotic factors are contributing to it. With the increasing population, which is expected to reach 9.2 billion by 2050 there is great demand to increase the crop productivity. Since ancient times people are selecting desired plants with higher biomass and yield which gave rise to the conventional plant breeding. But conventional is time consuming and laborious and there are several agronomic traits which are multigenic in nature and hence conventional breeding is not a handy tool to transfer those traits

(Jangra *et al.*, 2017). Since world population is increasing at alarming rates so, there is a need for fast and efficient methods to transfer trait of interest to the desired cultivar and genetic transformation can solve this problem. Genetic transformation or genetic engineering is a method through which it is possible to transfer foreign gene into desired crop plant. So, genetically engineered crop plants with higher yield and stress tolerance will ensure that there is enough food for the world. With the development of plant genetic transformation methods genetically modified crops have become commercially available and widely adopted.

The era of genetically modified crops began with the development of antibiotic resistant tobacco in 1982 (Fraley *et al.* 1983). The first field trails of genetically modified, herbicide resistance tobacco were conducted in France and USA in 1986. The involvement of private sector in GM crops began with the development of insect resistant tobacco using insecticidal gene from *Bacillus thuringiensis* (Bt) in 1987 by Plant Genetic Systems (Ghent Belgium), a company founded by Marc Van Montagu and Jeff Schell. The commercialization of genetically modified crops started, when FDA approved the delayed ripening tomato Flavr Svar for marketing in USA (Bruening and Lyon, 2000). Herbicide resistant tobacco was the first genetically modified crop to come to the market after gaining approval by European Union (MacKenzie, 1994). In 1995, US Environment Protection Agency approved Bt potato which was the first pest resistant crop to gain approval in USA. Another crops which got approval in 1995 were canola with modified oil composition (Calgene), glyphosate resistant soybean (Monsanto), virus resistant squash (Asgrow) and increased self-life tomato (DNAP, Zeneca/Peto and Monsanto). The field trials of genetically modified rice (Golden Rice) developed in 2000 are being carried out by IRRI in Bangladesh. Four varieties of Bt-brinjal (Bt-Uttra, Bt-Kajla, Bt-Nayantara and Bt-Iswardi) developed by Bangladesh Agriculture Research Institute (BARI) were approved for commercial cultivation in Bangladesh on October, 2013. Recently on January, 2015 USDA approved the first genetically modified apple (Arctic apple) developed by Okanagan for marketing in US. Since 1996, GM crops has become an important part of the agriculture to meet the global food demands. Farmers from all over the world has widely adopted GM crops.

Global Scenario of GM Crops

The global acreages under GM crops has been continuously increasing since 1996. According to International Service for Acquisition of Agri-Biotech Applications (ISAAA), 18 million farmers from 26 countries planted biotech crops on 185.1 million hectares in 2016, approximately 110-fold increase from 1996 and an increase of 3% from 2015 (equivalent to 5.4 million hectares). Of the total 185.1 million hectares 86.5, 75.4 and 23.1 were occupied by herbicide tolerant, stacked traits and insect resistant crops respectively. Principal biotech crops are soybean, cotton, maize and canola which were cultivated on 91.4, 60.6, 22.3 and 8.6 million hectares respectively. Biotech soybean has reached 50% of global biotech crop hectarage. Under GM crops, US continued to be the largest country with 72.9

million hectares under biotech crops, followed by Brazil with 49.1 million hectares under biotech crops. Out of the 26 countries planting biotech crops (table 1) in 2016, 19 were developing countries and out of the top 10 biotech countries 8 were developing (Brazil, Argentina, India, Paraguay, Pakistan, China, South Africa and Uruguay) and 2 were developed nations (USA and Canada). For the fifth consecutive year since 2012, developing countries planted more biotech crops than industrial countries in 2016.

Table 1. List of countries who planted biotech crops in 2016

Rank	Country	Area (million hectares)	Biotech Crops
	USA	72.9	Maize, Soybean, Cotton, Canola, Sugar Beet, Alfalfa, Papaya, Squash, Potato
	Brazil	49.1	Soybean, Maize, Cotton
	Argentina	23.8	Soybean, Maize, Cotton
	Canada	11.6	Canola, Maize, Soybean, Sugar beet, Alfalfa
	India	10.8	Cotton
	Paraguay	3.6	Soybean, Maize, Cotton
	Pakistan	2.9	Cotton
	China	2.8	Cotton, Papaya, Poplar
	South Africa	2.7	Maize, Soybean, Cotton
	Uruguay	1.3	Soybean, Maize
	Bolivia	1.2	Soybean
	Australia	0.9	Cotton, Canola
	Philippines	0.8	Cotton, canola
	Myanmar	0.3	Maize
	Spain	0.1	Maize
	Sudan	0.1	Cotton
	Mexico	0.1	Cotton, Soybean
	Colombia	0.1	Cotton, Maize
	Vietnam	<0.05	Maize
	Honduras	<0.05	Maize
	Chile	<0.05	Maize
	Portugal	<0.05	Maize
	Bangladesh	<0.05	Brinjal/eggplant
	Costa Rica	<0.05	Cotton, Soybean, Pineapple
	Slovakia	<0.05	Maize
	Czech Republic	<0.05	
	Burkina Faso	--	
	Romania	--	
	Total	185.1	

Source: ISAAA, 2016

GM Crops in India

Studies conducted by ISAAA indicates that India has enhanced farm income from insect resistant (IR) cotton by US$ 19.6 billion in thirteen years period 2002 to 2015 and US$ 1.3 billion in 2015 alone. A slight decrease of 7% in biotech cotton planting has been recorded in 2016. Nearly 7.2 million farmers in India were benefited from this technology. Another Bt crop, that is Bt-brinjal was approved by GEAC for commercial cultivation on October 2009, but due to the ethical issues raised by anti-GM activists, government banned the commercialization of Bt-brinjal on February 2010. Biotech mustard expressing the barnase-barstar gene has gained approval by GEAC in 2017 but again due to the activities of anti-GMO's, it has not been commercialized yet. Status of biotech crops trails conducted in 2016 are streamlined in table 2.

Table 2: Summary of biotech research trials conducted in the year 2016

Crop	Gene/Event	Developer	Status
Chickpea	Cry7Ac, cry1Aabc/IPCa2 & MP9	ICAR-Indian Institute of Pulses Research, Kanpur	BRL-I
Cotton	GHB 614 (Glytol)	Bayer biosciences Pvt. Ltd., Hyderabad	BRL-II
Cotton	WideStrike	Dow Agro Sciences Pvt. Ltd.	Hybrid Seed Production
Maize	NK603	Monsanto	BRL-II
Maize	Cry7F, cry1Ab and cp4EPSPS genes/ TC1507 x MON 810 x NK 603 (Das-01570-1 x MON-00810-6 x MoN-00003-6)	Pioneer Hi-Breed Pvt. Ltd., Hyderabad	BRL-I
Maize	TC1507 x MON 810	Pioneer Hi-Breed Pvt. Ltd., Hyderabad	BRL-I
Mustard	Bar, barnase & barstar/ events bn 3.6 & modbs 2.99	Delhi University	Environmental release
Pigeonpea	cry1Ac, cry1Aabc/IOCc2 & SS5	ICAR-Indian Institute of Pulses Research, Kanpur	Event selection
Rice	Drought & salinity tolerance and nutrition stress	Bioseed Research India, Pvt. Ltd., Hyderabad	Event selection
Rice	Cry2Aa2a	RasiSeeds Research Farm, Telangana	Event selection
Sugarcane	DREB	Sugarcane Research Institute, U.P. and Council of Sugarcane Research (UPCSR,) Shahjhanpur	Event selection

Source: ISAAA, 2016

Benefits of GM Crops

Since 1996, biotech GM crops have contributed to food and feed security and has increased productivity and economic gains. The market value of biotech seeds in US alone was about US$ 15.8 billion in 2016. Biotech crops has increased the productivity by 574 million tons and the economic gains of small scale farmers has reached up to US$ 167.8 billion (ISAAA, 2016). The increasing population has led to decrease in arable land and shelter, which has led to exploitation of several natural habitats but GM crops has increased the production thereby will help in conserving the biodiversity by stopping the anthropogenic activities. Apart from this, GM crops have eliminated the necessity of pre-emergence spraying, thus zero tilling or minimum tilling can be integrated with these crops. This will lead to reduced soil erosion and conservation of soil microfauna and flora (Marshall, 1998). Reports from Brookes and Barfoot (2017) has revealed that use of biotech crops in the last two decades has reduced the environmental footprints from agriculture by 619 million kgs of active ingredients. The most common herbicide resistance genes used are for glyphosate and glufosinate which are less recalcitrant than many of the chemical herbicides thus reduces groundwater contamination (Duke, 1998). Weed control using GM technology enables control of some herbicide resistant biotypes, like black grass, *Alopecurus myosuroides* (Duke, 1998). The use of genetically engineered insect resistant crops has also reduced the use of chemical insecticides(Peferoen, 1997). The pest control is also more effective as the gene expression is consecutive. This reduced herbicide and insecticide usage has resulted in reduced emission of greenhouse gases. In 2015 alone, a reduction in 26.7 billion kgs of CO_2 was recorded (ISAAA, 2016). Biotech crops has helped in mitigating challenges associated with climate changes like drought tolerant maize was released in 2013.

Demerits of GM Crops

Inspite of several advantages of GM crops, commercialization of GM crops is facing problems. The disputes related to commercialization of GM crops in the scientific communities and the activities of anti-GMO's have put a question about their impact on the environment. Issues are raised for several aspects like their effect on non-target species, superweed formation, effects on wildlife, their extent in reduction of chemical input and public acceptance (Dale *et al.*, 2002).

Gene Flow

The transgene can be transferred from the GM crops to the sexually compatible species and can impact environment by production of hybrids (Dale *et al.*, 2002). Gene transfer is mainly of two types; vertical gene transfer and horizontal gene transfer. In vertical gene transfer the genetic material is transferred from parents to offsprings, either through sexual or asexual reproduction. Two plants are crossed in case of plant breeding for vertical gene transfer to the next generation. In case of GM crops the genetically modified crop can cross with the non-transgenic crop and the resistance gene get transferred to the non-transgenic crop. In horizontal gene transfer pollen is the mediator or gene transfer from cultivated species to wild relative and vice-versa. Similar transfer can take place through genetically modified crops. This could be a hidden hazard of genetically modified crops and can lead to gene flow from species to species (Jangra *et al.*, 2016). Farmers in Canada found oilseed rape plants resistant to three herbicides (Orson, 2002). Management practices studies are going on in Europe to control such type of events. Today there is need to develop high standard weed control measures to control multiple resistance development in volunteer crops and these may affect the environment by use of high concentration of chemicals herbicides by farmers. In 2000, David Quist analyzed the fields of some farmers who cultivated organic maize in Mexico but he was surprised to find segments of DNA used in insect resistant maize in the DNA of the organic maize (Quist and Chapela, 2001). Similar study at same place was conducted by Piñeyro-nelson *et al.*, 2009, found the contamination of the same transgene in the fields. Another study conducted by Dyer *et al.*, 2009 found the seed from 1,765 households across Mexico contaminated with transgene.

Toxicity to Living Organisms

The undesirable effect of resistance gene incorporated in GM crops on ecofriendly organisms is referred to as non-target effects. Pest control without unwanted side effects is a major concern related to all control measures weather chemical or biological. The study which revolutionized the controversies against the GM crops was the study which showed the effect of Bt toxin on monarch butterfly larvae. The study was published in 1999 and it showed that when the leaves of Bt maize were consumed by the monarch butterfly larvae it was a potential hazard to them (Losey *et al.*, 1999). Studies carried out by Dively *et al.* 2004 showed that the development time of monarch butterfly larvae is increased when exposed to Cry1Ab expressing corn (Dively *et al.*, 2004). The Starlink corn expressing Cry9c was banned for animal feed by US Environment Protection Agency due to potential allergenicity (Bucchini and Goldman, 2002).

Effect on Biodiversity

The herbicide resistant GM crops may lead to reduction in number of weed species thereby reducing the weed diversity in the GM and neighboring fields. Broad spectrum herbicides are used in GM fields and their extensive use may result in loss of various weeds from the agricultural fields. However, the precise impact of herbicide resistant crops depends upon the crop rotation policies and agronomic practices adopted for their management (Dale *et al.*, 2002).

Effects on Soil Ecosystem

There are many concerns that Bt crops may affect natural and agricultural system (Rissler and Mellon, 1996; Conway, 2000; Stotzky, 2000 and 2002). Bt toxin enters soil through the residues incorporated after the crop is harvested (Tapp and Stotzky, 1998; Stotzky, 2000 and 2002) and from root exudates from some Bt plants (Saxena *et al.*, 1999, 2002a, b, 2004; Saxena and Stotzky, 2000, 2001a, 2003). The toxin is absorbed by the soil and bind to the surface-active particles and persist there in larvicidal form (Stotzky, 2000 and 2002). Twenty-five fields in the Vardha region of India were selected to study effect of Bt-cotton. In these fields where Bt-cotton was grown for three consecutive years, 8-9% decline in carbon biomass and nitrogen biomass was recorded in soil ecosystem. There was no alteration in fungal and nitrifiers count, esterase and alkaline phosphatase activities in Bt-cotton fields (Tarafdar *et al.*, 2012).

Formation of Superweeds

Continuous cultivation of GM crops may lead to the creation of superweeds because the weeds may evolve and may develop resistance to herbicide, creating a trouble to the farmers. Roundup ready resistant amaranth was first reported by Jay Holder in the Bt-cotton fields of Georgia, USA in 2004 and has become a pain to farmers as Palmer amaranth (*Amaranthus palmeri*) competes for water, sunlight and nutrients with the cotton plant (Gilbert, 2013). Ian Heap, Director, International Survey of Herbicide Resistant Weeds says that weeds with resistance to glyphosate has been reported in 18 countries with significant impact in Brazil, Argentina and Paraguay. A study conducted by Wang *et al.*, 2013 shows that weedy rice can acquire transgene (herbicide resistance) and the hybrid weedy rice is more efficient than the non-transformed and can be a problem to environment.

Conclusions

From the last 15 years, genetically modified crops are being cultivated and results have depicted that the impacts of genetically modified crops on biodiversity are balanced. Cultivation of GM crops have resulted in decreased use of herbicides and chemical fertilizers and have raised the agricultural productivity. Through studies on the impact of GM crops on crop diversity has not been conducted yet. However, small scale studies have been conducted on the impact of genetically modified cotton in US and India, genetically modified soybean in US. These studies have shown that there was no significant effect of GM crops on the agricultural biodiversity. Similarly, studies show that there is no or little effect of GM crops on the microbial community residing in the soil. Despite of some disputes and the activities of anti-GM activist the enormous potential of GM crops on agricultural productivity makes them a suitable candidate for attaining food security. Introduction of genetically modified crops have reduced the dependence on nitrogen fertilizers and the introduction of drought tolerant and salt tolerant varieties has made the cultivation possible on un-arable land. The development of herbicide resistant varieties of wheat and rice has resulted in the increased productivity and thereby will help in solving the problem of high amount of food requirements as the population is increasing at an alarming rate.

References

Brookes, G. and Barfoot, P. 2017. GM crops: global socio-economic and environmental impacts 1996-2015.

Tapp, H. and Stotzky, G.1998. Persistence of the insecticidal toxins from *Bacillus thuringiensis* subsp. Kurstaki in soil. *Soil Biol. Biochem.* 30: 471–476.

Tarafdar, J. C., Rathore, I. and Shiva, V. 2012. Effect of Bt-transgenic cotton on soil biological health. *Appl Biol Res.* 14(1): 1-9.

Wang, W., Xia, H., Yang, X., Xu, T., Si, H. J., Cai, X. X., Wang, F., Su, J., Snow, A. A. and Lu, B. R. 2013. A novel 5-enolpyruvoylshikimate-3-phosphate (EPSP) synthase transgene for glyphosate resistance stimulates growth and fecundity in weedy rice (*Oryza sativa*) without herbicide. *New Phytol.* 202(2): 679–688.

19

Omics in Agriculture: Applications, Challenges and Future Perspectives

Swati Shalini[1], Ankit Singla[2], Meenu Goyal[3], Vikender Kaur[4] and Pardeep Kumar[5*]

[1]National Institute of Biologicals, Ministry of Health and Family Welfare, Government of India, Noida-201309, India

[2]Regional Centre of Organic Farming, Department of Agriculture, Cooperation and Farmers Welfare, Ministry of Agriculture and Farmers Welfare, Govt. of India, Bhubaneswar-751021, India

[3]Department of Biotechnology, Central University of Haryana, Jant-Pali, Mahendergarh-123029, India

[4]Division of Germplasm Evaluation, [5*]Division of Plant Quarantine, ICAR- National Bureau of Plant Genetic Resources, New Delhi-110012, India

*Corresponding author: pardeep4yadav@gmail.com

Abstract

Omics technologies involving the study of the genome sequences, gene expression profiles, protein content and metabolites of an organism, are known as genomics, transcriptomics, proteomics and metabolomics, respectively. The analytical procedures involved in omics studies have been made easier, faster, precise and cost effective by revolutionary technological advancement in last few decades. Implementation of omics studies in crop biotechnology has resulted in identification of genomic sequences governing the crucial agronomic traits such as, grain yield, resistance to biotic and abiotic stresses etc. Integration of functional analysis with genomic sequence data provide more accurate gene annotations, functional markers, quantitative trait loci (QTL) maps etc. that can help in development of improved crop varieties. Multi-dimensional omics studies lead to better understanding of factors affecting the phenotypic traits, metabolic process and role of biomolecules in various biosynthetic pathways of crop plants.

Keywords: Genomics, Transcriptomics, Proteomics, Metabolomics, Crop plants

Introduction

There is an urgent need to enhance the yield and nutritive qualities of crop plants in order to meet the food requirements of ever-growing world population. With the advancement of high throughput omics technologies, it has become possible to improve the quantitative and qualitative traits of crop plants. Omics technologies imply the holistic approach for quantification and characterization of genes, transcripts, proteins and metabolites of a biological organism known as genomics, transcriptomics, proteomics and metabolomics, respectively (Varshney *et al.*, 2013). The crop traits are generally complex quantitative traits, which are controlled by multiple genes (Core *et al.*, 2008). The omics technologies have been successfully implied for studying and manipulating the economic traits of crop plants for enhancing yield and nutritional qualities. The development of next generation sequencing (NGS) technologies and accessibility to genomic information of model crop plants has helped to discover new genes and to study gene expression profiles (Mochida *et al.*, 2006. Massive sequencing and gene expression profiling are required for identification of key genes linked with desired phenotypic characters (Thruston *et al.*, 2005). The discovery of target loci associated with some phenotypic trait has enabled to explore complex biological phenomenon such as plant growth, development, photosynthesis, biotic and abiotic stress tolerance (Struik *et al.*, 2007). Application of biotechnological procedures in plant breeding programmes has enhanced the productivity and sustainability of crop plants. The crop traits are generally complex quantitative traits, which are controlled by multiple genes (Core *et al.*, 2008). Therefore, high throughput omics techniques assisted with bioinformatics tools are required to identify the factors affecting the growth and yield of food crops (Rhee *et al.*, 2006).

Genomics in Agriculture

Genomics involves collective and holistic approaches for studying the complete set of genes of an organism. The genomic studies of crop plants have provided the insights into total number of genes, gene organization, genetic mapping and role of genes in various metabolic processes. Genomic technologies study the genomes by gene mapping and DNA sequencing. First generation sequencing included automated Sanger's method of genome sequencing (Sanger, 1977). Sanger's method of sequencing involved incorporation of di-deoxy nucleotides (chain terminators) during *in vitro* DNA synthesis by DNA polymerase. However, this technique was very costly, time taking and laborious for whole genome sequencing purpose. NGS techniques have expanded the abilities and possibilities of studying the gene sequences. The second-generation technologies (Illumina and Roche 454) were highly efficient, cost effective and provided the in depth knowledge of genomic sequences (Metzker, 2010) as explained in Table 1. The third generation sequencing techniques include Helicos method, which is a single molecule synthesis method applying reverse terminators (Braslavsky, 2003). It was discovered in the lab of Stephen Quake and later commercialized by Helicos (Thompson and Steinmann, 2010). The other abundantly used third generation sequencing technique is single molecule real time (SMRT) procedure developed by Pacific Biosciences. Third generation sequencing technologies are cheaper and faster than the previous techniques and provide very long reads which lead to error free assembly of genomic sequences (Haque *et al.*, 2013).

Table 1: Genomic technologies used for sequencing of crop plants

Genomic techniques	Discovery	Analysis	Software tools
Di-deoxy or Chain termination method	Frederick Sanger, 1977	Exposer to UV light and X-ray	Analyzer
Automated DNA sequencing	Applied Biosystems, 1986	Laser light used to detect the fluorescence dye	Sequencing software
Shotgun sequencing method	Staden, 1979	The DNA fragment is ordered based on overlaps in genetic code	Assembler
Illumina	Barnes *et al.*, 2002	Solid phase bridge amplification	Sequencing software
Roche/454	Ronaghi *et al.*, 1998	Pyrosequencing	Sequencing software
SoLid	Landegren *et al.*, 1988	Sequencing by ligation	Sequencing software
Polonator	Shendure *et al.*, 2005	Emulsion PCR	Sequencing software
Helicos	Braslavsky, 2003	Single molecule sequencing system (Reversible terminators)	Sequencing software
SMRT	Eid *et al.*, 2009	Single polymerase molecule	Sequencing software

SMRT: Single Molecule Real-Time

Applications of Genomic Technologies

The genome wide association studies provide a broader view of working and interaction of all genes. Advances in genomic technologies have enabled us to design agricultural crops with promising economical traits. Genome sequencing, subsequent functional annotation and molecular analysis have been exploited in various fields of crop biotechnology.

Genome sequencing and gene prediction

The advancement of next generation sequencing technologies has helped in the prediction of genes by comparative genomics studies. The whole genome sequencing of first plant, *Arabidopsis thaliana* was completed in year 2000, which was subsequently annotated for identifying the functional 25,000 genes and further used as model crop for gene annotation of other crops. The comparative genomic study includes comparison of newly sequenced genome with model crops, which has resulted in discovering new genes (Mitchell *et al.*, 1998). The numbers of genes predicted in different agricultural crops by NGS are given in Table 2.

Table 2: The number of genes predicted in different agricultural crops by NGS techniques

Crops	Haploid number of chromosome	Number of genes predicted	Reference
Rice	12	37,544	Yu *et al.*, 2002
Tomato	12	34, 727	The tomato genome consortium, 2012
Potato	12	39, 031	The potato genome sequencing consortium, 2012
Soybean	20	46, 430	Schmutz *et al.*, 2010
Chickpea	8	28, 269	Varshney *et al.*, 2013
Banana	11	36, 542	D' Hont *et al.*, 2012
Maize	10	32,540	Schnable *et al.*, 2009
Wheat	21	94, 000	Brenchley *et al.*, 2012
Pigeonpea	11	48, 680	Varshney *et al.*, 2011
Cabbage	9	45, 758	Liu *et al.*, 2014
Mung bean	11	22, 427	Kang *et al.*, 2014
Pearl millet	7	38,579	Varshney *et al.*, 2017

Genetic diversity analysis and mapping of trait-specific markers

Molecular markers have been identified as important tools for early detection of desired characters in the progeny. The knowledge of molecular markers can now be applied to access and amplify the diversity of economically valuable traits of crop plants (Collard and Mackill, 2008). NGS technologies have enabled in the generation of massive sequences and detection the single nucleotide polymorphism (SNP) or simple sequence repeats (SSR) molecular markers located throughout the genome (Salgotra *et al.*, 2014) (Table 3). These molecular markers have been utilized for generating genetic and physical maps and identifying the regions responsible for adaptation of crop against various stress conditions (Varshney *et al.*, 2009). Genetic maps represent the location of markers in the linkage group based on their co segregation. NGS technologies have led to development of genetic maps with enhanced marker density. These enriched maps have been used to replace quantitative trait loci (QTL) mapping with association mapping (Perez-de-Castro *et al.*, 2012). The QTL mapping links a larger genomic area with a particular trait, but association mapping provide higher resolution since it uses more markers. Therefore, association maps are more precise and informative as molecular characterization tool.

Table 3: Markers discovered in crop plants using genome-sequencing approaches

Crops	Number of markers identified	Sequencing techniques	References
Rice	1,04,000 EST, 48,351 SSR	Random-fragment shotgun sequencing	International rice genome sequencing project, 2005
Pigeonpea	3,09,052 SSR	Illumina	Varshney *et al.*, 2011
Wheat	12,40,000 EST, 2,500 SSR	Solexa	International wheat genome sequencing consortium, 2008
Maize	36,000 SNP	Roche GS20	Barbazuk *et al.*, 2007
Grapes	17 million SNP	Roche 454	Novaes *et al.*, 2008
Arabidopsis	Epigenetic modification	Roche 454	Lister *et al.*, 2008
Chickpea	81,845 SSR	Illumina	Varshney *et al.*, 2013
Pearl millet	88,256 SSR	Illumina	Varshney *et al.*, 2017

EST: Expressed Sequence Tag, SSR: Simple Sequence Repeat, SNP: Single Nucleotide Polymorphism

Genetic improvement of crop plants

Omics studies have led to the advancement of agricultural research for improvement of food crops, feedstock and maintenance of the environment (Ahmad *et al.*, 2012). Genome sequencing and gene expression studies has helped in identification of the functional genes related to a particular trait and this information could be utilized for improvement of crop plants by addition of genes or post transcriptional gene silencing (Table 4). Genomics technologies have enabled to develop functional foods such as drought tolerant maize, rice with higher grain production (Ashikari *et al.*, 2005) and bananas with the longer shelf life (Mehrotra *et al.*, 2013). Plants are exposed to mutagenic reagents for the production of designer crops having desired economic traits, popularly known as mutation breeding (Burbank *et al.*, 1921). The progeny with desired character has been selected by marker-assisted breeding. The molecular markers such as SSR and SNPs discovered by genome sequencing techniques have been applied for improvement of agricultural crops (Salgotra *et al.*, 2014).

Table 4: The functional food crops developed by genomic techniques

Crops	Techniques	Functional food crops	Benefits
Plum and Apricot	Cross hybridization	Plumcot	More vitamin content
Broccoli and Cauliflower	Cross hybridization	Broccoflower	More vitamin content

Crops	Techniques	Functional food crops	Benefits
Tomato	Antisense RNA	Flavr savr	Longer shelf life and higher flavonoid content
Rice	Transgenic	Golden rice	Vitamin A fortified
Tomato and Banana	Transgenic	Edible vaccine	Vaccine against Diphtheria, pertussis, polio, Measles, tetanus, tuberculosis and Hepatitis B
Tobacco	Transgenic	Plantibodies	Monoclonal antibodies against West Nile Virus infection
Mustard	Transgenic	Reduced saturated fatty acid	Less saturated fatty acid

Challenges of Genomic technologies

The presence of highly repetitive DNA in plant genome sequences would produce hindrance during assembly of short fragments of genomic sequences. The gaps in genome sequences would lead to errors in final sequencing draft. The polyploidy and heterozygosity of agriculture crops produce hindrance during the assembly of their sequences. Functional annotation of many identified genes has still not been performed.

Future Perspectives of Genomic Technologies

The gene-based studies have shifted from single gene to whole genome, which would help in understanding the genetic interactions. The advancement in genomic technologies would help in exploring the evolutionary history of food crops, and identification of descendants from which the modern crop plants has been emerged. With application of genomics tools researchers are now equipped to develop engineered food crops, which could enhance the quality and quantity of agricultural production. The accuracy and efficiency of QTL mapping technique would increase with the help of modern genomic tools. Next generation sequencing techniques could applied for genome sequencing of new crops and it would lead to easier comparison of genomic data. The cross-genomic studies would help in better understanding of gene function.

Transcriptomics in Agriculture

Transcriptome provides the knowledge of expressed gene sequences in a specific tissue at a specific time. The transcriptome is dynamic in nature because the transcriptome of the cell keep on changing under different conditions and secondly, same gene can generate many transcripts because of alternative splicing. The mechanism and regulation of alternative splicing can be explored by sequencing transcripts. The study of transcriptome helps in understanding the function of genes, transcription levels and molecular mechanism of cellular metabolism (Borovitz and Chory, 2004). Expression analysis of genes helps in identification of candidate genes for a particular biological trait and can help in identifying key enzymes of a metabolic pathway (Table 5). Advancement in NGS technologies enabled to obtain cost effective transcriptome assemblies, which are useful for gene annotation (Mochida and Shinozaki, 2010). The analysis of transcriptome assemblies gives the information about various functional markers such as simple sequence repeats (SSR) and single nucleotide polymorphism (SNP) related with stress resistant response (Dunwell *et al.*, 2001). After obtaining the quantitative counts of each transcript, differential gene expression could be analyzed by normalizing the data with help of statistical modeling (Lowe *et al.*, 2017).

Table 5: List of transcriptomics techniques with their applications

Techniques	Analysis method	Application	Reference
DNA microarray	The chips are designed to visualize the mRNA profile during particular environmental condition	Study of circadian clock, plant defence, environmental stress response, fruit ripening	Aharoni and Vorst, 2001
EST	Short nucleotide sequence generated from a single RNA transcript	Sequence information for pre microarray design	Marra *et al.*, 1998
SAGE	Sequence expression profiling	Expression analysis of plants with less characterized genome	Velculescu *et al.*, 2000
Long SAGE	Gene expression technique	Derived transcriptome used for annotation of expressed genes	Saha *et al.*, 2002
MPSS	Generate sequences via complex series of hybridization	Identify and quantify the RNA transcript	Brenner *et al.*, 2000

EST: Expressed Sequence Tag, MPSS: Massive Parallel Signature Sequencing, SAGE: Serial Analysis of Gene Expression

Applications of Transcriptomics in Agriculture

The microarray technology have been propitiously utilized for explaining the variations in transcriptome data during seed germination, growth, development and various stresses (Poole *et al.*, 2007). Functional genomics is a combinatorial approach, which utilizes the genomic data along with expression analysis to identify putative genes linked with particular biological trait. By applying this approach, the region responsible for seed development has been located in wheat genome (Jordan *et al.*, 2007). QTLs related to grain development, resistance to biotic and abiotic stresses have been mapped on crop genomes; and have successfully been applied for improvement of crop varieties (Saha *et al.*, 2010). Functional techniques such as RNA interference (RNAi), mutagenesis and epigenetics can be applied as gene silencing tools for refinement of agricultural crops. Bulk segregate analysis (BSA) is a genome mapping technique for identification of the chromosomal region linked with expression of economically important crop traits. This technique was successfully applied for development disease resistant pigeon pea (Peng *et al.*, 2009). Molecular switches are sequence specific DNA binding transcriptional factors, which regulate the biological procedures. The transcriptional factors (TF) of newly sequenced plant species could be identified by comparing their genome sequence with TF sequences of model plants (*Aribidopsis thaliana*) (Riechmann *et al.*, 2000). The details of traits analysed by transcriptomics techniques are given in Table 6.

Table 6: The traits of plants analysed by transcriptomic techniques

Crop	Stress trait	Approach applied	Reference
Chickpea	Drought tolerance	cDNA library	Peng *et al.*, 2009
Pigeonpea	Fusarium and wilt resistant	BSA	Kotresh *et al.*, 2006
Soybean	Cyst nematode resistance	MAS	Cahill and Schmidt, 2004
Pea	Stemphylium blight resistant	QTL	Saha *et al.*, 2010
Cowpea	Striga resistant	SCAR markers	Omoigui *et al.*, 2012
Groundnut	Rust resistant	Back cross progenies	Varshney *et al.*, 2014
Alfalfa	Aluminium toxicity	Microarray	Zhou *et al.*, 2013

BSA: Bulk Segregation Analysis. QTL: Quantitative Trait Analysis, MAS: Marker Assisted Selection, SCAR: Sequence Characterized Amplified Regions

Challenges of Transcriptomic Technologies

The less availability of molecular markers has limited the ability for functional analysis of target traits. Limited access to high throughput and cost effective phenotyping platforms and lack of suitable infrastructure are the barriers for whole transcriptome analysis. The data of transcriptome is highly variable with respect to time and tissue, therefore, the sample collection and analysis should be done carefully.

Future Perspectives of Transcriptomic Technologies

The expression profile data could be generated based on cell type, developmental stage and physiological conditions. Gene expression analysis of mutant varieties would provide the information about genes involved in the particular metabolic phenomenon. Expression profile coupled with phenotypic characters would lead to functional annotations of uncharacterized genes. Analysis of multiple data sets simultaneously would enable for comparing the gene expression patterns among different species. The analysis of molecular and expression diversity would reveal the relationship between crop plants and their wild relatives.

Proteomics in Agriculture

Proteomics involves the identification and characterization of the whole set of proteins derived from a genome (Wilkins *et al.*, 1995). The genome code for the protein is required for maintaining the structure and sustaining essential regulatory functions (Whitelegge, 2002). Proteomics deals with the analysis of amino acid sequence for determination of their relative concentration and various post-translational modifications (Brygoo and Joyard, 2004). In contrast, of genomics, proteomics is dynamic in nature subjected to translation and post-translational modifications (Nat *et al.*, 2007). The study of proteomics helps in understanding the complex biological procedures and cellular response against environmental stress (Renaut *et al.*, 2006). Proteomics has emerged as an important tool for crop improvement since it explains the function of proteins maintaining homeostasis inside cells, involved in cell signalling pathways and required for structural maintenance.

Technologies Used for Proteome Analysis

Proteomic analyses include protein extraction, separation of proteins from a mixture followed by identification of proteins by mass spectrometer (MS) (Figure 1). The important technologies used in proteomics analysis are 2D-PAGE (2-D poly acrilamide gel electrophoresis) for separation of protein components of cellular extract (Klose, 1975) and MS for analysis of peptide sequence of proteins expressed at a particular time (Fenn *et al.*, 1989). Electron spray ionisation (ESI) is performed to convert peptides into ions by passing them through high voltage columns. Time of flight (TOF) is a procedure, which analyses the mass of peptide ions in mass spectrometry. Most abundantly used MS technique is MALDI TOF (Matrix-assisted laser desorption/ionization time-of-flight). These techniques have been used for peptide mapping, which allows the identification of protein and their complex interaction (Kersten *et al.*, 2002). In MS or MALDI, the peptides are fragmented by induced collisions and then the ionic peptides are quantified based on time of flight which varies with mass/charge ratio. Surface enhanced laser desorption (SELDI) is another MS method in which desired proteins are bound to the surface before the beginning of analysis (Hasiguchi *et al.*, 2010). The MS based quantification has been done based on labelled or labels free detection (Thelen *et al.*, 2007). The Combination of LC-MS (Liquid chromatography-mass spectrometry) and HPLC (High performance liquid chromatography) has been applied for multidimensional analysis of proteins since the high ionic charge of peptides enhances the separation of peptides (Washburn *et al.*, 2001). Genome wide protein interaction map has been prepared with help of yeast two-hybrid systems, these maps have been used to study the complexes involved in extracellular signalling (Chen and Harmon, 2006). Protein-protein interaction plays a vital role in maintaining the intracellular equilibrium (Ozbabacan *et al.*, 2011). Therefore, the study of proteome is essential to understand the mechanism of complex biological processes.

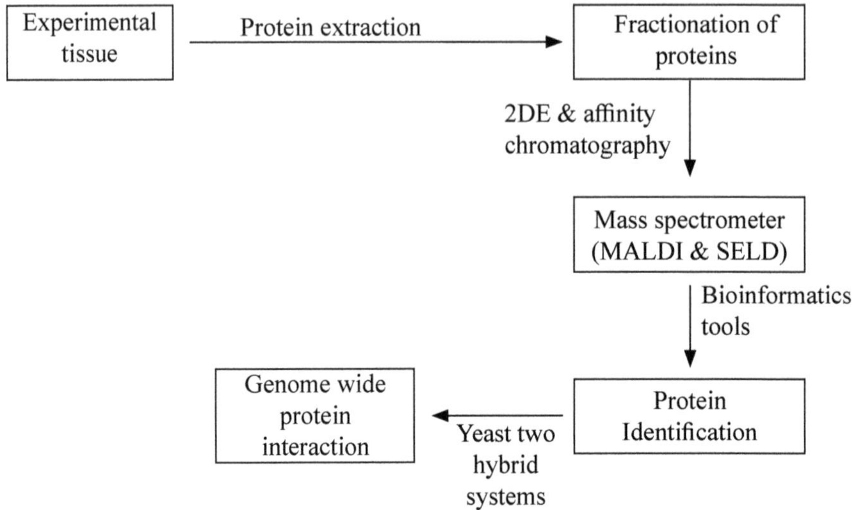

Fig. 1: The steps involved in proteomics studies

Applications of Proteomics in Agriculture

The attempts have been made to enhance the photosynthetic efficiencies and abiotic stress tolerance of crop plants. It was observed that C4 plants contain two types of chloroplasts and hence are more efficient with respect to energy conversion. A comparative proteomic study was performed for chloroplasts of C3 and C4 plants to identify the proteins responsible for more efficient light fixation (Zhao *et al.*, 2013). Wang *et al.* (2013) identified the proteins responsible for male sterility traits essentially required for hybrid selection. Heat stress conditions affect the growth and yield of crop plants. In order to figure out the proteins accumulating specifically during abiotic stress, the organ specific proteomic study was performed in numerous crop plants (Hossain and Komatsu, 2013). The knowledge of proteomic studies has been successfully applied for generation of transgenic plants with desired characteristics (Gong and Wong, 2013).

Challenges of Proteomic Technologies

Abundant amount of proteins are extracted from the samples, which would produce hindrance in analysis of desired protein. Currently there are no proper guidelines available for proteomic analysis and data interpretation procedure. The earlier technologies such as 2-DE (2-dimensional gel electrophoresis) are not considered reliable today, the advanced technologies are preferred for accurate characterization of proteins. The biological variations in protein are responsible for less reproducibility of proteomics data. Therefore, the analysis should be carefully done under controlled conditions. Proteomics also rely on *in silico* tools and software for protein-function prediction. The functions of proteins are often predicted by homology search with available dataset, which may lead to wrong predictions.

Future Perspectives of Proteomic Technologies

The recent advancement in proteomics and bioinformatics tools has helped to understand the functions of biomolecules and their roles in cellular metabolism. With the help of proteomics, the omics data are enriched and provide the insight of different biological procedures such as post-translational modifications and protein-protein interactions. The organ and time-specific proteomic analysis can identify the proteins accumulated during stress response, which would lead to development of stress resistant variety. The comparative proteome analysis of different organs can identify the proteins involved in organ-specific metabolic pathways and thus can help in identifying unique targets for inhibiting a particular pathway. The progress in technology would lead to resolution of less abundant proteins as well as finding their sub-cellular localization. In future, there is a possibility for development of protein based biosynthetic fungicide with the help of proteomics tools.

Metabolomics in Agriculture

The metabolite concentration is responsible for providing flavour, aroma, storage capacity and sustainability of crops (Memelink, 2005). The metabolomics involves identification and quantification of all metabolites in a biological sample. Metabolites are classified as primary and secondary based on their presence at the particular developmental stage. Analysis of metabolites becomes easier by the advancement of modern techniques such as mass spectrometry (MS), nuclear magnetic resonance (NMR) and fourier-transform infrared spectroscopy (FTIR). The metabolomics analysis depends on the extraction of the sample at a particular time (Hall, 2006). Metabolomics techniques provide us the insight of complex metabolic control and diversity of biological reactions involved in the regulation of plant growth, differentiation, stress resistance and defence mechanism against pathogens. Large datasets are produced and further processed by bioinformatics tools for analysing the components of the metabolic networks. Metabolomics study enhances the understanding of the complicated biosynthetic pathways and complex molecular networks. Computational techniques including three dimensional structural modeling and molecular dynamics simulations have helped in designing modern molecular breeding programs.

Applications of Metabolomics in Agriculture

Metabolomics studies have been performed in agriculture sector in order to generate metabolic profiles for the analysis of various stresses, enhancing economically important traits and impact of heredity. Metabolic profiles have been successfully applied for determining the impact of seasonal changes, geographical region and natural variation. Metabolic profiles have been evaluated for characterization of transgenic plants. Metabolic flux analysis has helped in the identification of orthologous enzymes with similar catalytic properties. Metabolic studies have also been performed for characterizing the growth profile, developmental stages and chemo taxis analysis.

The construction of biochemical network has been done by establishing the relative profiles of metabolites (Mendes, 2002). The integration of metabolome and transcriptome data has implied the understanding of regulatory network and association of genetic content with phenotypic characters (Urano *et al.*, 2009). The list of agricultural crops whose metabolic profile was studied for analysis of economically valuable traits is given in Table 7.

Table 7: Metabolic profiles of crop plants analysed by omics technologies

Crop	Metabolic profiles studied	Reference
Tomato	Introgression line	Overy *et al.*, 2005
Tomato	Analysis of carotonoids and isoprenoids	Fraser *et al.*, 2000
Tomato	Analysis of carbohydrate	Al-Babili *et al.*, 2001
Maize	Provitamin A, lycopene £-cyclase	Harjes *et al.*, 2008
Rice	β-Carotene	Oikawa *et al.*, 2008
Rice	Glycine betaine as marker for stress tolerance	Kasano *et al.*, 2007

Challenges of Metabolomic Technologies

The biological variability may lead to problems in analysis of metabolites linked with a particular trait. Most of the metabolites are part of multiple pathways, therefore it become difficult to analyse the metabolites related to a particular biosynthetic pathway. The difficulties occurring in traceability to specific pathways limit the applications of metabolites as biomarkers. Sometimes unknown metabolites have been identified during LC-MS (Liquid chromatography mass spectrometry) analysis, which cannot be utilized for any analysis. The data generated by metabolomics study is vast and complicated which need multivariate techniques for analysis.

Future Perspectives of Metabolomic Technologies

Metabolomics in association with other omics technologies can help in understanding different metabolic networks involved in biosynthesis procedures. Metabolome analysis has helped in identification of metabolites, which provides resistance to abiotic and biotic stresses. More such metabolites can be identified in future that can be used as biomarkers for the study of stress tolerance response. Metabolic contents are more closely associated with particular phenotypic trait of agricultural crops and could be used for QTL mapping in breeding programs. Metabolomics studies have been found useful for development of association map. Metabolic markers could be utilized for genetically engineering a metabolic pathway. Functional analysis of genes becomes easier by integration of genomic techniques with metabolomics.

Conclusions

To conquer the problems of growing population, changing climatic patterns and environmental stress there is a requirement of developing the novel crop varieties with higher yield, thermo-tolerance and less pesticide consumption. The advanced technologies have generated high dimensional biological data, which has been analysed by statistical methods in order to reveal the interrelationship between different omics studies. The interpretation of omics studies has increased the resolution of the biological assays, which have led the foundation for analysis of metabolic pathways and system biological studies. The expression analysis of agricultural crops by omics technologies have played very crucial role in genome function characterization, for example, gene ontology, pathway analysis and system modelling. However, the functional analysis of molecular markers and QTL mapping has enabled in restoration of environmentally favourable traits of wild varieties of crop plants and hence, enhanced their sustainability. The study of metabolic responses of crop plants against various biotic and abiotic stresses has helped in development of new resistant varieties. However, the transcriptome, proteome and metabolome data are highly variable over time, stages and environmental conditions, therefore, the sample collection and analysis has to be performed very carefully.

References

Aharoni, A. and Vorst, O. 2001. DNA microarrays for functional plant genomics. *Plant Mol. Biol.*, 48: 99-118.

Ahmad, P., Ashraf, M., Younis, M., Hu, X., Kumar, A., Akram, N.A. and Al-Quraіny, F. 2012. Role of transgenic plants in agriculture and biopharming. *Biotechnol. Adv.*, 30: 524–540.

Al-Babili, S., Ye, X., Lucca, P., Potrykus, I. and Beyer, P. 2001. Biosynthesis of beta-carotene (provitamin A) in rice endosperm achieved by genetic engineering. *Novartis Found Symp.*, 236: 219-228.

Ashikari M., Sakakibara H., Lin S., Yamamoto T., Takashi T., Nishimura A., Angeles E.R., Qian Q., Kitano H. and Matsuoka M. 2005. Cytokinin oxidase regulates rice grain production. *Science*, 309(5735): 741–745.

Barbazuk, W.B., Emrich, S.J., Chen, H.D., Li, L. and Schnable, P.S. 2007. SNP discovery via 454 transcriptome sequencing. *Plant J.*, 51: 910–918.

Barnes, C., Balasubramanian, S. Liu, X., Swerdlow, H. and Milton, J. 2002. Labelled nucleotides. US Patent 7,057,026. U. S. Patent and Trademarks Office, Department of Commerce, Washington, D.C., USA.

Borovitz, J.O. and Chory, J. 2004. Genomics tools for QTL analysis and gene discovery. *Curr. Opin. Plant Biol.*, 7: 132-136.

Braslavsky, I., Hebert B., Kartalov E. and Quake S.R. 2003. Sequence information can be obtained from single DNA molecules, *Proc. Natl. Acad. Sci.* U.S.A., 100: 3960–3964.

Brenchley, R., Spannag, M., Pfeifer, M., Barker, G.L.A., D'Amore, R., Allen, A.M. and McKenzie, N. 2012. Analysis of the bread wheat genome using whole-genome shotgun sequencing. *Nature,* 491(7426): 705–710.

Brenner, S., Johnson, M., Bridgham, J., Golda, G., Lloyd, D.H. and Johnson, D. 2000. Gene expression analysis by massively parallel signature sequencing (MPSS) on micro bead arrays. *Nat. Biotechnol,* 18: 630–4.

Brygoo, H. and Joyard, J. 2004. Introduction – focus on plant proteomics. *Plant Physiol. Biochem.,* 42: 913-917.

Burbank, L. 1921. Fruit Improvement. In How Plants are trained to Work for Man; P. F. Collier & Son: New York, 3: 85.

Cahill, D.J. and Schmidt, D.H. 2004. Use of marker assisted selection in a product development breeding program. In: Proceedings of the 4th International crop Science Congress, 26 September-1 October, Brisbane, Australia.

Chen, S. and Harmon, A.C. 2006. Advances in plant proteomics. *Proteomics,* 6: 5504-5516.

Collard, B.C.Y. and Mackill, D.J. 2008. Marker-assisted selection: an approach for precision plant breeding in the twenty-first century. *Philos. Trans. R. Soc. Lond. B. Biol. Sci.,* 363: 557–572.

Core, L.J., Waterfall, J.J. and Lis, J.T. 2008. Nascent RNA sequencing reveals wide spread pausing and divergent initiation at human promoters. *Science,* 322: 1845–1848.

D' Hont, A., Denoeud, F., Aury, J.M., Baurens, F.C., Carreel, F., Garsmeur, O. and Noel, B. 2012. The banana (*Musa acuminata*) genome and the evolution of monocotyledonous plants. *Nature,* 488 (7410): 213–217.

Eid, J., Fehr, A., Gray, J., Luong, K., Lyle, J. et al. 2009. Real-time DNA sequencing from single polymerase molecules. *Science,* 323: 133–138.

Fenn, J.B., Mann, M. and Meng, C.K. 1989. Electrospray ionization for mass spectrometry of large biomolecules. *Science,* 246: 64-71.

Fraser, P.D., Pinto, M.E., Holloway, D.E. and Bramley, P.M. 2000. Application of high-performance liquid chromatography with photodiode array detection to the metabolic profiling of plant isoprenoids. *Plant J.,* 24: 551-558.

Gong, C.Y. and Wang, T. 2013. Proteomic evaluation of genetically modified crops: current status and challenges. *Front. Plant Sci.,* 4: 41.

Hall, R.D. 2006. Plant metabolomics: from holistic hope to hype, to hot topic. *New Phytol.,* 169: 453-468.

Haque, F., Li, J., Wu, H.C., Liang, X.J. and Guo, P. 2013. Solid-state and biological nano-pore for real-time sensing of single chemical and sequencing of DNA. *Nano Today,* 8: 56–74.

Harjes, C.E., Rocheford, T.R., Bai, L., Brutnell, T.P. Kandianis, C.B. and Sowinski, S.G. 2008. Natural genetic variation in *Lycopene epsilon cyclase* tapped for maize biofortification. *Science,* 319: 330-333.

Hashiguchi, A., Ahsan, N. and Komatsu, S. 2010. Proteomics application of crops in the context of climatic changes. *Food Res. Int.*, 43: 1803–1813.

Hossain, Z. and Komatsu, S. 2013. Contribution of proteomic studies towards understanding plant heavy metal stress responses, *Front. Plant Sci.*, 3: 310.

International Rice Genome Sequencing Project. 2005. *International Rice Genome Sequencing Project. Nature*, 436: 793-800.

Jordan, M.C., Somers, D.J. and Banks, T.W. 2007. Identifying regions of the wheat genome controlling seed development by mapping expression quantitative trait loci. *Plant Biotechnology J.*, 5(3): 442–453.

Kang, Y.J., Kim, S.K., Kim, M.Y., Lestari, P., Kim, K.H., Ha, B.K. and Jun, T.H. 2014. Genome sequence of mung bean and insights into evolution within *Vigna* species. *Nature Communications*, 5: 5443.

Kersten. B., Bürkle, L. and Kuhn, E.J. 2002. Large-scale plant proteomics. *Plant Mol. Biol.*, 48: 133-141.

Klose, J. 1975. Protein mapping by combined isoelectric focusing and electrophoresis of mouse tissues. A novel approach to testing for induced point mutations in mammals. *Human Genetics*, 26: 231-243.

Kotresh, H., Fakrudin, B., Punnuri, S., Rajkumar, B., Thudi, M., Paramesh, H., Lohithswa H. and Kuruvinashetti, M.S. 2006. Identification of two RAPD markers genetically linked to a recessive allele of a Fusarium wilt resistance gene in pigeonpea (*Cajanus cajan* L.) Millsp.). *Euphytica*, 149:113–120.

Kusano, M., Fukushima, A., Kobayashi, M., Hayashi, N., Jonson, P. and Mortiz, T. 2007. Application of a metabolomics method combining one-dimensional and two-dimensional gas chromatography time of flight/ mass spectrometry to metabolic phenotyping of natural variants in rice. *J. Chromatogr. B. Anal Technol. Biomed. Life Sci.*, 855: 71-79.

Landegren, U., Kaiser, R., Sanders, J. and Hood, L. 1988. A ligase-mediated gene detection technique. *Science*, 241: 1077 – 1080.

Lister, R., O' Malley, R.C., Tonti-Filippini, J., Gregory, B.D., Berry, C.C., Millar, A.H. and Ecker, J.R. 2008. Highly integrated single-base resolution maps of the epigenome in *Arabidopsis*. *Cell*, 133: 395–397.

Liu, S., Liu, Y., Yang, X., Tong, C., Edwards, D., Parkin, I.A. and Zhao, M. 2014. The *Brassica oleracea* genome reveals the asymmetrical evolution of polyploid genomes. *Nature Communications*, 5: 3930.

Lowe, R., Shirley, N., Bleackley, M., Dolan, S. and Shafee, T. 2017. Transcriptomics technologies. *PLoS Comput. Biol.*, 13(5): 1005457.

Marra, M.A., Hillier, L. and Waterston, R.H. 1998. Expressed sequence tag-EST abolishing bridges between genomes. *Trends Genet.*, 14: 4–7.

Mehrotra, S. and Goyal, V. 2013. Evaluation of designer crops for biosafety- a scientist's perspective. *Gene*, 515: 241–248.

Memelink, J. 2005. Tailoring the plant metabolome without a loose stitch. *Trends Pl. Sci.*, 10: 305-307.

Mendes, P. 2002. Emerging bioinformatics for the metabolome. *Brief Bioinform.*, 3: 34-45.

Metzker, M.L. 2010. Sequencing technologies - the next generation. *Nature Reviews Genetics*, 11: 31 – 46.

Mitchell-Olds, T. and Pedersen, D. 1998. The molecular basis of quantitative genetic variation in central and secondary metabolism in *Arabidopsis*. *Genetic*, 149: 739-747.

Mittler, R. and Shulaev, V. 2013. Functional genomics, challenges and perspectives for the future. *Physiologia Plantarum*, 148: 317–321.

Mochida K and Shinozaki K. 2010. Genomics and Bioinformatics Resources for Crop Improvement. *Plant Cell Physiol.*, 51(4): 497–523.

Mochida, K., Kawaura, K., Shimosaka, E., Kawakami, N., Shin-I, T. and Kohara, Y. 2006. Tissue expression map of a large number of expressed sequence tags and its application to in silico screening of stress response genes in common wheat. *Mol. Gen. Genomics*, 276: 304 – 312.

Nat, N.V.K., Sanjeeva, S., William, Y. and Nidhi, S. 2007. Application of proteomics to investigate plant microbe interactions. *Curr. Proteomics*, 4: 28–43.

Novaes, E., Drost, D.R., Farmerie, W.G., Pappas, G.J., Grattapaglia, D., Sederoff, R.R. and Kirst, M. 2008. High throughput gene and SNP discovery in *Eucalyptus grandis*, an uncharacterized genome. *BMC Genomics*, 9: 312.

Oikawa, A. Matsuda, F., Kusano, M., Okazaki, Y. and Saito, K. 2008. Rice metabolomics. *Rice*, 1: 63-71.

Omoigui, L.O., Kamara, A.Y., Ishiyaku, M.F. and Boukar, O. 2012. Comparative responses of cowpea breeding lines to Striga and Alectra in the dry savannah of northeast Nigeria. *Afr. J. Agric. Res.*, 7: 747–754.

Ozbabacan, S.E., Engin, H.B., Gursoy, A. and Keskin, O. 2011. Transient protein-protein interactions. *Protein Eng. Des. Sel.*, 24: 635-648.

Pérez-de-Castro, A.M., Vilanova, S., Cañizares, J., Pascual, L., Blanca, J.M., Díez, M.J., Prohens, J. and Picó, B. 2012. Application of genomic tools in plant breeding. *Current Genomics*, 13(3): 179–195.

Poole, R., Barker, G., Wilson, I.D., Coghill, J.A. and Edwards, K.J. 2007. Measuring global gene expression in polyploidy; a cautionary note from allohexaploid wheat. *Functional and Integrative Genomics*, 3:207–219.

Renaut, J., Hausman, J.F. and Wisniewski, M.E. 2006. Proteomics and low temperature studies: bridging the gap between gene expression and metabolism. *Physiol. Plant*, 126: 97-109.

Rhee, S.Y., Dickerson J. and Xu, D. 2006. Bioinformatics and its applications in plant biology. *Annu. Rev. Plant Biol.*, 57: 335-360.

Riechmann, J.L., Heard, J., Martin, G., Reuber, L., Jiang, C., Keddie, J. 2000. *Arabidopsis* transcription factors: genome-wide comparative analysis among eukaryotes. *Science*, 290: 2105 – 2110.

Ronaghi, M., Uhlen M. and Nyrén P.I. 1998. A sequencing method based on real-time pyrophosphate. *Science*, 281: 363–365.

Saha, G.C., Sarker, A., Chen, W., Vandemark, G.J. and Muehlbauer, F.J. 2010. Inheritance and linkage map positions of genes conferring resistance to stemphylium blight in lentil. *Crop Sci.*, 50: 1831–1839.

Saha, S., Sparks A.B. and Rago, C. 2002. Using the transcriptome to annotate the genome. *Nat. Biotech.*, 20: 508–512.

Salgotra, R.K., Gupta, B.B. and Stewart, C.N. 2014. From genomics to functional markers in the era of next-generation sequencing. *Biotechnology Letters*, 36(3): 417–426.

Sanger, F., Nicklen, S. and Coulson, A.R. 1977. DNA sequencing with chain-terminating inhibitors. *PNAS*, 74(12): 5463–5467.

Schmutz, J., Cannon, S.B., Schlueter, J., Ma, J., Mitros, T., Nelson, W. and Hyten, D.L. 2010. Genome sequence of the palaeopolyploid soybean. *Nature*, 463(7278): 178–183.

Schnable, P.S., Ware, D., Fulton, R.S., Stein, J.C., Wei, F., Pasternak, S. and Liang, C. 2009. The B73 maize genome: Complexity, diversity and dynamics. *Science*, 326(5956): 1112–1115.

Shendure, J., Porreca, G.J., Reppas, N.B., Lin, X., Mccutcheon, J.P., Rosenbaum, A.M. and Wang, M.D. 2005. Accurate multiplex polony sequencing of an evolved bacterial genome. *Science*, 309: 1728 – 1732.

Staden, R. 1979. A strategy of DNA sequencing employing computer programs. *Nucleic Acids Res.*, **6** (7): 2601–10.

Struik, P.C., Cassman, K.G. and Koornneef, M. 2007. A dialogue on interdisciplinary collaboration to bridge the gap between plant genomics and crop sciences. p. 317-326. In: J.H.J. Spiertz, P.C. Struik and H. H.van Laar (eds), Scale and Complexity in Plant Systems Research: Gene-Plant-Crop Relations, Springer, The Netherlands.

The Potato Genome Sequencing Consortium. 2011. Genome sequence and analysis of the tuber crop potato. *Nature*, 475(7355): 189–195.

The Tomato Genome Consortium. 2012. The Tomato genome sequence provides insights into fleshy fruit evolution. *Nature*, 485(7400): 635–641.

Thelen, J.J. and Peck, S.C. 2007. Quantitative proteomics in plants: choices in abundance. *Plant Cell*, 19: 3339-3346.

Thompson, J. F. and Steinmann, K. E. 2010. Single-molecule sequencing with a HeliScope Genetic Analysis System. *Current Protocols in Molecular Biology*, 92: 1 – 14.

Thurston, G., Regan, S., Rampitsch, C., and Xing, T. 2005. Proteomic and phosphor proteomic approaches to understand plant–pathogen interactions. *Physiol. Mol. Plant Pathol.*, 66: 3–11.

Urano, K., Maruyama, K., Ogata, Y., Morishita, Y., Takeda, M. and Sakurai, N. 2009. Characterization of the ABA-regulated global responses to dehydration in Arabidopsis by metabolomics. *Plant J.*, 57: 1065 – 1078.

Varshney, R.K., Chen, W., Li, Y., Bharti, A.K., Saxena, R.K., Schlueter, J.A. et al. 2011. Draft genome sequence of pigeonpea (*Cajanus cajan*), an orphan legume crop of resource-poor farmers. *Nature Biotechnol.*, 30(1): 83–90.

Varshney, R.K., Nayak, S.N., May, G.D. and Jackson, S.A. 2009. Next-generation sequencing technologies and their implications for crop genetics and breeding. *Trends Biotechnol.*, 27(9): 522–530.

Varshney, R.K., Pandey, M.K., Janila, P., Nigam, S.N., Sudini, H., Gowda, M.V.C. et al. 2014. Marker-assisted introgression of a QTL region to improve rust resistance in three elite and popular varieties of peanut (*Arachis hypogaea* L.). *Theor. Appl. Genet.*, 127(8): 1771–1781.

Varshney, R.K., Shi, C., Thudi, M., Mariac, C., Wallace, J. et al. 2017. Pearl millet genome sequence provides a resource to improve agronomic traits in arid environments. *Nature Biotechnol.*, doi:10.1038/nbt.3943.

Varshney, R.K., Song, C., Saxena, R.K., Azam, S., Yu, S., Sharpe, A.G. and Cannon, S. 2013. Draft genome sequence of chickpea (*Cicer arietinum*) provides a resource for trait improvement. *Nature Biotechnol.*, 31(3): 240–248.

Velculescu, V.E., Vogelstein B. and Kinzler, K.W. 2000. Analysing uncharted transcriptomes with SAGE. *Trends Genet.*, 16: 423–425.

Washburn, M.P., Wolters, D. and Yates, J.R. 2001. Large-scale analysis of the yeast proteome by multidimensional protein identification technology. *Nature Biotechnol.*, 19: 242-247.

Whitelegge, J.P. 2002. Plant proteomics: blasting out of a Mud PIT. *PNAS*, 99:11564-11566.

Wilkins, M.R., Sanchez, J.C., Gooley, A.A., Appel, R.D., Humphery-Smith, I. and Hochstrasser, D. F. 1995. Progress with proteome projects: why all proteins expressed by a genome should be identified and how to do it. *Biotechnol. Genet. Eng. Rev.*, 13: 19–50.

Yu, J., Hu, S., Wang, J., Wong, K.S., Li, S., Liu, B. and Deng, Y. 2002. A draft sequence of the rice genome (*Oryza sativa* L. ssp. *indica*). *Science*, 296(5565): 79–92.

Zhao, Q., Chen, S. and Dai, S. 2013. C4 photosynthetic machinery: insights from maize chloroplast proteomics. *Front. Plant Sci.*, 4:85.

20

Crop Genomics for Food Security

Priti[1], Sumit Jangra[1*], Disha Kamboj[1], Neelam R. Yadav[1] and Ram C. Yadav[2]

[1]Department of Molecular Biology, Biotechnology & Bioinformatics, CCS Haryana Agricultural University, Hisar 125004, India

[2]Centre for Plant Biotechnology, CCS Haryana Agricultural University Campus, Hisar-125004, India

*Corresponding author: sumit.jangra712@gmail.com

Abstract

Making the world free of hunger and malnutrition is the ultimate goal of agricultural scientists. Decrease in the arable land and increasing global population has made it much more difficult. So, there is a need of advance techniques like genomics which will facilitate science based agricultural innovations such as development of nutrient rich and stress tolerant crop plants to eradicate hunger and malnutrition. Genomics or decoding the plant genome sequence using high-throughput techniques will allow scientific community to access agronomically important genes and will speed up the breeding programs for the development of superior varieties with higher yield and stress tolerance. Next generation genomics will help in the development of improved varieties which will be able to grow in harsh environmental and soil conditions without any compromise in the yield, which will lead to increase in farmers income and will make the world food secure.

Keywords: Genomics, food security, DNA sequencing, crop improvement and population.

Introduction

World population is increasing with high rate and is going to touch a mark of 9 billion by 2050 but the agricultural productivity is not able to cope with the increasing population. Although many countries have made significant progress in last few decades, but preponderance of famishment and malnutrition within the world population at dismaying rates challenges the global food security. According to FAO Hunger Report 2015, about 10.9% of the global population (one in nine people) is starving. Approximately 3.1 million children die every year due to hunger (http://www.worldhunger.org/world-child-hunger-facts/). Thus, seeing food security is the contiguous requisite for saving the global human potential, where the role of crop genomics is significant. Increasing agricultural productivity can increase access to food for people. So, the most populous countries like India has taken a step to increase land productivity. Green revolution, revolutionised the production of the major cereal crops with the introduction of semi-dwarf varieties of wheat and rice. But green revolution is not able to meet the demand of today's world. So, there is a need to focus on the development of high-yielding varieties. Agricultural productivity can be increased with the help of mechanisation, fertilisers, liming of soil acids to increase pH and to provide calcium and magnesium, herbicides, pesticides and development of high-yield varieties with the help of conventional breeding and genetic engineering. Conventional breeding has been contributing from a long time to crop improvement and development of high yielding varieties to meet the demand of increasing population. However, due to tapering of germplasm, identification of variability for incorporation into new cultivars is becoming more difficult. Therefore, there has been recourse to alternative approaches including mutagenesis, tissue culture and genetic transformation to aid breeding programs. Furthermore, close relatives of domesticated plants that is crop wild relatives represent gene pool that can be used for crop improvement by plant breeders. Since plant breeding is time-consuming and laborious, molecular breeding can speed up conventional breeding but still, there are some tasks which are unprocurable which can be achieved through genetic engineering. Genomics evolved with the advent of the invention of techniques of genetic engineering can generate data that can help to identify which region of genome needs to be targeted for crop improvement. Data generated from genomics will help in better understanding of gene expression, molecular and biochemical pathways which will aid plant breeders to entertain a different selection approach based on expression quantitative traits to maximise combinations of genes capable of conferring high performance. The advancements in high-throughput sequencing technologies have speed up the crop genomic research, role of genomic in ensuring food security is high lightened in this chapter.

Genomics

The term Genomics coined by Tom Roderick in 1986, is an interdisciplinary field of science which focuses on the study of genome of any organism. More specifically it refers to the analysis of the genome of organisms, both anatomically (sequences and organisation) and physiologically (expression and regulation) which further helps in deducing structure, function and mapping of genome (haploid set of chromosomes in a gamete or in each cell of multicellular organisms).

Research Areas

Structural Genomics

Branch of genomics that determines the three-dimensional structure of proteins encoded by a genome. This information can be helpful to find out where the desired gene resides in the genome.

Functional Genomics

Branch of genomics that determines the biological function of the genes and their products. The ability to find out gene and its function lay the basis of functional genomics. Various processes such as transcription, translation and epigenetics are investigated by functional genomics and answer when, where and how genes are expressed.

Epigenomics

It is the study of total gene expression that has not undergone any mutation in the DNA sequence. Various types of epigenetic changes like histone modification, DNA methylation, Nucleosome position occur in a cell. Epigenetic changes are reversible and heritable and change the overall chromatin structure.

Metagenomics

All the genetic material present in the environmental sample of the community of any organism is called metagenome. Metagenomics is the study of metagenome, genetic material recovered directly from the environment. Also called as environmental genomics, ecogenomics and community genomics. Metagenomics has made possible the study of non-culturable microbes.The phrase metagenome of soil first used by Handelsman *et al.*, (1998) to describe collective genome of soil microflora.

Genomics aid conventional breeding by gaining deeper insights into the biological mechanisms and can led to the development of new or improved screening methods for selecting superior genotypes more efficiently. These advances and development will provide an opportunity for efficient transfer of information systems from model species and major crops to orphan crops and will help in the development of crop plants with enhanced crop and productivity which will help in attaining food security.

Genome Sequencing Techniques

Background

DNA sequencing is the method of determining the actual arrangement of nucleotides within a DNA molecule. DNA sequencing includes any process or engineering which can be used to find out the order of the four bases: adenine, guanine, cytosine, and thymine within the DNA strand. Data generated from sequencing plays very crucial role in research perspectives as well as in a number of biological and its applied fields viz. forensics, biotechnology, virus studies, classification, medical science and many more. Initially, sequencing strategies were developed to sequence proteins but these were not applicable to nucleic acids. Because DNA molecule is larger and made up of fewer same units which make it difficult to sequence. Newer techniques were needed to sequence DNA. RNA was the first to be sequenced among nucleic acids, alanine tRNA from *Saccharomyces cerevisiae* was sequenced by Robert Holley and colleagues in 1965. The major milestone in RNA sequencing was the complete genome sequence of Bacteriophage in 1976 by Walter Fiers and coworkers. A huge number of efforts has been made to develop efficient DNA sequencing technology (fig. 1). Gilbert and Maxam in 1973 reported the sequence of Lac operator (24 bp) using wandering spot analysis. Frederick Sanger also followed this extension strategy and produced speedy method for sequencing DNA during his work at the MRC Centre, UK and released DNA sequencing with chain-terminating inhibitors in 1977 (Sanger *et al.*, 1977). Hood advanced the existing Sanger method of DNA sequencing which was becoming the common laboratory method and is the first semi-automated DNA sequencing machine in 1986. The first fully automated DNA sequencing machine (ABI 370) was marketed by Applied Biosystems in 1987. Revolution in DNA sequencing technology has brought down the cost of DNA sequencing and made the sequencing of an increased number of genomes both feasible and cost effective. The first plant genome Arabidopsis was completely sequenced in December 2000, and it was the third complete genome of a higher eukaryote. Subsequently, after Arabidopsis, several other crop plants have been sequenced (Sequence history of few important plants is represented in fig. 2).These genomes reveal numerous species-specific details, including genome size, gene number, patterns of sequence duplication, a catalogue of transposable elements, and syntenic relationships. To understand the complex instructions contained in all these raw sequence information of the plant genome, large-scale functional genomics projects are required. Progress towards a complete understanding of gene regulatory networks shared among many crop plants is important for improving cultivated species and for understanding different stress responsive mechanisms for crop improvement to increase productivity so that ample amount of food can be made available for growing population.

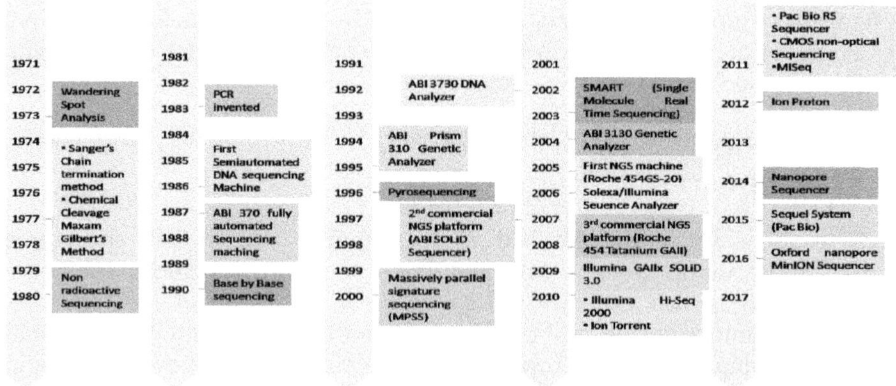

Fig. 1: Timeline representing the major breakthrough in DNA sequencing

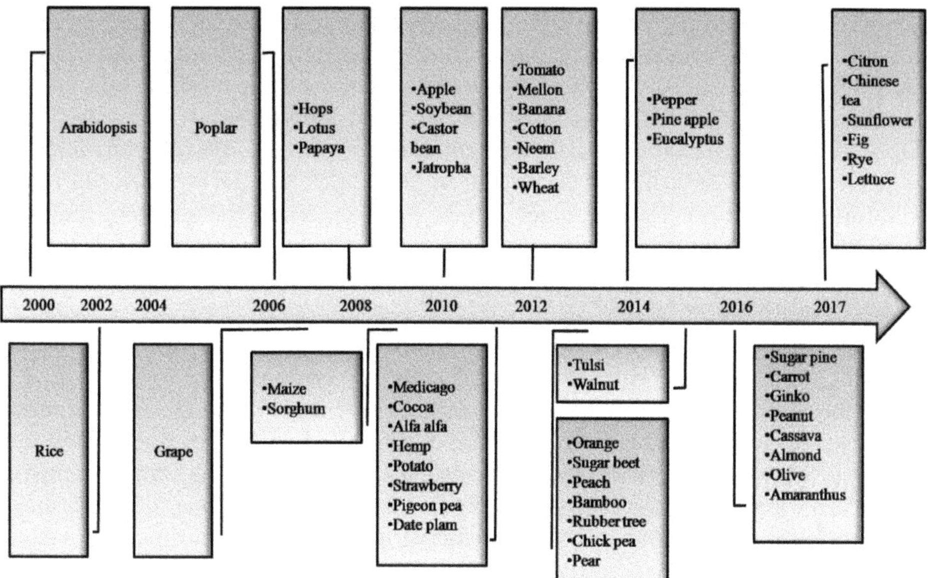

Fig. 2: Sequence history of some important plants

Types of DNA Sequencing Techniques

1. Sanger's method

The first DNA sequencing method devised by Sanger and Coulson in 1975 was called plus and minus sequencing that utilised *E. coli* DNA pol I and DNA polymerase from bacteriophage T4 with different limiting triphosphates. This method is also known as "Dideoxy Method". It prevailed from the 1980s until the mid-2000s. In this method, the synthetic nucleotides that lack the OH at the 3′ carbon atom are used for preparing DNA to be sequenced. When dideoxynucleotide added to the growing DNA, it stops chain elongation because there is no 3′ OH for the next nucleotide to be attached. The target DNA is assembled with the deoxynucleotides, tagged dideoxynucleotides and DNA polymerase I. These dideoxynucleotides are tagged such that each one fluorescence dissimilar colour. The elongation process goes on until it finds a dideoxynucleotide which leads to arrest the process. At the end, the fragments are separated according to length and the resolution is so good that a slight difference of single base pair is enough to separate that strand from the next shorter and next longer strand. When illuminated by laser beam, each of the four dideoxynucleotides fluorescence a give different colour and an automatic scanner provides a printout of the sequence.

2. Maxam-Gilbert method

This method required radioactive labelling at one 5′ end of the DNA (by kinase reaction using gamma-^{32}P ATP) and purification of the DNA fragment to be sequenced. Chemical treatment modifies the nucleotide base and then generates breaks at a small proportion of one or two of the four nucleotide bases in each of four reactions (G, A+G, C, C+T). The chemicals used for modification used in varied concentration such that they bring out only single change per molecule of DNA. As a result, a chain of labelled fragments is developed. Then, electrophoresis is performed for detachment according to size. Then the gel is disclosed under UV for visualisation. The sequence can be determined by the series of the band formation.

3. Pyrosequencing/Pal Nyren's Method

In 1996, Pal Nyren's group reported that natural nucleotide can be used to obtain efficient incorporation during a sequencing by synthesis protocol. Sequence detection is based on the release of pyrophosphate during the DNA polymerase activity, the pyrophosphate so generated is converted to ATP by sulfurylase and firefly luciferase converts this ATP to visible light which is recorded. Inclusion of dATPaS instead of dATP in the polymerization reaction was the first major improvement, which enables pyrosequencing to be done in real time. The second improvement was the introduction of apyrase to the reaction to make a four enzyme system. Apyrase allows nucleotides to be added sequentially without any intermediate washing step. In a cascade of enzymatic reaction, visible light is generated that is proportional to the number of incorporated nucleotides. The cascade starts with a nucleic acid polymerization reaction in which inorganic biphosphate (PPi) is released as a result of nucleotide incorporation by polymerase. The released PPi is subsequently converted to ATP by ATP sulfurylase, which provides the energy to luciferase to oxidise luciferin and generate light. The light so generated is captured by a CCD camera and recorded in the form of peaks known as pyrogram. Because the added nucleotide is known the sequence of the template can be determined.

3. Roche/454 FLX Pyrosequencer

The technology was commercialized in 2004 and is based on pyrosequencing. In this approach, the library fragments are mixed with agarose beads carrying oligonucleotides on the surface complementary to the 454-specific adapter sequences in the fragment library, so each bead is combined with a single fragment. The fragment:bead complexes isolated from oil:water-micelle comprises of the reacting agents of PCR. After thermal cycling, amplification of each DNA occurs on the bead surface. These single amplified DNA molecules are then sequenced. The process involves arrangement of beads into picotiter plate in such a way that each well has one bead in contact. Then, enzymes catalysing the reaction are added to each bead. The plate acts as a flow cell, which note downs every nucleotide's addition because of emission of light which is recorded by the CCD camera.

4. The Illumina genome analyzer

In this technique sequencing is performed by an automated device (Cluster Station), in this single molecule amplification step starts with an Illumina-specific adapter library. Sequencing takes place on the oligo-derivatized surface of a flow cell. The flow cell is an eight-channel sealed microfabricated device which let the bridge amplification of fragments to take place on its surface and employs DNA polymerase to make multiple copies or clusters of the single molecule that initiated the amplification. Same, separate or combination of both libraries can be added to each of the eight channels. Approximately one million copies of each fragment are present in a single cluster, which is enough for reporting bases at the required signal intensity for detection while sequencing. Sequencing-by-synthesis approach is utilised by Illumina system, in which all the four nucleotides are simultaneously added to the flow cell along with DNA polymerase. The nucleotides are labelled with a base-specific fluorescent label at 3-OH group. These steps are repeated for a specific number of cycles, as determined by user-defined instrument settings, which permits discrete read lengths of 25–35 bases. A base-calling algorithm assigns sequences and associated quality values to each read and a quality checking pipeline evaluates the Illumina data from each run, removing poor-quality sequences.

5. Applied Biosystems SOLiDTM system

The SOLiD (sequencing by oligonucleotide ligation and detection) platform uses an adapter-ligated fragment library other next-generation platforms, and uses an emulsion PCR approach with small magnetic beads to amplify the fragments for sequencing. This technology was developed by George Church in 2005 and was further improved and distributed by Applied Biosytem in 2007. The principle of this approach relies on DNA ligase to detect and incorporate bases in specific manner. Two flow cells are processed per instrument run, each of which can be divided to contain different libraries in up to four quadrants. Each Solid run requires 5 days and yield 2-4Gb DNA sequence data with average read length of between 25–35 bps. Once the reads are base called, have quality values, and low-quality sequences are removed, the reads are aligned to a reference genome to enable the second tier of quality evaluation called two-base encoding.

6. Helicos Heliscope TM

The Helicos sequencer, is based on work by Quake's group, which relies on cyclic interrogation of a dense array of sequencing features. No clonal amplification is needed in this method. A very sensitive fluorescence detection system is applied to directly find the DNA molecules through sequencing-by-synthesis method. The target DNA molecules libraries are devised by fragmenting randomly and addition of poly A tails. These are then hybridised with surface-attached poly T oligomers and are thus captured producing an irregular chain of single molecule sequencing templates. At each cycle, DNA polymerase and a single species of fluorescently labelled nucleotide are added, resulting in a template-dependent extension of the surface-immobilized, primer-template duplexes. Extension and imaging is permitted by the chemical cleavage and release of fluorescent labels after acquisition of images. An average read length of 25 bp or greater is generated by single base extension (i.e. A, G, C, T, A, G, C, T....) in several hundred cycles.

7. Pacific Biosciences SMRT

The Single Molecule Real Time Sequencing was developed by Pacific Biosciences. This approach involves the real-time monitoring of DNA polymerase activity. Nucleotide incorporations can potentially be detected through FRET (fluorescence resonance energy transfer) interactions between a fluorophore-bearing polymerase and gamma phosphate labelled nucleotides. This technique is ramped on two conceptions: zero-mode waveguides (ZMWs) and phosphor linked nucleotides. ZMWs helps in illuminating light at the bottom of the well where DNA polymerase-template complex is trapped. Phospho linked nucleotides grants recording of the trapped complex as DNA polymerase develops a DNA strand.

8. Nanopore-based Fourth Generation DNA Sequencing

This technology uses single-molecule technique which enables us to further examine the interaction between DNA and protein, as well as protein-protein interaction. Nanopore analysis opens a new door to molecular biology investigation at the single-molecule scale. The advantages of nanopores include label-free, ultra-longreads (104–106 bases), low cost, low material requirement and display results in real time, which simplifies theprocess and thus can be used for various DNA sequencing applications. Its principle is based on the fact that each base could produce different ionic current when DNA passed through nanopore, so it would be possible to distinguish different nucleotides. The nanometre-sized pores are in a biological membrane or formed in the solid-state film, which then separates two compartments containing conductive electrolytes and electrodes are emerged in each compartment. Upon applying voltage, electrolyte ions in solution move electrophoretically through the pore, thereby generating an ionic current signal. When the pore is blocked due to passage of biomolecule, current flowing through the nanopore would be blocked, interrupting the current signal. The physical and chemical properties of the target molecules can be determined by analyzing the amplitude and duration of current blockades from translocation events. The advantages and disadvantages of various DNA sequencing techniques is drafted in table 1.

Table 1: Comparison of various DNA sequencing techniques

Techniques	Read length	Accuracy	Reads per run	Time per run	Advantages	Disadvantages
Sanger Method	400-900bp	99.9%	NA	25 minutes to 3 hours	Long individual reads	Expensive and takes more time for plasmid cloning

Techniques	Read length	Accuracy	Reads per run	Time per run	Advantages	Disadvantages
Maxam and Gilbert Method	100 bp				DNA can be read directly, used to analyze DNA-protein interactions, nucleic acid structure and epigenetic modifications in DNA	extensive use of hazardous chemicals, cannot be used to analyze more than 500 base pairs, the read-length decreases due to incomplete cleavage reactions
Pyrosequencing (454)	700 bp	99.9%	1 million	24 hours	Fast and long read size	High cost, low throughput, homo-polymer errors
Illumina	50, 150, 250 or 300 bp	99.9%	1-25 million, 300 million -2 billion, 3 billion	6 hrs - 3 days	High sequence yield	Expensive, needs concentrated DNA
SOLiD	600 bases	99.9%	1.2-1.4 billion	1- 2weeks	Low cost	Slower, error during palindromic sequences
Helicose	32 nucleotides	99.5%	32 bases	8 days	Simplifies data analysis, avoids PCR bias	High error rate, time -consuming
SMRT	10,000 bp -15,000bp	87%	50,000 per cell	30 min - 4 hrs	Longest read length, fast, reliable	Costly, throughput is moderate
Nanopore-based sequencing	Depends on library size	92-97%	Selectedby user	1 min – 48 hrs	Portable, easy, very long read	Throughput is low, single read takes times

Genomics-Assisted Crop Improvement

Initially genomics research was focused on understanding the fundamentals of biology. To meet the demand of increasing population growth and to fight the climate changes, there has been growth in modernization of agricultural practices. Genomics is a rapidly expanding field of research fueled by reducing cost of DNA sequencing and genotyping (Edwards *et al.*, 2013). The complete and annotated DNA sequence of a plant genome canlocalise important traits that will be a valuable source for the plant breeders. The annotated DNA sequence of a plant genome informs about gene location, gene structure, distance between genes and locations of repetitive DNA sequence such as microsatellite and transposons. This information can further assist plant breeders to select effective breeding strategies. High throughput whole genome genotyping platforms such as VeraCode, Illumina Golden Gate, Infinium or DArT could be employed in breeding programs. High throughput screening methods could be developed based on genome sequence and by using genomics technologies, to identify desired progeny rapid and more efficiently than traditional methods. Accessibility of DNA sequence of the plant genome and transcriptome allows the development of DNA markers for marker-assisted breeding, allele mining, germplasm characterization, interpretation of gene function, analysis of genome evolution and study population genetics. Through these methods novel genes governing important traits can be found and could become new breeding targets for development of superior varieties to meet the global food requirement.

Genomics of Model Plants

Arabidopsis thaliana was the first plant and third multicellular organism whose complete genome was sequenced at the end of the year 2000. At that time, it was claimed that genome of *Arabidopsis thaliana* will help in the deeper understanding of plant development and environmental response and permits the structure and dynamics of plant genome to be accessed. The genome sequencing project began in 1996 with the formation of the Arabidopsis Genome Initiative, a consortium of international laboratories that utilised a BAC by BAC approach with Sanger sequencing to complete the genome sequence. The sequenced genome (135 Mb) is considered the gold standard for plant genomes due to the high quality and finished nature of the sequence.

Using the approach taken by Arabidopsis Genome Initiative, the genome of rice (*Oryza sativa*) was also sequenced by the International Rice Genome Sequencing Project (IRGSP). IRGSP was one of the few truly multinational plant genome projects with flags planted in chromosomes. The IRGSP was established in 1998 and taken up by ten nations [Chromosome 1 (Japan, Korea), Chromosome 2 (UK), Chromosome 3 (USA), Chromosome 4 (China), Chromosome 5 (Taiwan), Chromosome 6,7,8 (Japan), Chromosome 9 (Thailand& Canada), Chromosome 10 (USA), Chromosome 11 (USA & India) and Chromosome 12 (France &Brazil)] to obtain a complete finished quality sequence of rice genome (*Oryza sativa* L. sp.Japonica cv. Nipponbare). In December 2002 IRGSP released a high-quality map based draft sequence. This has permitted rice geneticists to identify several genes underlying traits and revealed very large and previously unknown segmental duplications that comprise 60% of the genome (Peterson *et al.*, 2014). The public sequence has also revealed new details about the syntenic relationships and gene mobility between rice, maize and sorghum(Paterson *et al.*, Salse *et al.*, and Lai *et al.*, 2004). Map based, finished quality sequence covers 95% of the 389 Mb genome, including all of the chromatin and two complete centromeres.

Nicotiana benthamiana a close relative of tobacco, widely used model for plant-microbe biology and other research applications. It is particularly useful because it is related to tomato, potato and will help in designing constructs for virus-induced gene silencing in order to reduce the possibility of 'off-target' gene silencing (Bombarley *et al.*, 2012). A new draft sequence of *N. benthamiana* genome has been released by researchers from the Boyce Thompson Institute for Plant Research (BTI) in 2012 and its estimated size was found to be 3Gb. Genome sequence of *Nicotiana slvestris* and *Nicotiana tomentosiformis* were released in 2013. Draft genomes of *N. sylvestris* and *N. tomentosiformis* were assembled to 82.9% and 71.6% of their expected size respectively.

Cajanus cajan, the first seed legume plant to have its complete genome sequenced in 2012. The sequencing efforts led by group of 31 Indian scientists from the Indian Council of Agricultural Research, it was then pursued by global research partnership, the International Initiative for Pigeon pea Genomics (IIPG), led by ICRISAT with partners such as BGI– Shenzhen (China), US research laboratories like University of Georgia, University of California-Davis, Cold Spring Harbor Laboratory and National Centre for Genome Resources, European research institutes like National University of Ireland, Galway, CGIAR Challenge Generation Programme and US National Science Foundation. Pigeon pea was considered an orphan crop but now significant amount of data has been generated, mainly because of the efforts by Indo-US Agricultural Knowledge Initiative (AKI), NSF and GCP-funded projects, (Varshney *et al.*, 2009, 2010a; Dutta *et al.*, 2011; Bohra *et al.*, 2011).

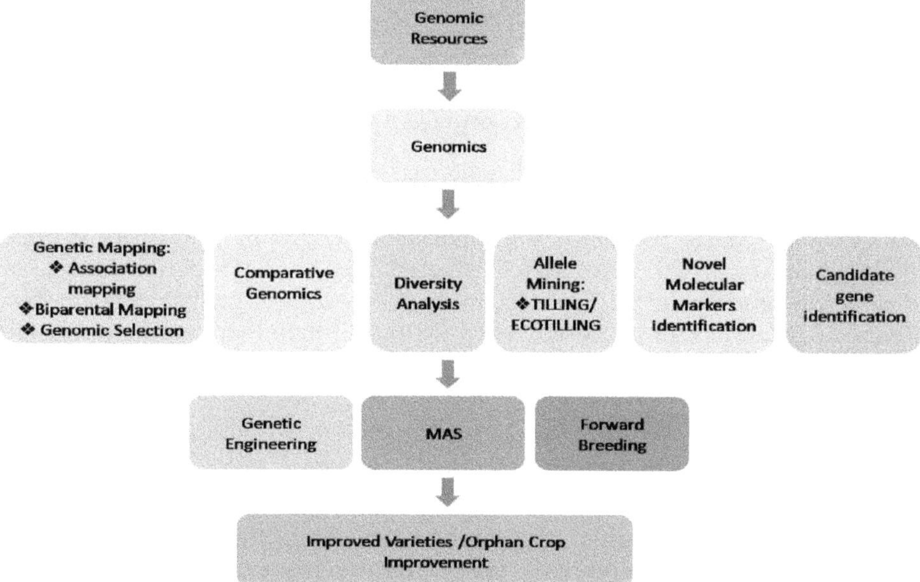

Fig. 3: Workflow of crop improvement through genomics

Medicago truncatula (~384 Mb) is a close relative of Alfalfa and is a preeminent model for the study of the processes of nitrogen fixation, symbiosis and legume genomics. The Medicago sequencing project began in 2003 with aim to elucidate sequences originates from the euchromatin portion of the genome. Among the eight chromosomes of Medicago, six were sequenced in US, chromosome no 5 was sequenced in France and chromosome no 3 was sequenced in the United Kingdom. In 2011 draft sequence was published based on BAC tilling path, supplemented with Illumina shotgun sequence (Mt3.5), together capturing 94% of all *M. truncatula* genes. In 2014 improved and more refined version of sequence was published based on *de novo* whole genome shotgun assembly of a majority of Illumina and 454 reads using ALLPATHS-LG.

Sequencing of model crop plants is not the end by itself but is just the beginning of a new venture to unravel genetic information and to gain better insights into the genetics of other species under investigation. The workflow of improvement of plant through genomics is depicted in fig. 3.

Mining of Genes for Agronomic Importance

With the progress in genome sequencing efforts for reducing cost, error rate and time have enabled to develop millions of novel markers, in non-model crop species, as well as identification of genes which are agronomically important. Identification of all genes within a species helps in understanding of how important agronomic traits are being governed, this information can further help in crop improvement. Biochemical function of these encoded proteins may be to gain greater understanding of the mechanism underlying the trait, whether change in gene structure or expression may further improve the trait. Knowledge of the gene responsible for trait may be further transferred to different cultivar by marker assisted selection or to species through genetic modification. While gene underlying the simple traits are simple to characterise at genome level, but there are many complex traits which are controlled by interacting gene networks.

Like related SNP-based functional molecular marker have been developed for SIDREB-2 gene involved in dehydration response in foxtail millet (Lata & Prasad, 2013). The EST-based SSR markers were developed for NBS-LRR R-gene in finger millet (Panwar *et al.*, 2011) to screen blast resistance. An allele-specific marker was developed based on the SNP detected in the eight exons of TaGW2 gene in wheat, to screen genotypes for increased kernel width and high thousand-kernel weight (Yang *et al.*, 2012)

Novel Molecular Markers

Various important and diverse agronomic traits coupled with genes and molecular markers which can be localized with availability of DNA sequence information creating new opportunities for crop improvement. Since 1980, with the discovery of very first molecular marker i.e. Restriction Fragment Length Polymorphism (RFLP), breeders employed various molecular markers in breeding programs like such as random amplified polymorphic DNA(RAPD), Amplified Fragment Length Polymorphism (AFLP), Sequence Tagged Sites (STS) and Simple Sequence Repeats (SSR). The major drawback with these random markers is that their prediction mainly depends on linkage between marker and target locus and for non-model crops it is difficult to develop sequence based markers due to lack of genomic information. Devlopment of next generation sequencing in transcrptomics creating coding sequence collection, which is much more cost effective has facilitated development of molecular markers from genic regions. As a result, functional molecular markers like genic SSRs, Single nucleotide polymorphism and other SSRs have been revealed and developed for non-model plant species. For example, genic SSRs are the functional molecular markers, derived from the primers which have been designed from the most conserved region of the gene (Varhsney *et al.*, 2005b).

Genic SSRs are electronically mined from publically available ESTs or gene sequence information using different bioinformatics softwares. The average frequency of SSRs in expressed regions of barley, sorghum, maize, rye, rice and wheat was reported to be 1 in every 6 kb (Varshney *et al.*, 2002, Varshney *et al.*, 2005c). The genic SSRs can be used in genetic mapping, functional diversity studies and these can be transferred among distantly related species (Varshney *et al.*, 2005c, Yu *et al.*, 2004). Like cssu45 a genic-SSR marker can distinguish the presence of Sr45 (stem rust resistance loci 45) across different genetic backgrounds in wheat (Periyannan *et al.*, 2014). Functional SNPs or insertions or deletions can cause direct phenotypic effect and such polymorphism are indispensable for the development of functional markers which can be further employed for crop improvement (Anderson & Lubberstedt, 2003; Bagge *et al.*, 2007; Domon *et al.*, 2004). Discovery of SNPs can further assist in GWAS using SNP array because we cannot afford whole genome sequencing in all individual plants. Traditionally SNP discovery involved PCR amplification of genes/genomic regions of interest from multiple individuals selected to represent diversity in the species or population of interest, followed by either direct sequencing of these amplicons, or the more expensive method of sequencing or cloning. Sequences are then aligned followed by screening for polymorphism. This approach is very expensive and time taking as large number of SNPs are required for most applications such as genetic mapping and association studies. SNP and SSR discovery through *in silico* methods are now being popular because of providing cheap and efficient methods for marker identification (Batley *et al.*, 2003; Robinson *etal.*, 2004; Jewell *et al.*, 2006; Duran *et al.*, 2009a,b). Large amount of sequencing data being produced in less time along more accuracy with latest sequencing technologies which provide a valuable resource for the mining of molecular markers (Imelfort *et al.*, 2009).

While next generation sequencing data produces a sequence data of quality, large amount of genome sequencing data makes it feasible to differentiate between true SNP and sequencing error. To identify genetic diversity in population and to study relationship between inherited genome and inherited traits, whole genome sequencing is the most potent method.

In plants one of the first application of next generation sequencing was identification of 36,000 maize SNPs using 260,000 and 280,000 EST, sequenced using Roche GS20. These SNP were identified between B73 and Mo17 inbred maize lines (Barbazuk *et al.,* 2007). Strict post processingof data reduced this number to >7000 putative SNP and over 85 % (94/110) of a sample of these SNPs were successfully validated by Sanger sequencing. Based on this validation rate, this pilot experiment conservatively identified >4900 valid SNP within >2400 maize genes, demonstrating the suitability and potential of the approach.

Genetic Mapping

Genetic mapping identify order and relative positions of molecular markers in linkage groups based on their pattern of inheritance. Genetic maps are prepared by analyzing populations derived from crosses of genetically diverse parents or from diverse line from natural environment. These markers can be further used to transfer gene of interest from donor to target genotypes.

Biparental QTL Mapping

Biparental mapping creates genetic maps on the basis of segregating population derived from biparental cross. The type of marker used for map construction should be present in enough density to increase resolution of the map. Genetic mapping now become robust method with the SNPs identified within whole genome sequence. This becomes possible to map a specific gene of interest and assist in the identification of linked or perfect markers for traits as well as increasing the density of markers on genetic maps (Rafalski, 2002).

The development of these markers also allows the integration of genetic and physical maps. By using molecular markers from related species allows the comparison of linkage maps. This allows the translation of information between model species with sequenced genomes and non-model species (Moore *et al.,* 1995). Furthermore, the integration of molecular marker data with phenomics, genomics and proteomics data allows researchers to link sequenced genome data with observed traits, bridging the genome to phenome divide. These markers can then be further employed in crop breeding programs.

Association Mapping

Association mapping (AM) identifies quantitative trait loci (QTL) by examining the marker-trait associations that can be attributed to the strength of linkage disequilibrium (LD) between markers and functional polymorphisms across a set of diverse germplasm. First reports of AM in plants emerged in 1996 in rice and 1997 in oat by Virk*et al.,* (1996) and Beer *et al.,* (1997) respectively. Association mapping is better than QTL mapping because of availability of broader genetic variations with wider background for marker trait correlation, high resolution, exploitation of historically measured trait data for association, time-saving and cost effective. In association mapping, unstructured populations represent many recombination events and are often many generations from common ancestor, providing the potential of great resolution for a particular set of population size. Sequencing led to development of large amount of molecular marker genotypic data which favours association studies over QTL mapping.

Conclusion

Recent advances in DNA sequencing technology has revolutionised crop improvement programs. Genomic tools reveal the information underlying plants responses to environmental stress and large inputs are needed to translate this information to climate resilient crops. Increased knowledge in crop genomics, along with reduced cost is expected to ease the sped-up betterment of orphan crops. Omics data generated from crop plants will allow us to address key agronomic traits that were difficult or impossible to address previously. Work on model plants has led to the refinement of technologies for later application to crop improvement. Genomics will help in identification of molecular markers linked to important agronomic traits which will help in the development of improved varieties with improved yield, quality, stress tolerance and disease resistance. All these will aid in making the world food secure.

References

Andersen, J. R. and Lübberstedt, T. 2003. Functional markers in plant. *Trends Plant Sci*, 8: 554-560.

Bagge, M., Xia, X. and Lubberstedt, T. 2007. Functional markers in wheat.*Curr Opin Plant Biol*, 10: 211–216.

Barbazuk, W.B., Emrich, S.J., Chen, H.D., Li, L. and Schnable, P.S. 2007. SNP discovery via 454 transcriptome sequencing. *Plant J*, 51: 910–918.

Batley, J., Barker, G., O'Sullivan, H., Edwards, K. J. and Edwards, D. 2003.Mining for single nucleotide polymorphisms and insertions/deletions in maize expressed sequence tag data. *Plant Physiol*, 132: 84–91.

Beer, S.C., Siripoonwiwat, W., O'Donoughue, L.S., Souza, E., Matthews, D. and Sorrells, M.E. 1997. Associations between molecular markers and quantitative traits in an oat germplasm pool: Can we infer linkages? *J. Agric. Genom*, 3: 1-16.

Bohra, A., Dubey, A., Saxena, R. K., Varma, P. R., Poornima, K. N., Kumar, N., Farmer, A. D., Srivani, G., Upadhyaya, H. D., Gothalwal, R., Ramesh, S., Singh, D., Saxena, K. B., Kavikishor, P. B., Singh, N. K., Town, C. D., May, G. D., Cook, D. R. and Varshney, R. K. 2011. Analysis of BAC-end sequences (BESs) and development of BES-SSR markers for genetic mapping and hybrid purity assessment in pigeonpea (*Cajanus* spp.). *BMC Plant Biol*, 11(56): 1-15.

Bombarely, A., Rosli, H.G., Vrebalov, J., Moffett, P., Mueller, L. A. and Martin, G.B. 2012. A draft genome sequence of *Nicotiana benthamiana* to enhance molecular plant- microbe biology research. *Mol Plant Microbe Interact*, 25(12): 1523-1530.

Domon, E., Yanagisawa, T., Saito, A. and Takeda, K. 2004. Single nucleotide polymorphism genotyping of the barley waxy gene by polymerase chain reaction with confronting two-pair primers. *Plant Breed,*123: 225–228.

Duran, C., Appleby, N., Clark, T., Wood, D., Imelfort, M., Batley, J. and Edwards, D. 2009a. Auto SNPdb: an annotated single nucleotide polymorphism database for crop plants. *Nucleic Acids Res,* 37: 951–953.

Duran, C., Appleby, N., Vardy, M., Imelfort, M., Edwards, D. and Batley, J. 2009b. Single nucleotide polymorphism discovery in barley using autoSNPdb. *Plant Biotechnol. J,* 7: 326–333.

Dutta, S., Kumawat, G., Singh, B. P., Gupta, D. K., Singh, S., Dogra, V., Gaikwad, K., Sharma, T. R., Raje, R. S., Bandhopadhya, T. K., Datta, S., Singh, M. N., Fakrudin, B., Kulwal, P., Wanjari, K. B., Varshney, R. K., Cook, D. R., Singh, N. K. 2011. Development of genic-SSR markers by deep transcriptome sequencing in pigeonpea [*Cajanus cajan* (L.) Millspaugh]. *BMC Plant Biol,* 11(17): 1-13.

Edwards, D., Batley, J. and Snowdon, R. J. 2013. Accessing complex crop genomes with next-generation sequencing. *Theor Appl Genet,*126: 1-11.

Gilbert, W. and Maxam, A. 1973. The Nucleotide Sequence of the lac Operator. *Proc. Natl Acad. Sci,*70(12): 3581–3584.

Guyot, R. and Keller, B. 2004. Ancestral genome duplication in rice. *Genome,* 47: 610–614.

Handelsman, J., Rondon, M. R., Brady, S. F., Clardy, J. and Goodman, R. M. 1998. "Molecular biological access to the chemistry of unknown soil microbes: A new frontier for natural products". *Chem. Biol,*5 (10): R245–R249.

Imelfort, M., Batley, J., Grimmond, S., and Edwards, D. 2009. Genome sequencing approaches and successes. *Methods Mol. Biol,* 513: 345–358.

Jewell, E., Robinson, A., Savage, D., Erwin, T., Love, C.G., Lim, G.A.C., Li, X., Batley, J., Spangenberg, G.C. and Edwards, D. 2006.SSR Primer and SSR taxonomy tree: Biome SSR discovery. *Nucleic Acids Res,* 34: 656–659.

Lai, J., Ma, J., Swigonova, Z., Ramakrishna, W., Linton, E., Llaca, V., Tanyolac, B., Park, Y.J., Jeong, O.Y., Bennetzen, J.L. and Messing, J. 2004. Gene loss and movement in the maize genome. *Genome Res,*14: 1924–1931.

Lata, C. and Prasad, M. 2013. Validation of an allele-specific marker associated with dehydration stress tolerance in a core set of foxtail millet accessions. *Plant Breed,* 132: 496–469.

Moore, G., Devos, K. M., Wang, Z. and Gale, M. D. 1995.Cereal genome evolution. Grasses, line up and form a circle. *Curr. Biol,*5: 737–739.

Panwar, P., Jha, A., Pandey P. K. and Kumar, A. 2011. Functional markers based molecular characterization and cloning of resistance gene analogs encoding NBS-LRR disease resistance proteins in finger millet (*Eleusine coracana*). *Mol Biol Rep,* 38: 3427–3436.

Paterson, A.H., Bowers, J.E. and Chapman, B.A. 2004. Ancient polyploidization predating divergence of the cereals, and its consequences for comparative genomics. *Proc. Natl Acad. Sci,* 101: 9903–9908.

Periyannan, S., Bansal, U., Bariana, H., Deal, K., luo, M. C., Dvorak, J. and Lagudah, E. (2014). Identification of a robust molecular marker for the detection of the stem rust resistance gene Sr45 in common wheat. *Theor. Appl. Genet,* 127 (4): 947-955.

Rafalski, A. 2002. Applications of single nucleotide polymorphisms in crop genetics. *Curr. Opin.Plant Biol,*5: 94–100.

Robinson, A.J., Love, C.G., Batley, J., Barker, G. and Edwards, D. 2004.Simple sequence repeat marker loci discovery using SSR primer. *Bioinformatics,* 20: 1475–1476.

Salse, J., Piegu, B., Cooke, R. and Delseny, M. 2004. New *in silico* insight in to the synteny between rice (*Oryza sativa* L.) and maize (*Zea mays* L.) highlights reshuffling and identifies new duplications in the rice genome. *Plant J,* 38(3): 396-409.

Sanger, F., Nicklen, S. and Coulson, A. R. 1977. DNA sequencing with chain-terminating inhibitors. *Proc. Natl Acad. Sci,* 74(12): 5463–5467.

The 3,000 Rice Genome Project: The Rice 3,000 Genome Project. GigaScience Database. 2014. *http://dx.doi.org/10.5524/200001*

Varshney, R. K., Close, T. J. Singh, N. K., Hoisington, D. A. Cook, D. R. 2009. Orphan legume crops enter the genomics era. Curr opin Plant Biol, 12:202-210.

Varshney, R. K., Graner, A. and Sorrells, M. E. 2005c. Genic microsatellitemarkers in plants: features and applications. *Trends Biotechnol,* 23:48–55.

Varshney, R. K., Sigmund, R., Borner, A., Korzunb, V., Steina, N., Sorrellsc, M. E., Langridged, P. and Graner, A. 2005b. Inter-specific transferability and comparative mapping of barley EST-SSR markers in wheat, rye and rice. *Plant Sci,* 168: 195–202.

Varshney, R. K., Thiel, T., Stein, N., Langridge, P. and Graner, A. 2002. In silico analysis onfrequency and distribution of microsatellites in ESTs of some cereal species. *Cell Mol Biol Lett,*7:537–546.

Virk, P. S., B. V. Ford-Lloyd, M. T. Jackson, H. S. Pooni, T. P. Clemeno *et al.,* 1996. Predicting quantitative variation within rice germplasm using molecular markers. *Heredity,*76: 296–304.

Yang, Z., Bai, Z., Li, X., Wang, P., Wu, Q., Yang, L., Li, L. and Li, X. 2012. SNP identification and allelic-specific PCR markers development for TaGW2, a gene linked to wheat kernel weight. *Theor Appl Genet,* 125: 1057–68.

Yu, J., Fleming, S.L., Williams, B., Williams, E.V., Li, Z., Somma, P., Rieder, C.L., Goldberg, M.L. 2004. Greatwall kinase: a nuclear protein required for proper chromosome condensation and mitotic progression in Drosophila. *J. Cell Biol,* 164(4): 487-492.

21

Energy Sustainability: Role of Omics Technologies in Bio-fuel Production

Surinder Singh*, Shivappa Hukkeri, Manu Pratap Gangola and Bharathi Rajaramadoss

Department of Plant Sciences, University of Saskatchewan, 51 Campus Drive, Saskatoon, Saskatchewan S7N 5A8, Canada.

**Corresponding author: singh.surinder13@gmail.com*

Abstract

The luxurious life style of modern society is increasingly dependent on the abundant supply of fuel energy. However, depletion of petroleum based fuel and their ill effects on environment have promoted renewable/sustainable alternative energy sources such as biofuels. The demand for sustainable energy ranks as one of the most demanding concern of the 21st century, necessary for the human civilization's energy future. Climate change and energy security are the major driving forces for worldwide development of biofuels which is also a potential source for agricultural economic returns.The key to the future sustainable biofuel production lies in the continuous supply of starting materials (plant material, oil seeds and microorganism) without competing with food production. In this chapter, we tried to elaborate on renewable energy sources by focusing on bioenergy crops, to microalgae,and to the utilization of high-throughput "omics" technologies in advancing our understanding of biofuel production. Among the "omics" technologies, we focussed on importance of the emerging genomics, transcriptomics, proteomics and metabolomics aspects to enhance and improve the biofuel production. With the dwindling fossil fuel reserves, rising crude oil prices and heightening fears over the effects of climate change, there is an urgent need to promote the use of biofuel energy for a sustainable future. Therefore,"omics"technologies will play a key role in achieving sustainable energy supply for future generations.

Keywords: Biofuels, Sustainability, Omics, Transgenic crops, Gene editing

Introduction

According to the United Nations, sustainability is defined as "meeting the needs of the present without compromising the ability of future generations to meet their own needs." The demand for sustainable energy is the most pressing concern of the 21st century world. There is an urgent need to promote the use of renewable/alternative energy sources for a sustainable future, especially due to the diminishing fossil fuel reserves, rising crude oil prices and their bad effects on climate.

Bioenergy, an environmentally safe and sustainable source of energy refers to the energy obtained from biological materials, specifically photosynthetic organisms such as green plants, grasses, and algae. Biofuels are the major category of sustainable bioenergy sources, readily available in this era. The initial surge of biofuels in industrial economies was driven by energy security and rising fossil fuel prices. The development of biofuels, emerging at the interface of agriculture and energy at the global level, has been one of the most significant agricultural development in recent years. It has been previously proposed that although a system that would combine energy derived from wind, water and sun will be attractive for global energy demands, however, fuels with high energy density like biofuels will still be a major component in the future to power large machinery, planes, and ships (Jacobson & Delucchi 2011). Currently, biofuel production is mainly in the form of bioethanol and biodiesel, derived from food crops. It is estimated that between 2013 and 2015, about 77% of produced bioethanol was based on the processing of maize and sugarcane; while 81% of biodiesel was obtained from vegetable oils (OECDFAO Agricultural Outlook 2016). Burning biofuels are more environmental friendly as they release carbon that has already been fixed by plants through photosynthesis and thus, theoretically, should not increase the net atmospheric CO_2 content (Naik 2010). While food and fuel production have often been being in conflict, there is actually substantial potential to boost food and fuel production simultaneously. This is because as food production expands to meet the nutritional needs of growing populations, there is also increased production of agricultural residues for biofuel production.

Apart from traditional breeding modern OMICs techniques is crucial in delineating the mechanisms and escalating the potential of biofuels from plants, oil seeds and algae. Researchers are continuously working in refining biofuel production using transgenics, as an efficient alternative to replace non-renewable fuels. This chapter will discuss the role and recent advances in contemporary techniques involving genomics, transcriptomics, proteomics and metabolomics in biofuel production and improvements.

Biofuels Overview

The imminent decline of the world's oil production, high market prices, irregularity in distribution and environmental impacts have made the production of biofuels to reach an unprecedented volume over last 20 years. Biofuels derived from biomass are either in the form of solid, liquid or gas. Solid fuels refer to numerous types of solid materials used as fuels that produce energy through combustion including peat, wood, coal, charcoal, and pellets. These solid fuels fall under the category of primary bio-fuels (Nigam & Singh, 2011, Lin *et al.*, 2011; Huang *et al.*, 2010) peroid on the other hand, the secondary fuels are liquid or gaseous fuels used for transportation enginesand are further classified as 1st, $2^{nd}$and 3^{rd}generation fuels (Nigam & Singh, 2011)

The "first generation" biofuels are made from sugar, starch, vegetable oil, or animal fats using conventional technology to produce biodiesel, bioethanol and biogas. The basic feedstocks to produce first-generation biofuels come from agriculture and food processing. Globally, in 2014 approximately 95 billion litres of bioethanol and 30 billionlitres of biodiesel was produced, equating to roughly 3.6% of gasoline supply and 1.5% of diesel supply, achieved by fermentation of sugars to produce bio-ethanol (feedstock: sweet Sorghum, sugar cane, sweet potato, sugar beet etc.), lipids derived from algae and other oil crops (feedstock: Jatropha, Pongamia, castor, sunflower, oil palm, etc.) and using syngas obtained from gasification of biomass.To date, liquid biofuels have been produced mainly in the USA, Brazil, and European nations, who have set up to diversify transport fuel supplies and improve energy security.

"Second-generation"biofuels use lignocellulosic feedstocks such as farm and forest residues, grasses and trees. These feedstocks have potentially high yields,sequester higher carbon, and can grow on land poorly suitable for food crops cultivation. These feedstocks are converted to biofuels using biochemical and thermochemical technologies.

"Third generation" biofuels, from algae, are the most advanced category of biofuels, and has been subjected to modern omics tools to enhance the production. They can grow on much less land area while producing a variety of other useful co-products. Due to the technical, economic and environmental promise, both second- and third-generation biofuels are the focus of intense research and development in several countries.

Role of OMICS in Modern Biofuels

Technologies that measure various characteristics of large family of cellular molecules, such as genes, transcripts, proteins, or small metabolites, have been named by appending the suffix "-omics," as in genomics- the study of genes and their function, transcriptomics- the study of the mRNA transcripts, proteomics- the study of proteins, metabolomics- study of metabolites. *Omics* refers to the collective technologies used to explore the roles, relationships, and actions of the various types of molecules that make up the cells of an organism. Omics studies aimed to understand the mechanisms of the activities of an organism, are important for modern biofuel industry. The traits important to biofuels can be improved using two approaches; either using conventional breeding which involvescrossing and selection, aided by identification of molecular markers for marker-assisted selection

(Price, 2006). The alternative approach utilizes transgenic or genetic modification to introduce new genes, modifyexisting genes, or interfere with gene expression (Torney *et al.*,2007). For both these approaches, advances in omics, whole-genome sequencing,molecular mapping, and bioinformatics providesa powerful strategy for gene discovery.

A considerable research has been carried out in the last decade to understand the molecular mechanisms of biofuel plants. Rapidly developing genomics and post-genomics, systems biology approaches such as transcriptomics, proteomics and metabolomics have become essential for understanding the response and adaptability of plants to changes in their environment and further in yield improvement. The utilization of such high-throughput approaches will lead to better and high-yielding biomass feedstocks which will eventually facilitate the acceleration in production andcommercialization of biofuels. Currently, genomic resources in the form of whole-genome information forseveral bioenergy crops which can be used as models, is available and accessible to public domains (Table. 1).

Table 1: Genome information for some important biofuel producing crops/ organisms.

Biofuel crop/organism	Genome size (Mbp)	Biofuel gen.	Web link for whole genome sequencing information
Sorghum bicolor (sorghum)	730	2nd	http://www.plantgdb.org/SbGDB/
Sachharumofficinarum (sugar cane)	~1000 (monoploid)	1st	http://sugarcanegenome.org/
Zea mays (maize)	2500	2nd	http://www.maizegdb.org/
Panicumvirga-tum (switchgrass)	1500 (haploid)	2nd	http://switchgrassgenomics.org/
Jatropha curcas(physic nut)	416	1st	http://www.kazusa.or.jp/jatropha/
Glycine max (soybean)	1150	1st	http://www.phytozome.net/soybean
Triticumaestivum(wheat)	17,000	2nd	http://www.wheatgenome.org/
Brassica rapa(canola)	530	1st	http://www.brassica.info/resource/sequencing.php
Helianthus annuus(sunflower)	3600	1st	https://www.sunflowergenome.org/
Botryococcusbraunii (Green microalga)	166	3rd	http://jgi.doe.gov/why-sequence-botryococcus-braunii/

The bio-resources that are involved in bioenergy production include plant and algal resources thatcan provide substrates for energy generation when utilized by appropriate microorganisms, which facilitate bioconversion of these substrates. Therefore, microorganisms have a key role to play in this new energy future.

Traditionally, this has been the way of fermentation, which is a relatively simple bioconversion of sugars to produce combustible solvents such as ethanol. Other roles include the production of various hydrolytic enzymes that may assist in conversion of non-fermentable and recalcitrant feedstock into fermentable products and the generation of new enzymes for biodiesel production. Furthermore, with developments in our understanding of cell metabolism and regulation, there is much greater capacity to manipulate microorganisms in a targeted way through the omics approaches. A range of tools are available to deepen our understanding of microbial molecular processes and create modifications to microorganisms that improve performance for anapplication. Starting from wild-type, modified strains can be created by cycles of mutation followed by screening, cross-breeding of haploid strains with desired features for cells with a sexual cycle (e.g. yeasts), genetic modification via plasmid transfection or genome editing to knock in/out particular genes (Marquis, 2016).

There are numerous studies on the engineering of bacteria, such as *E. coli, Z. mobilis,* and *P. putida,* and yeasts for applications in ethanol production (Neilsen *et al.,* 2013, Chen 2016). Compelling reasons to choose alternate hosts; the availability of a much wider range of genetic tools for modification and deeper understanding of their respective genomes.

Recently, the advancements in molecular assisted breeding, mutation breeding, genetic manipulation and/or the next generation sequencing methodologies has resulted in sequencing of whole genomes of various biofuel crops in short duration, precisely and inexpensively. Hence, Omics technologies enabled scientific community to simultaneously measure all the components of a biological system: genes (genomics), transcripts (transcriptomics), proteins (proteomics), metabolites (metabolomics), and phenotypes (phenomics)-to understand molecular and biochemical mechanisms that constrain biological functions. Together, these methods have revolutionized the study of organisms both in lab and in natural habitats. These biotechnological advancements have been complemented by developments in computer sciences, creating the new field of bioinformatics where powerful new databases and search algorithms are helping biologists share and build upon experimental results in ways and timescales that were never possible.

Genomics and Transcriptomics Studies to Improve Biofuels

Among the first-generation biofuels, storage oils derived from plant seeds in the form of triacylglycerols (TAGs) are excellent sources for generating biodiesel, due to their high chemical similarity to fossil fuels. Current first-generationbiofuel crops, such as soybean and *Jatropha curcas,* have either low or unpredictable oil yields (Fairless, 2007). Thus, increasing oil content in plants and redirecting the biosynthesis of fatty acids for accumulation of specific types are needed to achieve optimal biodiesel production. Genomic tools to overexpress fatty acid metabolism genes to increase the oil content have been applied earlier. For example, overexpression of a fungal DAGT2 (diacylglycerol acyltransferase) enzyme in soybean seeds led to a 1.5% increase in oil content (Lardizabal *et al.*, 2008). A similar phenotype was observed by overexpression of a *DGAT* cDNA in *Arabidopsis* (Jako *et al.*, 2001). Also, overexpression of the yeast glycerol 3-phosphate dehydrogenase (*ghpd1*) gene in canola seeds increased the lipid content by 40% (Vigeolas *et al.*, 2007). Advancement in genomics and availability of genetic data bases for various oil crops are helping the molecular breeders and biologists to explore the oil biosynthesis pathways. Recently, Soybean genome was sequenced in 2010 (Schmutz *et al.*,2010) and many genetic resources such as genetic maps, expression data, mutants and gene annotation tools are made available on https://www.soybase.org/ (accessed 27 June 2017). These tools are being used by researchers to further study and improve the genes involved in increasing oil content.

Apart from soybean, *Jatropha curcas* has been recently undergone an extensive research to functionally characterize the genes for increased oil production, and simultaneous profiling of gene expression products. In addition, with the advancement of genome sequencing in *J. curcas*, the expression of genes involved in the biosynthetic pathways have become more feasible, which should be beneficial for selective breeding of *J. curcas*, especially for oil quality, yield and low toxins (Carels, 2013). In addition to genomics, transcriptomic analyses of genes involved especially in fatty acid (FA) biosynthesis can provide fundamental molecular understanding of synthesis and storage of lipids and proteins in *Jatropha* seeds. Studies of the expression of *Jatropha* KAS III in different tissues revealed the highest expression in roots and in developing seeds (Li*et al.*, 2008). In addition, genes involved in the biosynthesis pathway of Triglycerides (TAGs) are of great interest to improve oil accumulation in *Jatropha* seeds throughgenetic modifications (Bahadur*et al.*,2013).

More recently a bioenergy crop *Camelina sativa*, which can grow well on marginal land with low inputs like less water, fertilizers and pesticides has emerged as a promising low cost renewable source for biodiesel and industrial bio-products. Its genome was sequenced in 2014, keeping in mind its potential as an industrial oil producing crop (Kagale *et al.*, 2014). The future studies, would aim at genetic engineering which can be achieved with relative ease and in a shorter time compared in *Camelina* as to other oilseed crops (Nguyen *et al.*,2013). The key areas of genomic investigation in optimizing the lignocellulosic biomass for 2nd generation biofuel production mainly depends onavailability of fuelstocks with modifiable traits, where enzymatic activities can be engineered for biorefining (Sarath.*et al.*, 2008).Grasses possessing C4 pathway, such as switch grass, sorghum, and *Miscanthus* are some of the best options for bioethanol feedstocks. For example, the U.S. Department of Energy's Bioenergy Feedstock Development Program identified switchgrass as one of the most promising energy feedstocks because of its yield rate, ability to grow from seed on marginal lands, capacity to protect soil from erosion, and potential to grow without fertilizer or replanting (Bouton, 2007). Recently, the Joint Genome Institute has sequenced the switch grass genome in 2013 (Phytozome v9.1: Panicum virgatum)and database is available for genetic improvement.

In yet another bioenergy feed stock for 2nd generation biofuel; Sorghum, the molecularmarkersystemhasbeenextensivelystudied(reviewedinMadhusudhana, 2015), with various linkage and QTL maps constructed for exploring the bioenergy production mechanism. The QTLs related to biofuel-traits wereintegrated with the whole genome sequence information providing a useful resource for designing efficient strategies for marker-assisted breeding. Further,mutant genes in the lignin biosynthesis (*bmr*genes) pathway was introgressed into elite sweet sorghum lines that results in the development of dual-purpose bioenergy sorghums, producing juice for the ethanol production and bagasse for the second-generation biofuel (Maehara *et al.*, 2011) Genomic tools has been further applied to traits like plant height, leaf morphology (photosynthetic rate), root architecture (water uptake efficiency) and other biotic and abiotic factors that impact productivity and traits for biofuel production. (Mathur *et al.*, 2017). These genomic tools are helping to elucidate the function of various genes using transcriptomics particularly in sugar/ starch metabolic pathways that could lead to a higher biomass.Corn (Zea mays), another C_4 plant has its genome well characterized, sequenced and mapped, and has potentially greater N-use and water–use efficiencies, thus also a promising biofuel feedstock. Corn stover provides a considerable source of lignocellulosic biomass that can be efficiently used as feedstock for ethanol and other biofuels while delivering the agronomic benefits. Modern genomic tools like marker assisted selection and transgenic approaches are the current strategies for improvement of these traits. Optimizing water and nutrient uptake using genomic tools will improve the sustainability of maize production as bioenergy crop. Further genetic diversity of maize provides many choices of target gene modification (such as cell wall genes) using contemporary gene editing tools (discussed later in this chapter).

Among third generation biofuels which include specially engineered energy crops such as algae, sometimes called oilgae have a potential 30 times the energy per unit area as compared to previous generation biofuels. Algae can play a significant role in providing biomass in areas not suitable to traditional agriculture or where unique resource utilization supports a mix of feedstocks. Phytozome (phytozome.net), a comparative hub for plant genome and gene family data and analysis, provides sequence and functional annotations for several strains of algae sequenced at the Joint Genome Institute, as well as selected species sequenced elsewhere. Recently two publicly available omics-based databases were developed: the Algal Functional Annotation Tool and 'The Greenhouse'. Further development of these tools will become vital as new species are sequenced.In order to aid with identifying algal proteins of unknown function, researchers at the University of California, Los Angeles developed the Algal Functional Annotation Tool (pathways.mcdb.ucla.edu/algal/index.html). This is a bioinformatics resource to visualize pathway maps, determine biological search terms, or identify genes to elucidate biological function *in-silico*. These types of analyses have been designed to support lists of gene identifiers, such as those coming from transcriptome gene-expression analysis.Similarly, 'The Greenhouse', developed by Los Alamos National Laboratory scientists, (greenhouse.lanl.gov/) provides a centralized website to display and share sequence-based data relevant to the improvement and advancement of algal biofuel feedstocks. This database seeks to provide consistent annotations across all algal species and strains; searchable data based on taxonomy, gene name, locus tag, protein function/families, pathways, Enzyme Commission numbers, and others; BLAST (Basic Local Alignment Search Tool)-searchable algal genomic databases; numerous comparative analyses and interactive visualizations; exportable sequences and information; data in a variety of formats; and user workspaces to allow manipulation of data within customized groups.

Apart, from the above-mentioned genomics resources for algae, a wide assortment of genetic tools has been developed specially for *Chlamydomona sreinhardtii*, including fluorescent tags and reporter genes for both nuclear and chloroplast expression (Rasala *et al.*, 2013; Kumar *et al.*, 2013), targeted expression constructs that link expression of the gene-of-interest to a selection gene (Rasala *et al.*, 2014), and nuclear gene editing and activation (Gao *et al.*, 2015 Jinkerson and Jonikas 2015). Transformation and expression in cyanobacteria is having no more impediments in regenerating the cells (Koksharova and Wolk 2002). Few robust molecular toolboxes have been developed for genetic engineering of algae strains, to avoid the difficulty involved in selecting native promoters and terminators sequences for expression and successful transformation (Georgianna *et al.*, 2013; Radakovits *et al.*, 2012). Although many transformation procedures werepublished, not all vectors or their sequence are publicly available (DOE, 2016) peroid while genome sequencing is an important component of any algal biofuels development, quantitative transcriptome profiling using high-throughput sequencing technologies are also become increasingly important sincetheyare not only aid with genome annotation (e.g., identifying coding regions of DNA), but also emerging as a robust approach for genome-wide expression analysis in response to environmental conditions. After gene identification, several transcriptomic profiling studies of potential production strains of microalgae have been completed to elucidate gene expression under various nutrients (such as nitrogen, phosphorous, or silicon) deprivation conditions (Jia *et al.*, 2015; Carpinelli *et al.*, 2014; Mansfeldt *et al.*, 2016), at different growth phases, under heterotrophic and autotrophic growth (Olivares, 2014), and many other conditions such as under varying light treatments, growth phase, nutrition, and heavy metal stress (Olivares, 2014). These transcriptomics studies aim at understanding the expression profiles and further improving the biofuel potential of an algae.

Modern Gene Editing Tool for Biofuels

Genome editing can accelerate biomass production required for biofuel feedstock by allowing the introduction of precise and predictable modifications through CRISPR/Cas9 system is particularly beneficial because multiple traits can be modified simultaneously (Congo *et al.*,2013). CRISPR-Cas9 is a ground breaking technique that enabled scientists to make precise targeted changes in living cells. Unlike traditional gene-editing methods (RNAi, TALENs, ZFNs and others; not to be discussed here), it is cheap, easy to use and effective in diverse organisms. The oleaginous (oil-producing) yeast *Yarrowia lipolytica*, which is known for converting sugars to lipids and hydrocarbons that are difficult to make synthetically have been edited using CRISPR-Cas9 technology (Schwartz *et al.*, 2016). *Saccharomyces cerevisia* strains has also been engineered using CRISPR-Cas9 for ethanol fermentation, which uses food biomass (corn starch, sugarcane) or lignocellulosic feedstocks (straw, corn stover, wood). Development of CRISPR-Cas9 in thermophilic fungi *Myceliophthora thermophile* (biomass-degrading fungus) has attracted industrial interest for the production of efficient thermostable enzymes such as lignocelluloses, required to breakdown plant biomass into fermentable sugars for biofuel production (Liu *et al.*,2017). Recently Ajjawi *et al* (2017) has demonstrated to double the production of lipid from *Nannochloropsisgaditana* using editing of a single transcriptional regulator that will enable the commercialization of microalgal derived biofuels.

Though genomics and transcriptomics reveals the genes and the transcripts involved in cellular regulatory mechanism, the post transcriptional regulations through proteins and the end products of cellular metabolism components such as metabolites are also important to understand the cellular mechanism of biofuel biosynthesis.

Proteomic Profiling of an Organism to Improve Biofuel Production

The productivity of biofuels especially second and third generation has improved in the recent-past due to proteomic analysis and genetic engineering. For example, proteomic analysis of nitrogen starved microalgae revealed that the up-regulation of hydroxyacyl-ACP dehydrogenase and enoyl-ACP reductase enzymes is required for lipid accumulation. The latter enzymes can be employed in engineering microalgae for algae based biofuels production (Rai *et al.*, 2017). Hence, the key enzyme responsible for triacylglycerols (TAGs) accumulation acetyl CoA carboxylase (ACCase) is a path forward for microalgae based biofuel production (Hu *et al.*, 2008). In yet another case, triglycerides producing yeast *Rhodosporidium toruloides* is an economical lipid organism but, not capable of surviving in hydrolysate conditions. Therefore, the transcriptomic and proteomic analysis of mutant strain of *R. toruloidium* revealed 57% of transcripts correlated to their proteins playing a role in hydrolysate tolerant mechanism (Qi *et al.*,2017). Further, the genes encoding these proteins can be explored to engineer the triacylglyceride metabolic pathways.

The genetic engineering of biofuel biosynthesis pathway enzymes in *E. coli* strains led to breakdown of cellulose and hemicellulose for advanced biofuels production (Bokinsky *et al.*, 2011). The major components of plant cell wall proteins constitute the glycosyl hydrolases, oxireductases, peroxidase and laccases necessary for cell wall polymer biosynthesis, identified through cell wall proteomic analysis. The *Panicumvirgatum* (switchgrass) is seen as a potential biofuel crop and the proteomic exploration on coleoptile endomembrane revealed 1750 proteins with plausible role in cellular membrane polymer biosynthesis (Lao, 2015). These proteins are best candidates to identify the genes and use of genome editing to enhance the biofuel production in switchgrass. The proteo-genomic analysis revealed *Paenibacillus* and *Gemmatimonadetes* are important bacterial groups to explore for genetic engineering and to deconstruct the switchgrass biomass for biofuel production (D'haeseleer *et al.,* 2013). Since, cellulose and hemi-cellulose are major plant cell wall components and, second generation ethanol major resources. Hence, the targeted modification of proteins involved in cell wall polymers biosynthesis will enhance the potential of biofuel production (Calderan-Rodrigues *et al.,* 2017).

Biofuel Metabolic Pathway Engineering

Metabolomics is a snap shot of cellular metabolism, while other "omics" studies are related to molecular regulatory processes of the cell. Metabolomics deals with systematic analysis and quantification of metabolic profiles of a biological system (Fiehn, 2002). Whereas, metabolome represents the collection of metabolic profiles in a biological cell, tissue, organ or organism and metabolites are small molecules with lower molecular weight (Oliver *et al.,* 1998). The accumulation of different metabolites at various cellular processes in-turn guides the molecular processes of the cell. Generally, nuclear magnetic resonance (NMR) and other mass spectrometry (MS) analysis like gas chromatography (GC)-MS and liquid chromatography (LC)-MS, have been in use to quantify the differential accumulation of these metabolites during various cellular metabolisms. Engineering of targeted metabolic pathways by modifying the genetic and regulatory processes within the cells, to increase the targeted bio products/biofuels accumulation is called as metabolic engineering (Kulkarni, 2016). Application of advanced omics technologies and engineering the biofuel producing metabolic pathways related in the renewable resources like microbes and plant biomass are essential.

Metabolomic Pathways for Plant Biofuels

Plant biomass is an eco-friendly and renewable resource for biofuel production (Liao *et al.,* 2016). The scope of modifying the metabolic pathways is gaining importance in the recent-past due to genomics evolution.Engineering of cuticular alkane biosynthetic metabolic pathway genes/enzymes are having greater potential of producing renewable hydrocarbon fuels from plants (Jetter and Kunst, 2008). Previously, the ectopic expression of rate limiting enzyme diacylglycerol *O*-acyltransferase 1 (DGAT1) from triacylglycerols pathway led to 1.5 to 2.0 fold increase in energy density of the biomass for improved and sustainable biofuel production in *Jatropha curcas* Maravi *et al.,* 2016). Also, availability of the whole genome sequence and high-throughput omics profiling for most of the cereal crops helped in engineering the metabolic pathways. For example, plant terpene metabolic pathways engineering was made possible through the availability of annotated genomes and genome editing tools. These plant derived terpenes were identified as specialty biofuels, having high energy densities and volumetric net heats of combustion (Mewalal *et al.,* 2017). The bottleneck problems in exploiting lignocellulosic substrates like birch, aspen, spruce, pine, wheat straw, sugarcane bagasse, corn stover, switch grass, Miscanthus, *Eicchorniacrassipes* and *Lantanacamara* for commercial biofuel production can be resolved through genetically modified metabolic pathway engineering (Chandel and Singh, 2011). To mitigate the scale-up problems of second generation biofuels, a third-generation biofuel resources like, microalgae were considered as promising and sustainable feedstocks for converting the carbon dioxide into carbon-rich lipids to generate biodiesels for the future (Wijffels and Barbosa, 2010). The algae and microalgae are also considered as potential sources of feedstock and cost effective for biofuel production (Salama *et al.*, 2017). However, current commercial exploration of biofuel production is focussing on metabolomic pathway engineering in oxygenic photosynthetic microorganisms and algae are also called as fourth generation biofuels (Lu *et al.*, 2011).

Microbial Metabolic Engineering for Biofuels

The second and third generation biofuels have been gaining importance due to genetic and metabolic engineering technologies. The microorganisms have been used to biosynthesise or modify the expression of a key enzyme encoding genes, to manifold increase in the accumulation of targeted metabolites within the cell. This increased performance in the microorganism to biosynthesize the metabolites requires deep understanding of metabolic pathways and genes encoding the key enzymes to finally increase the biofuel productivity. For example, the over expression of key enzyme *NudB*, responsible for conversion of isopentenyl pyrophosphate (IPP) to 3-methyl-3-buten-1-ol in mevalonate metabolic pathway, indicated 60% increase in accumulation of methylbutenol in *E. coli* based on LC-MS analysis (Martien and Amador-Noguez, 2017). Metabolic engineering of butanol production by *Clostridium acetobutylicum* was achieved through the addition of glycerol or methyl vilogen as redox potential decreasing co-substrate in the tricarboxylic acid (TCA) cycle (Lutke-Eversloh and Bahl, 2011). The advancement in the field of mass-spectrometry-based-metabolomics helped in increasing the biofuel productivity through metabolic pathway engineering in microbes. For example, ethanol production by using xylose as precursor through heterologous xylose reductase (XR), xylitol dehydrogenase (XDH), and xylulose kinase enzymes expressions led to increased flux through the oxidative pentose phosphate pathway (PPP) and tricarboxylic acid (TCA) cycle to meet increased NADPH and energy demands in yeast (Martien and Amador-Noguez, 2017). The synthetic biology comprising of all omics approaches with biological parts lead to the novel synthetic pathways combined with metabolic engineering in yeast species helped to generate biofuels from lignocellulosic biomass. However, exploring the non-conventional yeast systems is having greater opportunity to meet the biofuel requirements in coming future (Madhavan *et al.*, 2017). *E. coli* and *Saccharomyces cerevisiae* are promising organisms for biofuel production, due to their tolerance against heat rapid carbon assimilation and diverse substrate utilization (Clomburg and Gonzalez, 2010). Often, lack of critical engineering possibilities for genetic mutation and feedback inhibition of the metabolic pathways tend to decrease the biofuel production. Particular, the seamless supply of carbon, energy (such as, ATP and NAD(P)H) and microbial incapability of catabolism of higher sugar molecules present in the biomass (i.e. hexose and pentose). Also, cellular toxicity of the isobutanol accumulation hinders the biofuel production efficiency (Minty *et al.*, 2011).

Conclusion and Future Perspectives

Currently the trends in energy consumption are neither environmentally safe nor economically sustainable. An inevitable energy crisis projected in near future owing to shortage of petroleum fuels, will severely affect socio-economic growth with drastic environment implications. Therefore, an alternative clean fuel is an immediate need of the hour for energy suitability. Biofuels serves to be the most viable option to meet the current and future energy crisis, providing a sustainable alternative to other forms of renewable energy. The continuous supply of starting material (plant material, oil seeds and microorganism) without competing with food production can make biofuels a sustainable source of energy. Modern OMICS technology has the potential to harness this environmental friendly and sustainable form of energy. Based on current land-use requirements,transgenic cellulosic and lignocellulosic plants seem to be the most viable option because the cell wall is the most abundant renewable source of energy on the planet. In the years to come, a combination of current bioenergy crops and new biofuel sources like Algae, yeast and, *E.coli* along with advanced genome editing tools will fulfill the energy requirements. The omics research will ensure that the uninterrupted supply of third and fourth generation biofuels through genetically engineered plant biomass and microorganisms for our future generations.

References

Ajjawi, I., Verruto, J., Aqui, M., Soriaga, L. B., Coppersmith, J. and Kwok, K. 2017. Lipid production in *Nannochloropsis gaditana* is doubled by decreasing expression of a single transcriptional regulator. *Nat Biotechnol*, 35(7): 647-652.

Bahadur, B., Sujatha, M. and Carels, N. 2013. Jatropha, Challenges for a New Energy Crop: Springer.Volume 2: *Genetic Improve Biotechnol.* ISBN: 978-1-4614-4914-0 (Print) 978-1-4614-4915-7 (Online).

Bokinsky, G., Peralta-Yahya, P. P., George, A., Holmes, B. M. and Steen, E. J.2011. Synthesis of three advanced biofuels from ionic liquid-pretreated switchgrass using engineered *Escherichia coli., Proc Natl Acad Sci*, 108(50): 19949-19954.

Bouton, J. H. 2007. Molecular breeding of switchgrass for use as a biofuel crop. *Curr Opin Genet Dev*, 17(6): 553-558.

Calderan-Rodrigues, M. J., Fonseca, J. G., Labate, C. A., and Jamet, E. 2017. Cell Wall Proteomics as a Means to Identify Target Genes to Improve Second-Generation Biofuel Production *Frontiers in Bioenergy and Biofuels*: In Tech.

Carels, N. 2013. Towards the domestication of Jatropha: the integration of sciences Jatropha, challenges for a new energy crop (pp. 263-299): *Springer.*

Carpinelli, E. C., Telatin, A., Vitulo, N., Forcato, C., Angelo, D. and, Schiavon, M. 2014. Chromosome scale genome assembly and transcriptome profiling of *Nannochloropsis gaditana* in nitrogen depletion. *Mol plant,* 7(2): 323-335.

Chandel, A. K., and Singh, O. V. 2011. Weedy lignocellulosic feedstock and microbial metabolic engineering: advancing the generation of 'Biofuel'. *Appl microbiol biotechnol,* 89(5): 1289-1303.

Chen, R., and Dou, J. 2016. Biofuels and bio-based chemicals from lignocellulose: metabolic engineering strategies in strain development. *Biotechnol Lett,* 38(2): 213-221.

Clomburg, J. M., and Gonzalez, R. 2010. Biofuel production in *Escherichia coli*: the role of metabolic engineering and synthetic biology. *Appl microbiol biotechnol, 86*(2): 419-434.

Cong, L. 2013. Multiplex genome engineering using *CRISPR/Cas* systems. *Science,* 339(6121): 819-823.

D'haeseleer, P., Gladden, J. M., Allgaier, M., Chain, P. S., Tringe, S. G., Malfatti. 2013. Proteogenomic analysis of a thermophilic bacterial consortium adapted to deconstruct switchgrass. *PLoS One, 8*(7): e68465.

DOE (U.S. Department of Energy). 2016. National Algal Biofuels Technology Review. U.S. Department of Energy, *Office of Energy Efficiency and Renewable Energy, Bioenergy Technologies Office.*

Fairless, D. 2007. Biofuel: the little shrub that could-may be. *Nature News,* 449(7163): 652-655.

Fiehn, O. 2002. Metabolomics–the link between genotypes and phenotypes. *Plant Mol Biol, 48*(1-2): 155-171.

Gao, H., Wang, Y., Fei, X., Wright, D. A., and Spalding, M. H. 2015. Expression activation and functional analysis of HLA3, a putative inorganic carbon transporter in *Chlamydomonas reinhardtii. Plant J,* 82(1): 1-11.

Georgianna, D. R. 2013. Production of recombinant enzymes in the marine alga *Dunaliella tertiolecta. Algal Res,* 2(1): 2-9.

Hu, Q., Sommerfeld, M., Jarvis, E., Ghirardi, M., Posewitz, M., Seibert, M., and Darzins, A. 2008. Microalgal triacylglycerols as feedstocks for biofuel production: perspectives and advances. *Plant J, 54*(4): 621-639.

Huang, G., Chen, F., Wei, D., Zhang, X., and Chen, G. 2010. Biodiesel production by microalgal biotechnology. *Appl Energy,* 87(1): 38-46.

Jacobson, M. Z., and Delucchi, M. A. 2011. Providing all global energy with wind, water, and solar power, Part I: Technologies, energy resources, quantities and areas of infrastructure, and materials. *Energy Policy,* 39(3): 1154-1169.

Jako, C., Kumar, A., Wei, Y., Zou, J., Barton, D. L. and Giblin, E. M. 2001. Seed-specific over-expression of an Arabidopsis cDNA encoding a diacylglycerol acyltransferase enhances seed oil content and seed weight. *Plant physiol,* 126(2): 861-874.

Jetter, R., and Kunst, L. 2008. Plant surface lipid biosynthetic pathways and their utility for metabolic engineering of waxes and hydrocarbon biofuels.*Plant J, 54*(4): 670-683.

Jia, J., Han, D., Gerken, H. G., Li, Y., Sommerfeld, M., Hu, Q., and Xu, J. 2015. Molecular mechanisms for photosynthetic carbon partitioning into storage neutral lipids in *Nannochloropsis oceanica* under nitrogen-depletion conditions. *Algal Res,* 7: 66-77.

Jinkerson, R. E., and Jonikas, M. C. 2015. Molecular techniques to interrogate and edit the Chlamydomonas nuclear genome. *Plant J,* 82(3): 393-412.

Kagale, S., Koh, C., Nixon, J., Bollina, V., Clarke, W. E. and Tuteja, R. 2014. The emerging biofuel crop *Camelina sativa* retains a highly undifferentiated hexaploid genome structure. *Nat. commun.,* 5.

Koksharova, O., and Wolk, C. 2002. Genetic tools for cyanobacteria. *Appl microbio biotechnol,* 58(2): 123.

Kulkarni, R. 2016. Metabolic engineering. *Resonance, 21*(3): 233-237. doi:10.1007/s12045-016-0318-4.

Kumar, A., Falcao, V. R., and Sayre, R. T. 2013. Evaluating nuclear transgene expression systems in *Chlamydomonas reinhardtii. Algal Res,* 2(4): 321-332.

Lardizabal, K. 2008. Expression of Umbelopsis ramanniana DGAT2A in seed increases oil in soybean. *Plant physiol,* 148(1): 89-96.

Li, J., Li, M.-R., Wu, P.-Z., Tian, C.-E., Jiang, H.-W., and Wu, G.-J. 2008. Molecular cloning and expression analysis of a gene encoding a putative β-ketoacyl-acyl carrier protein (ACP) synthase III (KAS III) from *Jatropha curcas. Tree physiol,* 28(6): 921-927.

Liao, J. C., Mi, L., Pontrelli, S., and Luo, S. 2016. Fuelling the future: microbial engineering for the production of sustainable biofuels. *Nat. Rev. Microbiol., 14*(5): 288-304.

Lin, L., Cunshan, Z., Vittayapadung, S., Xiangqian, S., and Mingdong, D. 2011. Opportunities and challenges for biodiesel fuel. *Appl Energy,* 88(4): 1020-1031.

Liu, Q., Gao, R., Li, J., Lin, L., Zhao, J., Sun, W., and Tian, C. 2017. Development of a genome-editing *CRISPR/Cas9* system in thermophilic fungal *Myceliophthora* species and its application to hyper-cellulase production strain engineering. *Biotechnol Biofuels,* 10(1): 1.

Lu, J., Sheahan, C., and Fu, P. 2011. Metabolic engineering of algae for fourth generation biofuels production. *Energy Environ Sci,* 4(7), 2451-2466

Lutke-Eversloh, T., and Bahl, H. 2011. Metabolic engineering of *Clostridium acetobutylicum*: recent advances to improve butanol production. *Curr opin biotechnol,* 22(5): 634-647.

Madhavan, A., Jose, A. A., Parameswaran, B., Raveendran, S., Pandey, A., Castr, G. E., and Sukumaran, R. K. 2017. Synthetic biology and metabolic engineering approaches and its impact on non-conventional Yeast and Biofuel production. *Front Energy Res, 5*: 8.

Madhusudhana, R. 2015. Linkage mapping. In R. Madhusudhana, P. Rajendrakumar, and J. V. Patil (Eds.), Sorghum molecular breeding. New Delhi: *Springer.*

Maehara, T., Takai, T., Ishihara, H., Yoshida, M., Fukuda, K., Gau, M., and Kaneko, S. 2011. Effect of lime pretreatment of brown midrib sorghums. *Biosci Biotechnol Biochem*, 75. doi:10.1271/bbb.110573.

Mansfeldt, C. B., Richter, L. V., Ahner, B. A., Cochlan, W. P., and Richardson, R. E. 2016. Use of De Novo transcriptome libraries to characterize a novel oleaginous marine *Chlorella* species during the accumulation of triacylglycerols. *PLoS ONE*, 11(2): e0147527.

Maravi, D. K., Kumar, S., Sharma, P. K., Kobayashi, Y., Goud, V. V. 2016. Ectopic expression of AtDGAT1, encoding diacylglycerol O-acyltransferase exclusively committed to TAG biosynthesis, enhances oil accumulation in seeds and leaves of Jatropha. *Biotechnol.Biofuels, 9*(1): 226.

Marquis, C. 2016. Gene discovery in microbes for renewable energy development 2016 *pre-breeding and gene discovery for food and renewable energy security*, 109.

Martien, J. I., and Amador-Noguez, D. 2017. Recent applications of metabolomics to advance microbial biofuel production. *Curr.Opin.Biotechnol, 43*, 118-126.

Mathur, S., Umakanth, A., Tonapi, V., Sharma, R., and Sharma, M. K. 2017. Sweet sorghum as biofuel feedstock: recent advances and available resources. *Biotechnol.Biofuels*, 10(1): 146.

Mewalal, R., Rai, D. K., Kainer, D., Chen, F., Kulheim, C., Peter, G. F., and Tuskan, G. A. 2017. Plant-derived terpenes: A feedstock for specialty biofuels. *Trends Biotechnol, 35*(3): 227-240.

Minty, J. J., Lesnefsky, A. A., Lin, F., Chen, Y., Zaroff, T. A., Veloso, A. B. 2011. Evolution combined with genomic study elucidates genetic bases of isobutanol tolerance in *Escherichia coli. Microb. Cell Fact, 10*(1): 18.

Naik, S. N., Goud, V. V., Rout, P. K., and Dalai, A. K. 2010. Production of first and second generation biofuels: A comprehensive review. *Renewable Sustainable Energy Rev*, 14(2): 578-597.

Nguyen, H. T., Silva, J. E., Podicheti, R., Macrander, J., Yang, W., Nazarenus, T. J., et al., 2013. Camelina seed transcriptome: a tool for meal and oil improvement and translational research. *Plant Biotechnol. J*, 11(6): 759-769.

Nielsen, J., Larsson, C., van Maris, A., and Pronk, J. 2013. Metabolic engineering of yeast for production of fuels and chemicals. *Curr. Opin. Biotechnol*, 24(3): 398-404.

Nigam, P. S., and Singh, A. 2011. Production of liquid biofuels from renewable resources. *Prog. Energy Combust. Sci*, 37(1): 52-68.

OECD/FAO 2016 Biofuels, in OECD-FAO Agricultural Outlook 2016-2025. *OECD Publishing*, Paris.

Olivares, J. 2014. National Alliance for Advanced Biofuels and Bioproducts Synopsis (NAABB) Final Report. *US DOE-EERE Biotechnologies Office.*

Oliver, S. G., Winson, M. K., Kell, D. B., and Baganz, F. 1998. Systematic functional analysis of the yeast genome. *Trends Biotechnol, 16*(9): 373-378.

Price, A. H. 2006. Believe it or not, QTLs are accurate!*Trends Plant Sci,* 11(5) : 213-216.

Qi, F., Zhao, X., Kitahara, Y., Li, T., Ou, X., Du, W. 2017. Integrative transcriptomic and proteomic analysis of the mutant lignocellulosic hydrolyzate-tolerant *Rhodosporidium toruloides*. *Eng. Life Sci, 17*(3): 249-261.

Radakovits, R., Jinkerson, R. E., Fuerstenberg, S. I., Tae, H., Settlage, R. E., Boore, J. L., and Posewitz, M. C. 2012. Draft genome sequence and genetic transformation of the oleaginous alga *Nannochloropis gaditana*. *Nat. Commun,* 3: 686.

Rai, V., Muthuraj, M., Gandhi, M. N., Das, D., and Srivastava, S. 2017. Real-time iTRAQ-based proteome profiling revealed the central metabolism involved in nitrogen starvation induced lipid accumulation in microalgae. *Sci. Rep, 7.*

Rasala, B. A., Barrera, D. J., Ng, J., Plucinak, T. M., Rosenberg, J. N., Weeks, D. P., et al., 2013. Expanding the spectral palette of fluorescent proteins for the green microalga *Chlamydomonas reinhardtii. Plant J,* 74(4): 545-556.

Rasala, B. A., Chao, S.-S., Pier, M., Barrera, D. J., and Mayfield, S. P. 2014. Enhanced genetic tools for engineering multigene traits into green algae. *PLoS ONE,* 9(4): e94028.

Salama, E.-S., Kurade, M. B., Abou-Shanab, R. A., El-Dalatony, M. M., Yang, I.-S., Min, B., and Jeon, B.-H. 2017. Recent progress in microalgal biomass production coupled with wastewater treatment for biofuel generation. *Renewable and Sustainable Energy Rev, 79*: 1189-1211.

Sarath, G., Mitchell, R. B., Sattler, S. E., Funnell, D., Pedersen, J. F., Graybosch, R. A., and Vogel, K. P. 2008. Opportunities and roadblocks in utilizing forages and small grains for liquid fuels.*J. Ind. Microbiol. Biotechnol,* 35(5): 343-354.

Schmutz, J., Cannon, S. B., Schlueter, J., Ma, J., Mitros, T., Nelson, W., et al., 2010. Genome sequence of the palaeopolyploid soybean. *Nature,* 463(7278): 178-183.

Schwartz, C. M., Hussain, M. S., Blenner, M., and Wheeldon, I. 2016. Synthetic RNA polymerase III promoters facilitate high-efficiency *CRISPR–Cas9*-mediated genome editing in *Yarrowia lipolytica. ACS synth.Biol,* 5(4): 356-359.

Tatsis, E. C., and O'Connor, S. E. 2016. New developments in engineering plant metabolic pathways. *Curr.Opin. Biotechnol, 42*: 126-132.

Torney, F., Moeller, L., Scarpa, A., and Wang, K. 2007. Genetic engineering approaches to improve bioethanol production from maize. *Curr. Opin. Biotechnol,* 18(3): 193-199.

Vigeolas, H., Waldeck, P., Zank, T., and Geigenberger, P. 2007. Increasing seed oil content in oilseed rape (*Brassica napus* L.) by overexpression of a yeast glycerol-3-phosphate dehydrogenase under the control of a seedspecific promoter. *Plant Biotechnol J*, 5(3): 431-441.

Wijffels, R. H., and Barbosa, M. J. 2010. An outlook on microalgal biofuels. *Science, 329*(5993): 796-799.

22

Role of Metabolomics for Crop Improvement

Jyoti Taunk[1]* and Jaideep Kumar[2]

[1]Department of Molecular Biology, Biotechnology and Bioinformatics, CCS Haryana Agricultural University, Hisar- 125004, India

[2]Division of Animal Biotechnology, National Dairy Research Institute, Karnal- 132001, India

*Corresponding author: jyotibiotech86@gmail.com

Abstract

Metabolomics is a potent tool for quantitative measurement and identification of low molecular weight metabolites in a given sample, cell, tissue or organ. Metabolomics plays a significant role in plant biology. Globally, metabolomic analysis empowers phenotyping and diagnostic analysis of plants and is an important tool for functional annotation of genes and understanding cellular responses. Also, metabolomics has implementation in improving stress tolerance, plant breeding, association analysis etc. This chapter is aimed to comprehend application of metabolomics as a tool for crop improvement. Apart from general scheme of metabolomics and brief about analytical techniques used, the chapter also discusses bottleneck and future prospective of this technology.

Keywords: Analytical techniques, Crop improvement, Metabolomics, Stress tolerance

Introduction

Metabolites are the end products of cellular regulatory processes, and their levels can be regarded as the ultimate response of biological systems to genetic or environmental changes. The set of metabolites synthesized by a biological system constitute its "metabolome" (Fiehn, 2002). Plants produce large numbers of metabolites (approximately 200,000 to 1,000,000) that play crucial roles in plant growth, development and response to environments and also contribute to the color, taste, aroma and scent of fruits and flowers (Bino *et al.*, 2004).The metabolic richness comes not only from the number of genes present (20,000 to 50,000) but, also from multiple substrate specificities for many enzymes (Aharoni *et al.*, 2000). Generally, metabolites are classified into primary and secondary metabolites. The former are requisite for the growth and development of a plant, while the latter are not essential but are crucial for plant to survive under stress conditions by maintaining a delicate balance with the environment. Primary metabolites are highly conserved in their structures and abundances while secondary metabolites differ widely across plant kingdoms (Scossa *et al.*, 2016). Metabolites are highly dynamic in time and space and show an enormous range of structures raising the challenges for analytical procedures in their measurement (Stitt and Fernie, 2003). Therefore, implementing a metabolomic research strategy necessitates planning for considerable complexity (Watkins *et al.*, 2001).

Metabolomics is an advanced, specialised form of analytical biochemistry. It represents the comprehensive, nonbiased, high throughput analysis of complex metabolite mixtures allowing ideally the identification and quantification of every individual metabolite. Metabolomics has also unfolded as a functional genomics methodology that contributes to our understanding of the complex molecular interactions in biological systems (Hall *et al.*, 2002). Metabolomics was originally developed from metabolic profiling. In the early 1970's, Gas chromatography-Mass Spectrometry (GC-MS) technologies were used to analyse steroids, acids and neutral and acidic urinary drug metabolites (Devaux *et al.*, 1971; Horning and Horning, 1970, 1971). Soon afterwards, the concept of using metabolic profiles to screen, diagnose and assess health began to spread (Cunnick *et al.*, 1972;Mroczek, 1972). However, it was not until early 1990's that metabolite profiling was used to study plants (Roessner *et al.*, 2000). The term "*metabolic profiling*" is now used for identification and quantification of a selected number of pre-defined metabolites, generally related to a specific metabolic pathway(s). *Metabolite footprinting* is a similar approach that examines the signals associated with an organism's growth in a substrate or medium by analysis of chemical species in the substrate (Kaderbhai *et al.*, 2003). On the other hand, *metabolic fingerprinting* represents a high-throughput and rapid, global analysis of samples to provide sample classification, in which quantification and metabolic identification are generally not employed, but such screening enables to discriminate between samples of different biological status or origin. Qualitative and quantitative analysis of one or a few metabolites related to a specific metabolic reaction employed when low limits of detection are required is called *metabolite target analysis* (MTA) (Dunn and Ellis, 2005).

The financial cost of metabolomics experiment is approximately an order of magnitude lower than that of a transcriptome or microarray experiment, adding to its practical enactment (Sumner, 2010). However, attaining the broadest overview of metabolic composition is very complex and entails establishing a multifaceted, fully integrated strategy for optimal sample extraction, metabolite separation, detection, automated data gathering, handling, analysis and ultimately quantification (van Beek and Montoro, 2009; van Beek, 2005). Both analytical and computational developments are indispensable to achieve this goal. Currently, there is no single technology available to detect all the compounds in single analysis in any plant or other organism (Fernie *et al.*, 2005; Hall, 2006). Amalgamation of multiple analytical techniques, such as GC, liquid chromatography (LC), capillary electrophoresis (CE)-MS, Nuclear Magnetic Resonance (NMR) and rapid scanning time-of-flight (TOF) are generally performed to detect maximum compounds (Dettmer *et al.*, 2007; Krishnan *et al.*, 2005; Dunn, 2008). The *in silico* analysis of genes and metabolites using publicly available data can construct gene-to-metabolite networks. Such meta-analysis toward systems biology can accelerate the studies required to fill in the missing speck in our knowledge of cellular processes. However, due to technical limitations of metabolome detection, the set of publicly available metabolome data is generally smaller than that of transcriptome data (Fukushima *et al.*, 2009). Linkage of functional metabolomics information to messenger RNA (mRNA) and protein expression data makes it possible to envisage the functional genomics repertoire of an organism (Clark *et al.*, 2007; Lisec *et al.*, 2008; Wentzell *et al.*, 2007). Therefore, metabolomics plays a key role in understanding cellular system and decoding the functions of genes (Hagel and Facchini, 2008; Schauer and Fernie, 2006; Saito *et al.*, 2006).

Metabolomics also plays a key role in plant molecular biotechnology, where plant cells are modified by the expression of engineered genes, because we can obtain information on the metabolic status of cells *via* a snapshot of their metabolome (Oksman-Caldentey and Saito, 2005; Saito and Matsuda, 2010).Metabolomic information will not only assist in the establishment of deeper understanding of complex interactive nature of plant metabolic network and their response to environmental and genetic change but will also provide unique insight into the fundamental nature of plant phenotypes in relation to development, physiology, tissue identity, resistance, biodiversity etc. (Guimera and Amaral, 2005; Fukushima *et al.*, 2009).

Analytical Methods Used in Plant Metabolomics

For metabolomics analysis, metabolites can be extracted from a few milligrams of tissues. In plants, a major part of the large diversity in the metabolome is due to the presence of a wide range of secondary metabolites, which general greatly exceeds the number of primary metabolites. There are various methodologies for extracting compounds from biological materials: liquid extraction (temperature- or pressure-assisted); solid-phase extraction (SPE); solid-phase microextraction (SPME) and microwave-assisted extraction (MAE). Complex mixture of metabolomics sample then needs to be simplified by separating some analytes from others. This can be achieved by GC, high performance LC (HPLC) or CE. LC is probably the most versatile separation method, as it allows separation of compounds of a wide range of polarity with little effort in sample preparation.

MS is used to identify and quantify metabolites after separation. For this analysis, the analytes are imparted with a charge and transferred to the gas phase either by electron or chemical ionization or through electrospray ionization. The most extensively used methods for plant metabolite analysis are GC-MS (Lisec *et al.*, 2006; Kim *et al.*, 2011) and LC-MS (Chen *et al.*, 2007; De Vos *et al.*, 2007; Vorst *et al.*, 2005). Other important analytical techniques include *LC (photodiode array detection) coupled to MS* (LC-PDA/MS) (Huhman and Sumner, 2002), CE-MS (Harada and Fukusaki, 2009; Takahashi *et al.*, 2009), *Fourier transform – ion cyclotron resonance MS* (FT-ICR/MS) (Oikawa *et al.*, 2006), Raman spectroscopy and NMR spectroscopy (Ward *et al.*, 2003; Krishnan *et al.*, 2005). *FT–* Infrared Spectroscopy (FT-IR) has been inaugurated as a metabolic fingerprinting technique within the plant sciences (Baranska *et al.*, 2004; Kaderbhai *et al.*, 2003).

The NMR spectrum of a particular molecule is unique, and therefore, NMR is considered as one of the most selective techniques for compound elucidation. NMR is a quantitative technique, as the number of nuclear spins is directly related to the intensity of the signal (Pauli, 2001). Based on sample solubility, NMR-based methods can be broadly classified into solution NMR and insoluble or solidstate NMR. In one dimensional (1D) NMR, protons (^{1}H) are usually observed (^{1}H-NMR) due to the sensitivity and common occurrence of this magnetic nucleus. More detailed analyses, such as metabolite identification or flux assay, can be obtained with other nuclei, particularly ^{13}C and ^{15}N that are coupled with ^{1}H nuclei in two-dimensional (2D) or multi-dimensional NMR analysis (Kikuchi *et al.*, 2004; Sekiyama and Kikuchi, 2007). The advent of cryogenic probe-heads has escorted vital improvements in NMR sensitivity (Kovacs *et al.*, 2005). There are several software tools (ACD/Labs, Chem Office, and PERCH Solutions) that can help in ^{1}H NMR spectral analysis by providing NMR spectral predictions (Moco *et al.*, 2006). The combination of MS with NMR for deciphering the identity of a molecule is one of the most powerful strategies. Another technical advance that offers great promise for metabolomics is dynamic nuclear polarization (DNP) (Ardenkjaer-Larsen *et al.*, 2003) which uses low temperatures, high magnetic fields and microwaves to strongly polarize nuclear spins prior to NMR analysis. This approach yields enhanced high resolution NMR signals resulting in sensitivity increase up to 10,000 fold or substantial reduction in measurement times or both (Golman *et al.*, 2003). An additional layer of omics i.e. "Fluxome", is required between proteome and metabolome layers to consider the dynamic aspects of plant metabolism (Fernie *et al.*, 2005; Libourel and Shachar-Hill, 2008). Recent coupling of direct laser desorption (Nemes *et al.*, 2009) and matrix-assisted laser desorption ionization (Burrell *et al.*, 2007; Li *et al.*, 2008) with TOF-MS imaging have made the molecular imaging of metabolites a practical reality. These technologies have the ability to spatially resolve metabolites in tissues and cells and offer powerful new tools to better understand the function and segregated biochemistry of cells and organs. The extraction of valuable conclusions from the analysis of metabolomics data is equally important to analytical measurements. There are a variety of methods that allow the transformation of raw data, directly taken from the instrument, passing through different treatments and ultimately leading to a list of metabolites. Prior to any data analysis, it is important to be aware of the possible sources of variation present in the samples that can influence the final conclusions if these are not overseen. In principle, biological variance should surpass all analytical variance. XCMS (Smith *et al.*, 2006) and MZmine (Katajamaa*et al.*, 2006) are some of the available alignment toolboxes for MS applications, as HiRes is for NMR applications (Zhao *et al.*, 2006). These are relevant in reducing raw data to workable datasets. A generalized scheme of metabolite analysis in plants is represented in Fig. 1.

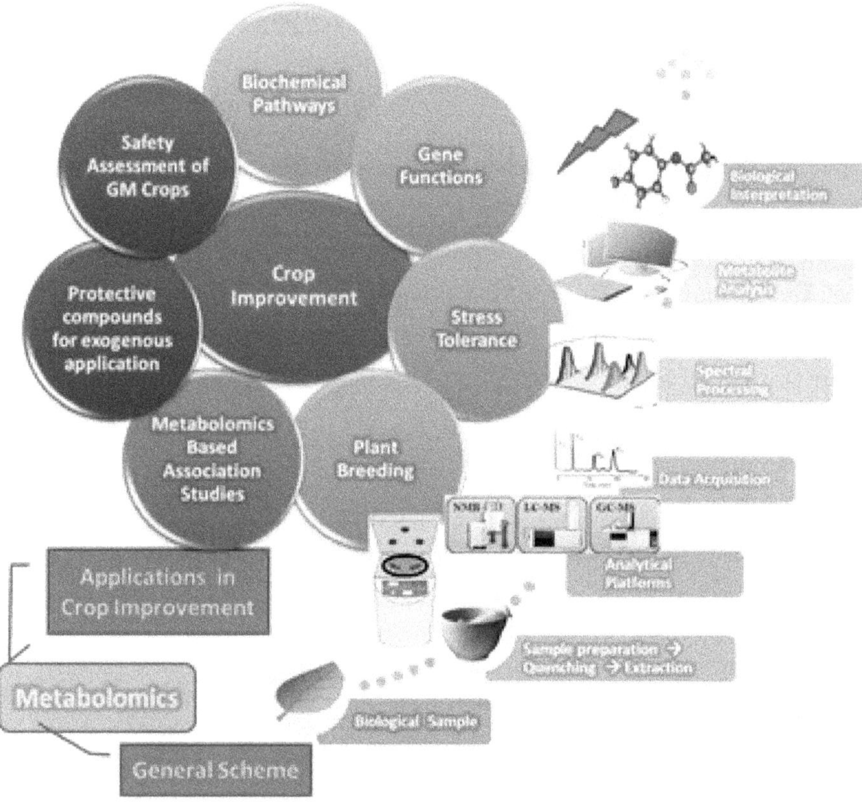

Fig. 1: General scheme for plant metabolite analysis and application of metabolomics for crop improvement.

Informatics for Plant Metabolomics Studies

Informatics is an essential tool for processing large metabolomic datasets (Fukushima *et al.*, 2009; Tohgeand Fernie, 2009, 2010; Go, 2010 Automated softwares can identify peaks from raw MS or NMR data and can also align the peaks among samples. These softwares can identify and quantify each metabolite.). The Golm Metabolome Database (http://gmd.mpimpgolm.mpg.de/) contains information about mass spectra from active metabolites quantified by GS-MS. The Madison Metabolomics Database is a resource for metabolomics research based on NMR and MS (http://mmcd.nmrfam.wisc.edu/). PRIMe, a platform for RIKEN metabolomics is a database that integrates genomic and metabolomics data. It contains information on metabolites obtained from NMR spectroscopy, GC-MS, LC-MS, and CE-MS (Akiyama *et al.*, 2008).

Several databases for plant species have also been developed, such as Plant metabolomics (http://plantmetabolomics.vrac.iastate.edu/ver2/), MassBank, the Plant Metabolic Network [(http://www.plantcyc.org), Wurtele *et al.*, 2007]; and the Madison Metabolomics Consortium Database (http://mmcd.nmrfam.wisc.edu).

Tokimatsu *et al.* (2005) reported a web based analysis tool known as KaPPA-view for integration of transcript and metabolite data on plant metabolic pathway maps. A related database is Terpmed (http://www.terpmed.eu/databases.html), which contains information about plant terpenoids, natural products and other secondary metabolites. There are different tools for visualization or databases that can be used to display the coupling of different omics data, for example, Kyoto Encyclopaedia of Genes and Genomes (KEGG) (www.genome.jp/kegg), MetaCyc(http://metacyc.org), MAPMAN (gabi.rzpd.de/projects/MapMan) and KappaView (kpv.kazusa.or.jp/kappaview).

Application of Metabolomics in Crop Improvement

Metabolomics is applied in a variety of ways to improve production and quality of crop plants. Brief overview of application of metabolomics in crop improvement is presented below (Fig. 1).

Understanding and Discovery of Biochemical Pathways

In a typical plant biochemical pathway, a single precursor synthesizes multiple metabolites. So, the presence of a precursor does not assure the synthesis of any specific metabolite (Dixon *et al.*, 2006). Several plant biochemical pathways cross-talk for effective regulation of cellular processes in an organism. An end product of one pathway becomes the precursor or intermediate substrate for another pathway. Furthermore, a single enzyme affects several biochemical pathways. Therefore, a single point mutation could lead to significant changes in a metabolome (Newell-McGloughlin, 2008). Alteration in one pathway could potentially change the flux in non-target metabolites (Davies, 2007). Currently, the goal of crop improvement is to produce changes in crops and foods that provide yield and health benefits, respectively. In the past few years, researchers have begun to use pathway-based approaches to identify the genetic determinants of crop compositional quality in several plant species. These approaches have led to a detailed dissection and an increase in our understanding of glucosinolate biosynthesis (Kliebenstein *et al.*, 2001), seed oil synthesis (Hobbs *et al.*, 2004) and oligosaccharine metabolism (Bentsink *et al.*, 2000) in *Arabidopsis* and flavonoid biosynthesis in *Arabidopsis* (Yonekura-Sakakibara *et al.*, 2007; Tohge *et al.*, 2005), tomato (Spencer *et al.*, 2005) and *Populus* (Morreel *et al.*, 2006).

Metabolomics is essential to understand fundamental biochemistry of plant metabolism and its direct relationship to nutrition. It has also been central to advancing the efficiency of metabolic engineering of plant nutrition and nutraceuticals in plants (Hall *et al.*, 2008). Examples include carotenoids/lycopene (DellaPenna and Pogson, 2006), vitamin A (Beyer *et al.*, 2002), folate (DellaPenna, 2007), anthocyanins (Xie *et al.*, 2006) and lignin (Vanholme *et al.*, 2008). Metabolomics has played a major role in production of Golden Rice (GR), which is genetically modified rice that accumulates β-carotene in the endosperm. (GR) allowed alleviating vitamin A deficiency, a major nutritional problem worldwide (Ye *et al.*, 2000). The nutritional value of GR was later improved by the overexpression of a phytoene synthase gene leading to GR2 variety, which accumulates higher amounts of carotenoids (84% of the total is β-carotene) (Paine *et al.*, 2005).

Metabolomic analysis of rice leaves infected with the fungus *Magnaporthe grisea,* which causes rice blast disease, presented a model for how this biotrophic/ hemi-biotrophic pathogen succeeds in suppressing the host's defenses and takes up the nutrients required to propagate in plant tissue (Parker *et al.*, 2009). In this study, metabolomic analysis revealed a modification of the shikimate path way that resulted in a reduction in lignified papillae formation and an increase of the mannitol content of susceptible hosts. Since mannitol was proposed to be an important carbohydrate for fungal growth, its increased concentration in susceptible hosts suggests the active metabolic re-programming of infected plants by pathogens (Parker *et al.*, 2009). In addition, RNA Sequencing (RNA-Seq) and high-throughput Super SAGE (Super Serial Analysis of Gene Expression) analysis based on next generation sequencing recently revealed upregulation of quinate permease upon infection (Soanes *et al.*, 2012).

Revealing Gene Functions in Plant Genomes

One of the greatest achievements of plant biology is the completion of the whole genome sequences of model plants such as *Arabidopsis thaliana* and rice. In *Arabidopsis* ~27000 genes were predicted based on nucleotide sequence information; however, only half of these genes have been functionally annotated based on sequence similarity to known genes, and among these the function of only ~11% has been confirmed with direct experimental evidence (MASC report, 2007). The elucidation of unknown genes functions is therefore currently a major challenge in plant research. During the last decade, metabolomics approaches combined with reverse genetic tools [such as RNA interference (RNAi) and gene knockout] expanded tremendously our understanding of biochemical reactions and metabolic pathways not reported in those model plants (Bringaud *et al.*, 2015). Direct measurement of the alteration in the metabolome or specific metabolic compositions of mutants can facilitate the functional annotation of the causing genes.

Exposing the gdh (glutamate dehydrogenase) triple mutant to continuous darkness demonstrated that providing 2-oxoglutarate for the tricarboxylic acid cycle (TCA) is the main physiological function of NADH-GDH (NADH-dependent glutamate dehydrogenase) and that NADH-GDH impacts remarkably on amino acid accumulation in both roots and leaves (Jean-Xavier *et al.*, 2012). Fukushima *et al.* (2014) established a database called Metabolite Profiling Database for Knock-Out Mutants in *Arabidopsis* (MeKO) based on the metabolomic analysis on 50 *Arabidopsis* mutants, which includes images of mutants, accumulation patterns of different metabolites, as well as their statistical results, facilitating significantly the related studies in *Arabidopsis*. In rice, constitutive overexpression of the *Arabidopsis* chloroplast Nicotinamide adenine dinucleotide kinase (NADK) gene enhanced NADK activity, accumulated the Nicotinamide adenine dinucleotide phosphate [NADP(H)] pool, increased electron transport and rates of CO_2 assimilation, and verified the critical role of NADP content in the photosynthetic electron transport rate (Takahara *et al.*, 2010). In *Arabidopsis*, the analysis of 369 recombinant inbred lines (RILs) and 41 introgression lines (ILs) indicated that the metabolite heterosis is primarily contributed by epistasis (Lisec *et al.*, 2009). Aromatic plant Expressed Sequence Tag (EST) databases from ginger and turmeric tissues and from sweet basil glandular trichomes have been assembled and used to identify several genes involved in the production of these specialized metabolites. The expression profiles of genes in various tissues and at various stages of development coupled with LC-MS and GC-MS metabolic profiling are now being pursued to characterize genes involved in the production and regulation of specialized metabolites (Jiang *et al.*, 2006a, *2006b*).

Quantitative trait locus (QTL) mapping of metabolic phenotypes is a powerful approach to unravel the genetic component associated with metabolic profiles and to identify genes associated with metabolic marker. Khan *et al.* (2012) reported mQTL hotspot for phenolic compound of apple (*Malus X domestica* Borkh) on linkage group (LG) 16. Untargeted metabolic QTL (mQTL) mapping of metabolites showed 669 mQTL in peel and flesh, spread over 17 LGs of apple genome. Not all the metabolites showed mQTL and 99 annotated metabolites belonged to phenolic compounds (phenylpropanoids and polyphenols). In tomato, metabolite profiling in seeds of 76 ILs in two consecutive harvest seasons revealed the presence of 30 mQTLs and dissected partial mechanisms, underlying the variational contents of main primary metabolites (Schauer *et al.*, 2006). With the advancement of genome editing techniques, such as Clustered Regularly Interspaced Short Palindromic Repeats [CRISPR/Cas9 (CRISPR associated gene 9)] (Belhaj *et al.*, 2013), our understanding of a specific enzyme in plant metabolism will be significantly promoted.

Improving Stress Tolerance

Plants frequently encounter various biotic and abiotic stresses during their development process processes and they have evolved a series of adaptive changes at both transcriptional and post-transcriptional levels, leading to the reconfiguration of regulatory networks to maintain homeostasis (Verslues *et al.*, 2006). Once plant receptors are stimulated by stress signals, expression of stress responsive genes is activated and subsequently specialized metabolites are biosynthesized to adapt to environmental stresses (Nakabayashi and Saito, 2015). Plant metabolomics can help plant breeders to identify resistant biomarker metabolites and the selected biomarker may be used as a diagnostic metabolite for plant stress (Gauthier *et al.*, 2015).

Abiotic Stresses

Abiotic stresses are produced by inappropriate levels of physical components of the environment and cause injury through unique mechanisms that result in specific responses. Plant responses to abiotic stresses were comprehensively summarised in review papers of Shulaev *et al.* (2008) and Genga *et al.* (2011).

Temperature Stress

In freeze stress metabolomics study, Guy *et al.* (2008) found that active reconfiguration of the metabolome is regulated in part by changes in gene expression initiated by temperature-stress-activated signalling and stress-related transcription factors. They reported that changes in membrane lipid composition play multiple roles in plant adaptation and survival in face of chilling and freezing damage.Metabolomic analysis of rice grown under high night temperature conditions has been applied to find the dysregulation of central metabolism in developing caryopses (Yamakawa and Hakata, 2010), as well as differences in metabolic profiles among 12 cultivars with differing sensitivity to this stress during the vegetative stage (Glaubitz *et al.*, 2015).

Kaplan *et al.* (2004) has explored the temperature stress metabolome of *Arabidopsis*. They aimed metabolome profiling to determine metabolome temporal dynamics associated with the induction of acquired thermotolerance in response to heat shock and acquired freezing shock in response to cold shock. They used GC-MS for metabolite analysis and were able to monitor 81 identified metabolites and 416 unidentified mass spectral tags. In their study, by comparing the heat and cold shock responses, they revealed a previously unknown relationship that majority of heat shock responses were shared with cold shock responses. To their further notice, they found that many metabolites that showed increased response to both heat and cold shock were previously unlinked with temperature stress. Cook *et al.* (2004) has examined the changes that occur in *Arabidopsis* metabolome in response to low temperature and proposed that C-repeat/dehydration responsive element-binding factor (CBF) cold responsive pathway has a prominent role in bringing about these modifications.

Salt Stress

Widodo *et al.* (2009) has studied the metabolic responses of barley genotypes to salt stress. They found that salt sensitive plants had elevated levels of amino acids, including proline, Gamma amminobutyric acid (GABA) and polyamine putrescine, which are indicators of general cellular damage in plants, whereas in salt tolerant, plants levels of hexose phosphatases, TCA cycle intermediates and metabolites involved in cellular protection were increased in response to salt stress. To assess the relationship between stress tolerance, metabolome, water homeostasis and growth performance under osmotic stress, Lugan *et al.* (2010) has compared a salt tolerant *Brassicaceae* species, *Thellungiella salsuginea* with *Arabidopsis thaliana*. Through metabolic fingerprinting and profiling they found that apart from few notable differences in raffinose and secondary metabolites, the same metabolic pathways were regulated by salt stress in both species. Significant differences in two species were observed in terms of physicochemical properties of their metabolomes, like water solubility and polarity.

Drought Stress

Charlton *et al.* (2008) profiled leaf metabolome of *Pisum sativum* L. using 1D and 2D NMR spectroscopy, and monitor the changes induced by drought stress, under both glasshouse and simulated field conditions. They attributed significant changes in resonances to a range of compounds, identified as both primary and secondary metabolites, highlighting metabolic pathways that are stress-responsive. The metabolites that were present at significantly higher concentrations in drought stressed plants under all growth conditions included proline, valine, threonine, homoserine, myoinositol, GABA and trigonelline (nicotinic acid betaine). In order to understand the metabolic mechanisms acting in response to water stress, Foito (2010) conducted metabolite profiling using GC-MS. A clear difference in the metabolic profiles of the leaves of perennial ryegrass plants grown under water stress was observed. Differences were principally due to a reduction in fatty acid levels in the more water stress-susceptible genotype Cashel and an increase in sugars and compatible solutes in the drought-tolerant PI462336 genotype. Raffinose was identified as the metabolite exhibiting the largest accumulation under water-stress in the more tolerant genotype and may represent a target for engineering superior drought tolerance or form the basis of marker-assisted breeding in perennial ryegrass.

Irradiation Stress

Global metabolic profiling of 'Granny Smith' apple peel was employed for evaluating metabolomic alterations resulting from prestorage irradiation by Ultra violet (UV)-white light. The profile, including more than 200 components, 78 of which were identified, revealed changes in the metabolome provoked by UV-white light irradiation and cold storage, distinct temporal changes, before and after cold storage, related to prestorage irradiation in a diverse set of primary and secondary metabolic pathways. The results demonstrated that metabolic pathways associated with ethylene synthesis, acid metabolism, flavonoid pigment synthesis and fruit texture were altered by prestorage irradiation, whereby many of the alterations were detectable after 6 months of cold storage in air (Rudell and Mattheis, 2009).

Cultivation method and Antioxidant Activity

The metabolomic approaches could be applied in the study of the effect of cultivation methods on chemical composition in plants and the relationship with antioxidant activity. Ku *et al.* (2010) investigated the metabolome changes in green tea and shade cultured green tea (tencha) by LC-MS and GC-MS coupled with a multivariate data set and found that green tea clearly showed higher levels of galloylquinic acid, epigallocatechin, epicatechin, succinic acid and fructose together with lower levels of gallocatechin, strictinin, apigenin glucosyl arabinoside, quercetin *p*-coumaroylglucosyl rhamnosyl galactoside, kaempferol*p*-coumaroyl glucosyl rhamnosyl galactoside, malic acid and pyroglutamic acid than tencha. In addition, green tea showed stronger antioxidant activity than tencha, indicating that the antioxidant activity of green tea samples were significantly correlated with their total phenol and total flavonoid contents.

Crop Protection Chemicals

Metabolomics has been so far a valuable tool for high-throughput screening of bioactive substances in order to discover those with high selectivity, unique modes of action and acceptable ecotoxicological/toxicological profiles (Aliferis and Chrysayi-Tokousbalides, 2011). Rapid classification and identification of the mode of action of bioactive compounds applied to plants can be achieved by a robust and easy-to-use metabolic-profiling method. This method uses artificial neural network analysis of 1D ^{1}H NMR spectra of aqueous plant extracts to classify rapidly changes in the total metabolic profile caused by application of crop protection chemicals (Aranibar *et al.*, 2001).

Zou *et al.* (2011) reported a novel approach to detection, analysis and characterization of low-abundant metabolites using herbicide clomazone. Their methodology derived from the *predictive multiple reaction monitoring* (pMRM) mode available on *triple-quadrupolelinear ion trap* (QqQ-LIT)-MS systems offered the highest sensitivity among various acquisition modes for studying trace levels of metabolites of the herbicide clomazone in plants and allowed the identification of positional isomers of metabolites. Unknown metabolites were further identified and validated by obtaining accurate masses and isotopic ratios using *selected ion monitoring* (SIM) and data-dependent MS/MS scans on LC-High Resolution Mass Spectroscopy (LC-HRMS).

Metabolite profiling based on GC-MS was used for study the nickel-rich latex of the hyperaccumulating tree *Sebertia acuminate* (Callahan *et al.*, 2008). More than 120 compounds were detected, 57 of them were subsequently identified. Le Lay *et al.* (2006) investigated the incorporation and localisation of ^{133}Cs in a plant cellular model and the metabolic response induced was analysed as a function of external potassium concentration. The cellular response to the caesium stress was analysed using metabolic profiling. ^{13}C and ^{31}P NMR analysis of acid extracts showed that the metabolome impact of the caesium stress was a function of potassium nutrition. These analyses suggested that sugar metabolism and glycolytic fluxes were affected in a way depending upon the medium content in K^{+} ions.

Multiple Abiotic Stresses

In nature, plants often encounter not only one single stress, since once a single stress occurs, it will be followed by other stresses. For example, salinity stress frequently causes osmotic stress and flooding often leads to low-oxygen stress (Obata and Fernie, 2012). Combined stresses normally results in more extreme condition than that of each individual stress alone and therefore has profound effect on central metabolisms such as sugars, their phosphates and sulfur-containing compounds (Caldana *et al.*, 2011;Rizhsky*et al.*, 2004;Wulff-Zottele *et al.*, 2010). To better dissect the plant metabolic regulatory networks and their functions in the response to complex abiotic stresses, integrated multiple-omics analysis is required (Urano *et al.*, 2010;Caldana *et al.*, 2011, Maruyama *et al.*, 2014; Kanani *et al.*, 2010). As global warming is approaching, heat and drought stresses become big challenges to sustain grain yields. A recent work on rice floral organ development provided mechanistic understandings of the responses of rice floral organs to combined stresses, in which integrative analyses on metabolomics and transcriptomic features of floral organs revealed that sugar starvation is the determinant of the failure of reproductive success under heat and drought stress in rice (Li *et al.*, 2015). Krasensky and Jonak (2012) has elaborately reviewed plant's metabolic adjustments in response to high salt stress, drought stress and extreme temperatures and concluded that metabolic networks are highly dynamic and metabolites can move between different cellular compartments, therefore metabolic analysis at subcellular level in specific tissues is a challenging need.

Biotic Stresses

Similar to distinctive responses to diverse abiotic stresses, metabolic responses of plants to biotic stresses also depend on tissues, species and plant-pathogen or pest interactions. Biotic stress or interaction with plants and other organisms greatly affect plant metabolism and sometimes the activation of specialized (secondary) metabolism can be implemented in the defense reaction against biotic stress (Okazaki and Siato, 2016). Consequently, the identified compounds from biotic stressed plants help in searching for novel defense compounds, and meanwhile serve as important plant defensive state markers (Balmer *et al.*, 2013). Metabolic profiling has been used to monitor strain-dependent differences in response of specialized metabolism in rice infected with the symbiotic rhizobacterium *Azospirillum* (Chamam *et al.*, 2013). When subjected to bacterial leaf blight (BLB) caused by *Xanthomonas oryzae* pv. oryzae (Xoo), sensitive and tolerant rice cultivars display contrasting changes in several specific metabolites, such as acetophenone, xanthophylls, alkaloids, carbohydrates and lipids (Sana *et al.*, 2010). Metabolomic analysis revealed identical changes in metabolic patterns in barley, rice and purple false brome grass, in which malate, polyamines, quinate, and non-polymerized lignin precursors accumulate during infection by *M. oryzae* (Parker *et al.*, 2009). In wheat, the Fhb1 (*Fusarium* head blight 1 resistance locus) QTL is believed to be responsible for resistance to the spread of *F. graminearum* within the spikes. This resistance is mainly attributed to the activation of the phenylpropanoid, terpenoid and fatty acid metabolic pathways in addition to the detoxification of deoxynivalenol (DON) to DON-3-O-glucoside (DON-3G) (Gunnaiah*et al.*, 2012). Using Ultra HPLC (UHPLC)-TOF-MS, Marti *et al.* (2013) took an unbiased approach to determine changes in the metabolite profile at the local and systemic level in maize plants infestated with *Spodoptera littoralis,* and revealed 32 differentially regulated compounds.

Protective Compounds for Exogenous Application

Protective compounds for exogenous application may contribute towards improving the functional quality of crops. Jeserpersen *et al.* (2015) examined the metabolic profiles altered by exogenous treatment of creeping bentgrass under heat stress. Through polar metabolite profiling they identified increase in the content of certain organic acids (i.e. citric and malic acid), sugar alcohols, disaccharides (sucrose) and decreased accumulations of monosaccharides (i.e. glucose and fructose) with exogenous treatment of nitrogen, aminoethoxyvinylglycine (AVG) or zeatin riboside (ZR). They found that the alleviation of heat-induced leaf senescence by N, AVG, and ZR could be due to changes in the accumulation of metabolites involved in osmoregulation, antioxidant metabolism, carbon and nitrogen metabolism, as well as stress signaling molecules. Kim *et al.* (2011) investigated the effects of exogenous methyl jasmonate (MeJA) on phytochemical production in buckwheat sprouts cultivated under dark conditions by metabolomic analysis, using ultra performance LC-quadrupole-TOF (UPLC-Q-TOF) MS and partial least-squares-discriminant analysis (PLS-DA). They found that the accumulation of chlorogenic acid, catechin, isoorientin, orientin, rutin, vitexin and quercitrin caused the phenolic compound content and antioxidant activity of the sprouts to increase. Shi *et al.*(2014) reported that the exogenous application of calcium chloride ($CaCl_2$) improved both chilling and freezing stress tolerances, while ethylene glycol-bis-(β-aminoethyl) ether-*N,N,N,N*-tetraacetic acid (EGTA) reversed $CaCl_2$ effects in bermudagrass [*Cynodon dactylon* (L.) Pers.]. They found that 42 metabolites including amino acids, organic acids, sugar and sugar alcohols were regulated by $CaCl_2$ treatment under control and cold stress conditions. Similarly, Shi *et al.* (2015) found that exogenous application of melatonin conferred improved salt, drought, and cold stress resistances in bermudagrass. Exogenous melatonin treatment alleviated reactive oxygen species (ROS) burst and cell damage induced by abiotic stress; this involved activation of several antioxidants. They found that melatonin-pre-treated plants exhibited higher concentrations of 54 metabolites, including amino acids, organic acids, sugar and sugar alcohol than non-treated plants under abiotic stress conditions. Metabolic analysis showed that the underlying mechanisms of melatonin could involve major reorientation of photorespiratory, carbohydrate and nitrogen metabolism.

Plant Breeding

Plant metabolomic technology can provide information not only on the numbers of identified metabolites but also their correlations with each other and with agronomic important traits, thus it could lead to the development of more rational models to link specific metabolite or pathway with yield or quality associated traits. Metabolomics can also be envisioned as metabolic marker analysis to the nth power with substantial predictive power (Sumner, 2010). Accordingly, metabolomics is emerging as a promising tool in metabolomics-assisted breeding. In plant breeding, metabolite profiling is able to support the trait discovery process. Metabolomics is helping to provide targets for plant breeding by linking gene expression, and allelic variation to variation in metabolite complement (functional genomics) and is also being deployed to better assess the potential impacts of climate change and reduced input agricultural systems on crop composition. Metabolomics-assisted breeding offers unique opportunities for enhancing traits of commercial value (Fernie and Schauer, 2009). The first study looking at agronomically important traits in a commercial population was done by Steinfath *et al.* (2010). They studied potato cultivars to predict browning and chip color quality based on metabolite profiling in harvested potato tubers. One measure of potato quality is determined by the level of darkening that occurs during the frying process, which is an important characteristic for the processing industry. High dry matter potatoes are more susceptible to black spot but are also more preferred in the chip industry. Black spots start to appear within 24 hours as a slight discoloration and the full damage is visible only days later. Thus, a screening tool that could quickly predict the potential for black spot formation would be of great benefit for the industry. Steinfath *et al.* (2010) have thus analysed metabolite profiles of several commercial potato cultivars showing a wide range of variation on the extent of chip coloring and black spot bruising. Full metabolite profiles were integrated with data from analyses on browning and chip color. In tomato, screening of carotenoid metabolites by matrix-assisted laser desorption ionization (MALDI) TOF-MS was recently demonstrated to be useful for the screening of large populations. For this purpose, selected lines from two tomato populations (*S. pennellii* ILs and saturated mutants) were profiled to identify germplasm that is likely to be of high utility in the breeding of fruit containing high levels of these important nutriceuticals (Fraser *et al.*, 2007). Metabolic profiling of phytomedicinal species has proven a valuable tool for quality control (van der Kooy *et al.*, 2008) and optimization of yield (Mavituna *et al.*, 2005; Shyur*et al.*, 2008).

The potential value of using metabolites as biomarkers in plant breeding is tremendous. Markers consisting of a single metabolite can be easily transferred to a low cost and high-throughput bioassay, for example, using colorimetric or enzymatic assays for fructose or glucose for chip coloring prediction. On the other hand, the determination of biomarkers constructed from full metabolite profiles or fingerprints is more expensive and often relies on MS lab analyses, though here also, costs can be cut if an automated and high-throughput platform can be established. Furthermore, when databases are built with metabolite profiles of breeding populations and cultivars, it would be possible to screen for novel phenotypes in existing datasets once those relationships are identified. Therefore, biomarker discovery in plant breeding will be the next tool used to select for new and improved traits (van Dongen and Schauer, 2010). Meyer *et al.* (2007) reported a highly significant canonical correlation between biomass and a specific combination of metabolites. Lisec *et al.* (2008) also showed that mQTL could be correlated with biomass accumulation. Thus, the predictive power of numerous qualitative and quantitative metabolite markers promises to be a powerful resource for breeding and biomass production.

Metabolomics Based Association Studies

Association mapping has only recently been adopted in plant genetic research. Compared with most artificial mapping populations that are constructed by crosses between two parental accessions, association analysis populations are composed of large numbers of natural accessions containing more genetic variants as well as potential for the identification of unknown phenotype-associated loci in the plant genome. Benefiting from the development of next-generation sequencing technologies, metabolome-based Genome Wide Association Study (mGWAS) has been used to understand genetic mechanisms underlying metabolic diversity and their associations with complex traits in plants. Such mapping approaches have pinpointed associations between genomic regions of maize and kernel composition as well as starch content in potatoes, pigment content in tomato and provitamin A content in maize (Wilson *et al.*, 2004; Palaisa*et al.*, 2003; Li *et al.*, 2008; Stich *et al.*, 2008; Zhao *et al.*, 2007). Wen *et al.* (2014) studied maize metabolism to dissect the genetic basis of metabolic diversity in maize kernels. They quantified 983 metabolite features in 702 maize genotypes planted at multiple locations and identified 1,459 significant locus–trait associations across three environments through mGWAS. In rice, a GWAS analysis using metabolomic data obtained from 175 rice accessions successfully identified 323 associations between 143 Single Nucleotide Polymorphisms (SNPs) and 89 secondary metabolites, which revealed two sorts of genetic machineries determining the natural variations in rice secondary metabolite compositions (Matsuda *et al.*, 2014).

Safety Assessment of Genetically Modified (GM) Crops

Many people are perturbed about transgenicall yaugmented foods due to the potential and perceived risks of epistatic transgenes (Schubert, 2008). Therefore, risk assessments of GM plants and derived products are very strict in many countries and regions. The World Health Organization (WHO) and the United Nations Food and Agriculture Organization (FAO) conceded in May 2000 (FAO and WHO, 2000) that some aspects in the safety assessment process for novel foods could be refined. They indicated that profiling techniques following validation could be considered as a useful appendage to current practices for compositional analysis of novel foods. The use of intragenic and famigenic techniques, in conjunction with metabolomics, offers huge potential for nutraceutical breeding to obtain greater acceptability in the market (Rommens *et al.*, 2007). Metabolomics provides an additional facet to GM crop analysis, allowing the detection of both intended and unintended effects that might take place in GM crops because of genetic modification at the metabolic level. This facilitates greatly the worthwhile equivalence evaluation of GM crops.

In such cases, metabolite profiles of the GM should not only be compared with the corresponding parental line, but should also be assessed mindful of natural variability of metabolic profiles of conventional crops. Case studies juxtaposing metabolic changes between GM crops and their non-transgenic counterparts have covered almost all important agricultural crops, such as wheat, rice, pea, maize, barely, soybean, tomato, potato, etc., which have confirmed detectable alterations of their metabolites due to transgenic modifications (Simó *et al.*, 2014; Chang *et al.*, 2012). However, results also show that other facets such as environmental conditions customarily employ greater impact on metabolic compositions than genetic modification (Zhou *et al.*, 2009).

Kuiper *et al.* (2003) deduced that the results from microarray technology, proteomics and metabolomics may be successfully exerted to screen for intended and unintended side effects of GM foods. ^{1}H NMR-based metabolic profiling of the GM Ailsa Craig tomatoes and its azygous control denoted that GM Ailsa Craig tomatoes contained 5-8 times higher concentrations of cyclopropyl-sterol (cycloartenol) compared to controls. This change may be considered as an intended effect. Currently, efforts are underway initiated by the Metabolomics Society to develop and execute minimum reporting standards and current best practice protocols in the "Metabolomics Standards Initiative" (MSI) (http://msi-workgroups.sourceforge.net/). Guidance regarding the choice of analytes to monitor is provided from organizations such as the Organization for Economic Cooperation and Development (OECD) that publish consensus documents on compositional considerations for new plants (http://www.oecd.org/document). These documents provide scientifically based guidelines that are admissible to national regulatory bodies for the relevant nutrients and antinutrients to be analyzed for new crop varieties.

Bottleneck and Challenge

Metabolomics is not any more seen just as a tool of functional genomics, but it has now became an intrinsic part of system biology (Mendes, 2006). However, metabolomics is not as advanced as other omics because there is a critical difference between metabolites and other molecules, i.e., DNA, RNA and proteins. Later are linear polymers composed of a limited numbers of monomers, and the interpretation of RNA and protein sequences can be facilitated by genome information according to the central dogma of molecular biology. On the other hand, metabolites comprise a more heterogeneous group than these polymers in terms of their physical and chemical properties, varying widely with respect to size, polarity, quantity, and stability (Okazaki and Saito, 2012). Therefore, there are substantial challenges posed by metabolomics which includes:

a) Identification of large number of metabolites that are detected but, whose chemical nature is unknown

b) Temporal and spatial analyses

c) Identification of the active areas of metabolism responsible for changes in metabolite profiles

d) Dynamic range limitations

e) Creating standards for data and metadata format and reporting

Conclusion and Future Perspective

Metabolomics has emerged as an important tool that is revolutionizing plant biology and crop breeding. Metabolomics offers us the opportunity to gain deeper insights into and have better control of the basic biochemical basis of plants. Allying metabolomics with other "omics" technology can provide ac oherent picture encompassing all aspects of information flow from genome to metabolome and resulting phenotype. Metabolomics can remarkbly contribute to design modified breeding programmes focussed at better quality production, identification of unknown genes as well as to better understand pathways responsible for biosynthesis of pertinent metabolites. As yet, no single analytical method can apprehend the whole metabolome. Currently, unification of high-resolution MS and NMR provides the necessary information for elucidation of compounds. The shift from single metabolite measurements to platforms that can provide information on hundreds of metabolites simultaneously can lead to the development of better models to describe the links both within metabolism itself and between metabolism and yield associated traits. The development of improved bioinformatic tools will facilitate the management of large amounts of data and help integrate different datasets by sieving the metabolite information. Future advancements can and will lead to greater efficacy of this approach.

References

Aharoni, A., Keizer, L.P., Bouwmeester, H.J., Sun, Z., Alvarez-Huerta, M., Verhoeven, H.A., Blaas, J., Houwelingen, A.V., De Vos, R.H., Voet, H., Jansen, R.C., Guis, M., Mol, J., Davis, R.W., Schena, M., Tunen, A.J. and O'Connell, A.P. 2000. Identification of the SAAT gene involved in strawberry flavor biogenesis by use of DNA microarrays. *Plant Cell*, 12: 647-661.

Akiyama, K., Chikayama, E., Yuasa, H., Shimada, Y., Tohge, T., Shinozaki, K., Hirai, M.Y., Sakurai, T., Kikuchi, J. and Saito, K. 2008. PRIMe: a Web site that assembles tools for metabolomics and transcriptomics. *InSilicoBiol*, 8:339-345.

Aliferis, K.A. and Chrysayi-Tokousbalides, M. 2011. Metabolomics in pesticide research and development: Review and future perspectives. *Metabolomics*. 7: 35-53.

Aranibar, N., Singh, B.K., Stockton, G.W. and Ott, K.H. 2001. Automated mode-of-action detection by metabolic profiling. *BiochemBiophys Res Commun*, 286:150-155.

Ardenkjaer-Larsen, J.H., Fridlund, B., Gram, A., Hansson, G., Hansson, L., Lerche, M.H., Servin, R., Thaning, M. and Golman, K. 2003. Increase in signal-to noise ratio of > 10,000 times in liquid-state NMR. *Proc Natl AcadSci, USA,*100:10158-10163.

Balmer, D., Flors, V., Glauser, G. and Mauch-Mani, B. 2013. Metabolomics of cereals under biotic stress: Current knowledge and techniques. *Front Plant Sci,* 4:82.

Baranska, M., Schulz, H., Rosch, P., Strehle, M.A. and Popp, J. 2004. Identification of secondary metabolites in medicinal and spice plants by NIR-FT-Raman microspectroscopic mapping. *Analyst,* 129:926-930.

Belhaj, K., Chaparro-Garcia, A., Kamoun, S. and Nekrasov, V. 2013. Plant genome editing made easy: Targeted mutagenesis in model and crop plants using the CRISPR/Cas system. *Plant Methods,* 9: 39

Bentsink, L., Alonso-Blanco, C., Vreugdenhil, D., Tesnier, K., Groot, S.P.C. and Koornneef, M. 2000. Genetic analysis of seed-soluble oligosaccharides in relation to seed storability of *Arabidopsis. Plant Physiol,* 124: 1595-1604.

Beyer, P., Al-Babili, S., Ye, X., Lucca, P., Schaub, P., Welsch, R. and Potrykus, I. 2002. Golden Rice: introducing the beta-carotene biosynthesis pathway into rice endosperm by genetic engineering to defeat vitamin A deficiency. *J Nutr,* 132:506-510.

Bino, R.J., Hall, R.D., Fiehn, O., Kopka, J., Saito, K., Draper, J., Nikolau, B.J., Mendes, P., Roessner-Tunali, U., Beale, M.H., Trethewey, R.N., Lange, B.M., Wurtele, E.S. and Sumner, L.W. 2004. Potential of metabolomics as functional genomics tool.*Trends Plant Sci,* 9: 418-425.

Bringaud, F., Biran, M., Millerioux, Y., Wargnies, M., Allmann, S. and Mazet, M. 2015. Combining reverse genetics and nuclear magnetic resonance-based metabolomics unravels trypanosome-specific metabolic pathways. *Mol Microbiol,* 96: 917-926.

Burrell, M.M., Earnshaw, C.J. and Clench, M.R. 2007. Imaging matrix assisted laser desorption ionization mass spectrometry: a technique to map plant metabolites within tissues at high spatial resolution. *J Exp Bot,* 58:757-763.

Caldana, C., Degenkolbe, T., Cuadros-Inostroza, A., Klie, S., Sulpice, R., Leisse, A., Steinhauser, D., Fernie, A.R., Willmitzer, L. and Hannah, M.A. 2011. High-density kinetic analysis of the metabolomic and transcriptomic response of *Arabidopsis* to eight environmental conditions. *Plant J,* 67: 869-884.

Callahan, D.L., Roessner, U., Dumontet, V., Perrier, N., Wedd, A.G., O'Hair, R.A.J, Baker, A.J.M. and Kolev, S.D. 2008. LC–MS and GC–MS metabolite profiling of nickel (II) complexes in the latex of the nickel-hyperaccumulating tree Sebertiaacuminata and identification of methylated aldaric acid as a new nickel (II) ligand. *Phytochemistry,* 69: 240-251.

Chamam, A., Sanguin, H., Bellvert, F., Meiffren, G., Comte, G., Wisniewski-Dyé F., Bertrand, C. and Prigent-Combaret, C. 2013. Plant secondary metabolite profiling evidences strain-dependent effect in the *Azospirillum-Oryza sativa* association. *Phytochemistry*, 87: 65-77.

Chang, Y., Zhao, C., Zhu, Z., Wu, Z., Zhou, J., Zhao, Y., Lu, X. and Xu, G. 2012. Metabolic profiling based on LC/MS to evaluate unintended effects of transgenic rice with *cry1Ac* and *sck* genes. *Plant MolBiol*, 78:477-487.

Charlton, A.J., Donarski, J.A., Harrison, M., Jones, S.A., Godward, J., Oehlschlager, S., Arques, J.L., Ambrose, M., Chinoy, C., Mullineaux, P.M. and Domoney, C. 2008. Responses of the pea (*Pisum sativum L.*) leaf metabolome to drought stress assessed by nuclear magnetic resonance spectroscopy. *Metabolomics*, 4:312-327.

Chen, G.D., Pramanik, B.N., Liu, Y.H. and Mirza, U.A. 2007. Applications of LC/MS in structure identifications of small molecules and proteins in drug discovery. *J Mass* Spectrom, 42:279-287.

Clark, R.M., Schweikert, G., Toomajian, C., Ossowski, S., Zeller, G., Shinn, P., Warthmann, N., Hu, T.T., Fu, G., Hinds, D.A., Chen, H., Frazer, K.A., Huson, D.H., Scholkopf, B., Nordborg, M., Ratsch, G., Ecker, J.R. and Weigel, D. 2007. Common sequence polymorphisms shaping genetic diversity in *Arabidopsis thaliana*. *Science*, 317:338-342.

Cook D., Fowler S., Fiehn O. and Thomashow M.F. 2004. A prominent role for the CBF cold response pathway in configuring the low-temperature metabolome of *Arabidopsis*. *Proc Nat Acad Sci, USA*, 101 (42):15243-15248.

Cunnick ,W.R., Cromie, J.B., Cortell, R., Wright, B., Beach, E., Seltzer, F. and Miller, S. 1972. Value of biochemical profiling in a periodic health examination program: analysis of 1,000 cases. *Bull N YAcad Med*, 18:5-22.

Davies, K.M. 2007. Genetic modification of plant metabolism for human health benefits. *Mutat Res*, 622:122-137.

De Vos, C.H.R., Moco, S., Lommen, A., Keurentjes, J.J.B., Bino, R.J. and Hall, R.D. 2007. Untargeted large scale plant metabolomics using liquid chromatography coupled to mass spectrometry. *Nat Protoc*, 2:778-791.

DellaPenna, D. 2007. Biofortification of plant-based food: enhancing folate levels by metabolic engineering. *Proc Natl Acad Sci, USA*, 104:3675-3676.

DellaPenna, D. and Pogson, B.J. 2006. Vitamin synthesis in plants: tocopherols and carotenoids. *Ann Rev Plant Biol*, 57:711-738.

Dettmer, K., Aronov, P.A. and Hammock, B.D. 2007. Mass spectrometry-based metabolomics. *Mass Spectrom Rev*, 26:51-78.

Devaux, P.G., Homing, M.G. and Homing, B.C. 1971. Benyzloxime derivative of steroids, a new metabolic profile procedure for human urinary steroids. *Anal Lett*, 4:151.

Dixon, R.A., Gang, D.R., Charlton, A.J., Fiehn, O., Kuiper, H.A., Reynolds, T.L., Tjeerdema, R.S., Jeffery, E.H., German, J.B., Ridley, W.P. and Seiber,

J.N. 2006. Applications of metabolomics in agriculture. *J Agric Food Chem,* 54: 8984-8994.

Dunn, W.B. 2008. Current trends and future requirements for the mass spectrometric investigation of microbial, mammalian and plant metabolomes. *Physical Bio,* 5:11001.

Dunn, W.B. and Ellis, D.I. 2005. Metabolomics: Current analytical platforms and methodologies. *Trends Anal Chem,* 24:285-294.

FAO and WHO. 2000. Safety Aspects of Genetically Modified Foods of Plant Origin; report of a Joint FAO/WHO Expert Consultationon Foods Derived from Biotechnology; WHO: Geneva, Switzerland.

Fernie, A.R. and Schauer, N. 2009. Metabolomics-assisted breeding: a viable option for crop improvement? *Trends Genet,* 25:39-48.

Fernie, A.R., Geigenberger, P. and Stitt, M. 2005. Flux an important, but neglected, component of functional genomics. *Curr Opin Plant Biol,* 8:174-182.

Fiehn, O. 2002. Metabolomics-the link between genotypes and phenotypes. *Plant Mol Biol,* 48:155-171.

Foito, A. 2010. A metabolomics-based approach to study abiotic stress in *Loliumperenne* PhD. Thesis, Uni. of Dundee.

Fontaine, J.X., Tercé-Laforgue, T., Armengaud, P., Clement, G., Renou, J.P., Pelletier, S., Catterou, M., Azzopardi, M., Gibon, Y., Lea, P.J., Hirel, B. and Dubois, F. 2012. Characterization of a NADH-dependent glutamate dehydrogenase mutant of *Arabidopsis* demonstrates the key role of this enzyme in root carbon and nitrogen metabolism. *Plant Cell,* 24:4044-4065.

Fraser, P.D., Enfissi, E.M., Goodfellow, M., Eguchi, T. and Bramley, P.M. 2007. Metabolite profiling of plant carotenoids using the matrix-assisted laser desorption ionization time-of-flight mass spectrometry. *Plant J,* 49:552-564.

Fukushima, A., Kusano, M., Mejia, R.F., Iwasa, M., Kobayashi, M., Hayashi, N., Watanabe-Takahashi, A., Narisawa, T., Tohge, T. and Hur, M. 2014. Metabolomic characterization of knockout mutants in *Arabidopsis*: Development of a metabolite profiling database for knockout mutants in *Arabidopsis*. *Plant Physiol,* 165: 948-961.

Fukushima, A., Kusano, M., Redestig, H., Aita, M.and Saito, K. 2009. Integrated omics approaches in plant systems biology. *Curr Opin Chem Biol,*13: 532-538.

Gauthier, L., Atanasova-Penichon, V., Chéreau, S. and Richard-Forget, F. 2015. Metabolomics to decipher the chemical defense of cereals against *Fusarium graminearum* and deoxynivalenol accumulation. *Int J MolSci,* 16:24839-24872.

Genga, A., Mattana, M., Coraggio, I., Locatelli, F., Piffanelli, P. and Consonni, R. 2011. Plant Metabolomics: A characterisation of plant responses to abiotic stresses. In: Shanker A, Venkateswarlu B, editors. Abiotic Stress in Plants - Mechanisms and Adaptations. Rijeka: InTech:309-350.

Glaubitz, U., Erban, A., Kopka, J., Hincha, D.K. and Zuther, E. 2015. High night temperature strongly impacts TCA cycle, amino acid and polyamine biosynthetic pathways in rice in a sensitivity-dependent manner. *J Exp Bot,* 66:6385-6397.

Go, E.P. 2010. Database resources in metabolomics: an overview. *J Neuroimmune Pharmacol,* 5:18-30.

Golman, K., Ardenkjaer-Larsen, J.H., Petersson, J.S., Mansson, S. and Leunbach, I. 2003. Molecular imaging with endogenous substances. *Proc Natl Acad Sci, USA,* 100:10435-10439.

Guimera, R. and Amaral, N.A. 2005. Functional cartography of complex metabolic networks. *Nature,* 433:895-900.

Gunnaiah R., Kushalappa A.C., Duggavathi R., Fox S. and Somers D.J. 2012. Integrated metabolo-proteomic approach to decipher the mechanisms by which wheat QTL (Fhb1) contributes to resistance against *Fusarium graminearum. PLoSOne,* 7:e40695.

Guy, C., Kaplan, F., Kopka, J., Selbig, J. and Hincha, D.K. 2008. Metabolomics of temperature stress. *Physiol Plant,* 132:220-235.

Hagel, J. and Facchini, P. 2008. Plant metabolomics: analytical platforms and integration with functional genomics. *Phytochem Rev,* 7:479-497.

Hall, R., Beale,M., Fiehn, O., Hardy, N., Sumner, L. and Bino, R. 2002. Plant metabolomics: the missing link in functional genomics strategies. *Plant Cell,* 14:1437-1440.

Hall, R.D. 2006. Plant metabolomics: from holistic hope, to hype, to hot topic. *New Phytol,* 169:453-468.

Hall, R.D., Brouwer, I.D. and Fitzgerald, M.A. 2008. Plant metabolomics and its potential application for human nutrition. *Physiol Plant,* 132:162-175.

Harada, K. and Fukusaki, E. 2009. Profiling of primary metabolite by means of capillary electrophoresis-mass spectrometry and its application for plant science. *Plant Biotechnol,* 26:47-52.

Hobbs, D.H., Flintham, J.E. and Hills M.J. 2004. Genetic control of storage oil synthesis in seeds of *Arabidopsis. Plant Physiol,* 136:3341-3349

Horning, B.C. and Horning, M.G. 1970. Metabolic Profiles: Chromatographic methods for isolation and characterization of a variety of metabolites in man. Year book Medical Publishers, Chicago,Ill, USA.

Horning, B.C. and Horning, M.G. 1971. Human metabolic profiles obtained by GC and GC/MS. *J. Chromatogr Sci,*9:129-140.

Huhman, D.V. and Sumner, L.W. 2002. Metabolic profiling of saponins in *Medicago sativa* and *Medicago truncatula* using HPLC coupled to an electrospray ion-trap mass spectrometer. *Phytochemistry,* 59:347-360.

Jespersen D., Yu J. and Huang B. 2015. Metabolite responses to exogenous application of nitrogen, cytokinin, and ethylene inhibitors in relation to heat-induced senescence in creeping bentgrass. *PLoSOne*, 10:e0123744.

Jiang, H., Somogyi, A., Jacobsen, N.A.,Timmermann, B.N. and Gang, D.R. 2006a. Analysis of curcuminoids by positive and negative electrospray ionization and tandem mass spectrometry. *Rapid Commun Mass Spectrom*, 20:1001-1012.

Jiang, H., Timmermann, B.N. and Gang, D.R. 2006b. Use of liquid chromatography-electrospray ionization tandem mass spectrometry to identify diarylheptanoids in turmeric (*Curcuma longa* L.) rhizome. *J Chromatogr*, 1111:21-31.

Kaderbhai, N.N., Broadhurst, D.I., Ellis, D.I., Goodacre, R. and Kell, D.B. 2003. Functional genomics *via* metabolic footprinting: Monitoring metabolite secretion by *Escherichia coli* tryptophan metabolism mutants using FT-IR and direct injection electrospray mass spectrometry. *Comp Func Gen*, 4:376-391.

Kanani, H., Dutta, B. and Klapa, M.I. 2010. Individual vs. combinatorial effect of elevated CO_2 conditions and salinity stress on Arabidopsis thaliana liquid cultures: Comparing the early molecular response using time-series transcriptomic and metabolomic analyses. *BMC Syst Biol*, 4:177.

Kaplan F., Kopka J., Haskell D.W., Zhao W., Schiller K.C., Gatzke N., Sung D.Y. and Guy C.L. 2004. Exploring the temperature-stress metabolome of *Arabidopsis*. *Plant Physiol*, 136:4159-4168.

KappaView database [kpv.kazusa.or.jp/kappaview].

Katajamaa, M., Miettinen, J. and Oresic, M. 2006. MZmine: toolbox for processing and visualization of mass spectrometry based molecular profile data. *Bioinformatics*, 22:634-636.

KEGG database [www.genome.jp/keg].

Khan, S.A., Schaart, J.G., Beekwilder, J., Allan, A.C., Tikuno, Y.M., Jacobsen, E. and Schouten, H.J. 2012. The mQTL hotspot on linkage group 16 for phenolic compounds in apple fruit is probably the result of leucoanthocyanidin reductase gene at the locus. *BMC Res notes*, 5:618.

Kikuchi, J., Shinozaki, K. and Hirayama, T. 2004. Stable isotope labelling of *Arabidopsis thaliana* for an NMR-based metabolomics approach. *Plant Cell Physiol*, 45:1099-1104.

Kim, H.K., Choi, Y.H. and Verpoorte, R. 2011. NMR-based plant metabolomics: where do we stand, where do we go? *Trends Biotechnol*, 29:267-275.

Kim, S., Shin, M.H., Hossain, M.A., Yun2, E.J., Lee, H. and Kim, K.H. 2011. Metabolite profiling of sucrose effect on the metabolism of *Melissa officinalis* by gas chromatography-mass spectrometry. *Anal Bioanal Chem*, 399:3519-3528.

Kim, H.J., Park, K.J., and Lim, J.H. 2011. Metabolomic analysis of phenolic compounds in buckwheat (*Fagopyrum esculentum* M.) sprouts treated with methyl jasmonate. *J Agric Food Chem*, 59:5707-5713.

Kliebenstein, D.J., Gershenzon, J. and Mitchell-Olds, T. 2001.Comparative quantitative trait loci mapping of aliphatic, indolic and benzylic glucosinolate production in *Arabidopsis thaliana* leaves and seeds.*Genetics*, 159:359-370.

Kovacs, H., Moskau, D. and Spraul, M. 2005. Cryogenically cooled probes-a leap in NMR technology.*ProgNuclMagnResonSpectrosc*, 46:131-135.

Krasensky J. and Jonak C. 2012. Drought, salt, and temperature stress-induced metabolic rearrangements and regulatory networks. *J Exp Bot*, 63:1593-1608.

Krishnan, P., Kruger, N.J. and Ratcliffe, R.G. 2005.Metabolite fingerprinting and profiling in plants using NMR. *J Exp Bot*, 56:255-265.

Ku, K.M., Choi, J.N., Kim, J., Kim, J.K., Yoo, L.G., Lee, S.J., Hong, Y.S. and Lee, C.H. 2010. Metabolomics analysis reveals the compositional differences of shade grown tea (*Camellia sinensis* L.). *J Agric Food Chem*, 58:418-426.

Kuiper, H.A., Kok, E.J. and Engel, K.H. 2003. Exploitation of molecular profiling techniques for GM food safety assessment. *CurrOpin Biotechnol*, 14:238-243.

Le Lay, P., Isaure, M.P., Sarry, J.E., Kuhn, L., Fayard, B., Le Bail, J.L., Bastien, O., Garin, J. and Roby, C. 2006. Metabolomic, proteomic and biophysical analyses of *Arabidopsis thaliana* cells exposed to a caesium stress. Influence of potassium supply. *Biochimie*, 88:1533-1547.

Li, F., Wu, X., Lam, P., Bird, D., Zheng, H., Samuels, L., Jetter, R. and Kunst, L. 2008. Identification of the wax ester synthase/Acyl- CoA: diacylglycerol acyltransferase WSD1 required for stem wax ester biosynthesis in *Arabidopsis thaliana*. *Plant Physiol*, 148:97-107.

Li, X., Lawas, L.M., Malo, R., Glaubitz, U., Erban, A., Mauleon, R., Heuer, S., Zuther, E., Kopka, J. and Hincha, D.K. 2015. Metabolic and transcriptomic signatures of rice floral organs reveal sugar starvation as a factor in reproductive failure under heat and drought stress. *Plant Cell Environ*, 38:2171-2192.

Li, Y., Shrestha, B. and Vertes, A. 2008. Atmospheric pressure infrared MALDI imaging mass spectrometry for plant metabolomics. *Anal Chem*, 80:407-420.

Libourel, I.G. andShachar-Hill, Y. 2008. Metabolic flux analysis in plants: from intelligent design to rational engineering. *Ann Rev Plant Biol*, 59:625-650.

Lisec, J., Meyer, R.C., Steinfath, M., Redestig, H., Becher, M., Witucka- Wall, H., Fiehn, O., Torjek, O., Selbig, J., Altmann, T. and Willmitzer, L. 2008. Identification of metabolic and biomass QTL in *Arabidopsis thaliana* in a parallel analysis of RIL and IL populations. *Plant J*, 53:960-972.

Lisec, J., Schauer, N., Kopka, J., Willmitzer, L. and Fernie, A.R. 2006. Gas chromatography mass spectrometry-based metabolite profiling in plants. *Nat Protoc*, 1:387-396.

Lisec, J., Steinfath, M., Meyer, R.C., Selbig, J., Melchinger, A.E., Willmitzer, L. and Altmann, T. 2009. Identification of heterotic metabolite QTL in *Arabidopsis thaliana* RIL and IL populations. *Plant J*, 59:777-788.

Lugan, R., Niogret, M.F., Leport, L., Guegan, J.P., Larher, F.R., Savoure, A., Kopka, J. and Bouchereau, A. 2010. Metabolome and water homeostasis analysis of *Thellungiella salsuginea* suggests that dehydration tolerance is a key response to osmotic stress in this halophyte. *Plant J*, 64:215-229.

Madison Metabolomics Consortium Database. [http://mmcd. nmrfam.wisc.edu].

MAPMAN database [gabi.rzpd.de/projects/MapMan].

Marti, G., Erb, M., Boccard, J., Glauser, G., Doyen, G.R., Villard, N., Robert, C.A., Turlings, T.C., Rudaz, S. andWolfender, J.L. 2013. Metabolomics reveals herbivore-induced metabolites of resistance and susceptibility in maize leaves and roots. *Plant Cell Environ*, 36:621-639.

Maruyama, K.,Urano, K., Yoshiwara, K., Morishita, Y., Sakurai, N., Suzuki, H., Kojima, M., Sakakibara, H., Shibata, D. and Saito, K. 2014. Integrated analysis of the effects of cold and dehydration on rice metabolites, phytohormones, and gene transcripts. *Plant Physiol*, 164:1759-1771.

MASC. 2007. The multinational coordinated *Arabidopsis thaliana* functional genomics project-Annual Report.

MassBank. [http://www.massbank.jp/en/about.html].

Matsuda, F., Nakabayashi, R., Yang, Z., Okazaki, Y., Yonemaru, J., Ebana, K., Yano, M. and Saito, K. 2014. Metabolome-genome-wide association study (mGWAS) dissects genetic architecture for generating natural variation in rice secondary metabolism. *Plant J*, 81:13-23.

Mavituna, F., Cenkci, S. and Walmsley, R. 2005.Modeling of the metabolism of Catharanthusroseus plant cell cultures for improved yields of pharmaceuticals. In: Proceedings of the 7th World Congress of Chemical Engineering, Glasgow, Scotland, 85701-85707.

Mendes, P. 2006. Metabolomics and the challenge ahead. *Brief Bioinformatics*, 7:127.

Metabolomics Standards Initiative (MSI) [http://msi-workgroups.sourceforge. net/].

MetaCycdatabase [http://metacyc.org].

Meyer, R.C., Steinfath, M., Lisec, J., Becher, M., Witucka-Wall, H., Törjék, O., Fiehn, O., Eckardt, A., Willmitzer, L., Selbig, J. and Altmann, T. 2007. The metabolic signature related to high plant growth rate in *Arabidopsis thaliana*. *Proc Natl AcadSci, USA*, 104:4759-4764.

Moço, S., Bino, R.J., Vorst, O., Verhoeven, H.A., de Groot, J., van Beek, T.A., Vervoort, J. and De Vos, C.H. 2006. A liquid chromatography-mass spectrometry-based metabolome database for tomato. *Plant Physiol*, 141:1205-1218.

Morreel, K., Goeminne, G., Storme, V., Sterck, L., Ralph, J., Coppieters, W., Breyne, P., Steenackers, M., Georges, M., Messens, E. and Boerjan, W. 2006. Genetical metabolomics of flavonoid biosynthesis in Populus: a case study. *Plant J*, 47:224-237.

Mroczek, W.J. 1972. Biochemical profiling and the natural history of hypertensive diseases. *Circulation*, 45:1332-1333.

Nakabayashi, R. and Saito, K. 2015. Integrated metabolomics for abiotic stress responses in plants.*CurrOpin Plant Biol*, 24:10-16.

Nemes, P., Barton, A.A. and Vertes, A. 2009.Three-dimensional imaging of metabolites in tissues under ambient conditions by laser ablation electrospray ionization mass spectrometry.*Anal Chem*, 81:6668-6675.

Newell-McGloughlin, M. 2008.Nutritionally improved agricultural crops.*Plant Physiol*, 147:939-953.

Obata, T. and Fernie, A.R. 2012.The use of metabolomics to dissect plant responses to abiotic stresses.*Cell Mol Life Sci*, 69:3225-3243.

OECD Consensus Documents (http://www.oecd.org/document/9/0, 2340, en_2649_201185_1812041_1_1_1_1, 00.html).

Oikawa, A., Nakamura, Y., Ogura, T., Kimura, A., Suzuki, H., Sakurai, N., Shinbo, Y., Shibata, D., Kanaya, S. and Ohta, D. 2006. Clarification of pathway-specific inhibition by Fourier transform ion cyclotron resonance/mass spectrometry-based metabolic phenotyping studies. *Plant Physiol*, 142:398-413.

Okazaki, Y. and Saito, K. 2016. Integrated metabolomics and phytochemical genomics approaches for studies on rice. *BMC Giga Science*, 5:11.

Okazaki, Y. and Saito, K. 2012. Recent advances of metabolomics in plant biotechnology. *Plant Biotechnol Rep*, 6:1-15.

Oksman-Caldentey, K.M. and Saito, K. 2005. Integrating genomics and metabolomics for engineering plant metabolic pathways. *Curr Opin Biotechnol*, 16:174-179.

Paine, J.A., Shipton, C.A., Chaggar, S., Howells, R.M., Kennedy, M.J., Vernon, G., Wright, S.Y., Hinchliffe, E., Adams, J.L., Silverstone, A.L. and Drake, R. 2005. Improving the nutritional value of Golden Rice through increased pro-vitamin A content. *Nature Biotechnol*, 23:482-487.

Palaisa, K.A., Morgante, M., Williams, M.and Rafalski, A. 2003. Contrasting effects of selection on sequence diversity and linkage disequilibrium at two phytoene synthase loci. *Plant Cell*, 15:1795-1806.

Parker, D., Beckmann, M., Zubair, H., Enot, D.P., Caracuel-Rios, Z., Overy, D.P., Snowdon, S., Talbot, N.J. and Draper, J. 2009. Metabolomic analysis reveals a common pattern of metabolic re-programming during invasion of three host plant species by *Magnaporthe grisea*. *Plant J*, 59:723-737.

Pauli, G.F. 2001. qNMR--a versatile concept for the validation of natural product reference compounds. *Phytochem Anal*, 12:28-42.

Plant Metabolic Network - Plant Metabolic Pathway Databases. [http://www.plantcyc.org].

Plantmetabolics database [http://plantmetabolomics.vrac.iastate.edu/ver2/].

Rizhsky, L., Liang, H., Shuman, J., Shulaev, V., Davletova, S. and Mittler, R. 2004. When defense pathways collide. The response of Arabidopsis to a combination of drought and heat stress. *Plant Physiol*, 134:1683-1696.

Roessner, U., Wagner, C., Kopka, J., Trethewey, R.N. and Willmitzer, L. 2000. Simultaneous analysis of metabolites in potato tuber by gas chromatography-mass spectrometry. *Plant J*, 23:131-142.

Rommens, C.M., Haring, M.A., Swords, K., Davies, H.V. and Belknap, W.R. 2007. The intragenic approach as a new extension to traditional plant breeding. *Trends Plant Sci*, 12:397-403.

Rudell, D.R. and Mattheis, J.R. 2009. Superficial scald development and related metabolism is modified by postharvest light irradiation. *Postharvest Biol Technol* , 51:174-182.

Saito, K. and Masuda, F. 2010. Metabolomics for functional genomics, system biology, and biotechnology. *Ann Rev Plant Biol*, 61:463-489.

Saito, K., Dixon, R.A. and Willmitzer, L. 2006. eds. Plant Metabolomics. 57. Berlin: Springer-Verlag.

Sana, T.R., Fischer, S., Wohlgemuth, G., Katrekar, A., Jung, K.H., Ronald, P.C. and Fiehn, O. 2010. Metabolomic and transcriptomic analysis of the rice response to the bacterial blight pathogen *Xanthomonas oryzae* pv. oryzae. *Metabolomics*, 6:451-465.

Sato, S., Soga, T., Nishioka, T. and Tomita, M. 2004. Simultaneous determination of the main metabolites in rice leaves using capillary electrophoresis mass spectrometry and capillary electrophoresis diode array detection. *Plant J*, 40:151-163.

Schauer, N. and Fernie, A.R. 2006. Plant metabolomics: Towards biological function and mechanism. *Trends Plant Sci*, 11:508-516.

Schauer, N., Semel, Y., Roessner, U., Gur, A., Balbo, I., Carrari, F., Pleban, T., Perez-Melis, A., Bruedigam, C., Kopak, J., Willmitzer, L., Zamir, D. and Fernie, A.R. 2006. Comprehensive metabolic profiling and phenotyping of interspecific introgression lines for tomato improvement. *Nature Biotechnol*, 24:447-454.

Schubert, D.R. 2008. The problem with nutritionally enhanced plants. *J Med Food*, 11:601-605.

Scossa, F., Brotman, Y., de Abreu e Lima, F., Willmitzer, L., Nikoloski, Z., Tonga, T. and Fernie, A.R. 2016. Genomics-based strategies for the use of natural variation in the improvement of crop metabolism. *Plant Sci*, 242:47-64.

Sekiyama, Y. and Kikuchi, J. 2007. Towards dynamic metabolic network measurements by multi-dimensional NMR-based fluxomics. *Phytochemisrty*, 68:2320-2329.

Shi, H., Ye, T., Zhong, B., Liu, X., Zhulong, C. 2014. Comparative proteomic and metabolomic analyses reveal mechanisms of improved cold stress tolerance in bermudagrass (*Cynodon dactylon* (L.) Pers.) by exogenous calcium. *J Integr Plant Biol*, 56:1064–1079.

Shi,H., Jiang,C., Ye,T., Tan,D.X., Russel, J., Heng,R., Liu, Z.R., Zhulong, C. 2015. Comparative physiological, metabolomic, and transcriptomic analyses reveal mechanisms of improved abiotic stress resistance in Bermuda grass [*Cynodondactylon* (L). Pers.] by exogenous melatonin. *J Exp Bot*, 66:681–694.

Shulaev, V., Cortes, D., Miller, G. and Mittler, R. 2008. Metabolomics for plant stress response. *Physiol Plant*, 132:199-208.

Shyur, L.F. and Yang, N.S. 2008. Metabolomics for phytomedicine research and drug development. *Curr Opin Chem Biol*, 12:66-71.

Simó, C., Ibáez, C., Valdés, A., Cifuentes, A. and García-Cañas, V. 2014. Metabolomics of genetically modified crops. *Int J MolSci*, 15:18941-18966.

Smith, C.A., Want, E.J., Maille, G.O., Abagyan, R. and Siuzdak, G. 2006. XCMS: processing mass spectrometry data for metabolite profiling using nonlinear peak alignment, matching, and identification. *Anal Chem*, 78:779-787.

Soanes, D.M., Chakrabarti, A., Paszkiewicz, K.H., Dawe, A.L. and Talbot, N.J. 2012. Genomewide transcriptional profiling of appressorium development by the rice blast fungus *Magnaporthe oryzae*. *PLoSPathog*, 8:e1002514.

Spencer, J.P., Kuhnle, G.G., Hajirezaei, M.,cMock, H.P., Sonnewald, U. and Rice-Evans, C. 2005. The genotypic variation of the antioxidant potential of different tomato varieties. *Free Radic Res*, 39:1005-1016.

Steinfath, M., Strehmel, N., Peters, R, Schauer, N., Groth, D., Hummel, J., Steup, M., Selbig, J., Kopka, J., Geigenberger, P. and van Dongen, J.T. 2010. Discovering plant metabolic biomarkers for phenotype prediction using an untargeted approach. *Plant Biotechnol J*, 8:900-911.

Stich, B. Piepho, H.P., Schulz, B. and Melchinger, A.E. 2008. Multi-trait association mapping in sugar beet (*Beta vulgaris* L.).*Theor Appl Genet*, 117:947-954.

Stitt, M. and Fernie, A.R. 2003. From measurements of metabolites to metabolomics: an "on the fly" perspective illustrated by recent studies of carbon-nitrogen interactions. *Curr Opin Biotechnol*,14:136-144.

Sumner L.W. 2010. Recent advances in plant metabolomics and greener pastures. *F1000 Biol Rep*, 2:7.

Takahara, K., Kasajima, I., Takahashi, H., Hashida, S.N., Itami, T., Onodera, H., Toki, S., Yanagisawa, S., Kawai-Yamada, M. and Uchimiya, H. 2010. Metabolome and photochemical analysis of rice plants over expressing *Arabidopsis* NAD kinase gene. *Plant Physiol*, 152:1863-1873.

Takahashi, H., Munemura, I., Nakatsuka, T., Nishihara, M. and Uchimiya, H. 2009. Metabolite profiling by capillary electrophoresis-mass spectrometry reveals aberrant putrescine associated with idiopathic symptoms of gentian plants. *J Horticult Sci Biotechn*, 84:312-318.

Terpmed database [http://www.terpmed.eu/databases.html].

Tohge, T. and Fernie, A.R. 2009. Web-based resources for mass-spectrometry based metabolomics: a user's guide. *Phytochemistry*, 70:450-456.

Tohge, T. and Fernie, A.R. 2010. Combining genetic diversity, informatics and metabolomics to facilitate annotation of plant gene function.*NatProtoc*, 5:1210-1227.

Tokimatsu, T., Sakurai, N., Suzuki, H.,Ohta, H., Nishitani, K., Koyama, T., Toshiaki, U., Misawa, M., Saito, K. and Shibata, D. 2005. KaPPA-View: A web-based analysis tool for integration of transcript and metabolite data on plant metabolic pathway maps. *Plant Physiol*, 138:1289-1300.

Urano, K., Kurihara, Y., Seki, M. and Shinozaki, K. 2010. 'Omics' analyses of regulatory networks in plant abiotic stress responses. *Curr Opin Plant Biol*, 13:132-138.

vanBeek, T.A. 2005. Ginkgolides and bilobalide: their physical, chromatographic and spectroscopic properties. *Bioorg Med Chem*,13:5001-5012.

vanBeek, T.A. and Montoro, P. 2009. Chemical analysis and quality control of Ginkgo biloba leaves, extracts, and phytopharmaceuticals. *J Chromatogr A*, 1216:2002-2032.

van der Kooy, F., Verpoorte, R. and Meyer, J.J.M. 2008. Metabolomic quality control of claimed anti-malarial *Artemisia afra* herbal remedy and *A. afra* and *A. annua* plant extracts. *S Afr J Bot*, 74:186-189.

van Dongen, J.T. and Schauer, N. 2010. Metabolic marker as selection tool in plant breeding ISB News Report.

Vanholme, R., Morreel, K., Ralph, J. and Boerjan, W. 2008. Lignin engineering.*Curr Opin Plant Biol*, 11:278-285.

Verslues, P.E., Agarwal, M., Katiyar-Agarwal, S., Zhu, J. and Zhu, J.K. 2006. Methods and concepts in quantifying resistance to drought, salt and freezing, abiotic stresses that affect plant water status. *Plant J*, 45:523-539.

Vorst, O., de Vos, C.H.R., Lommen, A., Staps, R.V., Visser, R.G.F., Bino, R.J. and Hall, R.D. 2005. A non-directed approach to the differential analysis of multiple LC-MS derived metabolic profiles. *Metabolomics*, 1:169-180.

Ward, J.L., Harris, C., Lewis, J. and Beale, M.H. 2003. Assessment of 1H NMR spectroscopy and multivariate analysis as a technique for metabolite fingerprinting of *Arabidopsis thaliana*. *Phytochemistry*, 62:949-957.

Watkins, S.M., Hammock, B.D., Newman, J.W. and German, J.B. 2001. Individual metabolism should guide agriculture toward foods for improved health and nutrition. *Am J ClinNutr*, 74:283-286.

Wen,W., Li, D., Li, X., Gao, Y., Li, W., Li, H., Liu, J., Liu, H., Chen, W., Luo, J. and Yan, J. 2014. Metabolome-based genome-wide association study of maize kernel leads to novel biochemical insights. *Nat Commun,* 5:3438.

Wentzell, A.M., Rowe, H.C., Hansen, B.G., Ticconi, C., Halkier, B.A. and Kliebenstein, D.J. 2007. Linking metabolic QTLs with network and cis QTLs controlling biosynthetic pathways. *PLoS Genet,* 3:1687-1701.

Widodo, Patterson J.H., Newbigin E., Tester M., Bacic A. andRoessner U. 2009. Metabolic responses to salt stress of barley (*Hordeum vulgare* L.) cultivars, Sahara and Clipper, which differ in salinity tolerance. *J Exp Bot,* 60:4089-4103.

Wilson, L.M., Whitt, S.R., Ibáñez, A.M., Rocheford, T.R., Goodman, M.M. and Buckler, E.S. 2004. Dissection of maize kernel composition and starch production by candidate gene association. *Plant Cell,* 16:2719-2733.

Wulff-Zottele, C., Gatzke, N., Kopka, J., Orellana, A., Hoefgen, R.,Fisahn, J. and Hesse, H. 2010. Photosynthesis and metabolism interact during acclimation of *Arabidopsis thaliana* to high irradiance and sulphur depletion. *Plant Cell Environ,* 33:1974-1988.

Wurtele, E.S., Li, L., Berleant, D., Cook, D., Dickerson, J.A., Ding, J., Hofmann, H., Lawrence, M., Lee, E.K., Li, J., Mentzen, W., Miller, L., Nikolau, B.J., Ranson, N. and Wang, Y. 2007. MetNet: Systems biology software for *Arabidopsis* In Concepts in Plant Metabolomics. Edited by Nikolau, B.J., Wurtele, E.S.: Dordrecht, The Netherlands: Springer, 145-58.

Xie, D.Y., Sharma, S.B., Wright, E., Wang, Z.Y. and Dixon, R.A. 2006. Engineering plants for introduction of health-beneficial proanthocyanidins. *Plant J,* 45:895-907.

Yamakawa, H. and Hakata M. 2010. Atlas of rice grain filling-related metabolism under high temperature: Joint analysis of metabolome and transcriptome demonstrated inhibition of starch accumulation and induction of amino acid accumulation. *Plant Cell Physiol,* 51:795-809.

Ye, X., Al-Babili, S., Kloti, A., Zhang, J., Lucca, P., Beyer, P. and Potrykus, I. 2000. Engineering the provitaminA (@-carotene) biosynthetic pathway into (carotenoid-free) rice endosperm. *Science,* 287:303-305.

Yonekura-Sakakibara, K., Tohge, T., Niida, R. and Saito, K. 2007. Identification of a flavonol 7-orhamnosyltransferase gene determining flavonoid pattern in *Arabidopsis* by transcriptome coexpression analysis and reverse genetics. *J Biol Chem,* 282:14932-14941.

Zhao, J., Paulo, M.J., Jamar, D., Lou, P., van Eeuwijk, F., Bonnema, G., Vreugdenhil, D. and Koornneef, M. 2007. Association mapping of leaf traits, flowering time, and phytate content in *Brassica rapa. Genome,* 50:963-973.

Zhao, Q., Stoyanova, R., Du, S.Y., Sajda, P. and Brown, T.R. 2006. HiRes--a tool for comprehensive assessment and interpretation of metabolomic data. *Bioinformatics*, 22:2562-2564.

Zhou W., Yasuor H., Fischer A.J. and Tolstikov V.V. 2011. Trace metabolic profiling and pathway analysis of clomazone using liquid chromatography coupled with triple Quadruple-Linear Ion Trap Mass Spectrometry in Predictive Multiple Reaction Monitoring Mode. *LCGC N Am*, 29:860-869.

Zhou, J., Ma, C., Xu, H., Yuan, K., Lu, X., Zhu, Z., Wu, Y., and Xu, G. 2009. Metabolic profiling of transgenic rice with *cryIAc* and *sck* genes: an evaluation of unintended effects at metabolic level by using GC-FID and GC-MS. *J Chromatogr Analyt Technol Biomed Life Sci*, 877:725-732.

23

Role of Biotechnology for Enhancing Yield Under Water Deficit Stress Conditions

Lovejot Kaur and C. Appunu*

ICAR-Sugarcane Breeding Institute, Coimbatore-641007

**Corresponding author: cappunu@gmail.com*

Abstract

Worldwide agriculture accounts for more than 80% of all freshwater used for various activities involved in crop production. **But, the water use** is not sustainable i.e more water is used then replenished. Water resources can be used more sustainably by using different biotechnological approaches along with various agronomic and physiological approaches for efficient and sustainable agricultural production. Sugarcane is one of the highest water demanding commercial crop. Therefore, it necessitates to develop sugarcane crops that require less water to produce sufficient sugar and biomass yield. Conventional breeding for production of drought tolerant sugarcane has a limited success due to complex ploidy of the crop. On the other hand transgenic crops have assisted in reducing the water requirement of several crops. In case of transgenic sugarcane the rate of loss of water through transpiration is reduced while yield is improved under drought conditions. This chapter discusses on different biotechnological approaches used to combat water stress in sugarcane and also effect of engineering drought tolerance in sugarcane on different growth parameters.

Keywords: Sugarcane, Water Deficit Stress, Tolerance, Genetic Engineering, Growth Parameters

Introduction

Sugarcane is a rainfed crop which depend heavily on the amount and duration of precipitation, humidity, moisture content, temperature and soil condition (Gawander, 2007). During its 270-365 days of growth the demand for water is high (1000-1500). Development of sugarcane plant is divided into four distinct phases i.e. germination, tillering, grand growth and maturity. Tillering and grand growth phases are most susceptible to water stress. Ironically, these phases are also critical for sugarcane productivity. During these critical phases, water stress directly affect final yield through reduction of growth, dry matter accumulation, cane yield and juice quality (Naidu *et al.*, 1987). Due to this, the shortage of water adversely affects the crop yield.

With green revolution mankind has witnessed tremendous yield gains to many crops in many parts of the developing world. But in current situation land availability is static/reduced, so there is a need to efficiently grow crops which can tolerate the risks of different abiotic and biotic stresses. Agricultural biotechnology is a tool for increasing food production and also making agriculture more sustainable from an environmental point of view (Hansson and Joelsson, 2013). Genetic modification reduces the water requirement of crops since traits that are used for genetic manipulation increase's the rate of photosynthesis and depth of root structure and decrease the rate at which water is lost through transpiration. Hence, biotechnology has a tremendous potential to reduce the amount of water required for improving productivity.

Typically, sugarcane respond to water stress in the form of reduced length of stalk, decrease in shoot branching and reduction in leaf senescence. Interestingly, a mild drought stress can have positive impact on sugarcane yield (Gentile *et al.*, 2015). Withholding the irrigation before the harvest is an important strategy to enhance sucrose content in stalk. But, if the water deficit becomes too severe photosynthesis is inhibited resulting in lowering of cane and sucrose yields (Robertson *et al.*, 1998). Sugarcane is a C_4 plant with high water use efficiency and has the capacity to maintain the leaf photosynthesis even with closed stomata. Studies undertaken with C_4 plants have shown that enzymes that comprise the metabolic CO_2 pump are more resistant to water deficit than the enzymes of C_3 photosynthesis (Ghannoum, 2009). Despite of the above facts sugarcane is prone to drought induced inhibition. Also, Andrade *et al.* (2015) suggested that patterns of gene expression in sugarcane vary in different genotypes classified as drought tolerant indicating that there is high degree of complexity in response of sugarcane to water stress. Large proportion of area under sugarcane production is susceptible to severe water stress thus affecting sugarcane at one or the other stage of growth. Above facts suggests that sugar yield and productivity are severely affected under limited moisture. In this case biotechnology can provide a sustainable solution by enabling us to make efficient use of limited water resources.

Different Approaches Adopted for Engineering Drought Tolerance

Most biotechnological approaches using transgenics to improve performance under drought have targeted survival and reductions in water use, not improved photosynthesis under a given supply of water (Parry *et al.*, 2005). A number of different biotechnological approaches can be used to produce sugarcane crop that are better able to cope with water scarcity in agriculture *viz.* marker assisted selection and genetic engineering. Chen *et al.* (1987) first reported the genetic transformation in sugarcane through PEG mediated transformation. This was followed by use of biolistic approach for sugarcane transformation (Bower, 1992). *Agrobacterium* mediated transformation in sugarcane was first reported by Enriquez *et al.* (1997) and Arencibia *et al.* (1998).

In the past years various genes has been tested for drought resistance in major crops. Yang *et al.*, (2010) classified these genes into the following three categories based on their biological functions: (1) stress-responsive transcriptional regulation (2) post-transcriptional RNA or protein modifications such as phosphorylation/de-phosphorylation and farnesylation (3) osomoprotectant metabolism or molecular chaperones. Summary of all the recent approaches used for engineering drought tolerance in sugarcane are listed in Table 1

Table 1: Summary of different genes used during the past few years for engineering drought tolerance in sugarcane

Gene name	Gene function	Promoter	Method of transformation	References
Arabidopsis Bax Inhibitor-1 (*AtBI-1*)	Anti - cell death	Ubi-1	Particle bombardment	Ramiro *et al.*, (2016)
Arabidopsis DREB 2A (*AtDREB2A CA)*	Binds to cis-acting DRE sequence to activate the expression of downstream genes that are involved in drought, salt and heat stress	Rab 17	Particle bombardment	Rafaela *et al.*,(2014)
Arabidopsis Vacuolar Pyrophosphatase (*AVP1*)	Confers resistance against higher concentrations of NaCl and water deprivation	CaMV 35S	*Agrobacterium*-mediated transformation	Kumar *et al.*, (2014)
Erianthus arundinaceus HSP70 (*EaHSP70*)	HSP70 stabilizes the proteins by binding to its substrates and prevents denaturation or aggregation	Port Ubi 2.3	*Agrobacterium*-mediated transformation	Augustine *et al.*, (2015)
E. arundinaceus DREB2 (*EaDREB2)* and pea DNA helicase 45 (*PDH45)*	DNA helicase	Port Ubi 2.3	*Agrobacterium*-mediated transformation	Augustine *et al.*, (2015)
Arabidopsis vacuolar H+-pyro-phosphatase (*H+-PPase*)	Acidifies vacuoles in plant cells by pumping H+ from the cytoplasm into vacuoles	35S promoter with double enhancer	Particle bombardment	Raza *et al.*, (2016)

Gene name	Gene function	Promoter	Method of transformation	References
D1-pyrroline-5-carboxylate synthetase (*P5CS)*	P5CS plays the key role in proline biosynthesis and catalyzes the major regulated step	AIPC	Particle bombardment	Molinari *et al.*, (2007)
Sugarcane drought-responsive (*Scdr1*)	Scdr1 encodes a protein of unknown function	CaMV 35S	*Agrobacterium*-mediated transformation	Begcy *et al.*, (2012)
Sugarcane ethylene responsive factor (*SodERF3*)	Transcriptional regulator of the Ethylene Responsive Factor (ERF) superfamily	CaMV 35S	-------	Trujillo *et al.*, (2009)
Sugarcane dirigent protein gene *(ScDir)*	Direct the outcome of bimolecular coupling reactions toward regio- and stereospecific product formation.	-	-	Jin-long *et al.*, (2012)
Sugarcane MYB (*SoMYB18*)	Activate the expression of downstream genes that are involved in stress	CaMV 35S	*Agrobacterium*-mediated transformation	Shingote *et al.*, (2015)
Trehalose synthase gene *(TSase)*	Catalyze the intramolecular rearrangement of maltose into trehalose	2 x CaMV 35S	*Agrobacterium*-mediated transformation	Wang *et al.*, (2003)

(A) Transcription factors (TFs)

Different stresses such as water stress, salinity, cold etc induces either ABA – dependent or ABA independent pathways on signal perception. These pathways mediate adaptation to stress by the activation of at least two different regulons (Lata and Prasad, 2011). TFs involved in ABA dependent pathway signalling are MYC/MYB and bZIP. DREB1/DREB2 and NAC TFs are involved in ABA independent pathways. However, it has been suggested that some DREB1/DREB2 TFs can respond to an ABA-dependent pathways (Haake *et al.*, 2002). Genetic analysis indicates that there is no clear line of demarcation between ABA-dependent and ABA–independent pathways (Knight and Knight, 2001). Of all the TFs involved in these pathways DREB has been most studied. Generally, in *Arabidopsis* expression of DREB1/CBF gene is induced by cold and DREB2 induced by dehydration, high-salinity and heat stresses (Liu *et al.*, 1998; Shinwari *et al.*, 1998; Nakashima *et al.*, 2000). It has been shown in other grasses like maize (Qin *et al.*, 2007), rice

(Bihani *et al.,* 2011; Chen *et al.,* 2008), wheat and barley (Morran *et al.,* 2011) that overexpression of DREB2a results in improved stress tolerance.

In sugarcane, Reis *et al.,* (2014) showed overexpression of AtDREB2A gene under the control of the *Rab17* promoter enhanced the tolerance of sugarcane variety RB855156 to water deficit as the plants had improved initial bud sprouting, increased culm linear and internode lengths and high sucrose content with no biomass penalty. Augustine *et al.,* (2015) reported that expression of EaDREB2 is enhanced by drought stress in *Erianthus arundinaceus*. Similar group showed that transformation of sugarcane with the EaDREB2 gene under the control of the Port Ubi 2.3 promoter enhanced the tolerance to water deficit and salinity stress through improved physiological adaptation and enhanced stress-related gene expression.

A nobel sugarcane ethylene responsive factor (SodERF3) was cloned from ethephon- treated young sugarcane leaves (Trujillo *et al.,* 2009). SodERF3 is induced in sugar cane leaves by ethylene, abscisic acid, salt stress and wounding. Transgenic tobacco plants expressing SodERF3 displayed increased tolerance to drought and osmotic stresses without any visible phenotypic change in growth and development. These results suggest that SodERF3, if overexpressed in sugarcane and other grasses to improve drought tolerance.

Recent reports of MYB TF family playing important role in stress signalling gaining momentum (Abe *et al.,* 2003; Cominelli *et al.,* 2005; Dubos *et al.,* 2010; Amwabat *et al.,* 2013). In rice, Yang *et al.,* (2012) reported a MYB likely to play role in regulation of diverse stresses like salt, cold and dehydration. In wheat, two MYB TFs namely TaMYB1 and TaMYB2 participate in response to salt, ABA and osmotic stress (Wang *et al.,* 2006). Prabhu and Prasad (2012) identified stress-associated MYB transcription factor ScMYBAS1 in sugarcane. Shingote *et al.,* (2015) isolated a MYB gene (*SoMYB18*) from sugarcane variety Co740. This gene was transferred in tobacco and stable transgenic plants were evaluated for salt, cold and drought tolerance. Two detoxification enzymes *i.e* SOD and CAT, proline accumulation and chlorophyll content was found elevated level of expression in transgenic plants. These reports suggest that MYB TFs has potential for engineering to achieve drought tolerance in sugarcane.

(B) Role of Different Osmolytes

Drought tolerance in plants can be improved by engineering the genes which accumulate metabolites that function as osmoprotectants or increase water retention through osmotic adjustments. Proline is one of the most common compatible solutes in water stressed plants and in many other organisms (Delauney and Verma, 1993). During drought, synthesis of proline is enhanced while degradation is reduced. In grasses there are very limited reports of using genes involved in proline biosynthesis for engineering drought tolerance. The rate limiting enzyme of proline biosynthesis *i.e.* P5CS was overexpressed, ability of transgenic rice to tolerate stress enhanced (Zhu *et al.,* 1998). You *et al.,* (2012) reported that transgenic rice plants overexpressing ornithine δ-aminotransferase gene *OsOAT* confers drought and oxidative stress tolerance. Encouraging results have been obtained from petunia (Yamada *et al.,* 2005) and soybean (De Ronde *et al.,* 2004) by over expressing osmolyte accumulating genes. In sugarcane, a study evaluated the role of proline as an osmoprotectant and as a component of antioxidative defense system in response to severe water-deficit conditions, using transgenic sugarcane plants expressing a heterologous P5CS gene controlled by the stress inducible promoter AIPC (Molinari et al., 2007). They reported that stress-induced proline production confers drought stress tolerance to P5CS-transformed sugarcane plants. They suggested that instead of proline acting as an osmotic adjustment mediator it increase the antioxidant system, thus protecting photosynthetic apparatus.

Several sugar alcohols including trehalose have been used to induce stress tolerance in plants. Different reports suggested that introduction of trehalose biosynthetic genes into plants can improve drought tolerance without any negative effect on plant growth or productivity. Trehalose has two major functions; it acts as reserve carbohydrates and protect proteins from denaturation. Trehalose 6-phospate synthase (TPS) and Trehalose 6-phosphate (TPP) phosphatase are involved in trehalose biosynthesis. In rice, a TPP/TPS fusion gene from the *Escherichia coli* trehalose biosynthetic genes (*otsA* and *otsB*) was used to engineer plants for increased drought tolerance (Garg *et al.,* 2002; Jang *et al.,* 2003). Positive results were obtained. Additionally, plants exhibited less photo oxidative damage under salt and low-temperature stresses. Trehalose synthase gene *(TSase)* from *Grifola frondosa* was transferred into sugarcane *(Saccharum officinarum* L.) to improve sugarcane drought-tolerance (Wang *et al.,* 2003). The results indicated that transformed sugarcane plants showed multiple phenotypic alterations and exhibited increase in tolerance of osmotic stress.

C) Role of Heat/Cold Shock Proteins

Different stresses cause denaturation/mis-folding of proteins which renders them dysfunctional. Organisms have mechanism to cope from this loss by encoding heat shock or cold shock proteins. Heat shock proteins are chaperones that plants require for quick adaptation to temperature changes and recovery from the loss. In unstressed cell they exist as monomers while in stressed conditions they exist as trimers. Role of HSPs in stress tolerance has been indicated in different plant species (Malik *et al.,* 1999; Li *et al.,* 2003; Sun *et al.,* 2001). Katiyar *et al.,* (2003) introduced *Arabidopsis thaliana* hsp101 (*Athsp*101) cDNA into the Pusa basmati 1 cultivar of rice (*Oryza sativa* L.). Adverse effect of over-expression of the transgene on overall growth and development of transformants were not seen and thermotolerance advantage appeared to be solely due to over-expression of HSP101. The overexpression of mitochondrial HSP70/DnaK has been linked to suppression of programmed cell death in transgenic rice (Qi *et al.,* 2011). Augustine *et al.,* (2015) isolated the gene coding for HSP70 from *E. arundinaceus* and incorporated it into the sugarcane. Overexpression of EaHSP70 enhances tolerance to drought and salt stresses in sugarcane.

Another important role of heat shock proteins that some of its members such as HSP90 and HSP100, can cross-link microfilaments (Koyasu *et al.*, 1986). It has been implicated in mammalian and animal systems (Haus *et al.*, 1993, Simard, 2011) that HSP70 is involved in actin polymerization. Augustine et al. (2015) reported that overexpression of *E. arundinaceus* HSP70 gene (EaHSP70) in sugarcane (*Saccharum* spp. hybrid) leads to the formation of anisotropic interdigitation under drought stress. Confocal microscopy was used to show anisotropic interdigitation with high cortical actin intensities in the epidermal cells of *E. arundinaceus* under soil moisture stress. They also developed a bioinformatic data mining protocol and identified a connection between growth-related auxin-binding protein (ABP) and HSPs.

D) Role of Vacuolar Pyrophosphatase

Higher proton electrochemical gradient can be generated by overexpressing the activity of an H+ pump on the vacuolar membrane so as to mobilize more of H+ into the vacuoles. This helps to energize secondary transporters including vacuolar Na+ /H+ antiporters and ultimately lead to increase in both salt and drought tolerance in transgenic plants. Gaxiola *et al.*, (2001) showed that in transgenic *Arabidopsis* plants on overexpression of the *Arabidopsis* gene AVP1 that encodes a vacuolar H+ pyrophosphatase, the drought and salt tolerance can be increased significantly. Similar reports were obtained from rice (Zhao *et al.*, 2006), maize (Li *et al.*, 2008) and barley (Schilling *et al.*, 2014).

Overexpression of the H-pyrophosphatase (H-PPase) AVP1 results in generation of salt and water stress tolerant sugarcane (Kumar *et al.*, 2014). They also reported that transgenic plants developed more vigorous root system as compared to the wild type plants. More vegetative development of roots thus helps in providing higher water absorption and retention capacity from the soil. Moreover it also enhances nutrient use efficiency. Li *et al.*, (2005) also reported that the AVP1 overexpressing in *Arabidopsis* greatly enhanced root development by facilitating recovery under water-deficit conditions. Raza *et al.*, (2016) showed that overexpression of *Arabidopsis* AVP1 gene in sugarcane improved growth and physiological parameters, such as photosynthesis rate, stomatal conductance, RWC, osmotic and turgor potentials in transgenic plants. Above mentioned reports clearly implies that AVP1 gene can be thought of as potential target for engineering drought tolerance in sugarcane.

E) Engineering Dirigent Proteins for Drought Tolerance

Dirigent (DIR) proteins are believed to mediate the free radical coupling of monolignol phenols in plants to yield lignans and lignins (Burlat *et al.*, 2001). The presence of these proteins has been reported in many plants ranging from lichens to angiosperms. They are mainly involved in lignin production and lignification as well as role in biotic/abiotic stresse response. Wu *et al.,* (2009) reported that a dirigent gene (BhDIR1) from a flower *Boea hygrometrica* was involved in the response to plant dehydration.

Jin-long *et al.,* (2012) obtained a full-length cDNA sequence of a dirigent like gene from sugarcane based on the stem full-length cDNA library. This sequence belonged to plant DIRd subfamily member of ScDir homolog from sugarcane. The ScDir transcript levels in sugarcane seedling increased under H_2O_2, PEG or NaCl stresses. The expression level of ScDir was significantly upregulated under PEG stress with maximum at 12 h after stress. Authors concluded that both the ScDir-hosted cell performance and the enhanced expressions in sugarcane imply that the ScDir gene is involved in the response to abiotic stresses of drought, salts and oxidation

F) Engineering Proteins Involved in Cell Death Inhibition for Drought Tolerance

If plant faces a severe stress, there is a boom in ROS (reactive oxygen species) production resulting in damage of molecular and cellular components in the plant. This ultimately leads to programmed cell death (PCD). PCD is an integral mechanism of plant, animal and microbes that functions in the self-destruction of those cells that are damaged by various stress factors (Lam, 2004). Cell suicide is controlled by a well regulated balance between anti-apoptotic proteins and pro-apoptotic proteins. These proteins tightly regulate the induction of ROS signalling. BAX is a type of pro-apoptotic protein. On the other hand the gene Bax inhibitor 1 is a an anti-cell death gene present in both animal and plant genomes. BAX protein induces cytochrome c release, stimulation of caspases and cleavage of essential proteins leading to cell death (Danial and Korsmeyer, 2004). The AtBI-1 gene encodes a cytoprotective protein which expresses during senescence and stress conditions and modulate the activation of PCD (Watanabe and Lam, 2009). *In planta* validation of the role of BI 1 was obtained when Kawai-Yamada *et al.,* (2001) reported that cell death was reduced in transgenic Arabidopsis expressing BAX when retransformed with AtBI-1. Ramiro *et al.,* (2016) showed that in transgenic sugarcane expression of a highly conserved cell death suppressor, BAX Inhibitor-1 from *Arabidopsis thaliana* (AtBI-1), has ability to confer increased tolerance to long-term (>20 days) water stress conditions. BI 1 gene also increased plants tolerance to endoplasmic stress.

G) Role of Helicase

Helicase are motor proteins involved in unwinding of either stable duplex DNA or duplex RNA secondary structures. In plants, the first DNA helicase was identified in Lily by Tuteja *et al.,* (2003). Ten different DNA helicase has been isolated in different plant species. They are present in three organelles of the plant cell:nucleus, mitochondria and chloroplast. Number of DNA helicases play important role in gene regulation at various developmental stages as well as in different stress conditions. Potential role of a pea DNA helicase *i.e* PDH45 has been explored. The PDH45 gene is up-regulated by different abiotic stresses like dehydration, wounding, and low temperature (Sanan *et al.*, 2005). Previous studies reported that overexpression of this gene leads to drought tolerance in peanut (Manjulatha *et al.*, 2014). Augustine *et al.*, (2015) reported that the overexpression of PDH45 in transgenic sugarcane, results in enhanced cell membrane thermostability and upregulation of stress-responsive genes leading to abiotic stress tolerance in sugarcane. It was also found that in PDH45 overexpressing sugarcane, the upregulation of DREB2, HSP70, LEA, RD29, ERD, ERF, COR15, and BRICK abiotic stress-responsive genes. This study suggests that overexpression of PDH45 may be involved in the expression of downstream stress-related genes and thus confers abiotic stress tolerance.

Role of MicroRNA

MicroRNAs (miRNAs) are small noncoding RNAs of usually 20-24 nucleotides length, which acts as regulators of genes at post-transcriptional levels in number of organisms. Function of some of the miRNAs which are regulated by drought stress are functionally conserved across plant species and these miRNAs mediate the responses by modulating the amount of themselves, the amount of mRNA targets or the activity/mode of action of miRNA–protein complexes (Ferdous *et al.*, 2015). In plants the drought responsiveness of miRNA is species dependent. It has also been studied that even members of the same family respond differently to drought stress. Zhou *et al.*, (2013) reported that in bent grass, drought stress down- and upregulates members of miR319 family. Same miRNA in the same plant species can show different responses to drought depending on the real time conditions.

Until today, in sugarcane, of all the reports that analyzed miRNA expression in response to drought stress only few have implicated miRNA those are related to drought stress (Ferreira *et al.*, 2012; Thiebaut *et al.*, 2012; Gentile *et al.*, 2013). Ferreira *et al.*, (2012) identified 18 miRNA families by deep sequencing of small RNAs. Seven of the miRNAs were differentially expressed during drought. Six of these miRNAs were differentially expressed at two days of stress, and five miRNAs were differentially expressed at four days. The expression levels of five miRNAs were validated by RT-qPCR. Also identified six precursors and the targets of the differentially expressed miRNA using an *in silico* approach. miRNA expression under drought stress in sugarcane suggest that plants miRNA expression may adjust their micro-transcriptome in different ways to overcome different phases and intensity of drought stress and these responses may vary in different genetic backgrounds (Gentile *et al.,* 2013). Schematic approaches for engineering/manipulating of different group of genes for improving drought tolerance in sugarcane (Fig. 1).

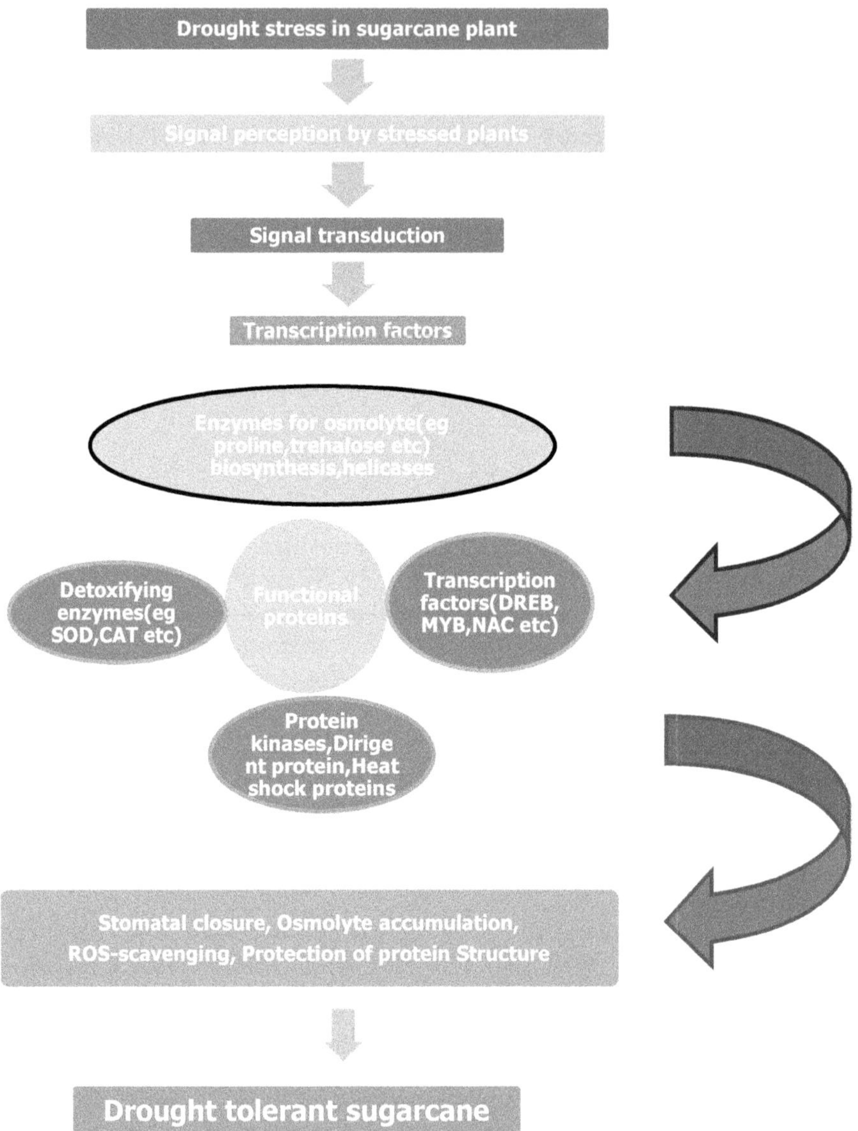

Fig.1: Schematic representation of approaches for engineering/manipulating of different group of genes for improving drought tolerance in sugarcane.

Effect of Engineering Drought Tolerance in Sugarcane on Different Growth Parameters

As a long term goal to improve stalk and sugar productivity under drought conditions, different physiological and biochemical parameters are used for evaluation of sugarcane clones. During phases of sugarcane growth most sensitive to drought is tillering and grand growth phases. Physiological and biochemical parameters are used to identify and distinguish genotypes tolerant to drought (Endres *et al.*, 2010). These parameters can be applied as base for research and development of drought tolerant cultivars. These are discussed as under

Physiological Parameters

C_4 plant like sugarcane can survive drought more efficiently by reducing the photorespiration rate and water loss. Even such mechanisms are present, sugarcane plant is susceptible to water deficit. Longer periods of drought on sugarcane crops can significantly decrease growth, productivity and quality of product (Wiedenfeld, 2000). However, C4 plants photosynthesis is highly sensitive to water deficit (Ghannoum, 2009). Different parameters that are used to select drought tolerant genotypes of sugarcane are: RWC, gas exchange parameters (stomatal conductance, photosynthetic rate, transpiration rate (Tr), electrolyte leakage (EL), cell membrane thermostabilty, intercellular CO_2 concentration (Ci), water use efficiency (WUE), leaf temperature, chlorophyll content (SPAD unit) and Photosynthetic efficiency (Fv/Fm).

Measuring Relative water content (RWC) is one of the efficient methods used to select and classify drought tolerant or susceptible genotypes. Both cellular and tissues hydration is indicated by RWC, which is important for the physiological plant metabolism (Silva *et al.*, 2007). Changes in RWC seem to directly affect all photosynthetic apparatus in sugarcane plants (Graca *et al.*, 2010). Plants tolerant to water stress have potential to successfully maintain the higher RWC by osmotic adjustment.

RWC is Calculated Using a Formula

$RWC(\%) = [(FM - DM)/(TM - DM)] \times 100$ where FW, DM and TM are the fresh, dry and turgid weights respectively. Raza *et al.*, (2016) reported that overexpression of AVP1 in transgenic sugarcane plants increased tolerance to drought stress this was indicated by increased relative water content (RWC) and leaf water (Ψw), osmotic (Ψs), and turgor potential (Ψp).

Photosynthetic efficiency: The maximum potential photochemical efficiency, is defined as the ratio of variable to maximum fluorescence emitted by chlorophyll (Fv/Fm). In healthy leaves, the Fv/Fm value is usually close to 0.8 in most plant species; A lower Fv/Fm value indicates that a proportion of PS II reaction centres are damaged or inactivated (Baker and Horton 1987; Baker and Rosenqvist, 2004;

Zlatev 2009). Generally, water deficit stress decreases the photochemical efficiency (PS II), and the ability of the cultivar to maintain a high level of *Fv/Fm*. Under drought stress the photosynthetic efficiency reduced to 0.22% as compared to unstressed with a final water potential -1.6MPa (Koonjah *et al.*, 2006).

Ability of the plant to tolerate drought is also indicated by the extent of cell-membrane injury under drought conditions. To estimate the cell membrane injury percentage, the cell-membrane thermostability test is carried out in V0 and V1 generations following the method described by Martineau *et al.*, (1979). Membrane Injury percentage can be computed using the following formula: Membrane injury % =1-[(1-T1/T2)/(1-C1/C2)]×100 where T and C refer to the values for treatment and control samples, and the subscripts 1 and 2 denote the initial and final conductance, respectively. Under high temperatures, the reduction of cell membranes stability is noted. According to Prasad *et al.*, (2008) and Sage and Kubien, (2007) damages in photosynthesis are more closely related to changes in membrane properties. In a study Augustine *et al.*, (2015) reported that transgenic sugarcane overexpressing EaDREB2 and the co-transformed events with PDH45 were found to have increased membrane stability under normal irrigated condition. Further increase in the membrane stability was observed under moisture stress.

Different studies in sugarcane have shown that gas-exchange parameters, such as stomatal conductance, transpiration rate and photosynthesis rate, were significantly higher under soil moisture stress. Chlorophyll content largely decides the photosynthetic capacity of the plant. Under stress, the functional chloroplast can be maintained by stable chlorophyll content and enables the plant to revive the photosynthesis after the stress. Chlorophyll content estimation can be used as one of the criteria's to select tolerant plants. Shingote *et al.*, (2015) reported that sugarcane plant transformed with a MYB transcription factor SoMYB18 had considerably high chlorophyll content compared to untransformed plants.

Biochemical Parameters

Under drought stress, the accumulation of different metabolites are used as an effective indicator to select drought tolerant cultivars, for example measurement of proline content, MDA –malonaldehyde, ABA activity , Antioxidant enzyme activity – SOD, CAT, POX etc. It has been reported in different studies that in sugarcane the proline content increases in both salt and water stress. In sugarcane, proline act more as ROS scavenger than as osmoprotectant (Molinari *et al.*, 2007).

Under drought stress signalling pathways regulated by ABA are highly responsive. ABA functions to close the stomata thus protecting the plant from dessication. Rodrigues *et al.*, (2011) reported that in sugarcane cultivars submitted to drought showed reduction in stomatal conductance and some genes also showed similarity to genes directly or indirectly involved in ABA biosynthesis in plants .

In photosynthesis reactions the accumulation of reactive oxygen species occurs naturally during the electron transport (Miller *et al.*, 2010). The generation of reactive oxygen species seems to stimulate mechanisms that reduce oxidative stress and so it may play an important role in drought tolerance (Arora *et al.*, 2002). This is associated with increase in the activity of ROS scavenging enzymes. Uncontrolled ROS production causes lipid peroxidation which results in membrane damage. Measurement of the concentration of thiobarbituric acid reactive substances (TBARS) such as malondialdehyde (MDA) is routinely used as an index of lipid peroxidation under stress conditions (Lokhande *et al.*, 2011). Shingote *et al.*, (2015) reported that when sugarcane plants were transformed with SoMYB18,a MYB transcription factor, a significant decrease in MDA content during late stress hours was observed in transformed sugarcane plants than the untransformed plants.

In plant cells, under stress conditions, harmful effects of ROS is prevented by increased accumulations of antioxidant enzymes such as catalase, ascorbate peroxidase (APX) and superoxide dismutase (SOD). Ramiro *et al.*, (2016) reported that induction of various antioxidant enzymes in WT sugarcane plants under the water stress treatment suggests that ROS management is part of the signal that mediates normal water stress response.

Future Prospects

Tolerance to water deficit stress involves number of gene regulatory networks and is a very complex trait. *Erianthus arundinaceus,* a wild relative of *Saccharum,* is well known for its high fibre, high biomass, tolerance to drought and water logging, pest and disease resistance with multi-ratooning ability (Augustine *et al.*, 2015). The extensive root system of *E. arundinaceus* is an indication of its adaptation to drought situations. Hence, this is a novel source for improvement/development of sugarcane varieties for water deficit stress tolerance. Considering sorghum genome as reference further research work can be carried out to identify miRNA targets that influence drought tolerance in sugarcane. Further, more genes that are involved in suppression of endoplasmic stress can be identified and characterized in sugarcane to improve its drought tolerance ability. Also further exploitation of genes from unconventional sources like microbes offers a great scope for development of water deficit stress tolerant varieties. Indonesia has approved drought tolerant sugarcane cultivars (NXI-1T, NXI-4T and NXI-6T) for commercial cultivation. These cultivars have been transformed with *betA* gene from *Rhizobium meliloti.*

More recently biotechnology has entered a new era precise editing of genomes. Application of CRISPR/Cas9 based genome editing in sugarcane is in infant stage (Shan et al., 2013). This technique has been successfully used for genome editing of several plant species, such as rice (Chen et al., 2017; Li et al., 2017; Ren et al., 2017), maize (Zong et al., 2017) and even in the hexaploid wheat (Gil-Humanes et al., 2017). However, only one study so far has reported the use of CRISPR/Cas9 technique to target a gene related to drought stress, the maize gene AGOS8, a negative regulator of ethylene response (Shi et al., 2017). Genome edited plants had higher levels of AGOS8 transcripts and increased grain yield under water deficit and no yield penalty under well-watered conditions. And also the use of CRISPR/Cas9 in crop plants is particularly interesting due to the regulatory issues involving the release of commercial products from genetically modified organisms (GMO).

Conclusion

With climate changing all around the globe, development of varieties tolerant to multiple stresses are essential. Drought is the most significant environmental stress in agriculture worldwide, and improving yield under drought conditions is a major goal of sustainable agricultural production. Development of transgenic crops modified for drought tolerance will provide increased yields in drier areas and increased average yields in rain-fed systems by reducing the effects of sporadic drought and by decreasing water requirements in irrigated systems (Edgerton, 2009). In the longer-term modifications aimed more specifically at stabilizing yields in stressed environments and increasing yields in more productive regions may help to offset the demand for the conversion of further forested lands to arable production, this seeming an inevitable consequence of the expanding world population (Raymond *et al.*, 2011). Sugarcane holds an important place as it is not only the source of sugar but also an important renewable energy source. Losses caused by drought can be reduced to a great extent by using genetic engineering and genome editing tools. CRISPR/Cas9 usefulness in plants suggests that genome editing in a complex polyploidy like sugarcane is also feasible. This would allow a new and promising paradigm shift in sugarcane productivity under drought conditions.

References

Abe, H., Urao, T., Ito, T., Seki, M., Shinozaki, K., and Yamaguchi-Shinozaki, K. 2003. Arabidopsis AtMYC2 (bHLH) and AtMYB2 (MYB) function as transcriptional activators in abscisic acid signaling. *The Plant Cell*, 15(1): 63-78.

Ambawat, S., Sharma, P., Yadav, N. R., and Yadav, R. C. 2013. MYB transcription factor genes as regulators for plant responses: an overview. *Physiology and Molecular Biology of Plants*, 19(3): 307-321.

Andrade, J. C. F. D., Terto, J., Silva, J. V., and Almeida, C. 2015. Expression profiles of sugarcane under drought conditions: Variation in gene regulation. *Genetics and molecular biology*, 38(4): 465-469.

Arencibia, A. D., Carmona, E. R., Tellez, P., Chan, M. T., Yu, S. M., Trujillo, L. E., and Oramas, P. 1998. An efficient protocol for sugarcane *(Saccharum* spp. L.) transformation mediated by *Agrobacterium tumefaciens. Transgenic Research*, 7(3): 213-222.

Arora, A., Sairam, R. K., and Srivastava, G. C. 2002. Oxidative stress and antioxidative system in plants. *Current Science-Bangalore-*, 82(10): 1227-1238.

Augustine, S. M., Cherian, A. V., Syamaladevi, D. P., and Subramonian, N. 2015. *Erianthus arundinaceus* HSP70 (EaHSP70) Acts as a Key Regulator in the formation of anisotropic inter-digitation in Sugarcane (*Saccharum* spp. hybrid) in Response to Drought Stress. *Plant and Cell Physiology*, pcv142.

Augustine, S. M., Narayan, J. A., Syamaladevi, D. P., Appunu, C., Chakravarthi, M., Ravichandran, V., and Subramonian, N. 2015. Introduction of Pea DNA Helicase 45 into Sugarcane (*Saccharum* spp. Hybrid) enhances cell membrane thermostability and up-regulation of stress-responsive genes leads to abiotic stress tolerance. *Molecular biotechnology*, 57(5): 475-488.

Augustine, S. M., Narayan, J. A., Syamaladevi, D. P., Appunu, C., Chakravarthi, M., Ravichandran, V., and Subramonian, N. 2015. *Erianthus arundinaceus* HSP70 (EaHSP70) overexpression increases drought and salinity tolerance in sugarcane (Saccharum spp. hybrid). *Plant Science*, 232: 23-34.

Baker, N. R., and Horton, P. 1987. Chlorophyll fluorescence quenching during photoinhibition. *Photoinhibition*, 9:145-168.

Baker, N. R., and Rosenqvist, E. 2004. Applications of chlorophyll fluorescence can improve crop production strategies: an examination of future possibilities. *Journal of experimental botany*, 55(403): 1607-1621.

Begcy, K., Mariano, E. D., Gentile, A., Lembke, C. G., Zingaretti, S. M., Souza, G. M., and Menossi, M. 2012. A novel stress-induced sugarcane gene confers tolerance to drought, salt and oxidative stress in transgenic tobacco plants. *PloS one*, 7(9): e44697.

Bihani, P., Char, B., and Bhargava, S. 2011. Transgenic expression of sorghum DREB2 in rice improves tolerance and yield under water limitation. *The Journal of Agricultural Science*, 149(01): 95-101.

Bower, R., and Birch, R. G. 1992. Transgenic sugarcane plants via microprojectile bombardment. *The Plant Journal*, 2(3): 409-416.

Burlat, V., Kwon, M., Davin, L. B., and Lewis, N. G. 2001. Dirigent proteins and dirigent sites in lignifying tissues. *Phytochemistry*, 57(6): 883-897.

Chen Y, Wang Z, Ni H, Xu Y, Chen Q and Jiang L. 2017. CRISPR/Cas9- mediated base-editing system efficiently generates gain-of-function mutations in *Arabidopsis*. *Sci. China Life Sci.* 60: 520–523.

Chen, J. Q., Meng, X. P., Zhang, Y., Xia, M., and Wang, X. P. 2008. Over-expression of OsDREB genes lead to enhanced drought tolerance in rice. *Biotechnology letters*, 30(12): 2191-2198.

Chen, W. H., Gartland, K. M. A., Davey, M. R., Sotak, R., Gartland, J. S., Mulligan, B. J., and Cocking, E. C. 1987. Transformation of sugarcane protoplasts by direct uptake of a selectable chimaeric gene. *Plant cell reports*, 6(4): 297-301.

Cominelli, E., Galbiati, M., Vavasseur, A., Conti, L., Sala, T., Vuylsteke, M., and Tonelli, C. 2005. A guard-cell-specific MYB transcription factor regulates stomatal movements and plant drought tolerance. *Current biology*, 15(13): 1196-1200.

Danial, N. N., and Korsmeyer, S. J. 2004. Cell death: critical control points. *Cell*, 116(2): 205-219.

De Ronde, J. A., Cress, W. A., Krüger, G. H. J., Strasser, R. J., and Van Staden, J. 2004. Photosynthetic response of transgenic soybean plants, containing an Arabidopsis P5CR gene, during heat and drought stress. *Journal of plant physiology*, 161(11): 1211-1224.

Delauney, A. J., and Verma, D. P. S. 1993. Proline biosynthesis and osmoregulation in plants. *The plant journal*, 4(2): 215-223.

Dubos, C., Stracke, R., Grotewold, E., Weisshaar, B., Martin, C., and Lepiniec, L. 2010. MYB transcription factors in Arabidopsis. *Trends in plant science, 15*(10), 573-581.

Edgerton, M. D. 2009. Increasing crop productivity to meet global needs for feed, food, and fuel. *Plant physiology*, 149(1): 7-13.

Endres, L., Silva, J. V., Ferreira, V. M., and Barbosa, G. D. S. 2010. Photosynthesis and water relations in Brazilian sugarcane. *Open Agric. J*, 4(3).

Enriquez-Obregon, G. A., Vazquez-Padron, R. I., Prieto-Samsonov, D. L., Perez, M., and Selman-Housein, G. 1997. Genetic transformation of sugarcane by *Agrobacterium tumefaciens* using antioxidant compounds. *Biotecnologia aplicada*, 14(3): 169-174.

Ferdous, J., Hussain, S. S., and Shi, B. J. 2015. Role of microRNAs in plant drought tolerance. *Plant biotechnology journal*, 13(3): 293-305.

Ferreira, T. H., Gentile, A., Vilela, R. D., Costa, G. G. L., Dias, L. I., Endres, L., and Menossi, M. 2012. microRNAs associated with drought response in the bioenergy crop sugarcane (*Saccharum* spp.). *PLoS One*, 7(10): e46703.

Garg, A. K., Kim, J. K., Owens, T. G., Ranwala, A. P., Do Choi, Y., Kochian, L. V., and Wu, R. J. 2002. Trehalose accumulation in rice plants confers high tolerance levels to different abiotic stresses. *Proceedings of the National Academy of Sciences*, 99(25): 15898-15903.

Gawander, J. 2007. Impact of climate change on sugar-cane production in Fiji. *World Meteorological Organization Bulletin*, 56(1): 34-39.

Gaxiola, R. A., Li, J., Undurraga, S., Dang, L. M., Allen, G. J., Alper, S. L., and Fink, G. R. 2001. Drought-and salt-tolerant plants result from overexpression of the AVP1 H+-pump. *Proceedings of the National Academy of Sciences*, 98(20): 11444-11449.

Gentile, A., Dias, L. I., Mattos, R. S., Ferreira, T. H., and Menossi, M. 2015. MicroRNAs and drought responses in sugarcane. *Frontiers in plant science*, 6.

Gentile, A., Ferreira, T. H., Mattos, R. S., Dias, L. I., Hoshino, A. A., Carneiro, M. S. and Menossi, M. 2013. Effects of drought on the microtranscriptome of field-grown sugarcane plants. *Planta, 237*(3), 783-798.

Ghannoum, O. 2009. C4 photosynthesis and water stress. *Annals of Botany*, 103(4): 635-644.

Gil-Humanes, J., Wang, Y., Liang, Z., Shan, Q., Ozuna, C. V., Sánchez-León, S., et al. 2017. High-efficiency gene targeting in hexaploid wheat using DNA replicons and CRISPR/Cas9. *Plant J.*, 89: 1251–1262.

Graça, J. P. D., Rodrigues, F. A., Farias, J. R. B., Oliveira, M. C. N. D., Hoffmann-Campo, C. B., and Zingaretti, S. M. 2010. Physiological parameters in sugarcane cultivars submitted to water deficit. *Brazilian Journal of Plant Physiology*, 22(3): 189-197.

Haake, V., Cook, D., Riechmann, J., Pineda, O., Thomashow, M. F., and Zhang, J. Z. 2002. Transcription factor CBF4 is a regulator of drought adaptation in Arabidopsis. *Plant physiology*, 130(2): 639-648.

Hansson, S. O., and Joelsson, K. 2013. Crop biotechnology for the environment?. *Journal of agricultural and environmental ethics*, 26(4): 759-770.

Haus, U., Trommler, P., Fisher, P. R., Hartmann, H., Lottspeich, F., Noegel, A. A., and Schleicher, M. 1993. The heat shock cognate protein from *Dictyostelium* affects actin polymerization through interaction with the actin-binding protein cap32/34. *The EMBO journal*, 12(10): 3763.

Jang, I. C., Oh, S. J., Seo, J. S., Choi, W. B., Song, S. I., Kim, C. H., and Kim, J. K. 2003. Expression of a bifunctional fusion of the Escherichia coli genes for trehalose-6-phosphate synthase and trehalose-6-phosphate phosphatase in transgenic rice plants increases trehalose accumulation and abiotic stress tolerance without stunting growth. *Plant physiology*, 131(2): 516-524.

Jin-long, G., Li-ping, X., Jing-ping, F., Ya-chun, S., Hua-ying, F., You-xiong, Q., and Jing-sheng, X. 2012. A novel dirigent protein gene with highly stem-specific expression from sugarcane, response to drought, salt and oxidative stresses. *Plant cell reports*, 31(10): 1801-1812.

Katiyar-Agarwal, S., Agarwal, M., and Grover, A. 2003. Heat-tolerant basmati rice engineered by over-expression of hsp101. *Plant molecular biology*, 51(5): 677-686.

Kawai, M., Pan, L., Reed, J. C., and Uchimiya, H. 1999. Evolutionally conserved plant homologue of the Bax inhibitor-1 (BI-1) gene capable of suppressing Bax-induced cell death in yeast. *FEBS letters*, 464(3): 143-147.

Kawai-Yamada, M., Jin, L., Yoshinaga, K., Hirata, A., and Uchimiya, H. 2001. Mammalian Bax-induced plant cell death can be down-regulated by overexpression of Arabidopsis Bax Inhibitor-1 (AtBI-1). *Proceedings of the National Academy of Sciences*, 98(21): 12295-12300.

Knight, H., and Knight, M. R. 2001. Abiotic stress signalling pathways: specificity and cross-talk. *Trends in plant science*, 6(6): 262-267.

Koonjah, S. S., Walker, S., Singels, A., Van Antwerpen, R., and Nayamuth, A. R. 2006. A quantitative study of water stress effect on sugarcane photosynthesis. In *Proceedings of the South African Sugar Technologists Association* ***(Vol. 80, pp. 148-158).***

KoYASU, S. H. I. G. E. O., Nishida, E., Kadowaki, T., Matsuzaki, F., Iida, K., Harada, F., and Yahara, I. 1986. Two mammalian heat shock proteins, HSP90 and HSP100, are actin-binding proteins. *Proceedings of the National Academy of Sciences*, 83(21): 8054-8058.

Kumar, T., Khan, M. R., Abbas, Z., and Ali, G. M. 2014. Genetic improvement of sugarcane for drought and salinity stress tolerance using Arabidopsis vacuolar pyrophosphatase (AVP1) gene. *Molecular biotechnology*, 56(3): 199-209.

Lam, E. 2004. Controlled cell death, plant survival and development. *Nature Reviews Molecular Cell Biology*, 5(4): 305-315.

Lata, C., and Prasad, M. 2011. Role of DREBs in regulation of abiotic stress responses in plants. *Journal of experimental botany*, 62(14): 4731-4748.

Li J, Sun Y, Du J, Zhao Y, Xia L. 2017. Generation of targeted point mutations in rice by a modified CRISPR/Cas9 System. *Molecular Plant*, 10(3): 526-529

Li, B., Wei, A., Song, C., Li, N., and Zhang, J. 2008. Heterologous expression of the TsVP gene improves the drought resistance of maize. *Plant Biotechnology Journal*, 6(2): 146-159.

Li, H. Y., Chang, C. S., Lu, L. S., Liu, C. A., Chan, M. T., and Charng, Y. Y. 2003. Over-expression of Arabidopsis thaliana heat shock factor gene (AtHsfA1b) enhances chilling tolerance in transgenic tomato. *Botanical Bulletin of Academia Sinica*, 44.

Li, J., Yang, H., Peer, W. A., Richter, G., Blakeslee, J., Bandyopadhyay, A., and Krizek, B. 2005. Arabidopsis H+-PPase AVP1 regulates auxin-mediated organ development. *Science*, 310(5745): 121-125.

Liu, Q., Kasuga, M., Sakuma, Y., Abe, H., Miura, S., Yamaguchi-Shinozaki, K., and Shinozaki, K. 1998. Two transcription factors, DREB1 and DREB2, with an EREBP/AP2 DNA binding domain separate two cellular signal transduction pathways in drought-and low-temperature-responsive gene expression, respectively, in Arabidopsis. *The Plant Cell*, 10(8): 1391-1406.

Lokhande, V. H., Srivastava, S., Patade, V. Y., Dwivedi, S., Tripathi, R. D., Nikam, T. D., and Suprasanna, P. 2011. Investigation of arsenic accumulation and tolerance potential of *Sesuvium portulacastrum* (L.) L. *Chemosphere*, 82(4): 529-534.

Malik, M. K., Slovin, J. P., Hwang, C. H., and Zimmerman, J. L. 1999. Modified expression of a carrot small heat shock protein gene, Hsp17. 7, results in increased or decreased thermotolerance. *The Plant Journal*, 20(1): 89-99.

Manjulatha, M., Sreevathsa, R., Kumar, A. M., Sudhakar, C., Prasad, T. G., Tuteja, N., and Udayakumar, M. 2014. Overexpression of a pea DNA helicase (PDH45) in peanut (*Arachis hypogaea* L.) confers improvement of cellular level tolerance and productivity under drought stress. *Molecular biotechnology*, 56(2): 111-125.

Martineau, J. R., Specht, J. E., Williams, J. H., and Sullivan, C. Y. 1979. Temperature tolerance in soybeans. I. Evaluation of a technique for assessing cellular membrane thermostability. *Crop Science,* 19(1): 75-78.

Miller, G., Suzuki, N., Ciftci-Yilmaz, S., and Mittler, R. 2010. Reactive oxygen species homeostasis and signalling during drought and salinity stresses. Plant, Cell andEnvironment .*Mol. Cell Biol,* 5: 305–315.

Molinari, H. B. C., Marur, C. J., Daros, E., De Campos, M. K. F., De Carvalho, J. F. R. P., Pereira, L. F. P., and Vieira, L. G. E. 2007. Evaluation of the stress-inducible production of proline in transgenic sugarcane (*Saccharum* spp.): osmotic adjustment, chlorophyll fluorescence and oxidative stress. *Physiologia Plantarum,* 130(2): 218-229.

Morran, S., Eini, O., Pyvovarenko, T., Parent, B., Singh, R., Ismagul, A., and Lopato, S. 2011. Improvement of stress tolerance of wheat and barley by modulation of expression of DREB/CBF factors. *Plant Biotechnology Journal,* 9(2): 230-249.

Naidu, K. M., and Sreenivasan, T. V. (1987, June). Conservation of sugarcane germplasm. In *Proceedings of the Copersucar International Sugarcane Breeding Workshop. Copersucar Technology Centre, Piracicaba-SP, Brazil* ***(pp. 33-53).***

Nakashima, K., Shinwari, Z. K., Sakuma, Y., Seki, M., Miura, S., Shinozaki, K., and Yamaguchi-Shinozaki, K. 2000. Organization and expression of two Arabidopsis DREB2 genes encoding DRE-binding proteins involved in dehydration-and high-salinity-responsive gene expression. *Plant molecular biology,* 42(4): 657-665.

Parry, M. A. J., Flexas, J., and Medrano, H. 2005. Prospects for crop production under drought: research priorities and future directions. *Annals of Applied Biology,* 147(3): 211-226.

Prabu, G., and Prasad, D. T. 2012. Functional characterization of sugarcane MYB transcription factor gene promoter (PScMYBAS1) in response to abiotic stresses and hormones. *Plant cell reports,* 31(4): 661-669.

Prasad, P. V., Pisipati, S. R., Mutava, R. N., and Tuinstra, M. R. 2008. Sensitivity of grain sorghum to high temperature stress during reproductive development. *Crop Science,* 48(5): 1911-1917.

Qi, Y., Wang, H., Zou, Y., Liu, C., Liu, Y., Wang, Y., and Zhang, W. 2011. Over-expression of mitochondrial heat shock protein 70 suppresses programmed cell death in rice. *FEBS letters,* 585(1): 231-239.

Qin, F., Kakimoto, M., Sakuma, Y., Maruyama, K., Osakabe, Y., Tran, L. S. P., and Yamaguchi-Shinozaki, K. 2007. Regulation and functional analysis of ZmDREB2A in response to drought and heat stresses in Zea mays L. *The Plant Journal,* 50(1): 54-69.

Ramiro, D. A., Melotto-Passarin, D. M., Barbosa, M. D. A., Santos, F. D., Gomez, S. G. P., Massola Júnior, N. S., and Carrer, H. 2016. Expression of Arabidopsis Bax Inhibitor-1 in transgenic sugarcane confers drought tolerance. *Plant biotechnology journal.*

Raymond Park, J., McFarlane, I., Hartley Phipps, R., and Ceddia, G. 2011. The role of transgenic crops in sustainable development. *Plant Biotechnology Journal*, 9(1): 2-21.

Raza, G., Ali, K., Ashraf, M. Y., Mansoor, S., Javid, M., and Asad, S. 2016. Overexpression of an H+-PPase gene from *Arabidopsis* in sugarcane improves drought tolerance, plant growth, and photosynthetic responses. *Turkish Journal of Biology*, 40(1): 109-119.

Reis, R. R., da Cunha, B. A. D. B., Martins, P. K., Martins, M. T. B., Alekcevetch, J. C., Chalfun-Júnior, A., and Yamaguchi-Shinozaki, K. 2014. Induced overexpression of AtDREB2A CA improves drought tolerance in sugarcane. *Plant science*, 221: 59-68.

Ren B, Yan F, Kuang Y, Li N, Zhang D, Lin H, Zhou H. 2017. A CRISPR/Cas9 toolkit for efficient targeted base editing to induce genetic variations in rice. Science China Life Sciences 60(5): 516-519

Robertson, M. J., Muchow, R. C., Donaldson, R. A., Inman-Bamber, N. G., and Wood, A. W. 1998. Estimating the risk associated with drying-off strategies for irrigated sugarcane before harvest. *Crop and Pasture Science*, 50(1): 65-78.

Rodrigues, F. A., Da Graca, J. P., De Laia, M. L., Nhani-Jr, A., Galbiati, J. A., Ferro, M. I. T., and Zingaretti, S. M. 2011. Sugarcane genes differentially expressed during water deficit. *Biologia Plantarum*, 55(1): 43-53.

Sage, R. F., and Kubien, D. S. 2007. The temperature response of C3 and C4 photosynthesis. *Plant, Cell and Environment*, 30(9): 1086-1106.

Sanan-Mishra, N., Pham, X. H., Sopory, S. K., and Tuteja, N. 2005. Pea DNA helicase 45 overexpression in tobacco confers high salinity tolerance without affecting yield. *Proceedings of the National Academy of Sciences of the United States of America*, 102(2): 509-514.

Schilling, R. K., Marschner, P., Shavrukov, Y., Berger, B., Tester, M., Roy, S. J., and Plett, D. C. 2014. Expression of the Arabidopsis vacuolar H+-pyrophosphatase gene (AVP1) improves the shoot biomass of transgenic barley and increases grain yield in a saline field. *Plant biotechnology journal*, 12(3): 378-386.

Shan Q, Wang Y, Li J, Zhang Y, Chen K, Liang Z, Zhang K, Liu J, Xi JJ, Qiu JL, Gao C. 2013. Targeted genome modification of crop plants using a CRISPR-Cas system. *Nature Biotechnology*, 31: 686–688

Shi, J., Gao, H., Wang, H., Lafitte, H. R., Archibald, R. L., Yang, M., Hakimi SM, Mo H, Habben JE. 2017. ARGOS8 variants generated by CRISPR-Cas9 improve maize grain yield under field drought stress conditions. *Plant Biotechnology Journal*, 15: 207-216.

Shingote, P. R., Kawar, P. G., Pagariya, M. C., Kuhikar, R. S., Thorat, A. S., and Babu, K. H. 2015. SoMYB18, a sugarcane MYB transcription factor improves salt and dehydration tolerance in tobacco. *Acta Physiologiae Plantarum*, 10(37): 1-12.

Shinwari, Z. K., Nakashima, K., Miura, S., Kasuga, M., Seki, M., Yamaguchi-Shinozaki, K., and Shinozaki, K. 1998. An Arabidopsis gene family encoding DRE/CRT binding proteins involved in low-temperature-responsive gene expression. *Biochemical and biophysical research communications*, 250(1): 161-170.

Silva, M. D. A., Jifon, J. L., Da Silva, J. A., and Sharma, V. 2007. Use of physiological parameters as fast tools to screen for drought tolerance in sugarcane. *Brazilian Journal of Plant Physiology*, 19(3): 193-201.

Simard, J. P., Reynolds, D. N., Kraguljac, A. P., Smith, G. S., and Mosser, D. D. 2011. Overexpression of HSP70 inhibits cofilin phosphorylation and promotes lymphocyte migration in heat-stressed cells. *J Cell Sci*, 124(14): 2367-2374.

Sun, W., Bernard, C., Van De Cotte, B., Van Montagu, M., and Verbruggen, N. 2001. At-HSP17. 6A, encoding a small heat-shock protein in Arabidopsis, can enhance osmotolerance upon overexpression. *The Plant Journal*, 27(5): 407-415.

Thiebaut, F., Grativol, C., Carnavale-Bottino, M., Rojas, C. A., Tanurdzic, M., Farinelli, L., and Ferreira, P. C. G. 2012. Computational identification and analysis of novel sugarcane microRNAs. *BMC genomics*, 13(1): 290.

Trujillo, L. E., Menéndez, C., Ochogavía, M. E., Hernández, I., Borrás, O., Rodríguez, R., and Hernández, L. 2009. Engineering drought and salt tolerance in plants using SodERF3, a novel sugarcane ethylene responsive factor. *Biotecnología Aplicada*, 26(2): 168-171.

Tuteja, N. 2003. Plant DNA helicases: the long unwinding road. *Journal of experimental botany*, 54(391): 2201-2214.

WANG, A. P., MAO, X. G., JING, R. L., CHANG, X. P., and YANG, W. D. 2006. Single nucleotide polymorphism of TaMyb2-◎ gene in common wheat (*Triticum aestivum* L.) and its relatives. *Acta Agronomica Sinica*, 12: 005.

Wang, Z. Z., Zhang, S. Z., Yang, B. P., and Li, Y. R. 2003. Trehalose synthase gene transfer mediated by *Agrobacterium tumefaciens* enhances resistance to osmotic stress in sugarcane. *Agricultural Sciences in China*, 2(1), 19-26. *Sugar Tech*, 7(1): 49-54.

Wang, Z. Z., Zhang, S. Z., Yang, B. P., and Li, Y. R. 2003. Trehalose synthase gene transfer mediated by *Agrobacterium tumefaciens* enhances resistance to osmotic stress in sugarcane. *Agricultural Sciences in China*, 2(1): 19-26.

Watanabe, N., and Lam, E. 2009. Bax inhibitor-1, a conserved cell death suppressor, is a key molecular switch downstream from a variety of biotic and abiotic stress signals in plants. *International journal of molecular sciences*, 10(7): 3149-3167.

Wiedenfeld, R. P. 2000. Water stress during different sugarcane growth periods on yield and response to N fertilization. *Agricultural Water Management*, 43(2): 173-182.

Wu, R., Wang, L., Wang, Z., Shang, H., Liu, X., Zhu, Y.,and Deng, X. 2009. Cloning and expression analysis of a dirigent protein gene from the resurrection plant *Boea hygrometrica*. *Progress in Natural Science*, 19(3): 347-352.

Yamada, M., Morishita, H., Urano, K., Shiozaki, N., Yamaguchi-Shinozaki, K., Shinozaki, K., and Yoshiba, Y. 2005. Effects of free proline accumulation in petunias under drought stress. *Journal of Experimental Botany*, 56(417): 1975-1981.

Yang, A., Dai, X., and Zhang, W. H. 2012. A R2R3-type MYB gene, OsMYB2, is involved in salt, cold, and dehydration tolerance in rice. *Journal of experimental botany*, 63(7): 2541-2556.

Yang, S., Vanderbeld, B., Wan, J., and Huang, Y. 2010. Narrowing down the targets: towards successful genetic engineering of drought-tolerant crops. *Molecular plant*, 3(3): 469-490.

You, J., Hu, H., and Xiong, L. 2012. An ornithine δ-aminotransferase gene OsOAT confers drought and oxidative stress tolerance in rice. *Plant science*, 197: 59-69.

Zhao, F. Y., Zhang, X. J., Li, P. H., Zhao, Y. X., and Zhang, H. 2006. Co-expression of the *Suaeda salsa* SsNHX1 and Arabidopsis AVP1 confer greater salt tolerance to transgenic rice than the single SsNHX1. *Molecular Breeding*, 17(4): 341-353.

Zhou, M., Li, D., Li, Z., Hu, Q., Yang, C., Zhu, L., and Luo, H. 2013. Constitutive expression of a miR319 gene alters plant development and enhances salt and drought tolerance in transgenic creeping bent grass. *Plant physiology*, 161(3): 1375-1391.

Zhu, B., Su, J., Chang, M., Verma, D. P. S., Fan, Y. L., and Wu, R. 1998. Overexpression of a Δ 1-pyrroline-5-carboxylate synthetase gene and analysis of tolerance to water-and salt-stress in transgenic rice. *Plant Science*, 139(1): 41-48.

Zlatev, Z. 2009. Drought-induced changes in chlorophyll fluorescence of young wheat plants. *Biotechnology and Biotechnological Equipment*, 23(sup1), 438-441.

Zong Y, Wang Y, Li C, Zhang R, Chen K, Ran Y, Qiu JL, Wang D, Gao C. 2017. Precise base editing in rice, wheat and maize with a Cas9-cytidine deaminase fusion. *Nature Biotechnology*, 35: 438-440

24

Proteomics: A Study Towards Stress Tolerance in Crop Plants

Minakshi Jattan[1], Nisha Kumari[2*] and Ram Avtar[1]

[1]Department of Genetics and Plant Breeding,
[2]Departments of Chemistry and Biochemistry, CCS Haryana Agricultural University, Hisar -125 004, India
*Corresponding author: nishaahlawat211@gmail.com

Abstract

Proteomics is the large scale study of the proteome i.e. protein complement of the genome. Proteome varies with time, and distinct requirements or stresses, that a cell or organism undergoes. Plants often experience various biotic and abiotic stresses during their life cycle. These biotic and abiotic stresses cause major losses to the crop plants. These stresses cause a multitude of changes in physiological, molecular and biochemical processes operating in plants. It is now widely reported that several proteins respond to these stresses at pre- and post-transcriptional and translational levels. By knowing the role of these stress inducible proteins, it would be easy to comprehensively explain the processes of stress tolerance in plants. The proteomics study offers a new approach to discover proteins and pathways associated with biotic and abiotic stress responses. Thus, studying the plants at proteomics level helps in understanding the pathways involved in stress tolerance and hence development of stress tolerant crop plants.

Keywords: Proteomics, proteome, biotic, abiotic, stresses

Introduction

As the world population is increasing exponentially, the agriculture sector worldwide is facing a major challenge of ensuring sufficient food supply to the masses through enhancing agricultural productivity (FAO, 2012). These challenges get further intensified with alterations in weather patterns due to changes in climate that impact crop productivity (Ahmad *et al.*, 2012, 2013; Lake *et al.*, 2012).

Adverse environmental conditions alter agro-ecological system that may affect the demand for increased agricultural production (Lake *et al.*, 2012; Ahmad *et al.*, 2013). Plants often experience various biotic and abiotic stresses during their life cycle. Biotic and abiotic stresses hamper the plant growth and development negatively thereby impacting crop production. It has been estimated that they may be responsible for over 50 % yield reduction in major crop plants. However, severity of losses depends on the plant developmental stage at which the stress occurs, its intensity and duration (Monneveux *et al.*, 2006; Lafitte *et al.*, 2007). Also, plants have developed stress-specific adaptations as well as simultaneous responses to a combination of various abiotic stresses. The efficiency of stress- induced adaptive responses is dependent on activation of molecular signalling pathways and intracellular networks by modulating expression or abundance, and/or post-translational modification (PTM) of proteins primarily associated with defence mechanisms. By knowing the role of these stress inducible proteins, it would be easy to comprehensively expound the processes of stress tolerance in plants. So far, breeding for biotic and abiotic stress tolerance has been achieved through crosses with tolerant donors, identification and introduction of QTLs, stress responsive genes such as ion transporters, transcription factors, late embryogenesis abundant proteins etc. Due to the low concentrations of ion transporters and transcription factors, they are not detected in two-dimensional gels. Therefore, the integration of emerging new technologies is essential and it assists in crop breeding programmes for stress tolerance. The proteomics offers a new approach to discover proteins and pathways associated with crop physiological and stress responses. This capability is especially useful for crops as it may give clues not only about nutritional value, but also about yield and how these factors are affected by adverse conditions.

Proteomics

The actual biological function of a gene, catalysis of reactions, maintenance of structural and chemical homeostasis is done by proteins. The number of proteins in a biological system exceeds the number of coding sequences due to various events i.e. alternate splicing etc. Furthermore, the chemical and physical diversity of proteins, from large to small and from acidic to basic, is much larger than the diversity of nucleotide-based macromolecules. All this makes the identification and sequence determination of proteins a task more difficult than the sequencing of polynucleotides. Nonetheless in recent years there have been large advances in the techniques and tools to study protein expression in plants. These advances in techniques have brought not only improvements in separation or identification methods, but also contributed to greater reproducibility, accuracy and reliability in the data provided. Proteomics is one such emerging new technique.

The terms "proteomics" and "proteome" were coined by Marc Wilkins and colleagues in 1994. The word "proteome" is derived from proteins expressed by a genome. Analogous to genomics, the term "proteomics" describes the study and characterization of the complete set of proteins present at a given time in the cell

(Wilkins *et al.*, 1995). Proteome varies with time, distinct requirements and various biotic and abiotic stresses. Although genomic studies are helpful in several ways, proteomic studies reveal the functional players for mediating specific cellular processes. The proteome, unlike the genome, which is static in nature, has dynamic capabilities. The study of proteins introduces post-translational modifications (PTMs) and provides the knowledge that is important for understanding the biological functions (Nat *et al.*, 2007). The PTMs that play important roles during the growth and development of a plant and/or in response to various stress conditions cannot be understood from the genome sequence projects and/or transcript abundance alone. Proteomics knowledge provides functional genomics with completeness towards understanding the process (Gygi *et al.*, 2000). Furthermore, the development of various advanced tools for bioinformatics and computational science are connecting proteomics to other "- omics," and the physiological data are further opening up new methods for crop improvement studies via the signalling, regulatory, and metabolic networks underlying plant phenotypes (Kitano, 2002; Langridge and Fleury, 2011).

Classification of Proteomics

Based on the protein response under stress conditions proteomics are classified into different groups i.e. expression, structural and functional proteomics.

Expression Proteomics

Expression proteomics is used to study the qualitative and quantitative expression of total proteins under two different conditions. Like the normal cell and treated or diseased cell can be compared to understand the protein that is responsible for the stress or diseased state or the protein that is expressed due to disease or stress of a given environment. Two-Dimensional gel electrophoresis and mass spectrometry techniques are used to observe the protein expressional changes, which is present or absent in a particular tissue under particular stress, when compared with normal tissue. The over expressed and under expressed proteins can be identified and characterized (Hinsby *et al.*, 2003).

Structural Proteomics

Structural proteomics helps to understand three dimensional shape and structural complexities of functional proteins. Structural proteomics gives detailed information about the structure and function of protein complexes present in a specific cellular organelle. Different technologies such as X-ray crystallography and NMR spectroscopy are used for structure determination (Junjie *et al.*, 2015).

Functional Proteomics

Functional proteomics explains the protein functions as well as unravels molecular mechanisms within the cell that depend on the identification of the interacting protein partners. The association of an unknown protein with partners belonging to a specific protein complex involved in a particular mechanism is in fact, strongly suggestive of its biological function. Furthermore detailed description of the cellular signalling pathways greatly benefits from the elucidation of protein-protein interactions (Kamal *et al.*, 2015).

Proteomics Technologies

The complete characterization of a proteome is a formidable task and the degree of success achieved depends on the methods available and their amenability to automation and high throughput formats (Junjie *et al.*, 2015). Parameters such as the complexity of the protein mixture, levels of expression and intracellular localization all impact the choice of proteomics technologies to be used (Holman *et al.*, 2013). The important proteomics technologies are two-dimensional polyacrylamide gel electrophoresis, mass spectrometery, chromatography, protein microarray etc.

Two-dimensional Polyacrylamide Gel Electrophoresis

To study the dynamic changes of proteome patterns in response to environmental stimuli. Two-dimensional gel electrophoresis (2-DGE) techniques have been used as a classical method. Ever since its advent in the late 1970s (Klose, 1975; O'Farrell, 1975), 2-DGE has remained one of the most important analytical techniques and is routinely applied for profiling the complex proteomes of various plant species (Van den Bergh and Arckens, 2005; Rossignol *et al.*, 2006). It resolves the proteins according to two independent parameters i.e. the isoelectric point in the first dimension and the molecular mass in the second dimension (Gorg *et al.*, 2004; Wittmann-Liebold *et al.*, 2006). These separated proteins are then visualized as spots on the gel by various staining methods. Stains such as Coomassie blue, silver, SYPRO Ruby and Deep Purple can be employed to visualize the proteins (Wu *et al.*, 2014). High-resolution 2-D PAGE can resolve up to 10,000 protein spots per gel. To overcome the limitations, such as gel-to-gel variations, and to increase the reproducibility, a new technique called difference in gel electrophoresis (DIGE) has been developed (Unlu *et al.*, 1997) (Fig. 1). In this technique a differential staining method is applied, in which each protein sample is labelled at a lysine residue with different fluorophores (i.e., Cy2, Cy3 and Cy5) (Viswanathan *et al.*, 2006). The samples are then mixed and separated on the same gel, allowing the high-sensitivity identification of differentially expressed proteins (Van den Bergh and Arckens, 2005).

Although gel-based proteomics techniques are widely used for protein profiling, due to certain limitations such as labour intensiveness, limited dynamic range and gel-to-gel inconsistencies, gel-free proteomics techniques have also been developed in recent years (Jorrin *et al.*, 2007; Thelen and Peck, 2007; Bachi and Bonaldi, 2008). The 2-D PAGE analysis is often coupled with the mass spectrometery technology and it has provided higher resolution, improved reproducibility, and higher loading capacity for preparative purposes with the intent of subsequent spot identifications (Kim *et al.*, 2013).

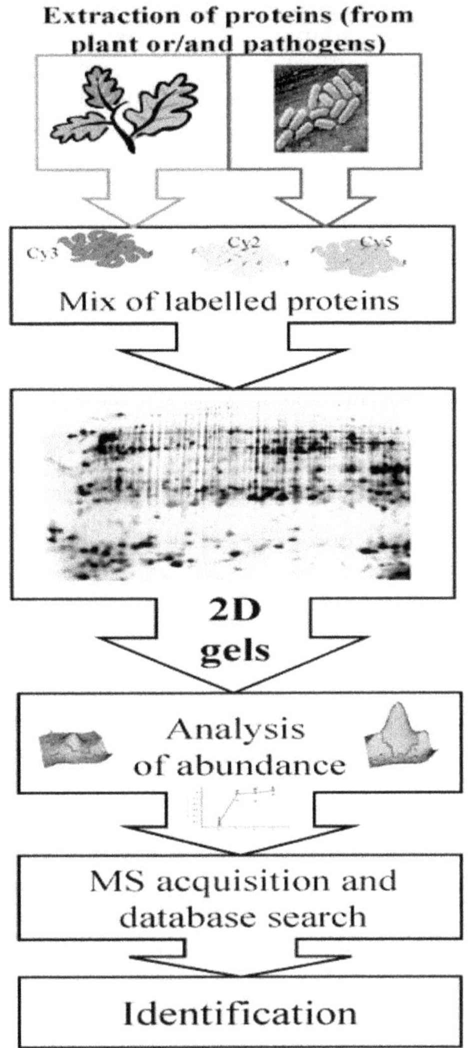

Fig.1: Schematic work-flow of complete 2DE-DIGE experiment (Unlu *et al.*, 1997)

Mass spectrometry

It is an analytical technique that in conjunction with the development of comprehensive protein databases (Kamal *et al.*, 2015) and advances in computational methods (Lahm and Langen, 2000), is being used for high-throughput characterization and identification of proteins. Mass spectrometer consists of an ion source, a mass analyzer and a detector. Mass spectrometry produces spectra of the masses of the atoms or molecules comprising a sample of material. The spectra are used to determine the elemental or isotopic signature of a sample, the masses of particles and of molecules, and to elucidate the chemical structures of molecules, such as peptides and other chemical compounds. It works by ionizing chemical compounds to generate charged molecules or molecule fragments and measuring their mass to charge ratios (Lodha *et al.*, 2013). Frequently used ionization methods include electro spray ionization (ESI), matrix-assisted laser desorption-ionization (MALDI), and surface-enhanced laser desorption/ionization (SELDI). Matrix-assisted laser desorption-ionization (MALDI) and electro spray ionization (ESI) are two technologies that are used for protein ionization (Yang, *et al.*, 2015).

MALDI-TOF-MS

The MALDI process has laser pulse as its energy source, as opposed to the electrostatic potential in ESI, to ionize peptides. Protein samples are being characterized by SDS PAGE by generating peptide maps. These peptide maps are used as fingerprints of protein or as a tool to know the purity of a known protein in a known sample. Mass spectrometry gives a peptide map when proteins are digested with proteolytic enzymes like trypsin. This peptide map can be used to search a sequence database to find a good match from the existing database (Jacoby *et al.*, 2013). Time-of-flight (TOF) and quadrupole mass analyzers have been developed for use in mass spectrometers (Wu *et al.*, 2014), and of these, TOF analyzers are more common because of their ease of operation. TOFs are commonly used with a MALDI ion source, whereas quadrupole analyzers are combined with an ESI source. The accurate determination of protein molecular weights is mainly achieved using a MALDI-TOF instrument. These peptide mass finger prints are compared with a Data base of virtual peptide mass finger prints generated by the theoretical digestion of known proteins by specific proteases (Kamal *et al.*, 2015).

Electro Spray Ionization (ESI)

This technology involves the production of gaseous ions by application of a potential to a flowing liquid, resulting in the formation of a spray of small droplets with solvent-containing analyte. Solvent is removed from the droplet by heat or another form of energy such as collision with a gas, and multiple charged ions are formed. Finally, electrostatic repulsion is sufficiently high to cause desorption of the analyte ions, which are then passed to the mass spectrometer. In ESI mass spectrometry the protein sample is in solution and the potential applied to it creates a fine mist of charged droplets that are subsequently dried and introduced into the mass analyzer (Lahm and Langen, 2000). In contrast to MALDI, ESI produces highly charged ions without fragmentation of the ions in the gas phase (Mann *et al.*, 2001).

Chromatography

Chromatography is utilized for the separation of closely related components in mixtures. The components to be separated are distributed between two phases, one of which is stationary (the stationary phase), while the other (the mobile phase). The phases are chosen such that components of the sample have differing affinities for each phase. The most commonly employed separation technique for Bioanalysis is high performance liquid chromatography (HPLC), also known simply as LC. Its increasing applications in proteomics research are due to its ability to analyze large, fragile biomolecules. With advancements in ionization methods and instrumentation, LC combined with MS has become a powerful technology for the characterization and identification of peptides and proteins in a complex mixture. This technology shows advantages over gel-based techniques in terms of speed, sensitivity, scope of analysis and dynamic range, since it can incorporate a wide choice of detection methods (Chandana *et al.*, 2013).

Protein Microarrays

Protein chips, protein biochips or protein micro-arrays are micro proteomic technologies for studying protein interactions and functions. This technology uses only a very small amount of crude sample (e.g. cell lysates) from the test individuals (Geho *et al.*, 2010). Protein chips are viewed as the most promising tools for proteome wide analysis. Thousands of proteins can be analysed and it also enables screening for specific types of post-translational modifications. Based on the comparisons of protein mass profiles from any two samples from different biological and pathological conditions, potential bio markers or disease-related protein targets can be identified. It is used as rapid profiling chip array system for a variety of studies, each exploiting the key strengths and advantages of this technology.

Proteome Analysis

Proteome analysis is based on two factors i.e. organ specific and cell specific proteome analysis.

Organ Specific Proteome Analysis

Proteome analysis is based on organs *viz.* leaves, roots, xylem, phloem etc. of plants.

i) Leaf Photosynthesis and Senescence

Leaf photosynthesis is the main source of plant biomass influencing potential crop yield. Chu *et al.*, (2015) analyzed changes in protein profiles upon the development of chlorophyll deficiency in *Brassica napus* leaves and provided new insights into the regulation of chlorophyll biosynthesis and photosynthesis in crops. Several proteomics studies have also been focused on the leaf senescence, mainly by analyzing the nitrogen mobilization from leaves during leaf senescence (Avice and Etienne, 2014).

ii) Root-to-leaf Signalling via Xylem and Phloem

Crop yield enhancement also depends on an optimal supply of nutrients to the leaves received from the root system via the xylem vessels. Xylem proteomic studies have recently become one of the major areas of interest in understanding plant development and responses to environmental perturbations, and illustrated several types of xylem sap containing proteins that participate in cell wall development and repair processes (Zhang *et al.*, 2014). Studies of protein and metabolite composition of xylem sap and apoplast in soybean (*Glycine max*) provide insights into the expression profiles and signalling roles of corresponding proteomes, and ultimately reveal contributions of the root towards pathogenic and symbiotic microbe interactions and root-to-shoot communications (Krishnan *et al.*, 2011). Other proteomic studies in agriculturally important crops like oilseed rape; predominantly focus on the analysis of phloem sap exudates (Froehlich *et al.*, 2012), identifying several hundred physiologically relevant proteins and ribonucleoprotein complexes. The phloem sap proteomes show enhanced presence of proteins involved in redox regulation, defence, stress responses, calcium regulation, RNA metabolism and G-protein signalling.

iii) Symbiotic Interactions in Legumes

The symbiosis between N-fixing bacteria and legumes results in formation of root nodules and is very important in agriculture. Differential plant and bacteroid responses to drought stress have been revealed by proteomic analysis. Two different groups reported protein expression studies in nitrogen fixing root nodules of soybean and white sweet clover, respectively (Reid *et al.*, 2012; Molesini *et al.*, 2013). The 2-demensional electrophoresis was used to identify differentially expressed proteins during the symbiotic interaction between the bacterium *Sinorhizobium meliloti* strain 1021 and white sweet clover. Over 250 proteins were induced or up-regulated in the nodule, compared with the root, and over 350 proteins were down regulated in the bacteroid form of the rhizobia, compared with cultured cells. Bacteroid cells showed down-regulation of several proteins involved in nitrogen acquisition, indicating that the bacteroid cells were nitrogen proficient. These studies are excellent examples of the potential of proteomics in plant symbiotic interactions.

iv) Abiotic Stress Responses

Stressful conditions often lead to delayed seed germination, reduced plant growth, and decreased crop yield. Under a wide range of abiotic stresses; organ-specific proteins are commonly accumulated in organs (Komatsu and Hossain, 2013). The emerging proteomic technology of multiplexed selective-reaction monitoring MS, having increased accuracy and throughput, provides an efficient method for identification of various stress-responsive proteins (Jacoby *et al.*, 2013). Freezing stress studies reveal that the plasma membrane plays significant roles in signal perception and cellular homeostasis, indicating that plasma membrane proteins are the most important factors in determining the environmental stress tolerance of plants (Takahashi *et al.*, 2013).

v) Plant-Pathogen Interactions or Biotic Stress Responses

Pathogens use different strategies to attack or interact with the plants like necrotrophic (e.g. rotting bacteria), biotrophic (e.g. viruses, nematodes, fungal mildews) or hemibiotrophic (e.g. *Phytophtora*). The outcome of any interaction of a plant with another biological entity ultimately depends on the compatibility or the incompatibility of molecular communication between the two. The perception and recognition of pathogens induces a defence response in the plant characterized by reactions such as cell wall thickening, apoplast acidification, signal transduction and transcription of genes considered as defence genes. This line of defence generally results in the resistance of plants to a wide array of pathogens. Different techniques at the proteomic level are used to unravel relationships between plants and pathogens. Studies show that most of the protein families identified are involved in defence mechanisms or protection of structures such as HSPs, chaperones, ascorbate peroxidases, catalases etc. (Sergeant and Renaut, 2010).

Cell Specific Proteome Analysis

Understanding how plant cells sense and respond to biotic and abiotic stresses is not only fundamental to our understanding of stress tolerance, but has the potential to yield novel approaches to improve crop productivity. Cellular proteomics plays an essential role in determining the functions of cellular compartments and the mechanisms underlying protein/gene targeting and trafficking. Currently, numerous organ-specific proteomic analyses of biotic and abiotic stresses in crops have contributed to our understanding of the response mechanisms of crops to abiotic stresses (Komatsu and Hossain, 2013). Obviously, the specifics of proteomic response to different stresses vary from tissue to tissue within a plant. Therefore, the crop stress responses should be analyzed at a cellular or subcellular level, integrated with studies on whole plants, organs or tissues, to discriminate the specific responses of different cell types to stresses.

At present, cell or organelle specific proteomic studies focus on relatively abundant, or easily isolated homogenous compartments (e.g., chloroplast, mitochondria, peroxisomes, and nuclei) mainly in Arabidopsis (Tanz *et al.*, 2013), but also in rice, wheat, barley, maize (Huang *et al.*, 2013; Millar and Taylor, 2014; Hu *et al.*, 2015). Chloroplast proteomic studies integrated with large scale genomic approaches have addressed the complexity of chloroplast proteolytic machinery during leaf senescence and identified different classes of senescence-associated proteases with unique physiological roles according to their expression profiles along the senescence progress (Roberts *et al.*, 2012).

The probability of identifying stress responsive proteins from specific cells or tissues increases with the development of appropriate and efficient sampling methods to obtain relatively pure subcellular fractions from the test plant material. A promising sampling method is laser capture microdissection (LCM), which can isolate specific cell types of interest from sectioned specimens of heterogeneous tissues under direct microscopic visualization with the assistance of a laser beam (Longuespée *et al.*, 2014). The LCM has been successfully used in transcriptome and microarray studies in maize (Nakazono *et al.*, 2003; Rajhi *et al.*, 2011) and rice (Suwabe *et al.*, 2008; Kubo *et al.*, 2013). When LCM is combined with more sensitive protein staining technologies and more advanced mass spectrometers;

it has the potential to promote crop stress proteomics at a cellular level. Another promising technique is free flow electrophoresis (FFE), which can isolate much purer membrane fractions or organelles. The FFE technique has been successfully applied in plant proteomics to isolate tonoplast, mitochondria, plasma membranes, and Golgi apparatus (Barkla *et al.*, 2013). The cell specific proteome analysis not only leads to a better definition of the specialized proteomic behaviour of certain cell types, but also illustrates that information gained at proteome levels based on whole plants or organs can be misleading. Therefore, cell specific proteome analysis surely helps to better understand the processes of crop stress acclimation and stress tolerance acquisition.

Proteomics in Abiotic Stresses

Proteins fulfil a vast diversity of functions. They act as enzymes, transcriptional factors, interact with other molecules, have protective functions, are involved in energy transfer or radicals scavenging and take part in signalling pathways and others. Specific composition of proteins present in cells in the defined environmental conditions reflects the true biochemical outcome of genetic information and indicates the biochemical pathways that may be involved. Various abiotic stresses are known to hinder plant growth and yield by causing a variety of adverse effects including disturbance in regulation of many proteins involved in protein folding, reactive oxygen species (ROS) scavenging, proteolysis, metabolic energy supply, biosynthesis of carbohydrates and nucleotides, signal transduction, RNA processing, redox homeostasis, energy metabolism, secondary metabolites, glycolysis, lipid peroxidation, ethylene biosynthesis and cell wall loosening etc. However, regulation of different proteins varies among different plant species. Therefore, complete dissection of all proteins involved in different metabolic processes in plant species under a variety of stresses needs to be carried out. Activation of stress-responsive pathways due to environmental adverse conditions causes significant changes in plant proteome. Explaining proteome changes is critical for understanding of rules with which cells work and adapt towards various stimuli. In recent years, much attention has been paid to study proteomics on crop plants. Examples of proteomics research on crop plants include, barley (Ashoub *et al.*, 2013), maize (Pechanova *et al.*, 2013) rice (Lee *et al.*, 2009), wheat (Peng *et al.*, 2009) and soybean (Cheng *et al.*, 2010) subjected to various abiotic stresses. Depending on the tolerance levels of the plant genotypes under various abiotic stresses, they may accumulate or enhance expression of particular proteins, which exhibit the protective functions. These stress responsive proteins include heat-shock proteins, osmolyte biosynthesis proteins or proteins belonging to ROS-scavenging systems and others.

Stress Responsive Proteins

Heat Shock Proteins

Proteins are to be maintained at proper conformation for normal functioning of plants. Abiotic stresses cause protein dysfunction. Therefore, plant survival under severe condition depends on plants ability to prevent incorrect folding and aggregation of the proteins. Heat shock proteins (HSPs) or chaperones are a large family of proteins playing roles in keeping cellular homeostasis both under optimal growth conditions and under stress. They are responsible for correct folding, translocation and degradation of proteins. They also prevent proteins from aggregation. In plants, five major families of HSPs are recognized. They are classified on the basis of their approximate molecular weight and according to their amino acid sequence homologies and functions: (1) HSP100 family; (2) HSP90 family; (3) HSP70 family; (4) HSP60 family; (5) small HSP family (sHSP; 15–40 kDa) (Kotak *et al.*, 2007; Gupta *et al.*, 2010). The HSP100/Clp takes part in the removal and reactivation of aggregated, misfolded or non-functional polypeptides (Agarwal *et al.*, 2001). Proteins belonging to HSP90 class play a role in protein folding and regulation of signal transduction (Buchner, 1999; Pratt and Toft, 2003). The HSP70 family prevents newly synthesized proteins from aggregation and helps them to fold correctly. They are also involved in transport and proteolytic degradation of unstable proteins (Su and Li, 2008). Proteins of HSP60 family are crucial to achieve native forms by newly synthesized proteins. For example, they assist in the proper folding of RUBISCO (Wang *et al.*, 2004). Among all HSPs or chaperonins in plants, small HSPs represent the most prevalent and diverse family with respect to their sequence homology, functions and cellular location (Waters *et al.*, 1996). They cannot refold non-native proteins; however, they bind to partially folded or denatured proteins and prevent aggregation or non-functional folding. They are also capable of degrading misfolded proteins (Sun *et al.*, 2002).

HSPs are known to be inducible by various abiotic stresses, including drought, salinity, high and low temperatures, light, ozone and metal stress (Wang *et al.*, 2003; Sabehat *et al.*, 1998; Lee *et al.*, 2000). The comparison of global expression of genes coding HSPs and their transcriptional factors (Hsf) in response to different abiotic stresses revealed that most Hsfs and HSPs have similar response and regulation patterns under different stresses; however, some of the genes show a highly specific response (Hu *et al.*, 2009).

Late Embryogenesis Abundant Proteins

Late embryogenesis abundant proteins (LEA) were originally discovered more than three decades ago in cotton seed, where they were synthesized and accumulated during late embryo development (Dure *et al.*, 1981). In plants, LEA proteins and their mRNAs accumulate to high concentrations in the embryo tissue during the last stages of seed development (Baker *et al.*, 1988). Dehydrins constitute one of the groups of LEA proteins (Roberts *et al.*, 1993). These proteins play a significant role in plant response and adaptation to abiotic stresses. The synthesis of dehydrins may be induced in the vegetative tissue by various stress factors. These proteins play a critical role in cellular protection during the osmotic shock. They stabilize the cell and protect the tissue from water loss.

Osmolyte Biosynthetic Enzymes

Osmolytes like glycine betaine (GB) or proline play an important role in enhancing plant resistance to abiotic stresses. Osmolytes exhibit protective functions on enzymes and membrane integrity and take part in regulation of osmotic pressure. During abiotic stresses their synthesis usually increases in plants (Quan *et al.*, 2004; Chen and Murata, 2011). Betaine aldehyde dehydrogenase (BADH) is an enzyme involved in the synthesis of glycine betaine through the conversion of betaine aldehyde. The overproduction of betaine glycine enhances photosynthesis and antioxidant activity (Wang *et al.*, 2010).

Enzymatic Reactive Oxygen Species (ROS) Scavengers

Partially oxidized oxygen particles, known as the reactive oxygen species (ROS), unlike atmospheric oxygen, have the capability of oxidizing various molecules and cell structures. Under normal environmental conditions, ROS are naturally present in plants; however, plants developed various antioxidant systems to balance their production and quantity (Gill and Tuteja, 2010). Main forms of ROS that arise in plants are the superoxide radical, hydrogen peroxide, singlet oxygen and hydroxyl radicals (Mittler, 2002). They are by-products of biological processes that take place in organelles with a high metabolic activity like chloroplasts (during photosynthesis) (Pfannschmidt, 2003), mitochondria (electron transport chain) (Rasmusson *et al.*, 2004) or peroxisomes (photorespiration, beta-oxidation) (del Rı´o *et al.*, 2002). Although ROS are considered to be very toxic for the plants, especially at higher concentrations, they also play a role in signalling and are involved in the activation of defence responses and in acclimation to changing environmental conditions (Dat *et al.*, 2000). Under abiotic stresses, diminished concentration of internal CO_2 due to stomata closing leads to the leakage of electrons to O_2 and is one of the main reasons for the excessive ROS production. Plants have developed various antioxidant mechanisms, both enzymatic and non-enzymatic, which very often form a common metabolic pathway. The ascorbate–glutathione cycle is one of the major ROS scavenging pathways. The enzymatic components of this cycle are ascorbate peroxidase (APX), monodehydroascorbate reductase (MDHAR), dehydroascorbate reductase (DHAR) and glutathione reductase utilize ascorbic acid and glutathione to scavenge H_2O_2 (Jimenez *et al.*, 1997). The SODs are found in various locations within the cell. They are classified by the metal cofactor they require: manganese (Mn-SOD), iron (Fe-SOD) and copper/ zinc (Cu/Zn-SOD) (Mittler, 2002).

Catalases (CAT) found in the peroxisomes are very efficient enzymes capable of dismutating H_2O_2 to H_2O and O_2 (Polidoros and Scandalio, 1999). Other enzymes involved in antioxidant defence include glutathione-S-transferases (GST) and various peroxiredoxins (PRX). The regulation of the expression of the ROS-scavenging enzymes differs between the plant species, depending on the tolerance level to the stress and its intensity. Analysis of the changes in the proteome of the roots of pea in response to salinity revealed an increased accumulation of SOD, but no other antioxidant enzymes were identified (Kav *et al.*, 2004).

Other proteins

This group comprises proteins involved in the secondary metabolism which respond to abiotic stresses. Phenylalanine ammonia lyase (PAL) is the first enzyme of the phenylpropanoid pathway, which catalyzes the conversion of phenylalanine to cinnamic acid. The phenylpropanoid pathway is crucial for the biosynthesis of lignins, phenolic acids and flavonoids, which play important roles in plant defence mechanisms (Vogt, 2010). Lignins, which are synthesized from phenylpropanoid compounds play a crucial role in conducting water in plant stem. Decreased resistance to drought was observed in *A. thaliana pal1 pal2* double mutants, which might have been the effect of the reduced lignin synthesis as the mutants contained only 30 % of the lignin level typical for the wild type. The reduced lignin contents in the double mutants might cause a reduced efficiency of water transport in the vascular tissue (Huang *et al.*, 2010). The protein synthesis elongation factor (EF-Tu protein) plays a role in the elongation phase of protein biosynthesis. This protein has been studied in several plants in response to abiotic stresses, such as extreme temperatures, salinity and drought. EF-Tu acts as a molecular chaperone and protects other proteins from thermal aggregation and degradation (Rao *et al.*, 2004).

Transcription factors (TF) are the specific group of proteins that play a regulatory role in gene control. TF factors have the ability to control the expression of target genes through binding to the specific sequence in these genes. Large plant TF families play significant roles in response to abiotic stresses (Nakashima *et al.*, 2009).

Proteomics in Drought Stress

Drought is a problem of great importance in world agriculture, especially in arid and semiarid regions. The water availability in agriculture is a key factor in agricultural production and non- availability of water cause severe losses. Ribas-Carbo *et al.*, (2005) reported that water stress inhibited CO_2 fixation in soybean leaves in just a few days, reducing carbon assimilation rates to almost zero. Plants respond to water deficit through a series of processes at the physiological, cellular and molecular levels, which can lead to stress tolerance. The effect of water stress on plant physiology has been extensively studied like stomatal closure, reduction in cell growth, decreased photosynthetic activity and increased respiration.

The most common responses to drought stress are the damage to chlorophyll, decreased antioxidant system, increase in the production of hydrogen peroxide (H_2O_2) and the O_2^- ion, lipid peroxidation, decreased photosynthesis (Qureshi *et al.*, 2007), increased oxidative stress and alterations in cell wall elasticity (Caruso *et al.*, 2009). At the cellular level, plants respond and adapt to a water deficit by accumulating osmolytes and synthesizing specific proteins (Fig. 2). The comparative analysis of barley leaves subjected to drought stress revealed the upregulation of HSP100, HSP90 in both sensitive and tolerant genotypes. A member of HSP70 was found to be down-regulated only in the sensitive genotype

(Ashoub *et al.*, 2013). Three HSP70 proteins were also down-regulated during the analysis of sugar beet leaves (Hajheidari *et al.*, 2005).

The HSP90 and HSP70 were found to be required for stomata closure and modulation of transcriptional responses to abscisic acid (Cle`ment *et al.*, 2011); thus, they may be crucial for enhanced drought-tolerance. The effects of drought stress on the activity of PAL in maize revealed an increased activity of this enzyme in leaves, while in the roots it remained constant. The antioxidant activity in leaves was also enhanced. It may suggest that PAL is a responsive antioxidative enzyme in the leaves, but not in roots (Ashraf, 2011).

Research on proteomics in wheat (*T. aestivum* and *T. durum*) (Caruso *et al.*, 2009) and rice (*O. sativa*) (Ke *et al.*, 2009) under water stress conditions, using the techniques of 2-DE and mass spectrometry, MALDI-TOF, showed changes in protein expression, with 12 proteins over-regulated and 24 sub-regulated in wheat, and 12 over-regulated and 6 sub-regulated in rice. In case of *T. durum*, 36 reproducible protein spots were detected in response to water stress. This study identified 21 different proteins, including some isoforms and subunits of enzymes that change their expression under water stress. The increased activities of APX and SOD were found in all three analyzed varieties of bean in response to drought, but the CAT activity was up-regulated only in two of them. Moreover, in the variety with the highest APX and CAT activity, the lowest H_2O_2 production and lipid peroxidation were observed (Zlatev *et al.*, 2006). Shotgun proteomic analysis of wheat cultivars subjected to drought revealed that Cu/Zn SOD and CAT were highly up-regulated in both tolerant and sensitive genotypes (Ford *et al.*, 2011). An increased activity of APX at mild drought was also reported in rice seedlings. However, the intensification of the stress diminished its activity (Sharma and Dubey, 2005).

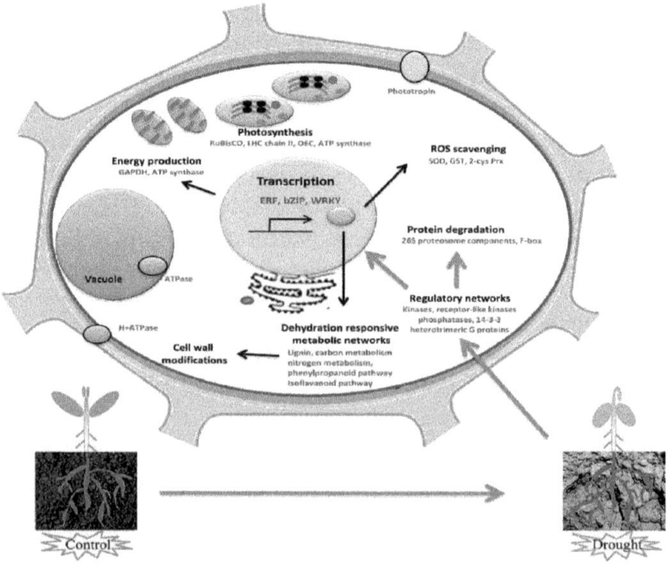

Fig. 2: Plant proteome under drought stress (Roy et al., 2011)

In the comparative study of dehydration sensitive and tolerant cultivars of chickpea, SOD, APX and GST were found to be induced only in the resistant genotype. Another enzyme, glyoxalase, involved in the detoxification of R-oxoaldehydes, was also up-regulated in the tolerant variety. The lack of induction of major ROS catabolic enzymes may lead to hypersensitivity of plants during dehydration-induced oxidative stress (Subba *et al.*, 2013).

Proteomics in Temperature Stress

Plants that are exposed to high temperatures show up-regulation of HSPs. Proteomic response of rice leaves to high temperatures was examined and out of 48 identified proteins, 18 proteins belonged to HSPs. Among them, members of HSP70, HSP100, DnaK-type, molecular chaperone BiP were identified and showed up-regulation. Seven induced sHSPs were also found. Such a result may reflect a rapid adaptation to changing environmental conditions (Lee *et al.* 2007). In the reports concerning barley (Su¨le *et al.*, 2004) and wheat (Majoul *et al.* 2003) also the upregulation and induction of HSPs were observed. Plants that are exposed to low temperatures also synthesize HSPs (Miura and Furumoto, 2013). In particular, the upregulation of HSP70, HSP100 and sHSPs is commonly reported (Hashimoto and Komatsu, 2007).

Quantitative differences in dehydrin gene expression and dehydrin protein accumulation under extreme temperatures was studied in wheat. The correlation between an increase of WCS120 LEA protein synthesis and the level of acquired frost tolerance was reported (Houde *et al.* 1992). Temperature stress-induced expression and accumulation of the elongation factor (EF) was first reported in maize. It indicated that there was a correlation between the increased accumulation of EF in plastids and the heat tolerance of maize lines (Ristic *et al.* 1991). The analysis of the proteome and transcriptome of maize indicated that the expression of EFTu factor was regulated differentially in heat-tolerant and sensitive cultivars. A significant increase in the accumulation of both the EF-Tu transcript and EF-Tu protein was observed in tolerant genotypes of maize. The up-regulation of glyoxylate was also observed in rice root tissue in response to cold treatment (Lee *et al.*, 2009). In another study concerning the analysis of cold induced changes of proteins in the roots of rice seedlings, APX was found to be up-regulated and also its phosphorylation pattern was changed (Chen *et al.*, 2012).

Proteomics in Salinity Stress

The salinity of the soil causes toxic levels of sodium, chloride and sulphates. Salts in the soil inhibit plant growth for two reasons; first, by an osmotic effect that reduces the capacity of plant water uptake and causes a slow plant growth as a result of water stress in response to salinity of the external environment, and second, by the toxic effect due to the entry of salts that cause tissue damage of the leaves (Khan *et al.*, 2007). The probable cause of damage is the high flux of ions and salts which exceeds the ability of the cell to compartmentalize them in the vacuole. With the rapid entry of salts and ions into the cytoplasm, enzyme activity is inhibited, the cell wall is altered and the cells become dehydrated and die (Munns, 2005). Plants under saline conditions generally have an osmotic imbalance. The osmotic imbalance mainly causes changes in ion concentration in the cytoplasm; particularly potassium and calcium are replaced by higher levels of Na+ and Cl^-, which have toxic effects on the membrane structure and the enzyme systems (Ashraf and Harris, 2004). Also, secondary stress, such as oxidative stress is caused by the production of toxic reactive oxygen which in turn leads to lipid peroxidation (LP) production (Qureshi *et al.*, 2005).

In the analysis of the changes in the proteome of grasspea exposed to salinity and drought, HSP 70 was found to be up-regulated in the plants exposed to increased salinity and drought (Chattopadhyay *et al.*, 2011). The over-expression of the wheat PMA80 dehydrin in rice enhanced its tolerance to drought and salinity (Cheng *et al.*, 2002). Several studies related to the comparative analysis of proteome between plants subjected to salt stress and control treatments have been performed in species like rice (*O. sativa*) (Yan *et al.*, 2005), wheat (*T. aestivum* sp.) (Huo *et al.*, 2004) and sorghum (*Sectarian italics* L.) (Veeranagamallaiah *et al.*, 2008), using the techniques of 2-DE and mass spectrometry, MALDI-TOF/TOF-MS. The results showed changes in expression of proteins related to the response to salt stress. Comparative proteome analysis of sorghum under different sodium chloride concentrations (100, 150 and 200 mM) and control treatments showed temporal changes in total protein profile (Veeranagamallaiah *et al.*, 2008). The results showed 175 reproducible and detectable protein spots. Through MS analysis, 29 different proteins were identified which expressed in response to salt stress, involved in various processes such as photosynthesis (31.0%), nitrogen metabolism (13.8%) lipid metabolism (6.8%), carbohydrate metabolism (6.8%), nucleotide metabolism (6.8%), cell wall biogenesis (6.8%) and signal transduction (10.3%), stress-related proteins (10.3%) and proteins with unknown functions (7.4%).

The study of the root proteome of barley genotypes response to salinity stress revealed that MDAR, APX and CAT showed a higher constitutional expression level in the sensitive genotype and two of them (MDAR, APX) were found to be more down-regulated during the salinity treatment in the tolerant genotype. However, two identified proteins involved in the glutathione-based detoxification of the reactive oxygen species (ROS) were more abundant in the tolerant cultivar. The higher constitutive expression of these proteins in the more tolerant genotype may represent a pre-formed tolerance mechanism (Witzel *et al.* 2009).

Combination of Multiple Abiotic Stresses

In nature, plants are usually exposed to multiple stresses. However, plant stress responses to combined stress treatments are seldom studied. A few proteomic studies dealing with combined stress treatments have shown that plant response to a combined stress treatment is specific when compared to the individual stress factors applied separately. Since no simple assumptions on an additive effect of individual stress treatments can be applied, plant response to combined stress treatments deserves to be studied. Peng *et al.*, (2009) compared the effects of drought and salinity as two separate treatments applied on a relatively sensitive bread wheat cv. Jinan 177 and its somatic hybrid with tall wheatgrass *Thinopyrum ponticum* named Shanrong 3. Comparison of proteome response to both treatments revealed that salinity induced significant alterations in a higher number of proteins than drought as a consequence of an ionic effect of salinity stress. Rollins *et al.*, (2013) studied the effects of drought (15% soil water content), heat (36 °C) and a combined drought and heat treatment in two relatively drought-tolerant barley genotypes originating from differential drought environments, Syrian landrace Arta and Australian cultivar Keel, differing in their drought response strategies. Drought led to a reduction in plant growth while maintaining relatively stable proteome composition. In contrast, heat led to enhanced protein damage, especially of PSII components, and thus an enhanced need for energy due to an increased protein turnover. An enhanced abundance of ROS scavenging enzymes and protective proteins (HSPs, a thermostable RubisCO activase B isoform) was found in heat-treated plants indicating an imbalance between the rates of primary (light-dependent) and secondary (light-independent) photosynthetic reactions and an enhanced risk of protein damage under heat stress. Yang *et al.*, (2011) investigated an effect of drought and heat (32 °C) treatment on proteome composition of wheat grain. Several proteins revealed a specific response to each stress treatment while only a few common proteins involved in redox metabolism, defence, carbohydrate metabolism, and storage, revealing an analogous response to multiple stress treatments. A comparison of contrasting water stresses i.e. drought and flooding was carried out by Oh and Komatsu, (2015) in soybean seedlings. The results showed differentially regulated stress responses. An increase in enzymes involved in regulation of redox homeostasis was found in drought-stressed plants while an increase in anaerobic metabolism-related (fermentation) enzymes was found in flooded plants. Moreover, an opposite pattern of changes in SAM synthetase (SAMS) was found in drought-treated plants versus flooded ones indicating an increase in SAMS under drought while a decrease in SAMS under flooding. Since SAM is a universal cell methylation agent involved in lignin biosynthesis, the differential pattern of SAMS may be related to changes observed in root cell wall lignification in soybean seedlings under the two treatments corresponding to an increased cell wall lignification under drought while a decreased cell wall lignification under flooding, respectively. Effects of drought, cold, and herbicide paraquat treatments on pea mitochondrial proteome were compared by Taylor *et al.*, (2005). The strongest adverse effects on mitochondrial proteins resulting in an oxidative damage were observed under paraquat treatment, followed by chilling while drought revealed the mildest effects.

Proteomics in Biotic Stresses

Biotic stresses are caused by the living organisms, such as fungi, bacteria, viruses, and insects etc. in the plants. The plant response to pathogen attack or microbial signals has been a long time interest of plant biologists and pathologists. Proteomics technology is playing a key role in developing inventory of proteins responsive to biotic stresses. The protein inventory has proved helpful in identifying common or specific proteins to biotic stresses. Different techniques are used to unravel relationships between plants and pathogens at the proteomic level e.g. 2DE, liquid chromatography etc. Most of the protein families reported, are involved in defence or protection of structures such as HSPs, chaperones, ascorbate peroxidases, catalases, pathogenesis- related proteins, and so on (Fig 3). Different relationships of crop plants with bacteria, fungi, insects, viruses or other parasitic plants have been reported to represent biotic stresses.

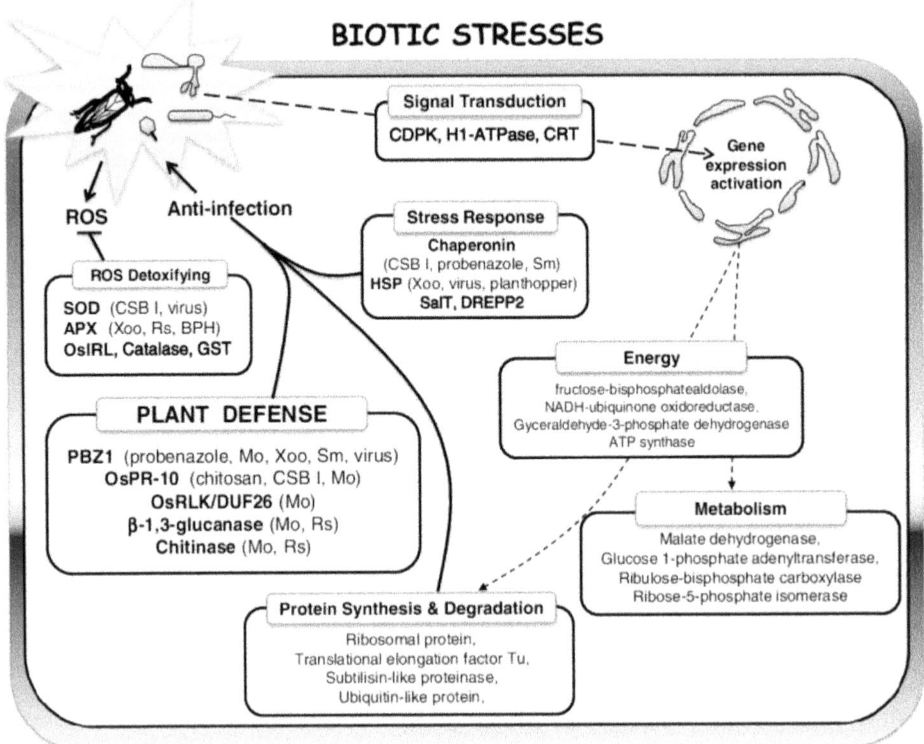

Fig.3: A model illustrating biotic stress-responsive proteins in rice (Wang *et al.*, 2011)

Insect Pest Stress

Examples of damage caused by insects in different crops are well-known. For instance, the ruin caused by Colorado potato beetles and the devastation caused locusts are two examples. Herbivorous insects can affect the health of plants either directly, through the damage they impose, or indirectly by serving as vector for the transmission of plant pathogenic microorganisms or viruses (Hooks and Fereres, 2006). The final response of the plants depends on the balance between the often antagonistic effects of phytohormones (Pieterse *et al.*, 2009). The production of toxic, repellent or anti digestive compounds and also the emission of volatiles that attract predators of the herbivores are the most important responses of plants during herbivore attack (Wu and Baldwin, 2009). The majority of plant viruses are spread by insects, in this situation the system that must be studied is composed of three biological entities; the plant, the insect and the microorganism/virus.

Despite the relative scarcity of examples of the use of proteome techniques in the study of the interaction of plants with insect herbivores; some research on the resistance of certain plants for this type of biological stress exists. The 2DE-PAGE combined with MALDI and/or ESI-MS/MS protein identification remains the workhorse. Several other techniques and tools have also been used in the study of plant-insect interactions. The study of the mid-gut content of leaf-fed insects revealed that potentially active plant proteins, known to be induced during herbivorous attack, could be identified. Both Kunitz-type proteinase inhibitor and amino acids degrading enzymes result in the depletion of essential amino acids (Philippe *et al.*, 2009), thereby limiting the fecundity of the herbivore.

Virus Stress

Viruses are characterized by a very small genome that encodes for a few proteins. The plant viruses belonging to Geminiviridae and Reoviridae, are among the most important causes of yield losses in many crops. They are spread in plantations either by the cultivation of contaminated seeds or during culture by normal agricultural practices (Fereres and Moreno, 2009; Rochon *et al.*, 2004). The detrimental impact of viruses on plant production can be limited by the control of biological vectors, crop trapping, the use of resistant plant material and the amelioration of agricultural practises to minimize spread. Apart from these, genetic engineering of virus-resistant crop plants is the current apex of applied research in viral phytopathology (Prins *et al.*, 2008). Although the study of plant virus interactions and the different forms of resistance/susceptibility are reviewed in few publications, the molecular mechanisms of viral diseases and resistance remain largely unknown (Kang *et al.*, 2005). Proteomics in plant and plant virus research is using subcellular fractionation to gain a more profound view on the interacting proteomes (Haynes and Roberts, 2007). The chloroplast proteome was targeted in peas infected with the Plum pox virus (Diaz-Vivancos *et al.*, 2008). The results of this study indicated that the virus disease is mainly caused by an increased imbalance of the antioxidant systems and by increased generation of reactive oxygen species, making the chloroplast a source of oxidative stress during the disease development. Although the study of the direct molecular interaction between viruses and plants during infection is an interesting field of research, relatively little studies on this topic have been done. Few studies have been done mainly applying yeast-two-hybrid screens and use of microarrays, but protein-based methods have also been used (Swatkoski *et al.*, 2007).

Pathogenic Fungi Stress

During their life cycle, plants have to cope with regular contacts with pathogenic fungi. Most of these fungi are able to penetrate the plants but their spread is rapidly stopped by the constitutive or induced defence strategies existing in plant cells (D'Ovidio *et al.*, 2004). Biotic and abiotic stresses are able to trigger almost similar pathways (Fujita *et al.*, 2006). Moreover, it is well-known that a plant weakened by an abiotic factor will be more susceptible to pathogen attack, and inversely. So, interconnected network triggered by one or the other type of stressor is activated in such a way that the plant invests a minimum of energy to enhance its survival chances. In a study, based on the interaction between canola and *Sclerotinia sclerotiorum* (Liang *et al.*, 2008), necrotic lesions were observed on the leaves as soon as after 12 hours of inoculation. After 2DE, spot picking and protein identification, 32 proteins (12 +ve and 20 -ve) responding to the infection after 6, 12, 24, 36 and 48 hours were classified according to their intracellular functions e.g. increases and decreases of abundance of proteins involved in photosynthesis, Calvin's cycle and hormone biosynthesis were detected upon *Sclerotinia* infection, while increased abundances were observed for proteins of energy metabolism, transport, nitrogen assimilation, several proteases, glyoxalase and peroxidase. Proteins showing less intensity upon infection have been identified as defence related proteins, probably explaining the susceptibility of *Brassica napus* to *Sclerotinia sclerotiorum*.

Bacterial Stress

Pathogenic bacteria use macromolecule delivery systems termed as type III and IV to inject microbial avirulence (avr) proteins and to transfer their DNA-protein complexes directly in the core of plant cells and to deliver effector molecules to the target. Type III secretion system can be activated by the contact between the host cell surface and the bacteria and proteins secreted facilitate the attack by interfering with host cellular processes, such as signal transduction. The type IV is mostly based on conjugation (transfer of DNA by cell-to-cell contact) or by a system called 'effector translocator' delivering protein substrates or DNA by direct contact with the host cells (Cascales and Christie, 2003). Pathogen-associated molecular patterns (PAMPs) are molecules used by plants or animals to detect invaders. Among these PAMPs, flagellin, a protein forming bacterial flagellum, or at least a part of this protein, is able to activate defence mechanism in plants (Zipfel *et al.*, 2004). Moreover, it has been established that phosphorylation events are commonly observed in response to bacterial attacks in plants. Several genomes of the most important plant pathogenic bacteria have been fully sequenced and made publically available (e.g. *Pseudomonas syringae, Xanthomonas campestris, Xylella fastidiosa* etc.) and these have been used tremendously for the completion of large-scale studies on the interaction between these pathogens and their hosts. The 'association' *Arabidopsis thaliana - Pseudomonas syringae* has become a model pathosystem to unravel the mechanisms specific to the association itself and the different reactions triggered by the association i.e. susceptibility, innate immune response or gene-for-gene resistance (Kaffarnik *et al.*, 2009).

Other Biotic Stresses

Fewer studies are currently available on other types of plant-pathogen interactions, for instance with nematodes or parasitic plants. Phytonematodes, parasites of plant roots, cause significant losses in plant production. Via a stylet (a protrusible spear in nematode's oral cavity), the nematodes inject secretions into root cells to pump host cell nutrients. Different proteins are secreted by the nematode to attack the roots, including proteins involved in e.g. cell wall modification, reception of signal transduction, cellular metabolism, cell cycle, selective protein degradation and localized defence response etc. A study with combined proteomic and metabolomic techniques helped to unravel differences between resistant and susceptible lines of soybean when roots were damaged by *Heterodera glycines* (Afzal *et al.*, 2009). Parasitic weeds have developed mechanisms to infest other crop plants. Small radicals attach the parasite to the host roots and a haustorium connects the parasitic plant to the vascular system of the host to use its resources. Relationships between parasitic plants and leguminous plants (e.g. alfalfa and pea) have been addressed in several studies to unravel mechanisms underlying the infestation and the resistance towards parasites, via the study of healthy and infested roots (Castillejo *et al.*, 2004).

Conclusions

Proteomics has emerged as an indispensable methodology and it will remain to be one of the fastest growing areas in research for large scale protein analysis in functional genomics. Proteomics is a powerful tool for investigating protein changes induced by various stresses. The global scale analysis of proteins is expected to yield more direct understanding of function and regulation than analysis of genes. Gel-based 2-DE proteomic approaches combined with gel-free MS-based quantitative proteomic techniques are widely used for crop proteome analysis. The proteomic studies have contributed to elucidation of complex relationships between stress tolerance and crop productivity, which would enable the development of novel breeding strategies resulting in an increase in crop productivity and environmental performance. Comparative proteomic studies on different tissues, organs, organelles and membranes, using different biological models allow monitoring of protein expression during different times, contributing significantly to the understanding of the mechanisms of adaptation of plants under certain stress conditions. Also combined stress treatments induce a unique plant stress response at the proteomic level and proteomic analyses can lead to an identification of proteins revealing common response to multiple stress treatments as well as proteins responding only to specific stress conditions.

References

Afzal, A.J., Natarajan, A., Saini, N., Iqbal, M.J., Geisler, M., El Shemy, H.A., Mungur, R., Willmitzer, L. and Lightfoot, D.A. 2009. The nematode resistance allele at the rhg1 locus alters the proteome and primary metabolism of soybean roots. *Plant Physiol*, 151: 1264-1280.

Agarwal, M., Katiyar-Agarwal, S., Sahi, C., Gallie, D.R. and Grover, A. 2001. *Arabidopsis thaliana* Hsp100 proteins: kith and kin. *Cell Stress Chaperon*, 6: 219-224.

Ahmad, F., Rai, M., Ismail, M. R., Juraimi, A. S., Rahim, H. A. and Asfaliza, R. 2013. Water logging tolerance of crops: breeding, mechanism of tolerance, molecular approaches, and future prospects. *Biomed Res. Int,* 963525.

Ahmad, P., Ashraf, M., Younis, M., Hu, X., Kumar, A. and Akram, N. A. 2012. Role of transgenic plants in agriculture and biopharming. *Biotechnol. Adv,* 30: 524–540.

Ashoub, A., Beckhaus, T., Berberich, T., Karas, M. and Bruggemann, W. 2013. Comparative analysis of barley leaf proteome as affected by drought stress. *Planta,* 237: 771-781.

Ashraf, G. 2011. Effects of drought on the activity of phenylalanine ammonia lyase in the leaves and roots of maize inbreds. *Aust. J. Basic Appl. Sci,* 5: 952-956.

Ashraf, M. and Harris, P.J.C. 2004. Potential biochemical induction of salinity tolerance in plants. *Plant Sci,* 166: 3-16.

Avice, J.C. and Etienne, P. 2014. Leaf senescence and nitrogen remobilization efficiency in oilseed rape (*Brassica napus* L.). *J. Exp. Bot.,* 65: 3813–3824.

Bachi, A. and Bonaldi, T. 2008. Quantitative proteomics as a new piece of the systems biology puzzle. *J. Proteomics,* 71: 357–367.

Baker, J., Steele, C. and Dure, L. 1988. Sequence and characterization of 6 Lea proteins and their genes from cotton. *Plant Mol. Biol,* 11: 277-291.

Barkla, B. J., Vera-Estrella, R. And Pantoja, O. 2013. Progress and challenges for abiotic stress proteomics of crop plants. *Proteomics,* 13:1801–1815.

Buchner, J. 1999. Hsp90 & Co.- a holding for folding. *Trends Biochem. Sci,* 24: 136-14.

Caruso, G., Cavaliere, C., Foglia, P., Gubbiotti, R., Samperi, R. and Laganà, A. 2009. Analysis of drought responsive proteins in wheat (*Triticum durum*) by 2D-PAGE and MALDI-TOF mass spectrometry. *Plant Sci,* 177: 570–576.

Caruso, G., Cavaliere, C., Guarino, C., Gubbiotti, R., Foglia, P. and Laganà, A. 2008. Identification of changes in *Triticum durum* L. leaf proteome in response to salt stress by two-dimensional electrophoresis and MALDI-TOF mass spectrometry. *Anal. Bioanal. Chem,* 391: 381–390.

Cascales, E. and Christie, P.J. 2003. The versatile bacterial type IV secretion systems. *Nat. Rev. Microbiol,* 1: 137-149.

Castillejo, M.A., Amiour, N., Dumas-Gaudot, E., Rubiales, D. and Jorrin-Novo, J.V. 2004. A proteomic approach to studying plant response to create broomrape (*Orobanche crenata*) in pea (*Pisum sativum*). *Phytochem,* 65: 1817-1828.

Chandana, N., Harithapavani, K., Deepa, R. N. and Maduri, J. 2013. A Review on Liquid Chromatography Tandem Mass Spectroscopy. *Journal of Chemical and Pharmaceutical Sciences,* 6:210.

Chattopadhyay, A., Subba, P., Pandey, A., Bhushan, D., Kumar, R. And Datta, A. 2011. Analysis of the grass pea proteome and identification of stress-responsive proteins upon exposure to high salinity, low temperature, and abscisic acid treatment. *Phytochemistry,* 72: 1293–1307.

Chen, J., Tian, L., Xu, H., Tian, D., Luo, Y., Ren, C., Yang, L. and Shi, J. 2012. Cold-induced changes of protein and phosphoprotein expression patterns from rice roots as revealed by multiplex proteomic analysis. *Plant Omics J,* 5: 194-199.

Chen, T.H. and Murata, N. 2011. Glycine betaine protects plants against abiotic stress: mechanisms and biotechnological applications. *Plant Cell Environ,* 34: 1-20.

Cheng, L., Gao, X., Li, S., Shi, M., Javeed, H., Jing, X., Yang, G. and He, G. 2010. Proteomic analysis of soybean [*Glycine max* (L.) Meer.] seeds during imbibition at chilling temperature. *Mol. Breed,* 26: 1-17.

Cheng, Z., Targolli, J., Huang, X. and Wu, R. 2002. Wheat LEA genes, PMA80 and PMA1959, enhance dehydration tolerance of transgenic rice (*Oryza sativa* L.). *Mol. Breed,* 10: 71-82.

Chu, P., Yan, G.X., Yang, Q., Zhai, L.N., Zhang, C. and Zhang, F.Q. 2015. ITRAQ-based quantitative proteomics analysis of *Brassica napus* leaves reveals pathways associated with chlorophyll deficiency. *J. Proteomics,* 113: 110–126.

Cle'ment, M., Leonhardt, N., Droillard, M.J., Reiter, I., Montillet, J.L., Genty, B., Laurie`re, C., Nussaume, L. and Noe¨l,L.D. 2011. The cytosolic/ nuclear HSC70 and HSP90 molecular chaperones are important for stomatal closure and modulate abscisic acid-dependent physiological responses in Arabidopsis. *Plant Physiol,* 156: 1481-1492.

Dat, J., Vandenabeele, S., Vranova,´ E., Van Montagu, M., Inze´ D. and Van Breusegem, F. 2000. Dual action of the active oxygen species during plant stress responses. *Cell Mol. Life Sci,* 57: 779-795.

del Rı´o, L.A., Corpas, F.J., Sandalio, L.M., Palma, J.M., Go´mez, M. and Barroso, J.B. 2002. Reactive oxygen species, antioxidant systems and nitric oxide in peroxisomes. *J. Exp. Bot.,* 53: 1255-1272.

Diaz-Vivancos, P., Clemente-Moreno, M.J., Rubio, M., Olmos, E., Garcia, J.A., Martinez-Gomez, P. and Hernandez, J.A. 2008. Alteration in the chloroplastic metabolism leads to ROS accumulation in pea plants in response to plum pox virus. *J. Exp. Bot,* 59: 2147-2160.

D'Ovidio, R., Mattei, B., Roberti, S. and Bellincampi, D. 2004 Polygalacturonases, polygalacturonase-inhibiting proteins and pectic oligomers in plant-pathogen interactions. *Biochim. Biophys. Acta,* 1696: 237-244.

Dure, L., Greenway, S.C. and Galau, G.A. 1981. Developmental biochemistry of cotton seed embryogenesis and germination: changing messenger ribonucleic-acid populations as shown by *in vitro* and *in vivo* protein-synthesis. *Biochemistry,* 20: 4162–68.

FAO. 2012. OECD-FAO *Agricultural Outlook,* 2012-2021. Paris: OECD Publishing. doi: 10.1787/agr_outlook-2012-en

Fereres, A. and Moreno, A. 2009. Behavioural aspects influencing plant virus transmission by homopteran insects. *Virus Res*, 141: 158-168.

Ford, K.L., Cassin, A. and Bacic, A. 2011. Quantitative proteomic analysis of wheat cultivars with differing drought stress tolerance. *Front. Plant Sci*, 2: 44.

Froehlich, A., Gaupels, F., Sarioglu, H., Holzmeister, C., Spannagl, M. and Durner, J. 2012. Looking deep inside: detection of low-abundance proteins in leaf extracts of Arabidopsis and phloem exudates of pumpkin. *Plant Physiol*, 159: 902–914.

Fujita, M., Fujita, Y., Noutoshi, Y., Takahashi, F., Narusaka, Y., Yamaguchi-Shinozaki, K. and Shinozaki, K. 2006. Crosstalk between abiotic and biotic stress responses: a current view from the points of convergence in the stress signalling networks. *Curr. Opin. Plant Biol*, 9: 436-442.

Geho, D.H., Lahar, N., Ferrari, M., Petricoin, E.F. and Liotta, L.A. 2010. Opportunities for nanotechnology-based innovation in tissue proteomics. *Biomed. Micro devices*, 6: 231–239.

Gill, S.S. and Tuteja, N. 2010. Reactive oxygen species and antioxidant machinery in abiotic stress tolerance in crop plants. *Plant Physiol. Biochem*, 48: 909-930.

Gorg, A., Weiss, W. and Dunn, M.J. 2004. Current two-dimensional electrophoresis technology for proteomics. *Proteomics*, 4(12): 3665-3685.

Gupta, S.C, Sharma, A., Mishra, M., Mishra, R. K. and Chowdhuri, D. K. 2010. Heat shock proteins in toxicology: how close and how far? *Life Sci*, 86: 377-384.

Gygi, S.P., Rochon, Y., Franza, B.R. and Aebersold, R. 1999. Correlation between protein and mRNA abundance in yeast. *Mol. Cell Biol*, 19(3): 1720-1730.

Hajheidari, M., Abdollahian-Noghabi, M., Askari, H., Heidari, M., Sadeghian, S.Y., Ober, E.S. and Salekdeh, G.H. 2005. Proteome analysis of sugar beet leaves under drought stress. *Proteomics*, 5: 950-960.

Hashimoto, M. and Komatsu, S. 2007. Proteomics analysis of rice seedling during cold stress. *Proteomics*, 7:1293-1302.

Haynes, P.A. and Roberts, T.H. 2007. Subcellular shotgun proteomics in plants: looking beyond the usual suspects. *Proteomics*, 7: 2963-2975.

Hinsby, A.M., Olsen J.V. and Bennett K.L. 2003. *Molecular and Cell Proteomics*, 2: 29-36.

Holman, J.D., Dasari, S. and Tabb, D.L. 2013. Informatics of protein and posttranslational modification detection via shotgun proteomics. *Methods Mol. Bio*, 1002: 167–179.

Hooks, C.R. and Fereres, A. 2006. Protecting crops from non-persistently aphid-transmitted viruses: a review on the use of barrier plants as a management tool. *Virus Res*, 120: 1-16.

Houde, M., Danyluk, J., Laliberte, J.F., Rassart, E., Dhindsa, R.S. and Sarhan, F. 1992. Cloning, characterization, and expression of a carrier DNA encoding a 50-kD protein specifically induced by cold-acclimation in wheat. *Plant Physiol*, 99: 1381-1387.

Hu, W., Hu, G. and Han, B. 2009. Genome-wide survey and expression profiling of heat shock proteins and heat shock factors revealed overlapped and stress specific response under abiotic stresses in rice. *Plant Sci*, 176: 583-590.

Hu, X.L., Yang, Y.F., Gong, F.P., Zhang, D.Y., Zhang, L. and Wu, L.J. 2015. Proteins HSP26 improves chloroplast performance under heat stress by interacting with specific chloroplast proteins in maize (*Zea mays*). *J. Proteomics*, 115: 81–92.

Huang, J., Gu, M., Lai, Z., Fan, B., Shi, K., Zhou, Y., Yu, J. and Chen, Z. 2010. Functional analysis of the Arabidopsis PAL gene family in plant growth, development, and response to environmental stress. *Plant Physiol*, 153: 1526-1538.

Huang, S., Shingaki-Wells, R.N., Taylor, N.L. and Millar, A.H. 2013. The rice mitochondria proteome and its response during development and to the environment. *Front. Plant Sci*, 4:16.

Huo, C.M., Zhao, B.C., Ge, R.C., Shen, Y.Z. and Huang Z. J. 2004. Proteomic analysis of the salt tolerance mutant of wheat under salt stress. *Acta Genetica Sinica*, 31: 1408–1414.

Jacoby, R.P., Millar, A.H. and Taylor, N.L. 2013. Application of selected reaction monitoring mass spectrometry to field-grown crop plants to allow dissection of the molecular mechanisms of a biotic stress tolerance. *Front. Plant Sci*, 4: 20.

Jimenez, A., Hernandez, J.A., del Rio, L.A. and Sevilla, F. 1997. Evidence for the presence of the ascorbate-glutathione cycle in mitochondria and peroxisomes of pea leaves. *Plant Physiol*, 114: 275-284.

Jorrin, J.V., Maldonado, A.M. and Castillejo, M.A. 2007. Plant proteome analysis: a 2006 update. *Proteomics*, 7: 2947–2962.

Junjie, H., Christof, R. and Natalian, V. 2015. Advances in plant proteomics toward improvement of crop productivity and stress resistance. *Front. Plant Sci*, 6: 209.

Kaffarnik, F.A., Jones, A.M., Rathjen, J.P. and Peck, S.C. 2009. Effector proteins of the bacterial pathogen Pseudomonas syringae alter the extracellular proteome of the host plant, *Arabidopsis thaliana*. *Mol. Cell Proteomics*, 8: 145-156.

Kamal, A.H M., Rashid, H., Sakata, K. and Komatsu, S. 2015. Gel-free quantitative proteomic approach to identify cotyledon proteins in soybean under flooding stress. *J. Proteomics*, 112:1–13.

Kang, B.C., Yeam, I. and Jahn, M.M. 2005. Genetics of plant virus resistance. *Annu. Rev. Phytopathol*, 43: 581-621.

Kav, N.N.V., Srivastava, S., Goonewardene, L. and Blade, S.F. 2004. Proteome-level changes in the roots of *Pisum sativum* in response to salinity. *Ann. Appl. Biol*, 145: 217-230.

Ke, Y., Han, G., He, H. and Li, J. 2009. Differential regulation of proteins and phosphor-proteins in rice under drought stress. *Biochem. Biophy. Res. Commun*, 379:133-138.

Kim, S.G., Wang, Y., Lee, K.H., Park, Z.Y., Park, J., Wu, J., Kwon, S.J., Lee, Y.H., Agrawal, G.K. and Rakwal, R. 2013. In-depth insight into *in vivo* apoplastic secretome of rice-Magnaporthe oryzae interaction. *J. Proteomics*, 78:58-71.

Kitano, H. 2002. Computational systems biology. *Nature*, 420: 206–210.

Klose, J. 1975. Protein mapping by combined isoelectric focusing and electrophoresis of mouse tissues. A novel approach to testing for induced point mutations in mammals. *Human genetics*, 26: 231–43.

Komatsu, S. and Hossain, Z. 2013. Organ-specific proteome analysis for identification of abiotic stress response mechanism in crop. *Front. Plant Sci*, 4:71.

Kotak, S., Larkindale, J., Lee, U., Von Koskull-Do¨ring, P., Vierling, E. and Scharf, K. D. 2007. Complexity of the heat stress response in plants. *Curr. Opin. Plant Biol*, 10:310-316.

Krishnan, H.B., Natarajan, S.S., Bennett, J.O. and Sicher, R.C. 2011. Protein and metabolite composition of xylem sap from field-grown soybeans (*Glycine max*). *Planta*, 233: 921–931.

Kubo, T., Fujita, M., Takahashi, H., Nakazono, M., Tsutsumi, N. and Kurata, N. 2013. Transcriptome analysis of developing ovules in rice isolated by laser microdissection. *Plant Cell Physiol*, 54:750–765.

Lafitte, H.R., Yongsheng, G., Yan, S. and Li, Z.K. 2007. Whole plant responses, key processes, and adaptation to drought stress: the case of rice. *J. Exp. Bot*, 58: 169-175.

Lahm, H.W. and Lagen, H. 2000. Mass spectrometry: a tool for the identification of proteins separated by gels. *Electrophoresis*, 21: 2105–14.

Lake, I.R., Hooper, L., Abdelhamid, A., Bentham, G., Boxall, A.B.A. and Draper, A. 2012. Climate change and food security: health impacts in developed countries. *Environ. Health Perspect*, 120: 1520–1526.

Langridge, P. and Fleury, D. 2011. Making the most of 'omics'for crop breeding. *Trends Biotechnol*, 29: 33–40.

Lee, B.H., Won, S.H., Lee, H.S., Miyao, M., Chung, W.I., Kim, I.J. and Jo, J. 2000. Expression of the chloroplast-localized small heat shock protein by oxidative stress in rice. *Gene*, 245: 283-290.

Lee, D.G., Ahsan, N., Lee, S.H., Kang, K.Y., Bahk, J.D., Lee, I.J. and Lee, B.H. 2007. A proteomic approach in analyzing heat-responsive proteins in rice leaves. *Proteomics*, 7:3369-3383.

Lee, D.G., Ahsan, N., Lee, S.H., Lee, J.J., Bahk, J. D., Kang, K.Y., Lee, B.H. 2009. Chilling stress-induced proteomic changes in rice roots. *J. Plant Physiol,* 166:1-11.

Liang, Y., Srivastava, S., Rahman, M.H., Strelkov, S.E. and Kav, N.N. 2008. Proteome changes in leaves of *Brassica napus* L. as a result of *Sclerotinia sclerotiorum* challenge. *J. Agric. Food Chem,* 56: 1963-1976.

Lodha, T.D., Hembram, P. and Basak, N. 2013. A Successful Approach to Understand the phylloclade of rice leaves free from cytosolic proteins: Application to study rice-Magnaporthe Oryzae interactions. *Physiol. Mol. Plant Pathol,* 88: 28-35.

Longuespée, R., Fléron, M., Pottier, C., Quesada Calvo, F., Meuwis, M.A., Baiwir, D. 2014. Tissue proteomics for the next decade? Towards a molecular dimension in histology. *OMICS,* 18:539–552.

Majoul, T., Bancel, E., Tribo, E., Ben Hamida, J. and Branlard, G. 2003. Proteomic analysis of the effect of heat stress on hexaploid wheat grain: characterization of heat-responsive proteins from total endosperm. *Proteomics,* 3: 175-183.

Mann, M., Hendrickson, R.H. and Pandey, A. 2001. Analysis of proteins and proteomes by mass spectrometry. *Annu. Rev. Biochem,* 70:437-73.

Millar, A.H. and Taylor, N.L. 2014. Subcellular proteomics-where cell biology meets protein chemistry. *Front. Plant Sci,* 5:55.

Mittler, R. 2002. Oxidative stress, antioxidants and stress tolerance. *Trends Plant Sci,* 7: 405-410.

Miura, K. and Furumoto, T. 2013. Cold signaling and cold response in plants. *Int. J. Mol. Sci,* 14: 5312-5337.

Molesini, B., Cecconi, D., Pii, Y. and Pandolfini, T. 2013. Local and systemic proteomic changes in *Medicago truncatula* at an early phase of *Sinorhizobium meliloti* infection. *J. Proteome Res,* 13: 408-421.

Monneveux, P., Sanchez, C., Beck, D. and Edmeades, G.O. 2006. Drought tolerance improvement in tropical maize source populations: evidence of progress. *Crop Sci,* 46: 180-191.

Munns, R. 2005. Genes and salt tolerance: Bringing them together. *New Phytol,* 167: 645-663.

Nakashima, K., Ito, Y. and Yamaguchi-Shinozaki, K. 2009. "Transcriptional regulatory networks in response to abiotic stresses in Arabidopsis and grasses,." *Plant Physiology,* 149 (1): 88–93.

Nakazono, M., Qiu, F., Borsuk, L. A. and Schnable, P. S. 2003. Laser- capture microdissection, a tool for the global analysis of gene expression in specific plant cell types: identification of genes expressed differentially in epidermal cells or vascular tissues of maize. *Plant Cell,* 15: 583–596.

Nat, N.V.K., Sanjeeva, S., William, Y. and Nidhi, S. 2007. Application of proteomics to investigate plant- microbe interactions. *Curr. Proteomics,* 4: 28-43.

O'Farrell P. H. 1975. High resolution two-dimensional electrophoresis of proteins. *J. Biol. Chem*, 250: 4007–21.

Oh, M. and Komatsu, S. 2015. Characterization of proteins in soybean roots under flooding and drought stresses. *J. Proteom, 114*: 161-181.

Pechanova, O., Takac, T.S., Amaj, J. and Pechan, T. 2013. Maize proteomics: an insight into the biology of an important cereal crop. *Proteomics*, 13: 637-662.

Peng, Z., Wang, M., Li, F., Lv, H., Li, C. and Xia, G.A. 2009. A proteomic study of the response to salinity and drought stress in an introgression strain of bread wheat. *Mol. Cell. Proteomics*, 8: 2676–2686.

Pfannschmidt, T. 2003. Chloroplast redox signals: how photosynthesis controls its own genes. *Trends Plant Sci*, 1: 33-41.

Philippe, R.N., Ralph, S.G., Kulheim, C., Jancsik, S.I. and Bohlmann, J. 2009. Poplar defense against insects: genome analysis, fulllength cDNA cloning, and transcriptome and protein analysis of the poplar Kunitz-type protease inhibitor family. *New Phytol*, 184: 865-884.

Pieterse, C.M., Leon-Reyes, A., Van der Ent, S. and Van Wees, S.C. 2009. Networking by small-molecule hormones in plant immunity. *Nat. Chem. Biol*, 5: 308-316.

Polidoros, A. N. and Scandalio, J.G. 1999. Role of hydrogen peroxide and different classes of antioxidants in the regulation of catalase and glutathione S-transferase gene expression in maize (*Zea mays* L.). *Physiol. Plant*, 106:112-120.

Pratt, W.B. and Toft, D.O. 2003. Regulation of signaling protein function and trafficking by the hsp90/hsp70-based chaperone machinery. *Exp. Biol. Med*, 228: 111-133.

Prins, M., Laimer, M., Noris, E. Schubert, J., Wassenegger, M. and Tepfer, M. 2008. Strategies for antiviral resistance in transgenic plants. *Mol. Plant Pathol*, 9: 73-83.

Quan, R., Shang, M., Zhang, H., Zhao, Y. and Zhang, J. 2004. Engineering of enhanced glycine betaine synthesis improves drought tolerance in maize. *J. Plant Biotechnol*, 2: 477-486.

Qureshi, M.I., M. Israr, M.Z. Abdin, and M. Iqbal. 2005. Responses of Artemisia annual L. to lead and salt-induced oxidative stress. *Environ. Exp. Bot*, 53: 185-195.

Qureshi, M.I., S. Qadir, and L. Zolla. 2007. Proteomics-based dissection of stress-responsive pathways in plants. *J. Plant Physiol*, 164: 1239-60.

Rajhi, I., Yamauchi, T., Takahashi, H., Nishiuchi, S., Shiono, K., Watanabe, R. 2011. Identification of genes expressed in maize root cortical cells during lysigenous paerenchyma formation using laser microdissection and microarray analyses. *New Phytol*, 190: 351-368.

Rao, D., Momcilovic, I., Kobayashi, S. and Callergari, E. 2004. Chaperone activity of recombinant maize chloroplast protein synthesis elongation factor, EF-Tu. *Eur J Biochem,* 271: 3684-3692.

Rasmusson, A.G., Soole, K.L. and Elthon, T.E. 2004. Alternative NAD(P)H Dehydrogenases of plant mitochondria. *Annu. Rev. Plant Biol,* 55: 23-39.

Reid, D.E., Hayashi, S., Lorenc, M., Stiller, J., Edwards, D. and Gresshoff, P.M. 2012. Identification of systemic responses in soybean nodulation by xylem sap feeding and complete transcriptome sequencing reveal a novel component of the auto regulation pathway. *Plant Biotechnology,* 10: 680–689.

Ribas-Carbo, M., Taylor, N.L., Giles, L., Busquets, S., Finnegan, P.M. and Day, D. A. 2005. Effects of water stress on respiration in Soybean leaves. *Plant Physiol,* 139: 466-473.

Ristic, Z., Gifford, D.J. and Cass, D.D. 1991. Heat shock proteins in two lines of Zea mays L. that differ in drought and heat resistance. *Plant Physiol,* 97: 1430-1434.

Roberts, I.N., Caputo, C., Criado, M.V. and Funk, C. 2012. Senescence associated proteases in plants. *Physiol. Plant,* 145: 130–139.

Roberts, J.K., DeSimone, N., Lingle, W.L. and Dure, L. 1993. Cellular concentrations and uniformity of cell-type accumulation of two LEA proteins in cotton embryos. *Plant Cell,* 5: 769-780.

Rochon, D., Kakani, K., Robbins, M. and Reade, R. 2004. Molecular aspects of plant virus transmission by olpidium and plasmodiophorid vectors. *Annu. Rev. Phytopathol,* 42: 211-241.

Rollins, J.A., Habte, E., Templer, S.E., Colby, T., Schmidt, J., and Von Korff, M. 2013. Leaf proteome alterations in the context of physiological and morphological responses to drought and heat stress in barley (*Hordeum vulgare* L.). *J. Exp. Bot, 64:* 3201-3212.

Rossignol, M., Peltier, J.B., Mock, H.P., Matros, A., Maldonado, A.M. and Jorrin, J.V. 2006. Plant proteome analysis: a 2004-2006 update. *Proteomics,* 6:5 529-5548.

Roy, A., Rushton, P.J. and Rohila, J.S. 2011. The potential of proteomics technologies for crop improvement under drought conditions. *Critical Reviews in Plant Sciences,* 30: 471–490.

Sabehat, A., Lurie, S., Weiss, D. 1998. Expression of small heat shock proteins at low temperatures. A possible role in protecting against chilling injuries. *Plant Physiol,* 117: 651-658.

Sergeant, K. and Renaut, J. 2010. Plant biotic stress and proteomics. *Current Proteomics,* 7: 1570-1646.

Sharma, P. and Dubey, R.S. 2005. Modulation of nitrate reductase activity in rice seedlings under aluminium toxicity and water stress: role of osmolytes as enzyme protectant. *J. Plant Physiol,* 162:854-864.

Su, P.H. and Li, H. M. 2008. Arabidopsis stromal 70-kD heat shock proteins are essential for plant development and important for thermotolerance of germinating seeds. *Plant Physiol,* 146: 1231-1241.

Subba, P., Kumar, R., Gayali, S., Shekhar, S., Parveen, S., Pandey, A., Datta, A., Chakraborty, S. and Chakraborty, N. 2013. Characterization of the nuclear proteome of a dehydration-sensitive cultivar of chickpea and comparative proteomic analysis with a tolerant cultivar. *Proteomics,* doi:10.1002/pmic.201200380

Süle, A., Vanrobaeys, F., Hajós, G., Van Beeumen, J. and Devreese, B. 2004. Proteomic analysis of small heat shock protein isoforms in barley shoots. *Phytochemistry,* 65: 1853-1863.

Sun, W., Van Motangu, M. and Verbruggen, N. 2002. Small heat shock proteins and stress tolerance in plants. *Biochim. Biophys. Acta,* 19: 1-9.

Suwabe, K., Suzuki, G., Takahashi, H., Katsuhiro, S., Makoto, E., Kentaro, Y. 2008. Separated transcriptomes of male gametophyte and tapetumin rice: validity of a laser micro dissection (LM) microarray. *Plant Cell Physiol,* 49:1407-1416.

Swatkoski, S., Russell, S., Edwards, N. and Fenselau, C. 2007. Analysis of a model virus using residue-specific chemical cleavage and MALDI-TOF mass spectrometry. *Anal. Chem,* 79: 654- 658.

Takahashi, D., Li, B., Nakayama, T., Kawamura, Y. and Uemura, M. 2013. Plant plasma membrane proteomics for improving cold tolerance. *Front. Plant Sci,* 4: 90.

Tanz, S.K., Castleden, I., Hooper, C.M., Vacher, M., Small, I. and Millar, H.A. 2013. SUBA3: a database for integrating experimentation and prediction to define the subcellular location of proteins in Arabidopsis. *Nucl. Acids Res,* 41: 1185–1191.

Taylor, N.L., Heazlewood, J.L., Day, D.A. and Millar, A.H. 2005. Differential impact of environmental stresses on the pea mitochondrial proteome. *Mol. Cell Proteom,* 4: 1122-1133.

Thelen, J.J. and Peck, S.C. 2007. Quantitative proteomics in plants: choices in abundance. *Plant Cell,* 19: 3339-3346.

Unlu, M., Morgan, M.E. and Minden, J.S. 1997. Difference gel electrophoresis: a single gel method for detecting changes in protein extracts. *Electrophoresis,* 18: 2071-77.

Van den Bergh, G. and Arckens, L. 2005. Recent advances in 2D electrophoresis: an array of possibilities. *Expert Rev. Proteomics,* 2: 243–252.

Veeranagamallaiah, G., Jyothsnakumari, G., Thippeswamy, M., Reddy, P.C.O., Sriranganayakulu, G., Mahesh, Y., Rajasekhar, B., Madhurarekha, C. and Sudhakar, C. 2008. Proteomic analysis of salt stress responses in foxtail millet (*Setaria italica* L. cv. Prasad) seedlings. *Plant Sci,* 175: 631-641.

Viswanathan, S., Unlu, M., and Minden, J.S. 2006. Two-dimensional difference gel electrophoresis. *Nature Protocols*, 1:1351–1358.

Vogt, T. 2010. Phenyl propanoid biosynthesis. *Mol. Plant*, 3:2-20.

Wang, G.P., Zhang, X.Y., Li, F., Luo, Y. and Wang, W. 2010. Over accumulation of glycine betaine enhances tolerance to drought and heat stress in wheat leaves in the protection of photosynthesis. *Photosynthetica*, 48: 117-126.

Wang, W., Vinocur, B. and Altman, A. 2003. Plant responses to drought, salinity and extreme temperatures: towards genetic engineering for stress tolerance. *Planta*, 218: 1-14.

Wang, W., Vinocur, B., Shoseyov, O. and Altman, A. 2004. Role of plant heat-shock proteins and molecular chaperones in the abiotic stress response. *Trends Plant Sci*, 9: 244-252.

Wang, Y., Kim, S.G., Kim, S.T., Agrawal, G.K., Rakwal, R. and Kang, K.Y. 2011. Biotic Stress-Responsive Rice Proteome: An Overview. *J. Plant Biol.*, 54: 219-226

Waters, E.R., Lee, G. J. and Vierling, E. 1996. Evolution, structure and function of the small heat shock proteins in plants. *J. Exp. Bot*, 47: 325-338.

Wilkins, M.R., Sanchez, J.C., Gooley, A. A., Appel, R.D., Humphery-Smith, I. and Hochstrasser, D.F. 1995. Progress with proteome projects: why all proteins expressed by a genome should be identified and how to do it. *Biotechnol. Genet. Eng. Rev*, 13:19-50.

Wittmann-Liebold, B., Graack, H.R. and Pohl, T. 2006. Two-dimensional gel electrophoresis as tool for proteomics studies in combination with protein identification by mass spectrometry. *Proteomics*, 6: 4688-4703.

Witzel, K., Weidner, A., Surabhi, G.K., Börner, A. and Mock, H.P. 2009. Salt stress-induced alterations in the root proteome of barley genotypes with contrasting response towards salinity. *J. Exp. Bot*, 60: 3545–3557.

Wu, C., Shi, T., Brown, J. N., He, J., Gao, Y. and Fillmore, T. L. 2014. Expediting SRM assay development for large-scale targeted proteomics experiments. *J. Proteome Res*, 13: 4479-4487.

Wu, J. and Baldwin, I.T. 2009. Herbivory-induced signalling in plants: perception and action. *Plant Cell Environ*, 32: 1161-1174.

Yan, S., Tang, Z., Su, W. and Sun, W. 2005. Proteomic analysis of salt stressresponsive proteins in rice root. *Proteomics*, 5: 235-244.

Yang, F., Jørgensen, A.D., Li, H., Søndergaard, I., Finnie, C., Svensson, B., Jiang, D., Wollenweber, B. and Jacobsen, S. 2011. Implications of high-temperature events and water deficits on protein profiles in wheat (*Triticum aestivum* L. cv. Vinjett) grain. *Proteomics*, 11: 1684–1695.

Yang, Y., Hu, M., Yu, K., Zeng, X. and Liu, X. 2015. Mass spectrometry-based proteomic approaches to study pathogenic bacteria-host interactions. *Protein Cell*, 6: 265-274.

Zhang, Z., Xin, W., Wang, S., Zhang, X., Dai, H. and Sun, R. 2014. Xylem sap in cotton contains proteins that contribute to environmental stress response and cell wall development. *Funct. Integr. Genomics*, 5: 17-26.

Zipfel, C., Robatzek, S., Navarro, L., Oakeley, E.J., Jones, J.D., Felix, G. and Boller, T. 2004. Bacterial disease resistance in Arabidopsis through flagellin perception. *Nature*, 428: 764-767.

Zlatev, Z.S., Lidon, F.C., Ramalho, J.C. and Yordanov, I.T. 2006. Comparison of resistance to drought of three bean cultivars. *Biol. Plant*, 50: 389-394.

25

Impact of Proteomic Studies for Biotic/ Abiotic Stress Tolerant Crops

Sumit Jangra[1*], Aakash Mishra[2], Neelam R. Yadav[1] and Ram C. Yadav[3]

[1]Department of Molecular Biology, Biotechnology & Bioinformatics, CCS Haryana Agricultural University, Hisar 125004, India

[2]4012, Frontera Drive, Davis, CA 95618, USA

[3]Centre for Plant Biotechnology, CCS Haryana Agricultural University Campus, Hisar-125004, India.

*Corresponding author: sumit.jangra712@gmail.com

Abstract

Biotic and abiotic stresses are the major factors limiting the crop production world-wide and to meet the needs of increasing global population there is a need to develop crop plants with higher productivity under stressed conditions. Various studies have been carried out to develop such varieties but no great success have been achieved. With the advancement in the omics technologies and availability of whole genome sequence of model crop will allow breeder to access a large number of genes but a major problem with this is that the number of genes do not correlate with the number of proteins produced and proteins also under go post-translational modifications and proteomics is the answer to it. Proteomic allows the study of all the proteins expressed inside the cell under stressed conditions and will help in better understanding of stress responsive mechanism and identification of genes and metabolites involved in stress adaptation, which will be helpful in the development of newer varieties of crop plants with improved stress resistance.

Keywords: Proteomics, protein, biotic/abiotic stress, resistance and functional genomics

Introduction

The term proteome refers to the entire set of proteins expressed by genome of an organism at a particular time and the study of proteome and its function is known as proteomics. The term proteome was coined by Marc Wilkins in 1994 and the term proteomics came in existence in late 1997 as an analog to genomics. Proteomics is the next level of study after genomics and is much more complicated than genomics because protein content varies for cell to cell and time to time. Plant response to stress is associated with enormous changes in the proteome content. Investigation of changes in plant proteome analysis during various biotic/abiotic stresses is highly important because changes in protein concentrations are directly involved in plant stress response and can add up to the knowledge of stress assimilation in crop plants. Various biomarkers tightly linked to target trait can be identified with the help of proteomics which could be undertaken by plant breeder to develop new variety with improved stress resistance. Novel genes involved in stress responsive mechanism can also be identified with the help of proteomics which can be utilized by pant biotechnologists to develop transgenic plant with improved stress tolerance.

Techniques for Proteome Analysis

2D-PAGE

2D-PAGE (Two-dimensional polyacrylamide gel electrophoresis) is commonly used high-resolution gel electrophoresis technique used to separate protein mixture on the basis of charge and mass. In this technology gel is run in two dimensions in the first-dimension gel is run in a pH gradient under non-denaturing conditions to resolve proteins on the basis of their pI values or isoelectric point this is called iso-electro focusing and in second dimension gel is run in denaturing conditions to resolve proteins by their molecular weight (MW). After this gel is stained usually with silver staining, which is highly sensitive, to locate the position of all the proteins. This results in a two-dimensional map; every single spot on the map represents the proteins being expressed. With the advancement in the technology, it is now possible to collect the 2D image data in digital format making it a practical mean for collection of proteomic information. Digitalization allows comparison of data, sharing of data among the research groups and cataloging of huge amount of generated data. Various softwares are used for this purpose. After the selection of protein of interest by differential analysis proteins can be excised out of the gel and analyzed and their molecular weight can be determined with the help of mass spectrophotometry and the use of bioinformatics tools has made high throughput protein identification possible. However, there are many challenges while using this technology, one is separation of hydrophobic proteins like membrane proteins which cannot be solubilized. To overcome this, detergents can be used to separate membrane proteins.

Mass Spectrometry

Once the proteins get resolved/separated through two-dimensional gel electrophoresis, further analysis and characterization can be done with the help of mass spectrometry. For this, the protein of interest is excised from the two-dimensional gel and are digested with a protease (trypsin). This leads to the generation of peptides of various molecular weight, which is named as protein fingerprint. The molecular weight of these protein fragments can be determined with the help of mass spectrometry a high-resolution mass analysis technique. At present electrospray ionization and matrix-assisted laser desorption ionization mass spectrometry are used. The only difference in both the techniques is the ionization protocol. In MALDI-MS a positive charge is applied to on the peptides and are allowed to pass through an analyzing tube with a magnetic field. Peptides are converted into gaseous phase before they are passed through analyzing tube. As the magnetic field will have more effect on the small peptides they will be deflected more than the larger peptides due to this peptide get separated according to their molecular mass and charge. The molecular weight data is displayed as a spectrum of ion density as function of mass to charge ratio. With further advancement in the technique, the peptides can be sequenced with sequential fragmentation and mass analysis, this is called tandem mass spectrometry, in which peptides are allowed to pass through two analyzers for sequence determination. In first analyzer, physical means is used to fragment the peptides of varying size differing by only one amino acid. The second analyzer uses these fragments to determine the molecular size and sequence of the fragment.

Differential in-gel Electrophoresis

Variation in protein expression can be studied in a way similar to fluorescently labelled DNA microarray using a technique known as differential in-gel electrophoresis (DIGE). Differently coloured fluorescent dyes are used to label the protein from the control and experimental sample. The labelled protein samples are mixed together and applying to two-dimensional gel electrophoresis. This leads to the separation and visualization of differentially expressed proteins in the same gel. This technique has several advantages over normal 2D-PAGE which includes reduction in the noise level and improvement in sensitivity and reproducibility. The major drawback of this technique is that different proteins take up fluorescent dyes to a different extent and labelling reduces the solubility of the proteins and the proteins may get precipitated before electrophoresis.

Protein Microarray/Protein Chip

High throughput protein function analysis can be carried out with the help of protein microarray chips which contains the entire proteome immobilized on them, similar to DNA microarray chips. Protein microarray chips are used to study protein function, protein-protein interactions, protein-DNA/RNA interactions and protein-ligand interactions. Protein array can be divided in two main types: (i) measures the abundance of protein (abundance based microarray) (ii) determines protein function. Abundance-based microarray can be further divided into two types: capture microarray and reverse phase microarray. Molecules like antibodies, peptides, DNA/RNA aptamers etc. are spotted on the surface to capture microarray to catch and assay target proteins from the complex mixture. The relative expression is then determined by comparing with the reference. In reverse phase microarray, unknown sample is spotted on the array and is probed with antibody or any other specialized molecule. In function based microarray the interest protein is blotted on the surface of microarray to its activity and biochemical properties. Lustiness, sensitivity and mechanization of protein microarray still need improvements to increase its use in proteomic studies.

Short Gun Proteomics

Mass spectrometry can be combined with analytical separation methods such as liquid chromatography (LC-MS) or gas chromatography GC-MS). In such cases mass spectrometry is used as detector after analytical separation. Nowadays this configuration is used by many laboratories because it combines high separation and increases detection capacities. Shotgun proteomics is a combination of one of these techniques, MS and multidimension chromatography and bioinformatics. In shotgun proteomics, protein sample is digested with a protease which gives rise to a more complex peptide mixture. The peptide developed after the proteolytic digestion are loaded in an LC/LC column placed in-line with MS/MS spectrophotometer. The spectra are generated as the peptides move from column to the mass spectrophotometer. Using specialized algorithms, the peptide sequence data generated is identified and compared with the databases. One such technique is multidimension protein identification technology (MudPIT), which has been used in proteomic studies of various organisms.

Proteomics of Model Plants

To identify genes responsible for important agronomical traits in crop plant is a major challenge for plant breeders, especially in crops like rice and wheat which plays a major role in global food production. Luckily all these challenges are brought forward when plant biotechnologists are witnessing noteworthy advancement in understanding key processes implied in growth and development of plants. Proteomic studies are a way towards better understanding of genes involved in biotic and abiotic stresses and in the development of stress tolerant plants. Keeping this in view proteomic studies has been carried out in the model plants in an established workflow to improve our fundamental knowledge about various metabolic, biochemical and molecular pathways involved in plant growth, development and stress tolerance.

1. Arabidopsis

Arabidopsis is recognized as a model plant for various biological studies worldwide. The availability of whole genome sequence, a large number of natural variations and advancement in the molecular technology has made it the most favourable model organism for post-genomic studies. Various key data platforms and databases like TAIR (http://www.arabidopsis.org/) already exists. The introduction of high-throughput techniques like shotgun proteomics has enabled the analysis of large number of samples at a very short time. High throughput proteomic techniques in combination with the genomics have readily become favourable tools of biotechnologists to get deeper insights into the molecular and biochemical pathways involved in plant growth and development and development of stress tolerant plants. An initiative was led by Weckwerth *et al.,* 2008 to coordinate international proteomic research on *Arabidopsis thaliana;* leading to the development of largest assembly of proteomic resource for a model plant. The Multinational Arabidopsis Steering Committee has made the major proteomics platform available online (MASCP; http://www.masc-proteomics.org/).

2. Rice

Rice is one of the major staple food crops and is an important model plant for various genetic and molecular biology research. Due to this substantial advancement, has been made in identification and cataloguing of rice proteins and the completion of rice genome sequencing has accelerated proteomic studies in this crop (Agrawal *et al.,* 2006). The knowledge of diverse proteins of rice has been increased by proteomic analysis of rice embryo (Fukuda *et al.,* 2003), endosperm (Komatsu *et al.,* 1993), root (Zhong *et al.,* 1997), leaves (Islam *et al.,*2004 and Zhao *et al.,* 2005) and etiolated shoot (Komatsu *et al.,* 1999), suspension cultured cells (Komatsu *et al.,* 1999), anther (Imin *et al.,* 2001and Kerim *et al.,*2003) and leaf sheath (Shen *et al.,* 2002) and other organs. 4892 proteins from nine tissues of rice (leaf, stem, root, germ, dark germinated seedlings, seed, bran, chaff and callus) were analyzed by Tsugita *et al.,* 1994. Different protocols were used to carry out all the above studies but data comparison was not possible. To do so, Rice Proteome database has been constructed (http://gene64.dan.affrc.go.jp/RPD/). Various rice organelles such as Golgi (Mikami *et al.,*2001), mitochondria (Heazlewood *et al.,*2003), chloroplast (Tsugita *et al.,*1994) and other subcellular compartments (Tanaka *et al.,*2004) have been analyzed.

3. Soybean

Soybean is an important source of essential fatty acids and proteins for both human and animal nutrition and also an important source of industrial feedstock and fuel (Thelen *et al.,* 2002). Soybean is an important part of US economy; accounts for nearly 12 billion dollars in annual crop value and the export value is nearly 5 billion dollars. Seed is the major economical important part of this crop and better understanding of the structure and storage proteins will help in crop improvement. Proteomic studies are a way towards better understanding to molecular and biochemical pathways involved in stress tolerance and improving seed quality. A total of 78 storage proteins were identified by Natarajan *et al.,*2006 by comparing protein profiles of wild and cultivated genotypes. Various high-throughput proteomic techniques were used to examine the expression scenario and determine hundreds of proteins involved in seed filling. During these studies, an overall decrease in metabolism related proteins and increase in seed filling proteins was observed. In addition to this, a soybean database (http://oilseedproteomics.missouri.edu) was developed (Hajduch *et al.,*2005). Sakata *et al.*,2009 developed another soybean proteome database (SPD) which contains proteome data from plant under flooding stress and is available at http://proteome.dc.affrc.go.jp/Soybean/. Soybean genome sequencing has been completed recently which will further supplement soybean proteomic studies and will help in the development of plants with desired characteristics.

4. Wheat

Wheat is the second most important cereal crop after rice and is an important ingredient of human diet, especially in the developing countries. 2D-PAGE was used to map the proteome of leaf (Bahrman *et al.*, 2004 and Donnelly *et al.*, 2005), flour (Mamone *et al.*,2000) and amyloplast (Andon *et al.*, 2002). A combination 2D-PAGE and MS was used by Vensel *et al.*, 2005 to identify over 250 proteins at early and late stages of grain development. These data put insights into molecular and biochemical events taking place during wheat development and highlights the scope of proteomic studies in better understanding of developmental pathways and will help biotechnologists to improve crop yield and develop plants with enhanced stress tolerance. Moreover, the development of proteome map will aid future studies and will help in addressing the effect of genetic and environmental factors on wheat growth and development. The draft of wheat genome sequencing has been released in 2014 and will supplement the proteomic studies in developing new wheat varieties with increased yield and enhanced stress tolerance.

5. Maize

Maize proteomics has been examined widely. 2D-PAGE and MS have been used to analyze maize egg cell and zygote (Okamato *et al.*, 2004). The studies indicated that egg cell contains ample amount of enzymes involved in energy metabolism. It was also found that the egg cell and zygote, annexin was involved in the exocytosis of cell wall materials. The foundation for the use of proteomics in understanding the molecular biology and biochemistry of various pathways involved in plant growth and development and stress tolerance was provided by these initial experiments. Comparative proteome analysis of maize primary roots and lateral roots revealed the proteins involved in the lignin biosynthesis, defense and citrate cycle (Liu *et al.*, 2006). It was also reported that the major backbone of root stock is formed by post-embryonically formed shoot-borne roots. Sauer *et al.*, 2006 described proteomic analysis of shoot-borne root introduction in maize and the protein identified were G-protein auxin-binding proteins. The cell wall proteome of maize primary root elongation zone was analyzed by Zhu *et al.*, 2006 and it was found that the proteins include endo-1,3:1,4-beta-D-glucanase and alpha-L-arabino furanosidase, which is the major polysaccharide present in type II cell walls. These experiments represent the first proteomic studies on root initiation and elongation and will add to the knowledge of better understanding of pathways involved in growth and development and biotic and abiotic stress tolerance. The availability of genome sequence of maize B73 line (http://maizesequence.org) has accelerated the advancement in maize research. Also, the availability of various genetic and genomic information resources such as Plant Proteome Database (http://ppdb.tc.cornell.edu), MaizeGDB (http://www.maizegdb.org/), TGIR Maize Database (http://maize.jcvi.org/) and Maize Assembled Genomic Island (http://magiplantgenomics.iastate.edu/) will aid plant biotechnologist in the development of high yielding and stress tolerant varieties of maize.

6. Barley

Barley is an important crop used for both feed and malting in the North European countries. There is a great difference in the characteristics of malting barley and feed cultivated barley and lot of studies are going on to improve the malting characteristics of barley and to develop new cultivar of barley with in improved malting properties and are tolerant to environmental stresses. Proteome analysis of barley seeds was reported by Ostergaard *et al.,* 2004. Moreover, analysis of water-soluble seed proteome has found a way towards identification of proteins by mass spectrometry in the major spots on 2D-PAGE. This led to the foundation for proteome studies at different developmental levels which will insights into the knowledge of mechanism involved in the growth and development of plants and to develop plants with higher yields and enhanced stress tolerance.

Proteomics in Biotic/Abiotic Stress Resistance

The concept of "stress" was introduced by Hans Selye in 1936, and afterward adopted in plant sciences and is now defined as any extrinsic factor that has a negative influence on plant growth, reproductive potential, survival and productivity. It constitutes of various types but can be grouped in two main categories: abiotic or environmental stress and biotic or biological stress. Both biotic and abiotic stress hampers plant growth and has negative impact on crop yield. Plants have developed various stress-related response mechanisms, which leads to upregulation and downregulation of various cellular proteins. The efficacy of stress-responsive pathway depends mainly on molecular signaling pathways and post-translational modification of stress related proteins. Study of these proteins and pathways will lead to better understanding of stress responsive mechanism and will be helpful in developing stress-resistant crop plants and proteomics is a way to it.

Proteomic Studies of Crop Under Biotic Stress

Biotic stress is the damage caused by living organism such as bacteria, fungi, viruses and insects. Biotic stress has a great historical background which includes the potato blight of Ireland, coffee rust in Brazil, maize rust in USA and the great Bengal famine in 1943. These are some the major event where biotic stress devastated food production and led to death of millions of people. Presently, the evolution of new pathotypes and insect biotypes is a major challenge that is further increased by the changing environment. Pathogen accounts for about 15% losses in global food production and disease or pest outbreak are expected to continue food production losses or even worsen by expanding the areas that were not prevalent before (Ijaz *et al.*, 2012). An array of morphological, genetic, biochemical and molecular processes is involved in plant resistance to various pathogens (Howe *et al.*, 2008). So, to overcome biotic stress the proper understanding of these pathways is must and proteomics is a way to it which will help in bringing forth more effective crop improvement strategies.

Lee *et al.*, 2006 carried out two-dimensional gel electrophoresis and tandem mass spectra analysis to detect the differentially expressed proteins in rice when infected with sheath blight caused by *Rhizoctonia solani.* The trigger of 3 β hydroxysteroid dehydrogenase/isomerase was detected for the first time in the resistant plants indicating the role of the enzyme in defense mechanism against *Rhizoctonia solani.* They were also able to found that four induced proteins were in close physical proximity with genetic marker for sheath blight resistance.

A proteomic study was carried out by Afroz *et al.*, 2009 to look into the molecular mechanism of bacterial wilt resistance in susceptible and resistant cultivars of tomato. Cultivar Roma and Riogarande and cultivar Pusa Ruby and Pant Bahr were selected as resistant and susceptible cultivars, respectively. 2-DE was used to separate proteins and a total of nine proteins were found to be differentially expressed. Amino acid sequencing of these proteins revealed that these proteins belong to energy, protein digestion, storage and defense categories. It was also found that a 60 kDa apical membrane antigen might be involved in defense signaling in tomato plants.

Sugarcane mosaic virus (SCMV) is a viral pathogen and causes a great loss in the yield, keeping this in mind a proteomic study was carried out in maize using 2D-DIGE by Wu *et al.*, 2013 to identify SCMV resistant proteins and to deduce the molecular mechanism of plant-SCMV interaction. They found that 93 protein spots showed differential expression after virus inoculation. Functional characterization of these proteins showed that the responsive proteins are mainly involved in energy, metabolism, photosynthesis, carbon fixation, stress and defense responses and the majority of them were located in chloroplast. This study provides further insights into the molecular events during compatible and incompatible interactions between viruses and host plants.

Approximately 1% of the flowering plants are parasitic on other plants and causes a great loss in yield, keeping this in mind a proteomic study in pea (*Pisum sativum*) was out by Castillejo *et al.*, 2004 to study the response of parasitic plant crenate broomrape (*Orobanche crenata*). Protein profiling of most resistant accession (Ps 624) was compared with a susceptible (Messire) cultivar. Out of 500 protein spots available, 22 protein spots differentiated control, non-infected, Messire and Ps 624 accessions. Both qualitative and quantitate differences were found among infected and non-infected root extracts. Proteins corresponding to enzymes of carbohydrate metabolism, nitrogen metabolism and electron transport chain decreased in susceptible check while proteins corresponding to nitrogen assimilation pathway (glutamine synthetase) or typical pathogen defense, PR proteins including β-1,3-glucanase and peroxidases increased in Ps 624.

Herbivores insects are another major biotic stress causing a great loss to yield by direct consumption of plant parts or contamination of harvested crops. So, to investigate the physiological factors affecting feeding behavior by larvae of the insect *Plutella xylostella* on herbivore-susceptible and herbivore-resistant *Arabidopsis thaliana*, a proteomic study was carried out by Collins *et al.*, 2010 using 2D-PAGE. The study showed that there was an increase in the production of hydrogen peroxide in the resistant line. Most of the proteins of resistant lines were found to be of ROS damaging category. Such proteins included carbonic anhydrases, malate dehydrogenases, glutathione S-transferase, isocitrate dehydrogenase-like protein (R1) and lipoamide dehydrogenase. The study suggested that enhances ROS may be a major pre-existing mechanism of *Plutella xylostella* resistance in *Arabidopsis thaliana.*

Proteomic Studies of Crops Under Abiotic Stress

At present, the important stress for the natural or cultivated crop plants is rapid climatic changes registered over the last years and is likely to follow. These climatic changes include changes in temperature, light, water supply and nutrient availability and have a negative impact on plant growth and productivity leading to losses in crop yield which is the major concern of agriculture. Estimates suggest that crop plants are reaching only 30% of their genetic potential under abiotic stress (Boyer, 1982) and over 90% of the global rural land area is affected by abiotic stress at some point of growing season (Cramer *et al.*, 2011). Alleviating the effect of not only crucial requisite to assure agriculture amelioration and to meet the demands of food for the increasing population (Tester et al, 2010) but improving plant fitness over a broader range of environmental conditions would allow for an increased use of degraded or marginal lands for agricultural production (Gisladottir *et al.*, 2005). For the fulfillment of these goals basic and fundamental understanding of physiological and molecular pathways is must. Genomic tools have helped to get insights into the molecular and physiological pathways of stress response. Proteomics has emerged as a powerful technique for identification of novel and stress responsive proteins to develop crop plants with improved stress tolerance.

1. Drought

Drought is a destiny of the changing climatic conditions throughout the world. Despite of several advancement in the technology, leading to early forecasting drought still remains a major threat to agriculture worldwide. Plant responses to drought has been studied extensively (Dinakar, Jogaiah and Abreu *et al.*, 2013) but till date there is no major break-through that could lead to the development of crop plants that can enhance production under drought stress. That is why the understanding and enhancing crop tolerance to drought is the need of todays. Drought induces a number of changes at phenotypic, physiological, molecular and biochemical level in the plant. Plants have evolved a number of ways to manage drought stress which includes drought avoidance and drought escape. Development in the "omics" technology has led to the development of lot of data which could be helpful in better understanding of drought responsive pathways but the level of proteins in the cell does not correlate with the mRNA and proteins also under post translational modifications so there is a need of another technique which could quantify all the cellular proteins to this proteomics is the answer which will be handy tool in the development of biomarkers tightly linked stress and development of stress tolerant crop plants.

To find new insight in the molecular mechanism of drought response a proteomic study of soybean roots subjected to severe drought stress at seedling stage was carried out by Alam *et al.,* 2010 using 2-DE. High resolution protein map demonstrated variation in about 45 protein spots, of these 28 were identified by mass spectrometry and level of 5 protein was increased, 21 was decreased and 2 were newly detected under drought stress. Protein expression analysis patterns revealed that protein associated with osmotic adjustment, defense signaling and programmed cell death plays an important role in plant adaptation to drought.

Drought is a major factor limiting the productivity of lowland and upland rice. Protein expression analysis of two-week-old seedling exposed to drought stress for 2 to 6 days was done by Ali *et al.,* 2006 using 2-D PAGE approach to investigate the initial response of rice to drought stress and it was observed that after 2 to 6 days of stress, level of 10 proteins was increased and that of 2 was decreased. These proteins were categorized defense, energy, metabolism, cell structure and signal transduction. It was also observed that actin depolarizing factor and light harvesting complex chain II were present at high level in drought tolerant cultivar before stress application. Drought stress application leads to expression of actin depolarizing factor in leaf blades, leaf sheaths and roots. Present investigation denotes that actin depolarizing factor is one of the targets proteins induced by drought stress.

2. Salinity

In the next coming 25 years, salinization of arable lands will lead to 30% land loss and up to 50% by 2050 (Yan *et al.,* 2003; Wang *et al.,* 2008; Abdel Latef and Chaoxing *et al.,* 2014; Ma *et al.,* 2015). Thus, development of salinity tolerant crop plant is of great importance to overcome the salinity problem world-wide. Advanced technologies are being applied in plant sciences to identify key proteins and metabolites, pathways and genes involved in stress tolerance. Proteomics is one such advanced technique which is used to study protein-protein interaction, post-translational modifications, protein expression profiling which will lead to identification of novel protein involved in stress response which will lead to development of stress tolerant varieties of crop plants.

To evaluate the changes in protein expression under salt stress proteomic approach was applied by Aghaei *et al.,* 2009. Soybean plants were exposed to 0, 50, 100 and 200 mM NaCl. 321 proteins spots were observed under salt stress. Under salt stress seven proteins were found to be up or down regulated by two to seven folds: LEA, β-conglycinin, elicitor peptide three precursor and basic/helix-loop-helix protein were up-regulated, while protease inhibitor, lectin and 31 kDa glycoprotein were down-regulated. This indicates that, these proteins may be involved in salt stress regulation in soybean.

Aghaei *et al.,* 2008 examined the molecular mechanism of salt stress in potato. When exposed to salt stress (90mM) a total of 322 and 305 proteins were detected in Kennebec (tolerant) and Concord (susceptible) cultivar respectively. Of these, 47 proteins were differentially expressed in shoot of both the cultivars under salt stress. Among differentially expressed proteins, photosynthesis and photosynthesis related proteins were drastically down-regulated, where osmotine-like proteins, TS-1 protein, heat shock protein, protein inhibitors and five novel proteins were up-regulated. This suggests that defense-related proteins may confer salt stress in potato.

3. Chilling

Low temperature is one of the major abiotic stress that limits the growth, productivity and geographical distribution of many crop plants. Temperature range from 0 to 12 ^{0}C is very common in the temperate region during growing seasons and considerably decreases plant growth and productivity (Allen *et al.,* 2001). Plant undergoes regulation of gene expression to combat the stress. Identification of determination of responsive genes will provide a molecular basis for efficient engineering strategies leading to enhanced stress tolerance. Transcriptomic studies do not correlate with the number of proteins produced under chilling stress because of post-translational modifications thus does not provide sufficient information for the development of chilling resistant plant this problem can be overcome with proteomics which gives the detailed information about the total proteins and hence is an important tool for development of chilling resistant plants.

To go deep into the understanding of chilling response in rice (*Oryza sativa* L. cv. Nipponbare) a comparative proteomic study was carried out by Yan *et al.,* 2006. Chilling stress was applied by treating the seedlings at 6 ^{0}C for 6 or 24 h. Two-dimensional gel electrophoresis was used to study the changes in protein level and it was found that 31 proteins spots out of 1000 were downregulated and 65 were upregulated. Mass spectrometry analysis identified 85 differentially expressed proteins, including well-known and novel cold-responsive proteins. Gene expression analysis of 44 different proteins by real-time PCR showed that mRNA level was not correlated well with the protein level. This study provides new insights in rice chilling stress response and exhibits the advantage of proteomics over transcriptomics.

Chilling stress has a devastating effect on Tea plant [*Camellia simesis* (L.) O. Kuntze]. So, to study the changes in protein expression under chilling stress a proteomic study was carried out by Wang *et al.,* 2016 using CBB stained 2-D gels. A treatment of 4 ^{0}C for 24 h was applied to one year old seedlings and then recovered for 24 h. Out of 1000 protein spots visualized 10 proteins spots were increased, 9 were decreased and 4 were induced. 18 differentially expressed proteins well known and novel cold responsive proteins were analyzed using MALDITOF/TOF mass spectrometry. This study provides new insights into molecular networks involved in chilling stress regulation and stress signaling.

4. Heat

Heat stress has become a major constraint impacting grain yield in crop plants and in near future the impacts of heat stress will be more severe due to greenhouse effect. By the end of this century vulnerability of crop plants will increase with increasing temperature variability. Therefore, the development of heat tolerant varieties of crop plants and understanding the molecular mechanism of heat stress has recently received much attention and to do so study of proteome is essential. Proteomics offers a mighty approach to disclose proteins and pathways that are crucial for heat stress responsiveness and tolerance.

Zhang *et al.*, 2013 exposed 30 days old seedlings of radish to a temperature of 40 ^{0}C to characterize heat stress response. Proteome analysis of leaf samples collected after 0 h, 12 h and 24 h exposure to heat stress was carried out using 2-DE and mass spectrometry and 11 differentially expressed proteins were found out of which four were heat shock protein (HSPs), four were related to energy metabolism, two were related to redox homeostasis and one was related to signal transduction.

Li *et al.*, 2013 exposed seedling of alfalfa to heat stress at 40 ^{0}C and samples were collected at 24, 48 and 72 h after exposure. Proteome analysis was carried out using 2-DE and mass spectrometry and a total of 81 differentially expressed proteins were found. These proteins were categorized into nine classes: including metabolism, energy, protein synthesis, protein destination/storage, transporters, intracellular traffic, cell structure, signal transduction and disease/defense.

Majority of the proteins expressed under high temperature stress fall under heat shock proteins (HSPs), and the relative expression differs in heat sensitive and heat susceptible cultivars. Identification and introgression of heat stress proteins from tolerant cultivars to susceptible cultivars using modern biotechnology tools could help in combating heat stress worl-wide.

Conclusion and Future Perspectives

Proteomics has come forward as worthful tool to identify proteins involved in biotic and abiotic stress response in crop plants. Proteins expressed in different cells and involved in specific response mechanism to stress have been analyzed and characterized. This omic approach will lead to better understanding of complex mechanism of stress adaptation on crop plants. Comparative proteomic studies of different plant parts will lead to the identification of different protein synthesized in different part and identifying their role in stress mitigation. Studies on protein-protein interactions and protein-ligand interactions and advances in genomics, transcriptomics and metabolomics will help in understanding the protein networking and identification of proteins and metabolites involved in stress responsive mechanisms. Proteomics will not only help in understanding the mechanism, proteins, pathways and metabolites underlying stress responsive mechanisms it will be helpful in development of biomarkers tightly linked to the trait of interest and will help plant breeders to develop new stress resistant varieties. It will also help in identification of novel genes involved in stress responsive mechanism which could be targeted by the plant biotechnologist to develop stress tolerant varieties with the help of transgenic technology.

References

Abdel Latef, A. A. and Chaoxing, H. 2014. Does the inoculation with *Glomus mosseae* improves salt tolerance in pepper plants? *J. Plant Growth Regul.*, 33: 644–653.

Abreu, I. A., Farinha, A. P., Negrao, S., Goncalves, N., Fonseca, C., Rodrigues, M., Batista, R., Saibo, N. J. M. and Oliveira, M. M. 2013. Coping with abiotic stress: Proteome changes for crop improvement. *J. Proteom.*, 93: 145–168.

Afroza, A., Khanb, M. R., Ahsana, N. and Komatsu, S. 2009. Comparative proteomic analysis of bacterial wilt susceptible and resistant tomato cultivars. *Peptides.*, 30: 1600–1607.

Aghaei, K., Ehsanpour, A. A. and Komatsu, S. 2008. Proteome Analysis of Potato under Salt Stress. *J. Proteome Res.*, 7(11): 4858–4868.

Aghaei, K., Ehsanpour, A. A., Shah, A. H. and Komatsu, S. 2009. Proteome analysis of soybean hypocotyl and root under salt stress. *Amino Acids.*, 36(1): 91-98.

Agrawal, G. K. and Rakwal, R. 2006. Rice proteomics: a cornerstone for cereal food crop proteomes. *Mass Spectrom. Rev.*, 25: 1– 53.

Alam I., Sharmin S. A., Kim K. H., Yang J. K., Choi M. S. and Lee B. H. 2010. Proteome analysis of soybean roots subjected to short-term drought stress. *Plant and Soil.* 333(1): 491-505.

Ali G. M. and Komatsu S. 2006. Proteomic analysis of rice leaf sheath during drought stress. *J. Proteome Res.*, 5(2): 396–403.

Allen, D. J., and Ort, D. R. (2001) Impacts of chilling temperatures on photosynthesis in warm-climate plants. *Trends in Plant Science,* 6: 36–42.

Andon, N. L., Hollingworth, S., Koller, A., Greenland, A. J., Yates J. R. and Haynes, P. A. 2002. Proteomic characterization of wheat amyloplasts using identification of proteins by tandem mass spectrometry. *Proteomics.,* 2(9): 1156–1168.

Bahrman, N., Negroni, L., Jaminon, O. and Le, G. J. 2004. Wheat leaf proteome analysis using sequence data of proteins separated by two-dimensional electrophoresis. *Proteomics,* 4(9): 2672–2684.

Boyer, J. S. 1982. Plant productivity and environment. *Science,* 218: 443–448.

Castillejo M. A., Amiour, N., Dumas-Gaudot, E., Rubiales, D. and Jorrin J. V. 2004. A proteomic approach to studying plant response to crenate broomrape (Orobanche crenata) in pea (Pisum sativum). *Phytochemistry,* 65: 1817–1828.

Collins, R. M., Afzal, M., Ward, D. A., Prescott, M. C., Sait, S. M., Rees, H. H. and Tomsett. A. B. 2010. Differential proteomic analysis of *Arabidopsis thaliana* genotypes exhibiting resistance or susceptibility to the insect herbivore, *Plutella xylostella, PLOS One.*5: 1-14.

Cramer, G. R., Urano, K., Delrot, S., Pezzoti, M. and Shinozaki, K. 2011. Effect of abiotic stress on plants: a systems biology perspective. *BMC Plant Biology,* 11: 163.

Dinakar, C. and Bartels D. 2013. Desiccation tolerance in resurrection plants: New insights from transcriptome, proteome, and metabolome analysis. *Front. Plant Sci.,* 2013 4: 1-14.

Donnelly, B. E., Madden, R. D., Ayoubi, P., Porter, D. R. and Dillwith, J. W. 2005. The wheat (Triticum aestivum L.) leaf proteome. *Proteomics,* 5: 1624–1633.

Fukuda, M.; Islam, N.; Woo, S.-H.; Yamagishi, A.; Takaoka, M.; Hirano, H. 2003. Assessing matrix-assisted laser desorption/ ionization-time of flight-mass spectrometry as a means of rapid embryo protein identification in rice. *Electrophoresis,* 24: 1319–1329.

Gisladottir, G. and Stocking, M. 2005. Land degradation control and its global environmental benefits. *Land Degrad. Dev.,*16: 99–112.

Hajduch, M., Ganapathy, A., Stein, J. W. and Thelen, J. J. 2005. A systematic proteomic study of seed filling in soybean. Establishment of high-resolution two-dimensional reference maps, expression profiles, and an interactive proteome database. *Plant Physiol,* 137: 1397-1419.

Heazlewood, J. L., Howell, K. A. and Millar, A. H. 2003. Mitochondrial complex I from Arabidopsis and rice: orthologs of mammalian and fungal components coupled with plant-specific subunits. *Biochim. Biophys. Acta,* 1604(3): 159–169.

Howe G. A. and Jander G. 2008. Plant immunity to insect herbivores. *Annual Review of Plant Biology,* 59: 41-66.

Ijaz, S. and Khan, A. I. 2012. Genetic pathways of disease resistance and plants-pathogens interactions. *Molecular Pathogens,* 3(4): 19–26.

Imin, N., Kerim, T., Weinman, J. J. and Rolfe, B. G. 2001. Characterisation of rice anther proteins expressed at the young microspore stage. *Proteomics,* 1: 1149–1161.

Islam, N., Lonsdale, M., Upadhyaya, N. M., Higgins, T. J., Hirano H. and Akhurst R. 2004. Protein extraction from mature rice leaves for two-dimensional gel electrophoresis and its application in proteome analysis. *Proteomics,* 4: 1903–1908.

Jogaiah, S., Govind, S. R. and Tran, L. S. P. 2013. Systems biology-based approaches toward understanding drought tolerance in food crops. *Crit. Rev. Biotechnol,* 33: 23–39.

Kerim, T., Imin, N., Weinman, J. J. and Rolfe, B. G. 2003. Proteome analysis of male gametophyte development in rice anthers. *Proteomics,* 3: 738–751.

Komatsu, S., Kajiwara, H. and Hirano, H. 1993. A rice protein library: a data-file of rice proteins separated by two-dimensional electrophoresis. *Theor. Appl. Genet,* 86: 935-942.

Komatsu, S., Muhammad, A. and Rakwal, R. 1999. Separation and characterization of proteins from green and etiolated shoots of rice (*Oryza sativa* L.): Towards a rice proteome. *Electrophoresis,* 20: 630–636.

Lee, J., Bricker, T. M., Lefevre, M., Pinson, S. R. M. and Oard, J. H. 2006. Proteomic and genetic approaches to identifying defence-related proteins in rice challenged with the fungal pathogen *Rhizoctonia solani. Molecular Plant Pathology,* 7(5): 405–416

Li, W., Wei, Z., Qiao, Z., Wu, Z., Cheng, L. and Wang, Y. 2013. Proteomics analysis of alfalfa response to heat stress. *PLOS One.* 8(12): 1-11.

Liu, Y., Lamkemeyer, T, Jakob, A., Mi, G., Zhang, F., Nordheim, A. and Hochholdinger, F. 2006. Comparative proteome analyses of maize (Zea mays L.) primary roots prior to lateral root initiation reveal differential protein expression in the lateral root initiation mutant rum1. *Proteomics,* 6: 4300–4308.

Ma, H., Yang, R., Song, L., Yang, Y., Wang, Q. and Wang, Z. 2015. Differential proteomic analysis of salt stress response in jute (*Corchorus capsularis & olitorius* L.) seedling roots. *Pak. J. Bot,* 47: 385–396.

Mamone, G., Ferranti, P., Chianese, L., Scafuri, L., Addeo, F., 2000. Qualitative and quantitative analysis of wheat gluten protein by liquid chromatography and electrospray mass spectrometry. *Rapid Commun. Mass Spectrom,* 14(10): 897–904.

Mikami, S., Hori, H. and Mitsui, T. 2001. Separation of distinct compartments of rice Golgi complex by sucrose density gradient centrifugation. *Plant Sci,* 161: 665–675.

Natarajan, S. S., Xu, C., Bae, H., Caperna T. J. and Garrett W. M. 2006. Characterization of storage proteins in wild (*Glycine soja*) and cultivated (*Glycine max*) soybean seeds using proteomic analysis. *J Agric Food Chem,* 54: 3114-3120.

Okamoto, T., Higuchi, K., Shinkawa, T., Isobe, T., Lorz H., Koshiba, T.and Kranz E. 2004. Identification of major proteins in maize egg cells. *Plant Cell Physiol,* 45: 1406–1412.

Ostergaard, O., Finnie, C., Laugesen, S., Roepstorff, P. and Svennson, B. 2004. Proteome analysis of barley seeds and malt and identification of major proteins from two-dimensional gels (pI 4-7). *Proteomics,* 4: 2437–2447.

Sakata, K., Ohyanagi, H., Nobori, H., Nakamura, T., Hashiguchi, A., Nanjo, Y., Mikami, Y., Yunokawa, H., and Komatsu, S. 2009. Soybean proteome database: a data resource for plant differential omics. *J. Proteome Res,* 8: 3539–3548.

Sauer, M., Jakob, A., Nordheim A. and Hochholdinger F. 2006. Proteomic analysis of shoot-borne root initiation in maize (*Zea mays* L.). *Proteomics,* 6: 2530–254.

Selye, H., 1936. A syndrome produced by diverse nocous agent. *Nature,* 138: 32-34.

Shen, S., Matsubae, M., Takao, T., Tanaka, N. and Komatsu, S. A. 2002. proteomic analysis of leaf sheaths from rice. *J. Biochem,* 132(4): 613–620.

Tanaka, N., Fujita, M., Handa, H., Murayama, S., Uemura, M., Kawamura, Y., Mitsui, T., Mikami, S., Tozawa, Y., Yoshinaga, T. and Komatsu, S. 2004. Proteomics of the rice cell: systematic identification of the protein populations in subcellular compartments. *Mol. Genet. Genomics,* 271: 566–576.

Tester, M. and Langridge, P. 2010. Breeding technologies to increase crop production in a changing world. *Science,* 327: 818–822.

Thelen, J. and Ohlrogge, J. 2002. Metabolic engineering of fatty acid biosynthesis in plants *Metab. Eng.,* 4: 12–21.

Tsugita, A., Kawakami, T., Uchiyama, Y., Kamo, M. Miyatake, N. and Nozu, Y. 1994. Separation and characterization of rice proteins. *Electrophoresis,* 15: 708–720.

Vensel, W. H., Tanaka, C. K., Cai, N., Wong, J. H., Buchanan B. B. and Hurkman W. J..2005. Developmental changes in the metabolic protein profiles of wheat endosperm. *Proteomics,* 5(6): 1594–1611.

W. Weckwerth, S. Baginsky, K. van Wijk, J.L. Heazlewood, H. Millar. 2008. The multinational Arabidopsis steering subcommittee for proteomics assembles the largest proteome database resource for plant systems biology. *J Proteome Res.* 7 (10): 4209–4210.

Wang, J., Wang, Y., Ding, Z. T., Wang, Y., Zhao, L., Pei, Y. H., Duan, F. M. and Sun, H. 2016. Differential proteomic analysis of chilling stress responses in tea plant [*Camellia sinensis* (L.) O. Kuntze]. *Journal of Microbiology and Biotechnology,* 5(2): 84-90.

Wang, X., Yang, P., Gao, Q., Liu, X., Kuang, T., Shen, S. and He, Y. 2008. Proteomic analysis of the response to high-salinity stress in Physcomitrella patens. *Planta,* 228: 167–177.

Wu, L., Hana, Z., Wanga, S., Wanga, X., Sund, A., Zua, X. and Chen Y. 2013. Comparative proteomic analysis of the plant–virus interaction in resistant and susceptible ecotypes of maize infected with sugarcane mosaic virus. *Journal of Proteomics,* 89: 124-140.

Yan, J., Tang, H., Huang, Y., Shi, Y., Li, J., and Zheng, Y. 2003. Dynamic analysis of QTL for plant height at different developmental stages in maize (Zea mays L.). *Chin. Sci. Bull,* 48: 2601-2607.

Yan, S. P., Zhang, Q. Y., Tang, Z. C., Su W. A. and Sun W. N. 2006. Comparative proteomic analysis provides new insights into chilling stress responses in rice. *Molecular & Cellular Proteomics,* 5: 484–496.

Zhang, Y., Xu, L., Zhu, X., Gong, Y., Xiang, F., Sun, X. and Liu, L. 2013. Proteomic analysis of heat stress response in leaves of radish (*Raphanus sativus* L.). Plant Molecular Biology Reporter. 31: 195-203.

Zhao, C., Wang, J., Cao, M., Zhao, K., Shao J., Lei, T., Yin, J., Hill, G. G, Xu, N. and Liu S. 2005. Proteomic changes in rice leaves during development of field-grown rice plants. *Proteomics,* 5: 961–972.

Zhong B., Karibe H., Komatsu S., Ichimura H., Nagamura Y., Sasaki T. and Hirano H. 1997. Screening of rice genes from a cDNA catalog based on the sequence data-file of proteins separated by two-dimensional electrophoresis. *Breeding Science,* 47: 245–251.

Zhu, J., Chen, S., Alvarez, S., Asirvatham, V. S., Schachtman, D. P., Wu, Y. and Sharp, R. E. 2006. Cell wall proteome in the maize primary root elongation zone. I. Extraction and identification of water-soluble and lightly ionically bound proteins. *Plant. Physiol.* 140: 311–325.

26

Transcriptomic Analysis for Abiotic Stress Resistance for Crop Plants

Bharti Aneja[1], Vishal Sharma[1*], Neelam R. Yadav[2] and Ram C. Yadav[3]

[1]ICAR-National Dairy Research Institute, Karnal -132001, India
[2]Department of Molecular Biology, Biotechnology & Bioinformatics, CCS Haryana Agricultural University, Hisar - 125004, India
[3]Centre for Plant Biotechnology, Hisar - 125004, India
*Corresponding author: vetvishal319@gmail.com

Abstract

Abiotic stresses like drought, high salinity, and extreme temperatures are common hostile environmental conditions that significantly reduce the crop productivity. Plants have the capability to sense and adjust to abiotic stresses, although the amount of adaptability to specific stresses varies from species to species. The adaptability to different environmental stresses is controlled by either simple or complex cascades of molecular networks. The responses of plants to different stresses are regulated by multiple signaling pathways. In response to different stresses, significant overlap between the gene expression patterns is observed in plants. Stress-related genes are induced primarily at the level of transcription, and regulation of the temporal and spatial expression patterns of different specific stress genes is vital part of the plant stress response. Far beyond the manipulation of single functional gene, genetic engineering of certain regulatory genes has emerged as an effective strategy for controlling the expression of many stress-responsive genes. Transcription factors (TFs) are good candidates for genetic engineering to breed stress-tolerant crop because of their role as master regulators of many stress-responsive genes. Many TFs belonging to families AP2/EREBP, WRKY, NAC, MYB, bZIP have been found to play role in various abiotic stresses and some genes have also been engineered to improve stress tolerance in model and crop plants.

Keywords: Abiotic stress, Regulatory genes, Transcription factors, Stress-responsive, Stress tolerance

Introduction

Plants as sessile organisms are regularly exposed to a variety of abiotic or biotic stresses, such as high salt, water deficiency, extreme temperatures, oxidative stress, chemical pollutants, pathogens, nematodes, arthropods, and herbivores (Guo *et al.*, 2016). Abiotic stress is the major cause of crop loss worldwide, and reduces crop productivity by around 50% annually (Wang *et al.*, 2004). In contrast to animals, plants could not change their sites to escape from the hostile stresses. But plants have adapted to certain changing stresses during the course of evolution, such as the presence of leaf epidermis with stomata for gas exchange, dominance of sporophyte enclosing the sensitive gametophyte, formation of stress resistant dormant organs, and presence of conducting tissues in big and long lived plants for long-distance transport of nutrient and water (Baniwal *et al.*, 2004; Al-Whaibi, 2011). An interconnected network of cellular stress response systems is a primary requirement for plant survival and productivity (Scharf *et al.*, 2012). Understanding of the mechanisms involved in the stress response is important for developing new methods to enhance plant stress tolerance.

During the last few decades, a great deal of efforts has been dedicated in breeding different crop cultivars with various stress-tolerant traits. Two main approaches have been employed to achieve success in this process. First is traditional breeding methods such as wide-cross hybridization and mutation breeding, which frequently brings about unpredictable results (Wang *et al.*, 2016). Second is transgenic technology by introducing novel exogenous genes or modifying the expression levels of different endogenous genes to improve stress tolerance. Since the conventional breeding approaches have resulted in limited success due to the complexity of stress tolerance traits, the transgenic approach is now being preferred to develop transgenic crops tolerant to abiotic stresses (Yamaguchi and Blumwald, 2005). Therefore, understanding the molecular mechanisms by which plants recognize the stress signals and induce the cellular machinery to recruit adaptive responses to the stress is an essential prerequisite for the identification of key genes and pathways to engineer stress-tolerant crop plants (Ray *et al.*, 2009).

Abiotic stress tolerance is multigenic in nature as it is governed by multiple loci, therefore adapting to different environmental cues is an extremely complex phenomenon (Yadav *et al.*, 2014). Furthermore, in response to abiotic stress hundreds of genes are activated (Seki *et al.*, 2002). These genes are categorized into two groups: regulatory genes and functional genes. The regulatory genes includes genes encoding for various transcription factors that regulate different stress-inducible genes. The functional genes are the genes that encode metabolic components like sugar, sugar alcohols, and amines that play a vital role in stress tolerance (Tran *et al.*, 2010).

Transcription factors (TFs) generally act as key regulators of gene expression. In general, the transcription factors with two distinct domains, viz., a DNA binding domain and a transcriptional activation/repression domain, regulate different cellular processes via controlling the transcriptional rates of different target genes

(Roy, 2016). Regulon is a transcription factor that controls the expression of a group of genes (Mizoi *et al.*, 2012). In response to abiotic stresses, four different regulons are identified in plants: (i) the NAC (NAM, ATAF, and CUC)/ZF-HD (zinc finger homeodomain) regulon, (ii) the CBF/DREB (C-repeat binding factor/dehydration-responsive element binding) regulon, (iii) the MYC (myelocytomatosis oncogene)/ MYB (myeloblastosis oncogene) regulon, and (iv) the AREB/ABF (ABA-responsive element-binding protein/ ABA-binding factor) regulon. The first two are ABA independent regulons while the last two are ABA dependent regulons (Saibo *et al.*, 2009). This suggests that there exists ABA independent and -dependent signal transduction pathways that convert the initial stress signal into cellular responses (Fig. 1). ABA is produced under water-deficit conditions, and it plays an important role in plants in the drought and salinity stress tolerance. This broad-spectrum phytohormone is also involved in coordinating various stress signal transduction pathways in plants during abiotic stresses, in addition to other commonly known functions like regulating stomatal opening, growth, and development (Agarwal and Jha, 2010).

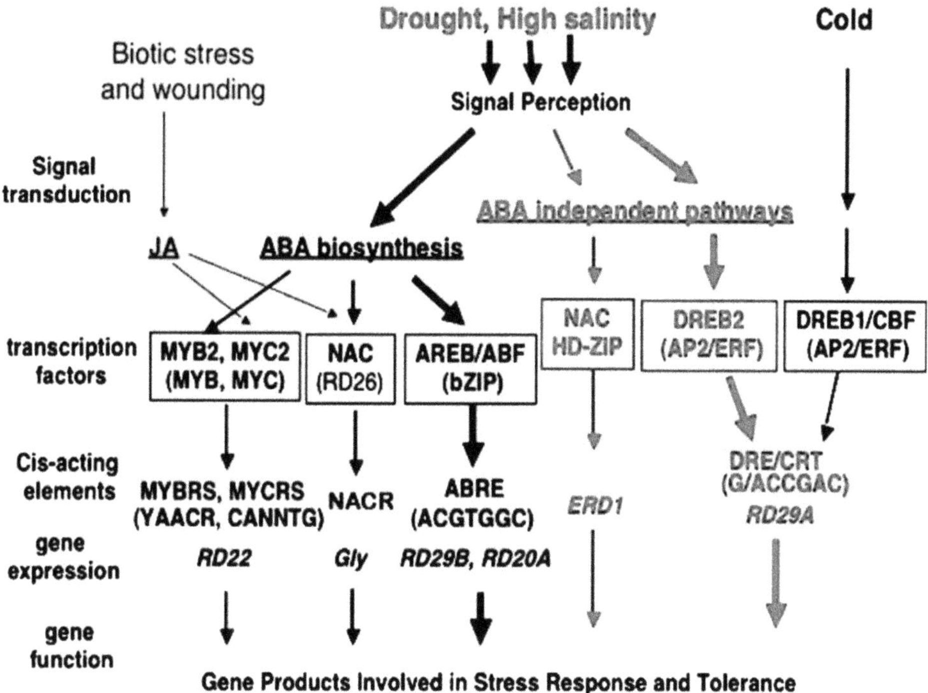

Fig.1: Abiotic Stress Signaling Network (Gujjar *et al.*, 2014)

NAC Transcription Factors

NAC TFs comprise a large plant-specific superfamily present in a wide range of plant species, similar to the transcription factor families mentioned earlier. The typical features of NAC TFs is that they contain a highly conserved NAC domain in the N-terminal region and a variable transcriptional regulatory region in the C-terminal region (Wang *et al.*, 2016). NAC was derived from the first letter of the names of the first three described proteins containing the DNA-binding domain, namely NAM (no apical meristem), ATAF1-2, and CUC2 (cup-shaped cotyledon) (Souer *et al.*, 1996; Aida *et al.*, 1997). NAM was the first NAC gene isolated from petunia. NACs regulate the abiotic stress response in plants through both the ABA-dependent and -independent pathways (Fujita *et al.*, 2004; Tran *et al.*, 2004). One of the NAC proteins, RD26, is probably involved in ABA-dependent stress signaling pathway. TaNAC2 is another NAC transcription factor that was isolated from wheat. Transgenic experiments, in *Arabidopsis*, indicated that TaNAC2 increases tolerance to drought, salt, and freezing stresses (Mao *et al.*, 2012).

NAC TFs interacts with NAC recognition sequence (NACRS) with the CACG core-DNA binding motif in the promoter region of the target genes and regulate the transcription of these target genes. NAC TFs are involved in various processes including shoot apical meristem formation, flower development, formation of secondary walls and cell division, leaf senescence, as well as biotic and abiotic stress responses (Olsen *et al.*, 2005; Tran *et al.*, 2010; Nakashima *et al.*, 2012; Nuruzzaman *et al.*, 2013; Banerjee and Roychoudhury, 2015).

Over-expression of another NAC TF, OsNAC10, under the control of the constitutive promoter GOS2 and the root-specific promoter RCc3 increased the tolerance of rice to drought, high salinity and low temperature at the vegetative stage (Jeong *et al.*, 2010). While over-expression of NAC TF, ZmSNAC1, in *Arabidopsis* led to hypersensitivity to ABA and osmotic stress, and increased tolerance to dehydration (Lu *et al.*, 2012). Different transgenic experiments indicated that TaNAC2 increases the tolerance of *Arabidopsis* to drought, salt, and freezing stresses, without any obvious negative effects on morphology by the over-expression of TaNAC2, suggesting a possible potential for utilization of such NAC TF gene in crop improvement (Mao *et al.*, 2012). A full-length cDNA of DgNAC1 was isolated from chrysanthemum and it contained a typical NAC domain. Using transgenic experiment it was observed that 35S:DgNAC1 transgenic tobacco exhibited a significantly greater tolerance to salt with no detectable phenotype defects under normal growth conditions (Liu *et al.*, 2011). Differential expression patterns such as tissue-specific, developmental stage- or stress-specific expression, of these stress-responsive NAC genes indicates their involvement in the complex signaling networks during plant stress responses (Wang *et al.*, 2016).

DREB family of transcription factors

DREB1 genes are the best studied group of TFs involved in abiotic stress, particularly in drought and cold tolerance (Reddy and Reddy 2008). A conserved drought responsive element (DRE) is present in the promoters of many osmotic stress inducible genes (Gujjar *et al.*, 2014). From *Arabidopsis thaliana*, several cDNAs encoding the DRE binding proteins, DREB1A and DREB2A have been isolated and they have been shown to specifically bind and activate the transcription of genes containing DRE sequences. In fact, in a number of studies, overexpression of stress inducible DREB transcription factor has resulted in the increased expression of many target genes having DRE elements in their promoters and that resulted in improved stress tolerance in the resulting transgenic plants (Table 1).

CBF/DREB1 transcription factors, controlling the level of COR (cold-regulated) gene expression, are considered as important regulators of the cold acclimation response and promotes tolerance to freezing (Gilmour *et al.*, 2000). Transformation of CBF/DREB1 genes has been observed to improve environmental stress tolerance to many plants, some of these have the potential to be used as freezing-tolerant varieties.

Table 1: Function of DREB genes in various abiotic stresses in plants (Lata and Prasad, 2011)

DREB TFs	Species	Accession no.	Stress response	References
DREB1A	Arabidopsis thaliana	AB007787	Cold	Liu *et al.*, 1998
DREB2A	Arabidopsis thaliana	AB007790	Drought, Salt, ABA	Liu *et al.*, 1998
DREB2C	Arabidopsis thaliana	At2g40340	Salt, Mannitol, Cold	Lee *et al.*, 2010
CBF1	Arabidopsis thaliana	U77378	Cold	Gilmour *et al.*, 1998
CBF2	Arabidopsis thaliana	AF074601	Cold	Gilmour *et al.*, 1998
CBF3	Arabidopsis thaliana	AF074602	Cold	Gilmour *et al.*, 1998
CBF4	Arabidopsis thaliana	AB015478	Drought, ABA	Haake *et al.*, 2002
OsDREB1A	Oryza sativa	AF300970	Cold, Salt, Wounding	Dubouzet *et al.*, 2003
OsDREB1B	Oryza sativa	AF300972	Cold	Dubouzet *et al.*, 2003
OsDREB1C	Oryza sativa	AP001168	Drought, Salt, Cold, ABA, Wound	Dubouzet *et al.*, 2003
OsDREB1D	Oryza sativa	AB023482	None	Dubouzet *et al.*, 2003
OsDREB2A	Oryza sativa	AF300971	Drought, Salt, faintly to Cold, ABA	Dubouzet *et al.*, 2003
OsDREB1F	Oryza sativa		Drought, Salt, Cold, ABA	Wang *et al.*, 2008
OsDREB2B	Oryza sativa		Heat, Cold	Matsukura *et al.*,2010
OsDREB2C	Oryza sativa	AK108143	None	Matsukura *et al.*, 2010

DREB TFs	Species	Accession no.	Stress response	References
OsDREB2E	Oryza sativa		None	Matsukura *et al.*, 2010
OsDREBL	Oryza sativa	AF494422	Cold	Chen *et al.*, 2003
TaDREB1	Triticum aestivum	AAL01124	Cold, Dehydration, ABA	Shen *et al.*, 2003
WCBF2	Triticum aestivum		Cold, Drought	Kume *et al.*, 2005
WDREB2	Triticum aestivum	BAD97369	Drought, Salt, Cold, ABA	Egawa *et al.*, 2006
HvDRF1	Hordeum vulgare	AY223807	Drought, Salt, ABA	Xue and Loveridge, 2004
HvDREB1	Hordeum vulgare	DQ012941	Drought, Salt, Cold	Xu *et al.*, 2009
ZmDREB2A	Zea mays	AB218832	Drought, Salt, Cold, Heat	Qin *et al.*, 2007
PgDREB2A	Pennisetum glaucum	AAV90624	Drought, Salt, Cold	Agarwal *et al.*, l., 2007
SbDREB2	Sorghum bicolor	ACA79910	Drought	Bihani *et al.*, 2011
SiDREB2	Setaria italica	HQ132744	Drought, Salt	Lata *et al.*, 2011a
CaDREB-LP1	Capsicum annum	AY496155	Drought, Salt, Wounding	Hong and Kim, 2005
AhDREB1	Artiplex hortensis		Salt	Shen *et al.*, 2003b
GmDREBa	Glycine max	AY542886	Cold, Drought, Salt	Li *et al.*, 2005
GmDREBb	Glycine max	AY296651	Cold, Drought, Salt	Li *et al.*, 2005
GmDREBc	Glycine max	AY244760	Drought, Salt, ABA	Li *et al.*, 2005
GmDREB	Glycine max	AF514908	Drought, Salt	Shiqing *et al.*, 2005
GmDREB2	Glycine max	ABB36645	Drought, Salt	Chen *et al.*, 2007
PpDBF1	Physcomitrella patens	ABA43697	Drought, Salt, Cold, ABA	Liu *et al.*, 2007
PNDREB1	Arachis hypogea	FM955398	Drought, Cold	Mei *et al.*, 2009
CAP2	Cicer arietinum	DQ321719	Drought, Salt, ABA, Auxin	Shukla *et al.*, 2006
DvDREB2A	Dendrathema	EF633987	Drought, Heat, ABA, Cold	Liu *et al.*, 2008
DmDREBa	Dendronthema 3 moriforlium	EF490996	Cold, ABA	Yang *et al.*, 2009

DREB TFs	Species	Accession no.	Stress response	References
DmDREBb	Dendronthema 3 moriforlium	EF487535	Cold, ABA	Yang *et al.*, 2009
PeDREB2	Populus euphratica	EF137176	Drought, Salt, Cold	Chen *et al.*, 2009
SbDREB2A	Salicornia brachiata	GU592205	Drought, Salt, Heat	Gupta *et al.*, 2010

Using both cDNA and GeneChip microarrays, more than 40 target genes of DREB1/CBF have been identified (Seki *et al.*, 2001; Fowler *et al.*, 2002; Maruyama *et al.*, 2004; Vogel *et al.*, 2005). In order to analyze DREB regulon in monocots, Dubouzet *et al.*, (2003) isolated five cDNAs for DREB homologs (OsDREB1A, OsDREB1B, OsDREB1C, OsDREB1D, and OsDREB2A). They found that cold induces the expression of OsDREB1A and OsDREB1B, while dehydration and high-salt stresses induces the expression of OsDREB2A. They observed that on over-expression of OsDREB1A in transgenic *Arabidopsis* induced over-expression of target stress-inducible genes of *Arabidopsis* DREB1A leading to higher tolerance to drought, high-salt, and freezing stresses. They concluded that OsDREB1A is potentially useful for producing transgenic monocots that are tolerant to drought, high-salt, and/or cold stresses.

MYB transcription factors

The MYB family of proteins is large, functionally diverse, and represented in all eukaryotes. MYB proteins are important factors involved in various regulatory networks controlling development, metabolism, and responses to biotic and abiotic stresses. Numerous MYB TFs play role in many significant physiological and biochemical processes including cell development and cell cycle, hormone synthesis and signal transduction, primary and secondary metabolism, as well as in plant responses to various biotic and abiotic stresses (Dubos *et al.*, 2010; Ambawat *et al.*, 2013). Members of this family were first identified to be involved in the regulation of anthocyanin biosynthesis (Goodrich *et al.*, 1992). Most MYB proteins function as transcription factors and possess varying numbers of MYB domain repeats conferring their ability to bind DNA (Dubos *et al.*, 2010). The DNA-binding domain of MYB proteins usually contain two imperfect repeats of about 50 residues (R2 and R3) in plants, while in animals it contains three repeats (R1, R2, and R3) in a helix–turn–helix structure. In plants, the MYB family has selectively expanded through the large family of R2R3-MYB. Among them, R2R3-MYBs are the most prevalent in plants (Dubos *et al.*, 2010; Ambawat *et al.*, 2013; Li *et al.*, 2015). The smallest class is the 4R-MYB group, containing four R1/R2-like repeats. A single 4R-MYB protein is encoded in several plant genomes. Members of this family play an important role in a variety of plant-specific processes, and are also involved in the regulation of many aspects of plant growth, development, metabolism, and stress responses as revealed by their extensive characterization in *A. thaliana*.

MYB factor regulates different target genes involved in several physiological processes under abiotic stresses. In addition, *OsMYB2* from rice was induced by salt, cold, and dehydration stress. The transgenic plants over-expressing *OsMYB2* exhibited increased tolerance to different stresses by bringing change in the expression levels of genes involved in different functions in response to stress (Yang *et al.*, 2012). Overexpression of either *GmMYB76* or *GmMYB177*, from soybean, in *Arabidopsis* resulted in significantly enhanced salt and freezing tolerance (Liao *et al.*, 2008). Cominelli *et al.*, (2005) reported that the transcription factor, AtMYB60, is responsible for stomatal movement regulation. AtMYB60 has guard cell specific expression and its expression is negatively controlled during drought stress conditions. AtMYB44 belongs to the R2R3 MYB subgroup 22 transcription factor family in *Arabidopsis*. AtMYB44 gene expression was also activated under various abiotic stresses, such as dehydration, low temperature, and salinity. Transgenic *Arabidopsis* carrying an AtMYB44 promoter-driven GUS construct, showed strong GUS activity in the vasculature and leaf epidermal guard cells. Jung *et al.* (2008) showed that *Arabidopsis* overexpressing AtMYB44 upon treatment with ABA induced AtMYB44 transcript accumulation within 30 min. and show more rapid ABA-induced stomatal closure response in comparison to wild-type and AtMYB44 knockout plants. Liang *et al.* (2005) have shown that AtMYB61 (At1g09540), a member of the R2R3-MYB family of transcription factors, is specifically expressed in guard cells and its expression is consistent with the manner of involvement in the control of stomatal aperture in *A. thaliana*.

AREB/ABF Regulon

ABA is a plant hormone that plays an important role in regulating different processes such as seed germination and development, root and stem growth, and biotic and abiotic stress responses (Busk and Pages, 1998). When plants are threatened to abiotic stresses, two independent signal transduction pathways are triggered: ABA-independent and -dependent signaling cascades (Bray, 1997; Shinozaki and Yamaguchi-Shinozaki, 1996). Considering the tissue specificity of ABA signaling pathways, various factors mediating ABA-dependent stress responses during the vegetative growth phase may have been unidentified so far. Many genes induced in response to drought and salinity stress, also respond to ABA and in their promoter regions they contain a conserved ABA-responsive, cis-acting element, designated as ABRE (PyACGTGG/ TC). Most known coupling elements are similar to ABREs. Narusaka *et al.*, (2003) showed that in *Arabidopsis* DRE and ABRE are interdependent in the ABA-responsive expression of the rd29A gene, whose expression is induced by dehydration, high-salinity, and low-temperature, in response to ABA. They concluded that DRE/CRT sequence may serve as a coupling element of ABRE in response to ABA in *Arabidopsis*, suggesting interaction between the DREB regulons and the ABRE-related regulons.

The factors, isolated by a yeast one-hybrid system using a prototypical ABRE are named as ABFs, and they belong to a distinct subfamily of basic leucine zipper (bZIP) proteins (Yadav *et al.*, 2014). A specific group of bZIP transcription factors have been shown to bind to ABRE, and are known as AREBs or ABFs (Uno *et al.*, 2000; Choi *et al.*, 2000). Around 75 distinct bZIP transcription factors exist in the genome of *Arabidopsis* and they have been divided into 11 groups. The ABFs/AREBs belong to group A, and usually play role in ABA signaling during seed maturation or stress conditions. The bZIP class ABRE-binding factor OSBZ8 (38.5 kDa) was identified by Mukherjee *et al.*, (2006). Like other TFs, the bZIP TFs respond to various abiotic stresses such as drought, high salinity, and cold stresses in addition to playing pivotal roles in the developmental processes (Jakoby *et al.*, 2002). Now, many members of the bZIP TF family have been identified or predicted at genome-wide level in some species. However, only a small portion of bZIP TFs has been well studied and most studies have shown that bZIP TFs are induced by ABA in response to stress and regulate the expression of stress-related genes in ABA-dependent manner by interacting with specific ABA-responsive cis- acting elements (ABRE) in their promoter region (Zou *et al.*, 2008).

WRKY Transcription Factor

WRKY transcription factors possess a conserved WRKYGQK sequence in their DNA binding domain. They bind to the W-box (TTGAC) region on the promoter of target gene. WRKY TFs encode for a wide family of transcription factors, characterized by the presence of approximately 60 amino acids having amino acid sequence WRKY at its N terminal end and a putative zinc finger motif at its C terminal end (Gujjar *et al.*, 2014). Till date, approximately 74 members of WRKY gene family in *Arabidopsis* and 112 members in rice have been reported. Recently, in response to fungal pathogens and hormone treatments, 46 WRKY genes have been isolated from canola (*B. napus* L.) (Yang *et al.*, 2009). The knowledge about the role of WRKY TFs in abiotic stress response remains murky and little information is available in limited crops only. Since the cloning of the first cDNA encoding a WRKY protein (SPF1), isolated from sweet potato (Ishiguro and Nakamura, 1994), an increasing number of WRKY TFs have been identified in various plants. *Arabidopsis* contains 72 WRKY genes which can be divided into three groups with several subgroups, on the basis of their WRKY domains. Rizhsky *et al.* (2002) reported the expression of a WRKY TF that responds to a combination of drought and heat stress in tobacco (*N. tabacum*).

ZF-HD Regulon

The ZF-HD regulons are characterized by the presence of conserved ZF domain, at N-terminal, containing several cysteine and histidine residues that determine the ZF structure. The ZF domain is responsible for potential protein–protein interaction and the C-terminal conserved HD is able to bind DNA. Using yeast one-hybrid screen, four ZF-HD proteins were first identified in *Flaveria trinervia*. These proteins bind to the regulatory regions of C4 phosphoenol pyruvate carboxylase (Windhovel *et al.*, 2001). There are 14 AtZF-HD genes in the *Arabidopsis* genome, and they encode for a group of transcriptional regulators that play overlapping regulatory roles in *Arabidopsis* floral development (Tan *et al.*, 2006). The expression of ZFHD1 was induced by drought and salt stress and ABA. Seven novel transcription factors (four ZF-HD and three C2H2- type transcription factors) bind to the promoter to repress the expression of OsDREB1B and are involved in the response to different abiotic stresses (Figueiredo *et al.*, 2012).

Conclusions

This review emphasizes the promising roles of TFs as vital tool to improve plant responses to multiple abiotic stresses. Although a great deal of information about the role of TFs has been accumulated on their involvement in response to various abiotic stresses and the role of a large number of candidate TF genes have been validated. Understanding the molecular mechanisms by which the plants respond to different abiotic stresses such as heat, drought and salinity is a primary requirement for the manipulation of plants to improve stress tolerance and productivity (Guo *et al.*, 2016). In response to these stresses, mainly TFs regulate the expression of most of the genes, and the gene products function in providing stress tolerance to plants (Lata and Prasad, 2011). TFs respond to specific stresses and they activate or repress genes through cis-acting sequences. In plants, a large portion of their genome capacity is dedicated towards the transcription, and it often belongs to large gene families that are unique to plants in certain cases (Yadav *et al.*, 2014). In plants, the response to abiotic stress is highly complicated because of the involvement of huge gene families and the complex interactions between TFs and cis-elements on the promoters of target genes. Moreover, a single TF may regulate large number of target genes with the corresponding cis-elements on the promoters, whereas a single gene with several types of cis- elements may be regulated by different families of TFs. Thus, the stress-responsive TFs not only function independently but also cross-talk between each other in response to various abiotic stress responses indicating the complexity of signaling networks involved in plant stress responses (Prasch and Sonnewald, 2015). The TFs can be genetically engineered to produce transgenic plants having greater tolerance against abiotic stresses, using different promoters.

References

Agarwal, P.K. and Jha, B. 2010. Transcription factors in plants and ABA dependent and independent abiotic stress signalling. *Biologia Plantarum,* 54: 201–212.

Aida, M., Ishida, T., Fukaki, H., Fujisawa, H. and Tasaka, M. 1997. Genes involved in organ separation in Arabidopsis: an analysis of cup-shaped cotyledon mutant. *Plant Cell,* 9: 841–857.

Al-Whaibi, M.H. 2011. Plant heat-shock proteins: a minireview. *J. King Saud Univ.-Sci,* 23: 139–150. doi:10.1016/j.jksus.2010.06.022.

Ambawat, S., Sharma, P., Yadav, N.R. and Yadav, R.C. 2013. MYB transcription factor genes as regulators for plant responses: an overview. *Physiology and Molecular Biology of Plants,* 19(3): 307-321.

Banerjee, A. and Roychoudhury, A. 2015. WRKY proteins: signaling and regulation of expression during abiotic stress responses. *Science World Journal* doi:10.1155/2015/807560.

Baniwal, S.K., Bharti, K., Chan, K.Y., Fauth, M., Ganguli, A., Kotak, S., Mishra, S.K., Nover, L., Port, M., Scharf, K.D., Tripp, J., Weber, C., Zielinski, D. and von Koskull-Döring, P. 2004. Heat stress response in plants: a complex game with chaperones and more than twenty heat stress transcription factors. *Journal of Biosciences* 29: 471–487. doi: 10.1007/BF02712120.

Bray, E.A. 1997. Plant responses to water deficit. *Trends in Plant Sciences* 2: 48–54.

Busk, P.K. and Pages, M. 1998. Regulation of abscisic acid-induced transcription. *Plant Molecular Biology,* 37: 425–435.

Choi, H.I., Hong, J.H., Ha, J.O., Kang, J.Y. and Kim, S.Y. 2000. ABFs, a family of ABA-responsive element binding factors. *Journal of Biological Chemistry,* 275: 1723–1730.

Cominelli, E., Galbiati, M., Vavasseur, A., Conti, L., Sala, T., Vuylsteke, M., Leonhardt, N., Dellaporta, S.L. and Tonelli, C. 2005. A guard-cell-specific MYB transcription factor regulates stomatal movements and plant drought tolerance. *Current Biology,* 15: 1196–1200.

Dubos, C., Stracke, R., Grotewold, E., Weisshaar, B., Martin, C. and Lepiniec, L. 2010. MYB transcription factors in Arabidopsis. *Trends in Plant Sciences* 15: 573–582.

Dubouzet, J.G., Sakuma, Y., and Ito, Y. 2003. OsDREB genes in rice, Oryza sativa L., encode transcription activators that function in drought-, high-salt- and cold responsive gene expression. *Plant Journal,* 33: 751–763.

Figueiredo, D.D., Barros, P.M., Cordeiro, A.M., Serra, T.S., Lourenco, T., Chander, S., Oliveira, M.M. and Saibo, N.J.M. 2012. Seven zinc-finger transcription factors are novel regulators of the stress responsive gene OsDREB1B. *Journal of Experimental Botany,* 63: 3643–3656.

Fowler, S. and Thomashow, M.F. 2002. Arabidopsis transcriptome profiling indicates that multiple regulatory pathways are activated during cold acclimation in addition to the CBF cold response pathway. *Plant Cell,* 14: 1675–1690.

Fujita, M., Fujita, Y., Maruyama, K., Seki, M., Hiratsu, K., Ohme-Takagi, M., Tran, L.S.P.; Yamaguchi-Shinozaki, K. and Shinozaki, K. 2004. A dehydration induced NAC protein, RD26, is involved in a novel ABA-dependent stress-signalling pathway. *Plant Journal,* 39: 863–876.

Gilmour, S.J., Sebolt, A.M., Salazar, M.P., Everard, J.D. and Thomashow, M.F. 2000. Over-expression of the Arabidopsis CBF3 transcriptional activator mimics multiple biochemical changes associated with cold acclimation. *Plant Physiology,* 124: 1854–1865.

Goodrich, J., Carpenter, R. and Coen, E.S. 1992. A common gene regulates pigmentation pattern in diverse plant species. *Cell,* 68: 955–964.

Gujjar, R.S., Akhtar, M. and Singh, M. 2014. Transcription factors in abiotic stress tolerance. *Indian Journal of Plant Physiology,* 19(4): 306–316.

Guo, M., Liu, J.H., Ma, X., Luo, D.X., Gong, Z.H. and Lu, M.H. 2016. The Plant Heat Stress Transcription Factors (HSFs): Structure, Regulation, and Function in Response to Abiotic Stresses. *Frontiers in Plant Science,* 7:114. doi: 10.3389/fpls.2016.00114

Ishiguro, S. and Nakamura, K. 1994. Characterization of a cDNA encoding a novel DNA-binding protein, SPF1, that recognizes SP8 sequences in the 5′ upstream regions of genes coding for sporamin and beta-amylase from sweet potato. *Molecular Genetics and Genomics,* 244: 563–571. doi:10.1007/BF00282746

Jakoby, M., Weisshaar, B., Dröge-Laser, W., Vicente-Carbajosa, J., Tiedemann, J., Kroj, T., Parcy, F. 2002. bZIP transcription factors in *Arabidopsis*. *Trends in Plant Sciences,* 7: 106–111. doi: 10.1016/S1360-1385(01)02223-3

Jeong, J.S., Kim, Y.S., Baek, K.H., Jung, H., Ha, S.H., Choi, Y.D., Kim, M., Reuzeau, C. and Kim, J.K. 2010. Root-specific expression of OsNAC10 improves drought tolerance and grain yield in rice under field drought conditions. *Plant Physiology,* 153: 185–197.

Jung, C., Seo, J.S., Han, S.W., Koo, Y.J., Kim, C.H., Song, S.I., Nahm, B.H., Choi, Y.D. and Cheong, J.J. 2008. Overexpression of AtMYB44 enhances stomatal closure to confer abiotic stress tolerance in transgenic Arabidopsis. *Plant Physiology,* 146: 623–635.

Lata, C. and Prasad, M. 2011. Role of DREB in regulation of abiotic stress responses in plants. *Journal of Experimental Botany,* 62: 4731–4748. doi:10.1093/jxb/err210

Li, Z., Tian, Y., Zhao, W., Xu, J., Wang, L., Peng, R. and Yao, Q. 2015. Functional characterization of a grape heat stress transcription factor VvHsfA9 in transgenic *Arabidopsis*. *Acta Physiologia Plantarum,* 37: 1–10. doi:10.1007/s11738-015- 1884-x

Liang, Y.K., Dubos, C., Dodd, I.C., Holroyd, G.H., Hetherington, A.M. and Campbell, M.M. 2005. AtMYB61, an R2R3-MYB transcription factor controlling stomatal aperture in Arabidopsis thaliana. *Current Biology,* 15: 1201–1206.

Liao, Y., Zou, H.F., Wang, H.W., Zhang, W.K., Ma, B., Zhang, J.S. and Chen, S.Y. 2008. Soybean GmMYB76, GmMYB92, and GmMYB177 genes confer stress tolerance in transgenic Arabidopsis plants. *Cell Research,* 18: 1047–1060.

Liu, Q.L., Xu, K.D., Zhao, L.J., Pan, Y.Z., Jiang, B.B., Zhang, H.Q. and Liu, G.L. 2011. Over-expression of a novel chrysanthemum NAC transcription factor gene enhances salt tolerance in tobacco. *Biotechnology Letters,* 33: 2073–2082.

Lu, M., Ying, S., Zhang, D.F., Shi, Y.S., Song, Y.C., Wang, T.Y. and Li, Y. 2012. A maize stress responsive NAC transcription factor, ZmSNAC1, confers enhanced tolerance to dehydration in transgenic Arabidopsis. *Plant Cell Reports,* 31: 1701–1711.

Mao, X., Zhang, H., Qian, X., Li, A., Zhao, G. and Jing, R. 2012. TaNAC2, a NAC type wheat transcription factor conferring enhanced multiple abiotic stress tolerances in Arabidopsis. *Journal of Experimental Botany,* 63: 2933–2946.

Maruyama, K., Sakuma, Y., Kasuga, M., Ito, Y., Seki, M., Goda, H., Shimada, Y., Yoshida, S., Shinozaki, K. and Yamaguchi- Shinozaki, K. 2004. Identification of cold inducible downstream genes of the Arabidopsis DREB1A/CBF3 transcriptional factor using two microarray systems. *Plant Journal,* 38: 982–993.

Mizoi, J., Shinozaki, K., and Yamaguchi-Shinozaki, K. 2012. AP2/ERF family transcription factors in plant abiotic stress responses. *Biochimica et Biophysica Acta,* 18: 86–96.

Mukherjee, K., Choudhury, A.R., Gupta, B., Gupta, S. and Sengupta, D.N. 2006. An ABRE-binding factor, OSBZ8, is highly expressed in salt tolerant cultivars than in salt sensitive cultivars of indica rice. *BMC Plant Biology,* 6: 18.

Nakashima, K., Takasaki, H., Mizoi, J., Shinozaki, K. and Yamaguchi-Shinozaki, K. 2012. NAC transcription factors in plant abiotic stress responses. *Biochimica et Biophysica Acta,* 19: 97–103 doi:10.1016/j.bbagrm.2011.10.005

Narusaka, Y., Nakashima, K., Shinwari, Z.K., Sakuma, Y., Furihata, T., Abe, H., Narusaka, M., Shinozaki, K. and Yamaguchi-Shinozaki, K. 2003. Interaction between two cis-acting elements, ABRE and DRE, in ABA dependent expression of Arabidopsis rd29A gene in response to dehydration and high salinity stresses. *Plant Journal,* 34: 137–148.

Nuruzzaman, M., Sharoni, A.M. and Kikuchi, S. 2013. Roles of NAC transcription factors in the regulation of biotic and abiotic stress responses in plants. *Frontiers in Microbiology,* 4: 248 doi:10.3389/fmicb.2013.00248

Olsen, A.N., Ernst, H.A., Leggio, L.L. and Skriver, K. 2005. NAC transcription factors: structurally distinct, functionally diverse. *Trends in Plant Science,* 10: 79–87. doi:10.1016/j.tplants.2004.12.010

Prasch, C.M. and Sonnewald, U. 2015. Signaling events in plants: stress factors in combination change the picture. *Environmental and Experimental Botany* 114: 4–14. doi: 10.1016/j.envexpbot.2014.06.020

Ray, S., Dansana, P.K., Bhaskar, A., Giri, J., Kapoor, S., Khurana, J.P. and Tyagi, A.K. 2009. Emerging trends in functional genomics for stress tolerance in crop plants. *Plant Stress Biology,* ed H.Hirt (Weinheim: Wiley- VCH Verlag GmbH & Co. KGaA), Ch 3: 37–63.

Reddy, G.L. and Reddy, A.R. 2008. Rice DREB1B promoter shows distinct stress-specific responses, and the overexpression of cDNA in tobacco confers improved abiotic and biotic stress. *Plant Cell,* 16: 2481–2498.

Rizhsky, L., Liang, H. and Mittler, R. 2002. The combined effect of drought stress and heat shock on gene expression in tobacco. *Plant Physiology,* 130: 1143–1151.

Roy, S. 2016. Function of MYB domain transcription factors in abiotic stress and epigenetic control of stress response in plant genome. *Plant Signaling & Behaviour* 11(1): e1117723, http://dx.doi.org/10.1080/15592324.2015.1117723.

Saibo, N.J., Lourenco, T. and Oliveira, M.M. 2009. Transcription factors and regulation of photosynthetic and related metabolism under environmental stresses. *Annals of Botany,* 103: 609–623.

Scharf, K.D., Berberich, T., Ebersberger, I. and Nover, L. 2012. The plant heat stress transcription factor (Hsf) family: structure, function and evolution. *Biochimica et Biophysica Acta,* 1819: 104–119. doi: 10.1016/j.bbagrm.2011.10.002.

Seki, M., Narusaka, M., Abe, H., Kasuga, M., Yamaguchi-Shinozaki, K., Carninci, P., Hayashizaki, Y. and Shinozaki, K. 2001. Monitoring the expression pattern of 1300 Arabidopsis genes under drought and cold stresses by using a full-length cDNA microarray. *Plant Cell,* 13: 61–72.

Seki, M., Narusaka, M., Ishida, J., Nanjo, T., Fujita, M., Oono, Y., Kamiya, A., Nakajima, M., Enju, A. and Sakurai, T. 2002. Monitoring the expression profiles of 7000 Arabidopsis genes under drought, cold and high-salinity stresses using a full-length cDNA microarray. *Plant Journal,* 31: 279–292.

Shinozaki, K. and Yamaguchi-Shinozaki, K. 1996. Molecular responses to drought and cold stress. *Current Opinion in Biotechnology,* 7: 161–167.

Souer, E., van Houwelingen, A., Kloos, D., Mol, J. and Koes, R. 1996. The No Apical Meristem gene of petunia is required for pattern formation in embryos and flowers and is expressed at meristem and primordia boundaries. *Cell,* 85: 159–170.

Tan, Q.K.G. and Irish, V.F. 2006. The Arabidopsis zinc finger- homeodomain genes encode proteins with unique biochemical properties that are coordinately expressed during floral development. *Plant Physiology,* 140: 1095–1108.

Tran, L.S.P., Nakashima, K., Sakuma, Y., Simpson, S.D., Fujita, Y. and Maruyama, K. 2004. Isolation and functional analysis of Arabidopsis stress-inducible NAC transcription factors that bind to a drought responsive cis-element in the early responsive to dehydration stress 1 promoter. *Plant Molecular Biology,* 68: 533–555.

Tran, L.S.P. Nishiyama, R., Yamaguchi-Shinozaki, K. and Shinozaki, K. 2010. Potential utilization of NAC transcription factors to enhance abiotic stress tolerance in plants by biotechnological approach. *GM Crops,* 11: 32–39.

Uno, Y., Furihata, T., Abe, H., Yoshida, R., Shinozaki, K. and Yamaguchi-Shinozaki, K. 2000. Arabidopsis basic leucine zipper transcription factors involved in an abscisic acid dependent signal transduction pathway under drought and high-salinity conditions. *Proceedings of National Academy of Sciences,* USA 97: 11632–11637.

Vogel, J.T., Zarka, D.G., VanBuskirk, H.A., Fowler, S.G. and Thomashow, M.F. 2005. Roles of the CBF2 and ZAT12 transcription factors in configuring the low temperature transcriptome of Arabidopsis. *Plant Journal,* 41: 195–211.

Wang, H., Wang, H., Shao, H. and Tang, X. 2016. Recent Advances in Utilizing Transcription Factors to Improve Plant Abiotic Stress Tolerance by Transgenic Technology. *Frontiers in Plant Science,* 7: 67. doi: 10.3389/fpls.2016.00067

Wang, W., Vinocur, B., Shoseyov, O. and Altman, A. 2004. Role of plant heat- shock proteins and molecular chaperones in the abiotic stress response. *Trends in Plant Sciences,* 9: 244–252. doi:10.1016/j.tplants.2004.03.006

Windhövel, A., Hein, I., Dabrowa, R. and Stockhaus, J. 2001. Characterization of a novel class of plant homeodomain proteins that bind to the C4 phosphoenolpyruvate carboxylase gene of Flaveria trinervia. *Plant Molecular Biology,* 45: 201–214.

Yadav, N.R., Taunk, J., Rani, A., Aneja, B. and Yadav, R.C. 2014. Role of Transcription Factors in Abiotic Stress Tolerance in Crop Plants. *Climate Change and Plant Abiotic Stress Tolerance,*. First Edition. Wiley-VCH Verlag GmbH & Co. KGaA. doi: 10.1002/9783527675265.ch23

Yamaguchi, T. and Blumwald, E. 2005. Developing salt-tolerant crop plants: challenges and opportunities. *Trends in Plant Sciences,* 10: 615–620. doi: 10.1016/j.tplants.2005.10.002

Yang, A., Dai, X. and Zhang, W-H. 2012. A R2R3-type MYB gene, OsMYB2, is involved in salt, cold and dehydration tolerance in rice. *Journal of Experimental Botany,* 63: 2541–2556.

Yang, B., Jiang, Y., Rahman, M.H., Deyholos, M.K. and Kav, N.V. 2009. Identification and expression analysis of WRKY transcription factor genes in canola (Brassica napus L.) in response to fungal pathogens and hormone treatments. *BMC Plant Biology,* 9: 1–19.

Zou, M., Guan, Y., Ren, H., Zhang, F. and Chen, F. 2008. AbZIP transcription factor, OsABI5, is involved in rice fertility and stress tolerance. *Plant Molecular Biology,* 66: 675–683. doi: 10.1007/s11103-008-9298-4.

27

Breeding for Biotic Stress Resistance Plants

Chhagan Lal

Division of Plant Breeding & Genetics, Rajasthan Agricultural Research Institute, Jaipur

Corresponding author: chhagangpb123@yahoo.com

Abstract

Goal of crop improvement for biotic stress tolerance inplants is an important objective of plant breeders. Plants must continuously defend themselves against attacks from bacteria, viruses, fungi, and even other pests. This chapter will therefore summarize the mechanism of resistance in plants. Attempts will be made to provide a description on the effective genetic mechanisms that plants have developed to recognize and respond to infection by a number of pathogens and pests, such as non-host resistance, constitutive barriers and race-specific resistance. This chapter also covers the most relevant problems in breeding for resistance to pest and will include aspects related to specificity of defence mechanisms, specificity of parasitic ability, inheritance of resistance, gene-for-gene interaction, and durability of resistance. Major considerations in breeding for resistance to pest, conventional sources of resistance and possible alternatives, namely mutation breeding, genetic manipulations, tissue cultures, and molecular interventions to develop plants resistant to pests and pathogens will also be dealt.

Keywords: Biotic stress, Pest

Introduction

Most of the problems facing agriculture in the twenty- first century relate to the growing world population, which is expected to stabilize at around 10–12 billion during the next 70 years (Heszky, 2008). During the last decade, world grain yield increased around 0.5% per year, which is three-fold lower than the population growth rate in the same period. The main task for breeders and agronomists will therefore be to increase yields while reducing the use of chemicals. Because more than 42% of the potential world crop yield is lost owing to biotic stresses (15%

attributable to insects, 13% to weeds, and 13% to other pathogens), a reduction in this incidence will be one of the more important possibilities for improving plant production (Pimentel, 1997). In this context, the development of tolerant plants to biotic stresses is therefore an important objective of plant breeding strategies with relevant implications for both farmers and the seed and agrochemical industries. In fact genetic resistance has several obvious advantages over the use of chemical pesticides or other methods for pestcontrol. These include nominal genetic permanency, negligible cost once cultivars are developed, and quite high efficiency. The major downside of genetic resistance to biotic stresses is the fact that selection pressure is placed on pest populations to develop means of overcoming the resistance, thus practically limiting the time of effectiveness.

Table 1: Overview of Potential and Actual Losses.

	Pest and Pathogens				
	Fungi and Bacteria	Virus	Animal Pests	Weeds	Total
Loss potential (%) a	14.9	3.1	17.6	31.8	67.4
Actual losses (%) a	9.9	2.7	10.1	9.4	32.0
Efficacy (%) b	33.8	12.9	42.4	70.6	52.5

Source: Modified from Oerke and Dehne (2004)

a) As percentage of attainable yields

b) As percentage of loss potential prevented

Breeding for Resistance to Biotic Stresses

Disease Resistance

Disease is an abnormal condition in the plant produced by an organism of an environmental factor. More specially, disease may be defined as 'the series of invisible and visible responses of plant cells and tissues to a pathogenic microorganism of an environmental factor that result in adverse changes in form, function or integrity of plants and may lead to partial impairment or death of the plant or its parts. But in this chapter, we shall consider only such disease that are induced by organism and the abnormalities produced by the non-biological environment or by a disease is known as host, while the organism that produces the disease is termed as pathogen. Diseases are produced by a variety of organisms from plant and animal kingdom *viz.*, fungi, bacteria, viruses, nematodes and insects.

Different crops are attacked to different degree by the different kinds of pathogens, but it may be emphasized that all the crop science are attacked by them. For example, cereals suffer from epidermises of air-borne fungi, most solanaceous crops are infected severely viruses, cotton, is damaged by many insects and so on. In the order of their importance, the pathogen may be listed as fungi

>bacteria>viruses>nematodes= insects. Much of the breeding efforts have been directed against diseases caused by fungi, which may be greater than the effort against all other pathogens put together. Therefore, our discussion would be primarily foccused on the knowledge of fungal disease.

Mechanism of Disease Resistance

A variety of mechanisms are involved in disease resistance. In some cases, the basis of resistance is better known than in others. The various mechanisms of disease resistance are as follows: 1) Mechanical, 2) Hypersensitivity and 3) Nutritional.

(1) Mechanical

Certain mechanical or anatomical features of the host may prevent infection. For example, closed flowering habit of wheat and barley prevents infection by the spores of ovary infecting fungi.

(2) Hypersensitivity

In a large number of cases, immune reaction is due to the hypersensitive reaction of the host. This mechanism is found in case of biotrophic organism or obligate parasites.

(3) Nutritional

The reduction is growth and in spore production is generally supposed to be due to an unfavourable physiological conditions within the host. Most likely, a resistant host does not fulfil the nutritional requirements of the pathogen and thereby limits its growth and reproduction. However, more precise information is not available on this aspect.

Breeding for Resistance to Biotic Stresses

Insect Resistance

Like disease, insects are important causal factors of biotic stress in crop plants. Insects attack all the crop plants and lead to considerable losses in yield as well as quality. Insect attack leads to various types of damages. In crop plants such as (1) Reduction in plant growth or stunting, (2) Damage of vegetative and reproductive parts, (3) Premature defoliation, and (4) Wilting of plants. Insect cause 14% estimated yield loss of all important crops on global bases. Insect cause yield loss directly either by sucking cell sap or by eating away various plant parts. Insects also cause yield loss through transmission of various diseases.

Mechanisms of Insect Resistance

There are four mechanisms of insect resistance *viz.*, 1) non-preference, 2) antibiosis, 3) tolerance, and 4) avoidance or escape. The first three mechanisms were given by Painter (1951) and the fourth one was added subsequently. A resistant variety may have one, two or more of these mechanisms.

1. Non – preference

It refers to various features of host plant that make the host undesirable for unattractive to insects for food, shelter, or reproduction. This type of insect resistance is also known as non acceptance and antixenosis. Non acceptance appears to be more accurate term, because in most known examples of this type of resistance, insects will not accept a resistant host plant even if there is no alternative source of food. Various plant characters which are associated with non preference include colour, light penetration, hairiness, leaf angle, odour and taste. For example, in cotton red plant body, smooth leaves, okra, leaf, open canopy, nectarilessness, fregobract, thickness and hardness of boll rind and long pedicel are examples of non-preference to bollworms, and hairiness of leaf and stem is non-preference for jassids. In pea, yellow green genotypes are less preferred by pea aphid than blue green genotypes. In soybean, hairy genotypes are less preferred by potato is related to odour. In maize and sorghum resistance to grass hoppers appears to be related to differences in taste. Non preference involves various morphological and chemical features of host plants.

Non Preference Mechanism of Insect Resistance in Some Crop Plants:

Host Crop	Insect Pest	Non-preference	Preference
Wheat	Stem Sawfly	Solid stem	Hollow stem
Rice	i) Rice stem borer	Lignified stem	Non-lignified stem
	ii)Brown plant hopper	Low asparagines	High asparagines
Maize	i)Corn earworm	Toughness of husk	Soft husk
	ii)Corn leaf aphid	High DIMBOA	Low DIMBOA
Soybean	Potato leaf hopper	Hairiness	Smoothness
Pea	Pea aphid	Yellow green leaves	Blue grren leaves
Cabbage	Cabbage aphid	Leaves with high light reflection	Leaves with low light refelction
Sugarbeet	Aphid	Low free sugar	High free sugar
Brassica	Cabbage aphid	Low sinigrin	High sinigrin

2. Antibiosis

It refers to the adverse effect of host plant on the development and reproduction of insect pests which feed on resistant plant. Resistant plants retard the growth and rate of reproduction of insect pest. In some cases, antibiosis may lead even to death of an insect. An antibiosis is considered as the true form of resistance to insect pests. In cotton, antibiosis is related with high level of gossypol, tannins ,heliocides and silica contents, antibiosis may involve morphological, physiological and biochemical features of the host plant.

3. Tolerance

It refers to the ability of a variety to produce greater yield than susceptible variety at the same level of insect attack. In other words, a tolerant variety will give higher yield than susceptible one despite the insect attack. The tolerance is measured in terms of rejuvenation potential, healthy leaf growth, flowering compensation potential and superior plant vigour. Hybrid cottons, by virtue of their very high potential, show tolerance to insect pest. Tolerant cultivars have greater recovery of damaged parts than susceptible ones.
Antibiosis Mechanism of Insect Resistance in Some Crop Plants:

Host Crop	Insect Pest	Cause of Antibiotics
Wheat & Barley	Green bugs	High benzyl alcohol
Rice	Rice stem borer	High silica content
Cotton	Bollworms	High gossypol
		High tannis
Sugarbeet	Aphid	Low free sugar
Alfalfa	Spotted aphid and pea aphid	High saponin
Brassica	Cabbage aphid	Low sinigrin
Potato	Aphid	Gummy Trichome exudates
Tobacco	Mites	Exudates of glandular leaf hairs
Medicago	Alfalfa weevil	Exudates of secondary trichomes on leaves.

4. Avoidance or Escape

It refers to escape or avoid of a variety from insect attack either due to earliness or its cultivation in the season where insect population is very low. For example, early maturing cotton varieties escape pink bollworm infestation which occurs late in the season. Avoidance is also an effective means of protecting crop from the damage of insect pests.

Genetic Basis of Resistance

Genetic analysis of disease resistance in plants began over 100 years ago when Biffin (1905) reported that resistance in wheat to stripe rust (*Puccinia striiformis*) was inherited as a single recessive Mendelian trait. Since this initial work, many genes conferring resistance to pathogens in crop plants have been characterized, and the genetic basis of pathogenicity (virulence/avirulence) has been studied in many plant pathogens. This knowledge culminated in the development of the gene- for- gene hypothesis by Flor (1971) based on genetic studies of the interaction between flax and the flax rust pathogen, which has provided a framework for much if not all of the work on disease resistance in the years since. In genetic terms, resistance is generally defined by the mode of inheritance, with broad distinctions between oligogenic (controlled by one or few genes of major effect) and polygenic (controlled by many genes of low individual phenotypic effect) resistance.

Qualitative Resistance

Evidence made it clear that many cases of resistance were inherited in a simple way. Most characterized resistance genes are dominant in action; for example the *Hm1*gene of maize conferring resistance to *Cochlioboluscar bonum*race 1, a causal agent of northern leaf spot of maize, However, some recessive resistance genes have proven important sources of durable resistance – e.g. gene *Sr2*conferring resistance to stem rust in wheat (McIntosh *et al.* 1995) ; gene *mlo* for mildew resistance in barley (Jorgensen 1994). There are also many examples of resistance genes that display partial dominance (gene dosage dependence; e.g. the resistance gene *Lr9* in wheat to *Puccinia recondita*). Dominance or recessiveness of resistance genes is, however, not absolute andcan even be governed by the attribute used to measure the disease phenotype(Johnson 1992), genetic background, pathogen isolate or environment. Examples of oligogenic resistance are known in which additive and non-additive interaction occurs between genes at separate loci. The genes *Lr13* and *Lr34* in wheat interact in an additive manner to confer resistance to leaf rust, not only with each other, butalso with other genes for resistance to leaf rust (Kolmer1992). Non-additive geneinteraction occurs when two genes in the host are only effective when presenttogether. In such cases, the genes in the host are referred to as complementary. Resistance to bacterial infections is not well developed as virus and fungal resistance,partly because bacterial diseases are a main problem only in crop plants likepotato, rice, and some fruit trees. Similarly to fungal diseases the most effective typeof protection is genetic resistance, which is based on single dominant or semi-dominantgenes. Different classes of *R* genes cloned from various plant species were characterized and tested for their ability in conferring resistance against bacterial pathogens. For example, among these, a map-based cloned *Xa21* gene from rice, gave resistance to bacterial blight, a serious disease in rice caused by *Xantomonas oryzae.* Infiltration of different maize lines with a variety of bacterial pathogens of maize, rice and sorghum has permitted to identify a maize gene, *Rxo1*, which conditions a strong HR to the non-host bacterialpathogen*X. o.* pv. *Oryzicola*(Zhao *et al.* 2004). The same locus carries a gene(designated *Rba1*) controlling resistance to the maize and sorghum bacterial stripe pathogen *Burkholderia andropogonis*. It was surprising that the same locus controlled resistance to two of only four bacterial pathovars tested. This suggests that this locus may condition defence reactions to other bacterial pathogens

Quantitative Resistance and QTLs

Quantitative resistance, in contrast to qualitative resistance, is generally considered as partial resistance in a particular cultivar (Young 1996). This type of disease resistance is controlled by multiple loci, referred to as polygenes or quantitative traitloci (QTLs), and does not comply with simple Mendelian inheritance. Examples of such polygenetically inherited resistance are the partial resistance in potato to

Phytopathora infestans, in maize to *Puccinia sorghi*, and in barley to *Puccinia hordei* (Parlevliet and Zadoks, 1977).

Although genetically complex forms of disease resistance are still poorly understood an effective strategy for studying complex and polygenic forms of disease resistance is known as QTL mapping, which is based on the use of DNA markers (see Young 1996, for a review). With QTL mapping, the roles of specific loci ingenetically complex traits can be described; this has also permitted insight to begained into fundamental questions that have puzzled researchers in the field of plantpathology for decades. Although results of QTL mapping indicate that it is generallynot the case, there are examples of several (>10) QTLs involved in quantitativeresistance; however, it is much more common to find only three to five loci: frequently, 1 or 2 QTLs predominate.

QTL mapping may also help to determine whether individual QTLs are racespecific or not, and when there is an indication of specificity, the degree to whichpartial resistance differs between races. For example, quantitative resistance to *P. Infestans* in potato was initially described as race non-specific (Van der Plank1982). Dissecting the contributions of individual QTLs, it was clearly demonstratedthat loci show distinctly different resistance effects against different pathogen races (Leonards-Schippers *et al.* 1994). Indeed, only 5 of the 11 statistically significantgenomic regions showed no specificity against just two races tested, while the others were significant against just one. Moreover, genetic mapping with DNA markersmakes it possible to ask whether homologous resistance genes exist in related planttaxa and may help to test the hypothesis that QTLs are simply variants of qualitativeresistance loci that have been (partially) overcome by their respective pathogen. Forinstance, in rice blast, 3 of the QTLs mapped to the same marker intervals as previously identified qualitative blast resistance genes. It is conceivable that these QTLsrepresent allelic variants of the known qualitative resistance genes, though onlymore precise mapping and gene cloning can resolve this definitely. In potato lateblight, 1 QTL coincided in location with a dominant, race-specific gene known as *R1*, as well as *Rx2*, a gene for resistance to potato virus.

References

Biffin, R.H. 1905. Mendel's laws of inheritance and wheat breeding. *J AgricSci,* 1: 4–48.

Heszky, L. 2008. Challengee of plant breeding early in 21th century. *Hungarian Agric Res* 4: 4–8.

Jørgensen, J.H. 1994. Genetics of powdery mildew resistance in barley. *Crit Rev Plant Sci,* 13: 97–119.

Flor, H.H. 1971. Current status of the gene-for-gene concept. *Annu Rev Phytopathol,* 9: 275–296.

Kolmer, J.A. 1992. Enhanced leaf rust resistance in wheat conditioned by resistance gene pairs withLr13. *Euphytica,* 61: 123–130.

McIntosh, R.A., Wellings, C.R. and Park, R.F. 1995. Wheat rusts: an atlas of resistance genes. CSIRO Publishing, Melbourne.

Leonards-Schippers, C., Gieffers, W., Schiifer-Pregl, R., Ritter. E., Knapp, S.J., Salamini, F. Gebhardt, C. 1994. Quantitatve resistance to *Phytophtorainfestans* in potato: a cause study of QTL mapping in an allogamous plant species. *Genetics,* 137: 67–77.

Painter, R.H. 1958. Insect resistance in crop plants. University Press of Kansas, Lawrence.

Pimentel, D. 1997. Techniques for reducing pesticide use. Wiley, Hoboken.

Parlevliet, J.E. and Zadoks, J.C. 1977. The integrated concept of disease resistance. A new view including horizontal and vertical resistance in plants. *Euphytica,* 26: 5–21.

Van der Plank, J.E. 1982. Disease resistance in plants. Academic, New York.

Young, N.D. 1996. QTL mapping and quantitative disease resistance in plants. *Annu Rev Phytopathol,* 34: 479–501.

Zhao, B.Y., Ardales, E., Brasset, E., Clafl, L.E., Leach, J.E. and Hulbert, S.H. 2004. The *Rxo1/ Rba1* locus of maize controls resistance reactions to pathogenic and non-host bacteria. *Theor Appl Genet,* 109: 71–79.

28

Commonly Used Computational Tools of Genomics and Proteomics

Anil Panwar[1*], Ajay Pal[2] and Hemant Poonia[3]

[1]Department of Molecular Biology, Biotechnology and Bioinformatics
[2]Department of Chemistry and Biochemistry
[3]Department of Mathematics, Statistics and Physics, CCS HAU Hisar
**Corresponding author: anilpanwar@hau.ernet.in*

Abstract

Enormous data has been generated with the evolution of new high throughput sequencing methods like pyrosequencing and illumina sequencing. This gigantic data captivate the workers to stand on a single platform known as bioinformatics. The use of information technology and computer science to retrieve, collect and analyse biological data with the help of statistical tools is known as bioinformatics. A number of algorithms have been designed to extract the information from this uncooked data. These algorithms are being used in the form of online and stand alone tools. But, many a times user is confused with the type of genomics/ proteomics approaches to be used. Considering this issue we have tried to flash a light on the operation of various tools in this chapter.

Keywords: Bioinformatics, Genomics, Proteomics, Sequencing data

Introduction

Bioinformatics is the application of information technology, computer science and statistics to retrieve, store, organize and analyse biological data using *in-silico* tools. Voluminous biological data is available in the form of sequence and structure databases. Biological databases are libraries of life sciences information collected from scientific experiments, high throughput experiment technology like sequencing and published literature. These databases contain the information from different research areas like genomics, proteomics, metabolomics, phylogenetics and also of experiment and organism specific. The highlight of the last decade in the life science was the production of massive amount of data and therefore, the objective of approaching decade is to analyse this data to extract some useful information which can lead to new discoveries. In order to turn the available data into knowledge there is a need to formulate the hypothesis, validate them and convert into computer programs and tools. So, for knowledge discovery, a numbers of bioinformatics tools are developed and some commonly used tools are discussed in this chapter.

Genomics tools: The branch of biotechnology which is concerned with applying the techniques of genetics and molecular biology to genetic mapping and DNA sequencing of sets of genes or the complete genomes of selected organism, with organising the result in databases and with the applications of the data is called genomics. The tools which are applied on these data sets are called genomics tools.

Sequence similarity searching: Many times, it happens that a researcher has a sequence but he does not know the source of the sequence for example while working on metagenomics. So, in these cases, a user has numerous nucleotide or protein sequences but he does not know the organism to which these sequences belong, as the genetic material is recovered directly from environmental sample like soil. Under these conditions, it becomes necessary to find similarity searching to identify the organisms that are present in his environmental samples.

1. Similarity Searching tools:

a) BLAST

BLAST (**B**asic **L**ocal **A**lignment **S**earch **T**ool) comes under the category of homology and similarity tools. It is a set of search programs designed for the Windows platform and is used to perform fast similarity searches regardless of whether the query is for protein or DNA. Comparison of nucleotide sequences in a database can be performed. Also, a protein database can be searched to find the match against the queried protein sequence. NCBI has also introduced the new queuing system to BLAST (Q BLAST) that allows users to retrieve results at their convenience and format their results multiple times with different formatting options.

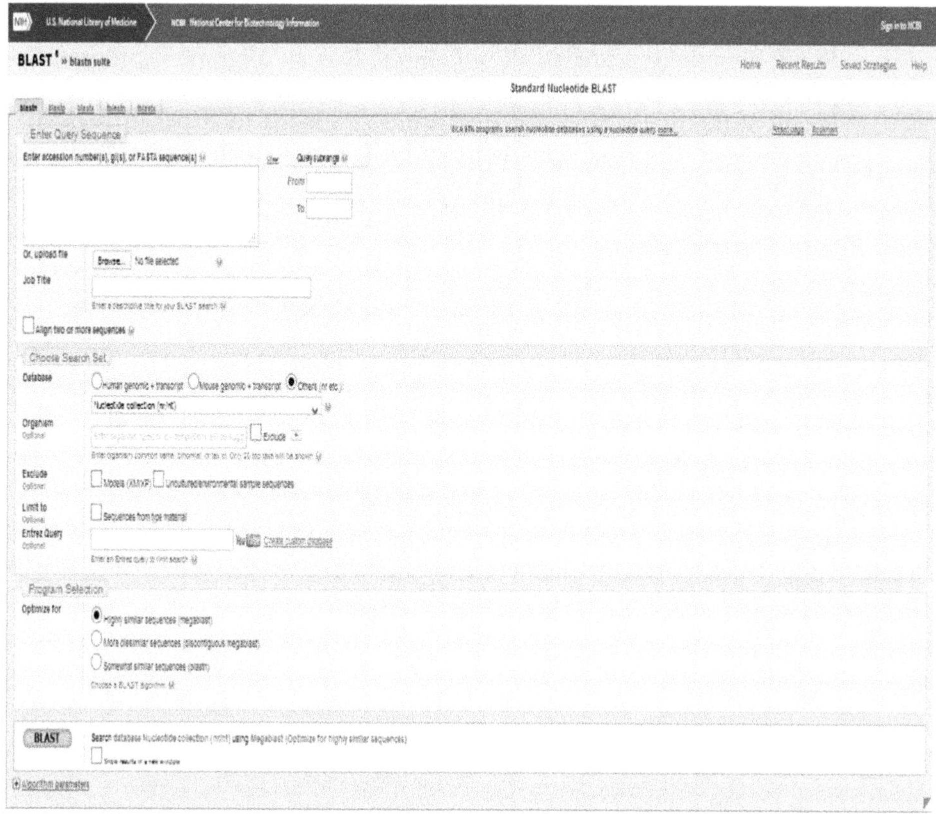

Types of BLAST:

i) Nucleotide BLAST: The Query sequence is of nucleotide and Searching Database is also of nucleotide.

ii) Protein BLAST: The Query sequence is of Protein and Searching Database is also of protein.

iii) BLASTX: The Query sequence is translated nucleotide and Searching Database is also of protein.

iv) TBLASTN: The Query sequence is of protein and Searching Database is of translated nucleotide.

v) Primer-BLAST: Designs primers specific to your PCR template.

vi) MOLE-BLAST: Establishes taxonomy for uncultured or environmental sequence.

vii) Ig-BLAST: Search immunoglobulins and T-cell receptor sequences.

b) FASTA:

FAST homology searches all sequences. It is an alignment program for protein sequences created by Pearsin and Lipman in 1988. The program is one of the many heuristic algorithms proposed to speed up sequence comparison. The basic idea is to add a fast prescreen step to locate the highly matching segments between two sequences, and then extend these matching segments to local alignments using more rigorous algorithms such as Smith-Waterman.

FASTA

Protein Similarity Search

Multiple sequence alignment: Multiple sequence alignment (**MSA)** of many nucleotides or amino acids is an important tool in bioinformatics. The multiple sequence alignment technique identifies diagnostic patterns or motif to characterize protein families. It can also detect or demonstrate homology between new sequences and existing families of sequences. Thus, it helps predict the secondary and tertiary structures of the new sequences which are an essential prelude to molecular evolutionary analysis. Many multiple sequence alignment tools have been proposed to reduce the high computation time of fully performing alignment of all sequences.

Biologists produce high quality multiple sequence alignments by hand using expert knowledge of protein sequence evolution. Importance factors include: specific sorts of columns in alignments, such as highly conserved residues or buried hydrophobic residues; the influence of secondary and tertiary structure; and expected patterns of insertions and deletions. Furthermore, the phylogenetic relationships between sequences dictate constraints on the changes that occur in columns and in the patterns of gaps. Manual multiple alignment is tedious. Automatic multiple sequence alignment methods are a topic of extensive research in computational biology. In general, an automatic method must have a way to assign a score so that better multiple alignments get better scores.

In a multiple sequence alignment, homologous residues among a set of sequences are aligned together in columns. Homologous is meant in both the structural and evolutionary sense. Ideally, a column of aligned residues occupy similar three-dimensional structural positions and all diverge from a common ancestral residue. However, except for trivial cases of highly identical sequences, it is not possible to unambiguously identify structurally or evolutionarily homologous positions and create a single correct multiple alignment. Since protein structures also evolve, we do not expect two protein structures with different sequences to be entirely super imposable. An evolutionary correct alignment can be even more difficult to infer than a structural alignment. While structural alignment has an independent point of reference (superposition of crystal or NMR structures), the evolutionary history of the residues of a sequence family is not independently known from any source; it must itself be inferred from sequence alignment.

Thus, our ability to define a single 'correct' alignment will vary with relatedness of the sequences being aligned. An alignment of very similar sequences will generally be unambiguous, but these alignments are not of great interest to us. For cases of interest, e.g. for a family of proteins sharing perhaps only 30% average pairwise sequence identity, there is no objective way to define an unambiguously correct alignment. We should focus attention on the subset of columns corresponding to key residues and core structural elements that can be aligned with more confidence.

Multiple Sequence Alignment Methods

Multiple sequence alignment is an extension of pairwise alignment to incorporate more than two sequences at a time. Multiple alignment methods try to align all of the sequences in a given query set. Multiple alignments are often used in identifying conserved sequence regions across a group of sequences hypothesized to be evolutionarily related. Such conserved sequence motifs can be used in conjunction with structural and mechanistic information to locate the catalytic active sites of enzymes. Alignments are also used to aid in establishing evolutionary relationships by constructing phylogenetic trees. Varities of methods are developed some are;

Dynamic Programming

The technique of dynamic programming is theoretically applicable to any number of sequences; however, because it is computationally expensive in both time and memory, it is rarely used for more than three or four sequences in its most basic form. This method requires constructing the n-dimensional equivalent of the sequence matrix formed from two sequences, where n is the number of sequences in the query. Standard dynamic programming is first used on all pairs of query sequences and then the "alignment space" is filled in by considering possible matches or gaps at intermediate positions, eventually constructing an alignment essentially between each two-sequence alignment.

Progressive Methods

Progressive, hierarchical, or tree methods generate a multiple sequence alignment by first aligning the most similar sequences and then adding successively less related sequences or groups to the alignment until the entire query set has been incorporated into the solution. The initial tree describing the sequence relatedness is based on pairwise comparisons that may include heuristic pairwise alignment methods similar to FASTA. Progressive alignment results are dependent on the choice of "most related" sequences and thus can be sensitive to inaccuracies in the initial pairwise alignments.

Iterative Methods

Iterative methods attempt to improve on the heavy dependence on the accuracy of the initial pairwise alignments, which is the weak point of the progressive methods. Iterative methods optimize an objective function based on a selected alignment scoring method by assigning an initial global alignment and then realigning sequence subsets. The realigned subsets are then themselves aligned to produce the next iteration's multiple sequence alignment.

Motif Finding

Motif finding, also known as profile analysis, constructs global multiple sequence alignments that attempt to align short conserved sequence motifs among the sequences in the query set. This is usually done by first constructing a general global multiple sequence alignment, after which the highly conserved regions are isolated and used to construct a set of profile matrices. The profile matrix for each conserved region is arranged like a scoring matrix but its frequency counts for each amino acid or nucleotide at each position are derived from the conserved region's character distribution rather than from a more general empirical distribution. The profile matrices are then used to search other sequences for occurrences of the motif they characterize.

2. Tool For Multiple Alignment:

a) **ClustalW:** ClustalW is a general purpose multiple alignment program for DNA or proteins. ClustalW works by a three step process: All sequences are aligned and compared to each other and a score or distance is calculated between each pair of sequences; this matrix of distances between pairs of sequences is used to create a dendrogram (guide phylogenetic tree) among the included sequences; the dendrogram is used as a basis for constructing the real multiple sequence alignment whereby the most closely related pairs of sequences are aligned first. The quality of the alignment is determined by assigning a positive score to each pair of identical aligned residues, and a lower or negative score is assigned to mismatches. The scores are read off from whatever substitution matrix is being used. The parameters most likely to affect the quality of the alignment are the gap penalty (GapOpen), the gap-extension penalty (GapExt), and to a lesser extent, the substitution matrix.

b) **T-Coffee** (**T**ree-based **Co**nsistency **O**bjective **F**unction **F**or alignment **E**valuation) is a multiple sequence alignment software using a progressive approach. It generates a library of pairwise alignments to guide the multiple sequence alignment. It can also combine multiple sequences alignments obtained previously and in the latest versions can use structural information from PDB files (3D-Coffee). It has advanced features to evaluate the quality of the alignments and some capacity for identifying occurrence of motifs (Mocca). It produces alignment in the aln format (Clustal) by default, but can also produce PIR, MSF, and FASTA format. The most common input formats are supported (FASTA, PIR).

c) **MUSCLE:** Abbreviation stand for **mu**ltiple **s**equence **c**omparison by **l**og-**e**xpectation, is public domain, multiple sequence alignment software for protein and nucleotide sequences. The MUSCLE algorithm proceeds in three stages: the 'draft progressive', 'improved progressive' and 'refinement' stages. In the 'draft progressive' stage, the algorithm produces a draft multiple alignment, with the emphasis on speed rather than accuracy. In the 'improved progressive' stage, the Kimura distance is used to re-estimate the binary tree used to create the draft alignment, in turn producing a more accurate multiple alignment. The final 'refinement' stage refines the improved alignment produced in the second step. Multiple alignments are available at the end of each stage. MUSCLE is often used as a replacement for Clustal, since it typically (but not always) gives better sequence alignments, depending on the chosen options. In addition, MUSCLE is significantly faster than Clustal, especially for larger alignments.

3. Gene finding tools:

Gene finding refers to identifying stretches of sequences (genes) in genomic DNA that are biologically functional. Gene finding is crucial in understanding the genome of a species.Approaches Computational methodology for finding genes in a genome has evolved significantly over the last 20 years. Many approaches have been proposed to find genes in both prokaryotes and eukaryotes. These approaches mainly fall into three categories: homology-based approaches, Ab Initio approaches and comparative genomics approaches.

a) **GENSCAN:**GENSCAN is a program to identify complete gene structures in genomic DNA. It is a GHMM-based program that can be used to predict the location of genes and their exon-intron boundaries in genomic sequences from a variety of organisms. The GENSCAN Web server can be found at MIT. GENSCAN was developed by Christopher Burge, department of Mathematics, Stanford University.

b) **FGENESH:** One of the most popular and accurate ab initio gene finders available. It is also the fastest gene prediction program - 50 to 100 times faster than Genscan. FGENESH can be supplied with data sets specifically trained for several taxonomic groups. Custom parameter sets for specific organism or group can also be created. Recently released ver. 3 of FGENESH supports non-standard GC-donor splice sites and has various output options.

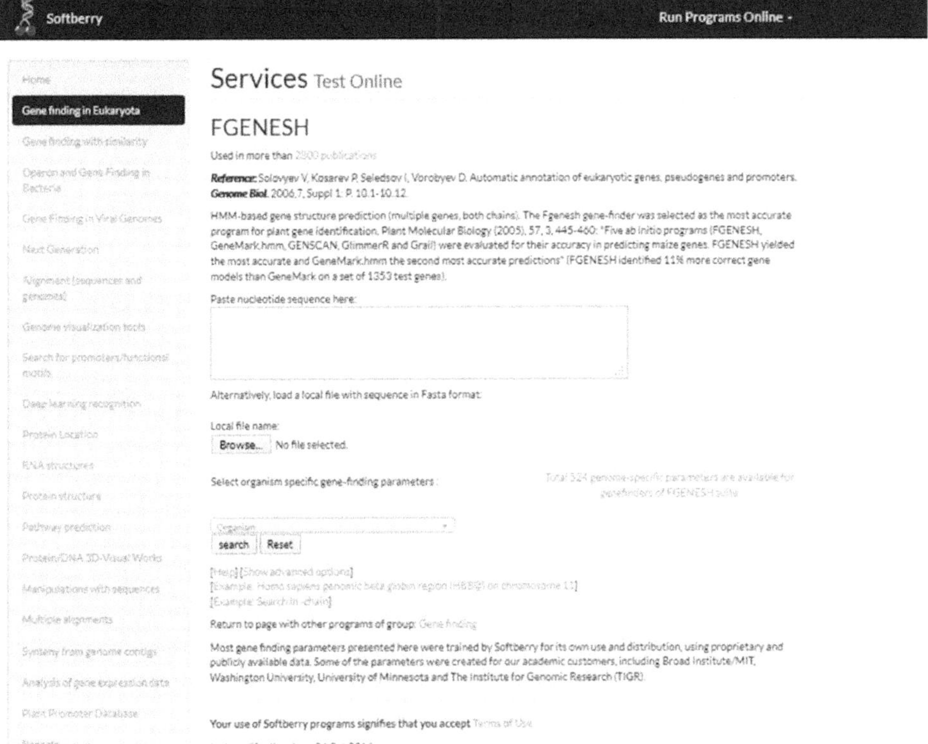

c) **GLIMMER:** Gene Locator and Interpolated Markov Modeller. Glimmer is a tool for finding genes in microbial DNA, especially the genomes of bacteria, Achaea, and viruses. Glimmer is the system of choice for genome annotation efforts on a wide range of bacteria, archaeal, and viral species due to high accuracy. Glimmer was used by the DNA Databank of Japan (DDBJ) to re-annotate all bacterial genomes in the International Nucleotide Sequence Databases. It is also being used by this group to annotate viruses.

d) **GeneID**: Geneid is a program to predict genes in anonymous genomic sequences designed with a hierarchical structure. Three steps it include to predict gene. In the first step, splice sites, start and stop codons are predicted and scored along the sequence using Position Weight Arrays (PWAs). In the second step, exons are built from the sites. Exons are scored as the sum of the scores of the defining sites, plus the log-likelihood ratio of a Markov Model for coding DNA. Finally, from the set of predicted exons, the gene structure is assembled, maximizing the sum of the scores of the assembled exons.

e) **ORF FINDER:** An open reading frame, or ORF, is a genetic sequence that begins with a start codon, ends with a stop codon. An ORF is called a coding sequence (CDS) if it is actually translated into a protein product by the cell. ORF-finder identifies ORF's in a genetic sequence in six reading frames, three for each DNA strand. Protein BLAST is used to determine if the translated ORF matches known protein sequences. Failure of any ORF in your genetic sequence to match the target protein will indicate that mutations in the sequence have disrupted the CDS; the gene is now a pseudogene!

4. SNP detection: Single nucleotide polymorphisms (SNPs) are a valuable resource for investigating the genetic basis of disease. These variants can serve as markers for fine-scale genetic mapping experiments and genome-wide association studies. Certain of these nucleotide polymorphisms may predispose individuals to illnesses such as diabetes, hypertension, or cancer, or affect disease progression. Bioinformatics techniques can play an important role in SNP discovery and analysis.

a) **Poly phred:** It is an integrated program to identify heterozygous locations for SNP. The program together with three other programs is used to compare florescence based sequence across DNA traces obtained from different individuals to detect heterozygous sites for SNPs. The contributed programs are as Phred (is responsible for detection of potential heterozygous sites for SNP using base call and peak characterization), Phrad (provides assemblies or sequence alignment) and Consed (responsible for editing, scanning, interpretation and visualization of results)

b) **Polybays:** The PolyBayes software is used to detect SNPs in redundant DNA sequences. By three steps, Creation of multiple sequence alignment, Identification of paralog and SNP detection.

c) **novoSNP:**novoSNP is a program used for detection of SNP in SNP reconstructing/resequencing based researches. It's input is a reference sequence and a number of sequencing trace files and output will come in the form of a list of possible variations with a quality score.

d) **QualitySNPng:** It is another software tool for detection of SNPs, designed in such a way that uses haplotype based scheme, it uses DNA sequences derived from next generation sequencing data. It performs on diploid as well as polyploidy species. Complete sequenced reference genome is not required by the tool.

Proteomics Tools

1. Homology and Similarity search Tools:

Homologous sequences are sequences that are related by divergence from a common ancestor. Thus the degree of similarity between two sequences can be measured while their homology is a case of being either true of false. This set of tools can be used to identify similarities between novel query sequences of unknown structure and function and database sequences whose structure and function have been elucidated.

a) **PSI-BLAST**: The emphasis of this tool is to find regions of sequence similarity, which will yield functional and evolutionary clues about the structure and function of your novel sequence. Position specific iterative BLAST (PSI-BLAST) refers to a feature of BLAST 2.0 in which a profile is automatically constructed from the first set of BLAST alignments. PSI-BLAST is similar to NCBI BLAST2 except that it uses position-specific scoring matrices derived during the search , this tool is used to detect distant evolutionary relationships. PHI-BLAST functionality is available to use patterns to restrict search results.

b) **COPIA:** COPIA (COnsensus Pattern Identification and Analysis) is a protein structure analysis tool for discovering motifs (conserved regions) in a family of protein sequences. Such motifs can be then used to determine membership to the family for new protein sequences, predict secondary and tertiary structure and function of proteins and study evolution history of the sequences.

2. Protein Function Analysis

This group of programs allow you to compare your protein sequence to the secondary (or derived) protein databases that contain information on motifs, signatures and protein domains. Highly significant hits against these different pattern databases allow you to approximate the biochemical function of your query protein.

a) **Pfam Scan**: PfamScan is used to search a FASTA sequence against a library of Pfam HMM.

b) **Pratt**: Pratt - Pattern Matching. An important problem in sequence analysis is to find patterns matching sets or subsets of sequences. This tool allows the user to search for patterns conserved in sets of unaligned protein sequences. The user can specify what kind of patterns should be searched for, and how many sequences should match a pattern to be reported.

c) **Prosite Scan**: PS_SCAN - compare a protein sequence against the signatures in PROSITE.

d) **RADAR**: RADAR stands for Rapid Automatic Detection and Alignment of Repeats in protein sequences. RADAR identifies gapped approximate repeats and complex repeat architectures involving many different types of repeats.

3. Structure Prediction

a) **MODELLER:** MODELLER is a computer program for comparative protein structure prediction. In the simplest case, the input is an alignment of a sequence to be modeled with the template structures, the atomic coordinates of the templates and a short script file. MODELLER then automatically calculates a model containing all non-hydrogen atoms, without any user intervention and within minutes on a Pentium processor. MODELLER implements comparative protein structure modeling by satisfaction of spatial restraints .The spatial restraints include: (i) homology-derived restraints on the distances and dihedral angles in the target sequence, extracted from its alignment with the template structures (ii) stereochemical restraints such as bond length and bond angle preferences, obtained from the CHARMM-22 molecular mechanics forcefield (iii) statistical preferences for dihedral angles and non-bonded inter-atomic distances, obtained from a representative set of known protein structures (iv) optional manually curated restraints, such as those from NMR spectroscopy, rules of secondary structure packing, cross-linking experiments, fluorescence spectroscopy, image reconstruction from electron microscopy, site-directed mutagenesis and intuition.

b) **MODWEB:** MODWEB is a web server for automated comparative protein structure modelling. MODWEB accepts one or many sequences in the FASTA format and calculates models for them based on the best available template structures from the Protein Data Bank.Alternatively, MODWEB also accepts a protein structure as an input and calculates models for all its identifiable sequence homologs in the non-redundant SWISS-PROT protein sequence database . The latter mode is a useful tool for various structural genomics efforts to assess the impact of a newly determined structure on the modeling coverage of the sequence space.

ModWeb
· Sali Lab Home · ModWeb · ModBase · ModEval · PCSS · FoXS · IMP · MultiFit · ModPipe ·
Login ModWeb Home Current ModWeb queue Help Contact News
ModWeb: A Server for Protein Structure Modeling
Additional functionality for registered users:
• Model leverage calculations
• Access to all user's ModWeb datasets
Developers:
Eswar Narayanan
Ursula Pieper
Ben Webb
Acknowledgements:
David Eramian
Mallur S. Madhusudhan
Marc A. Marti-Renom
Min-Yi Shen
Andrej Sali
ModWeb version r186
General information
Name
Email address (optional)
Modeller license key
Dataset name (optional)
Availability
Add to academic dataset
Input data
Input protein sequences
or upload sequences file (FASTA Format)
Browse... No file selected.
Calculate Models Reset
Model selection criteria
Best scoring model
Longest well scoring model
Fold assignment methods
Slow (Seq-Prf, PSI-Blast)
Upload models to ModBase

c) **PROSPECT:** PROSPECT (PROtein Structure Prediction and Evaluation Computer ToolKit) is a protein-structure prediction system that employs a computational technique called protein threading to construct a protein's 3-D model.

4. Structural Analysis

This set of tools allows us to compare structures with the known structure databases. The function of a protein is more directly a consequence of its structure rather than its sequence with structural homologs tending to share functions. The determination of a protein's 2D/3D structure is crucial in the study of its function.

a) **CASP**: CASP refers to Critical Assessment of protein Structure Prediction experimental methods to establish the current state of the art in protein structure prediction with identification of the progress made so far and highlight future efforts to be focused. CASP provides research groups with an opportunity to objectively test their structure prediction methods and delivers an independent assessment of the state of the art in protein structure modeling to the research community and software users. Prediction methods are assessed on the basis of the analysis of a large number of blind predictions of protein structure.

b) **DALI**: DALI stands for Distance Alignment Matrix Method. DALI is used to align two structures by generating a comparison matrix of intra-molecular distances and optimizes that matrix using a Monte Carlo procedure. DALI is a common and popular method that breaks down the protein that is inputted into hexapeptide fragments and then calculates a distance matrix through the understanding of the contact pattern between successive fragments. Secondary structure features with residues that are contiguous in sequence are shown on the matrix's main diagonal. The other diagonals represent residues that are not next to each other in sequence. When the diagonals are parallel to the main diagonal, the features they represent are also parallel. When the diagonals are perpendicular, however, their features are anti parallel. If two proteins' distance matrices are the same or share similar features in almost the same positions, they can be said to have similar folds and length loops connecting the secondary structure elements.

c) **MAMMOTH**: algorithm, also implemented in a multiple structure alignment method, MAMMOTH-Mult used in the DBAlidatabaseis a fast method for aligning two structures based on a vector representation of intra-molecular distances compared by a dynamic programming optimizer. Also, SALIGN command of the MODELLER package which is also used in the DBAli database, compares structure and properties calculated from the 3D coordinates of two or more proteins that are then aligned by a dynamic programming optimizer.

References

Altschul, S.F., Gish, W., Miller, W., Myers, E.W. and Lipman, D.J. 1990. Basic local alignment search tool. *J. Mol. Biol.,* 215: 403-410.

Pearson, W.R. 1991. Searching protein sequence libraries: comparison of the sensitivity and selectivity of the Smith-Waterman and FASTA algorithms. *Genomics,* 11 (3): 635-50.

Chenna, R., Sugawara, H., Koike, T., Lopez, R., Gibson, T.J., *Higgins, D.G.* and Thompson, J.D. 2003. Multiple sequence alignment with the Clustal series of programs. *Nucleic Acids Res.* 31 (13): 3497–3500.

Paolo, D. T., Sebastien, M., Ioannis, X., Miquel, O., Alberto, M., Jia-Ming, C., Jean-François, T. and Cedric N. 2011. T-Coffee: a web server for the multiple sequence alignment of protein and RNA sequences using structural information and homology extension. *Nucleic Acids Res.,* 1; 39(Web Server issue): W13–W17

Solovyev, V., Kosarev, P., Seledsov, I. And Vorobyev, D. 2006. Automatic annotation of eukaryotic genes, pseudogenes and promoters. *Genome Biol.,* 7 (1): 10.1-10.12.

Eswar, N., Marti-Renom, Webb, M. A. B., Madhusudhan, M. S., Eramian, D., Shen, M., Pieper, U. and Sali, A. 2006. Comparative Protein Structure Modeling With MODELLER. *Current Protocols in Bioinformatics, John Wiley & Sons, Inc., Supplement,* 15: 5.6.1-5.6.30.

Arnold, K., Bordoli, L., Kopp, J., and Schwede, T. 2006. The SWISS-MODEL Workspace: A web-based environment for protein structure homology modelling. *Bioinformatics,* 22: 195-201.

Benkert, P., Tosatto, S.C.E. and Schomburg, D. 2008. QMEAN: A comprehensive scoring function for model quality assessment.*Proteins: Structure, Function, and Bioinformatics,* 71(1): 261-277.

Buchan, D.W., Ward, S.M., Lobley, A.E., Nugent, T.C., Bryson, K. & Jones, D.T. 2010. Protein annotation and modelling servers at University College London.*Nucl. Acids* Res., 38 Suppl, W563-W568. (PSIPRED)

Kelley, L.A. and Stemberg, M.J.E. 2009. Protein structure prediction on the web: a case study using the Phyre server. *Nature Protocols,* 4: 363 – 371.

Wass, M.N., Kelley, L.A. and Sternberg, M.J. 2010. 3DLigandSite: predicting ligand-binding sites using similar structures.*NAR* 38 Suppl: W469-73.

29

Role of Information Technology in Agriculture

KS Nehra[1*], Mukesh R Jangra[1], Sumit Jangra[2] and Raj Kumar[3]

[1]Deptt. of Biotechnology, Govt College, Hisar- 125 001, India

[2]Deptt. of Molecular Biology, Biotechnology & Bioinformatics, CCSHAU, Hisar- 125 001, India

[3]Deptt. of Botany, Govt College, Hisar- 125 001, India

*Corresponding author: ksnehra@gmail.com

Abstract

Information technology can be defined as a set of various technical tools and resources used to communicate, broadcast, deposit and handle information. Information technology include computers, internet, networking hardware and software, satellites, broadcasting technologies (radio and television), and telephony (land lines and cellular). In addition to this, it requires services and functions linked with it for instance web portals, email, SMS, video-conferencing, etc. In short information technology is helpful to communicate the knowledge. In developing countries like India where Agriculture is a major contributor to GDP, it cannot be ignored in an era of rapid transformation. Information technology refers to how we use information, compute and communicate information to the people The role of information technology is, users need with the right information, in right form, in right time. The generation and application of agricultural knowledge is progressively important, particularly for small and marginal farmers, who require relevant information in order to improve, sustain, and diversify their farm enterprises.

Keywords: Agriculture, farming, information technology, rural, population and internet.

Introduction

The potentiality of agriculture in economic growth of a nation was recognized long back in 2009 by Byerlee and Sadoulet. Despite tremendous efforts, the agricultural production and productivity in the developing countries has declined in the last few decades. One of the potential clarifications for this reduced yield is the ignorance of farmers about advanced agricultural techniques in the developing countries since 1970. The impetus of digital technology in the modernization of agricultural practices has been determined by various studies (Feder, Just and Zilberman, 1985; Foster and Rosenzweig, 1995; Foster and Rosenzweig, 2010). Whereas the particular bases of technology acceptance depend on the setting and type of technology, usual features recognized in academic and observed fictions comprise of education, wealth, tastes, risk preferences, complementary inputs and access to information learning. Among all these, the role of asymmetric and expensive information has received the highest attention. Agriculture extension services have been started by government and international institutions to overcome the information failures regarding technology implementation (Anderson and Feder, 2007). In 2005, there were 5, 00,000 agriculture extension personals out of which 95% were working in public agriculture extension services. Insipte of huge amount of investment in extension programmes which has gone through a lot of criticism and its existence is facing huge problems still the evidences of their influence on agricultural technology and knowledge adoption remains limited (Anderson and Feder, 2007). A unique opportunity to spread information by public and private collaboration is provided by rapid development of ICTs in developing nations. Increased efficiency, reduced costs and enhanced productivity determines the role of ICTs in agriculture. Development of information systems based on farmer's requirement should be given highest priority and focus of systems should be on new challenges addressed from deregulation and globalization of agriculture sector (Samah *et al.*, 2009).

Informational Needs of Farmers

Access to reliable, timely and relevant information can help significantly and in many ways to reduce farmers' risks and uncertainty, empowering them to make good decisions. However, whether or not this access leads to an impact often depends on issues related to markets, institutions, policies and resource availability. Several studies have shown that the wide availability and multiple sources of information have not significantly changed farm-

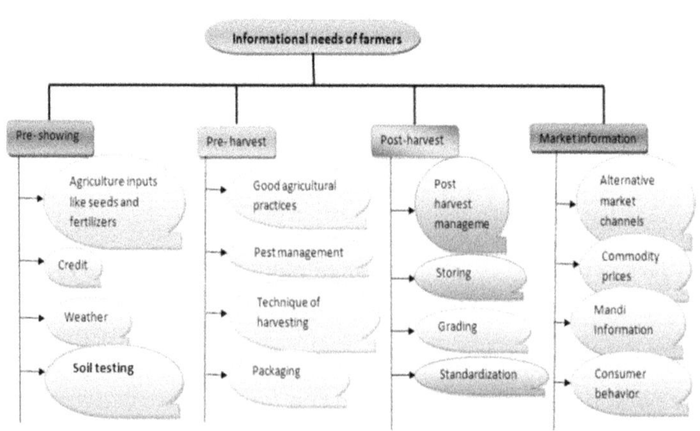

Fig. 1: Summary of informational needs of farmers

ers' behavior towards new technologies and information - a fact that is often attributed to lack of knowledge or understanding of farmers' perspectives and needs on the part of information providers. Informational needs of farmers are summarized below (fig. 1)

Classification of ITs'

Even if farmers are generally having a great knowledge of room conditions and significant practical knowledge or expertise of exploiting to environment to get the best out of it but they still require appropriate and novel information gathered from research and development to tackle with harsh weather and diseases (Correa *et al.* 1997). It was observed (Olawoye, 1996) that, farmer's agricultural productivity can be improved by agricultural knowhow if they have proper access to it. Jonston (1986) studied that extension programmes have been mainly personalized to deliver enough information that is related to rural farmers. Thus mass media plays an important role in spreading of agricultural information. With the help of mass media same message can be delivered simultaneously and repeatedly to a large number of populations in a given period of time. To do so, mass media can take use of printing press, radio transmitters, movie, exhibitions and audio-visuals. The last few years has recognized the importance of mass media in enhancement of agricultural productivity with time (Davies 1992). The information technology was classified into two types (fig.2)

- **Traditional IT**: It includes radio, television and print media. Based on their educational requirements, different countries can take advantage of radio and television in terms of informal education. Television is acknowledged as the most important medium for communicating with the rural populations of developing countries (FAO, 2001). For valuable transmission of agricultural technology to the agricultural community, TV seems to be an effective means of mass media. Presently, it is one of the most valuable tools available for communication. Teaching techniques comprising audio-visual seems to be the most effective. According to Carpenter (1983) coupling of audio visual can change the human behavior and eventually improved farmers learning skills. It is having the capability to spread the information to large number of audiences and highly diverse and geographical area where personal contact is not possible (Calvert, 1990). Radio is another important devices of mass media and is an effective communication tool. The initial experiments of ICT in agriculture in India after independence started with radio. Programmes in local languages were broadcasted with the help of a network of All India Radio (AIR) stations that were established post-independence. Prasar Bharti (earlier AIR) has been playing considerable role since many decades in improving agriculture by bringing latest technological information related to agriculture and other allied subjects to the farmers. Recently with the relaxation in licensing policy related to broadcasting rules has received a new motivation in India. This type of communication has proven to be very handy tool in socio-economic development at initial level. The rural population which is left behind by the main stream media can be taken up by radio community. Even farmer to farmer extension can be achieved through ample capacity building as the HAM radio experience underway in Tamil Nadu and Andhra Pradesh shows. Transferring new findings and technologies to rural farmers remain a promising strategy for increasing agricultural productivity. The new idea must reach farmers' farms and homes through effective extension the and mass media channels, so new technique can be adopted and put to use (Ekoja, 2003). Using the mass media has caused an increase in the knowledge level and the output of educational system in recent decades. Apart from it, print media play an important role in agriculture production. Picture and diagrams are used by print media to convey detailed and accurate information on mass scale. Printed material can be used by farmers for longer time and can be used repeatedly. Print media is more effective when the contents are tailored according to the target audience. Attention and popularity is gained by print media when it addresses the problems faced by agriculturist and come up with a feasible solution. Printed media can be used by extension workers along with media channels to strengthen the farmers learning process. In agriculture extension printed media is considered as a permanent message (Oakley and Garforth, 1985). However, illiteracy is a major problem with print media thus only educated or literate farmers are benefited with such kind of information.

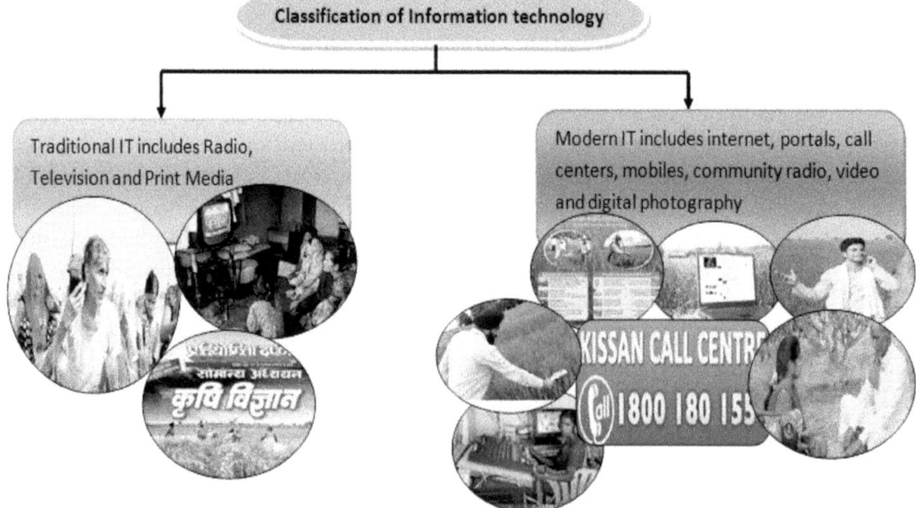

Fig. 2: Types of information technology

◈ New or Modern ITs

It includes internet, portals, mobiles, community radio, video, digital photography and call centers. Internet services, in conjunction with existing and more widely used communication media such as rural radio, will enable the broadest enhancement of information and communication resources for rural people. For example, national or regional agricultural market information systems or extension information systems hosted on the Internet can be excellent information sources for the staff of rural radio stations throughout a region or nation. Using information on current market prices broadcast by rural radio stations (including national variations and international figures), farmers can negotiate better prices from local buyers. Improved horizontal communication and improved information resources can improve the quality of the decisions and interventions that impact upon rural people. Apart from it, selling or buying online began to become popular in the world. However, it's most important role remains communication, and the internet has provided us with an ideal opportunity to do so. Central, state governments and private organizations have taken ICT measures for agriculture extension which include ITC- e-choupal, Kisan Kerala, Aaqua, Rice knowledge management portal, e-krishi, Mahindra Kisan Mitra, IFFCO Agri-portal, Village knowledge centers (VKCs)- M.S. Swaminathan Research Foundation, Village resource centers (VRCs)- ISRO etc. The Department of Agriculture and Cooperation, Ministry of Agriculture has implemented a National e-Governance Programme (NeGP) in Agricultural Sector as Mission Mode Project

The major aims of the project are:

a) Linking farmer centricity and facility alignment of the programs.

b) Boosting spread and influence of extension facilities.

c) Efficient management of DAC schemes motivating a mutual framework within states.

d) Enhancing entry of farmers to information and services.

Internet of Things (IoT)

Now a days many agricultural industries have adopted IoT technology for smart agriculture to increase proficiency and production, global market and other features such as minimum human intervention, time and cost etc. The advancement in the technology ensures that the sensors are getting smaller, sophisticated and more economic. The networks are also easily accessible globally so that smart farming can be achieved with full pledge. Focusing on promoting novelties in agriculture, smart agriculture is the solution to the problem which is faced by the industries presently. Smart phones and IoT devices are handy tool for this. Farmer can get any required data or information as well can monitor his agricultural sector. The Internet of things (IoT) is the most efficient and important techniques for development of solutions to the problems. IoT evolve from different building blocks which includes lots of sensors, software's, network components and other electronic devices. Also it makes data more effective. IoT allows exchanging the data over the network without human involvement. In Internet of things, we can represent things with natural way just like normal human being, like sensor, like car driver etc. This thing is assigned an ip address so that it can transfer data over a network. As per the report generated by Garner, at the end of 2016 there will be 30% rise in count of connected devices as compared to 2015. He further says that, this count will increase to 26 billion by 2020 (Jim, 2013). The IoT technology is more efficient due to following reasons:

1. Global Connectivity through any devices.
2. Minimum human efforts.
3. Faster Access.
4. Time Efficiency.
5. Efficient Communication.

The smart agriculture market is expected to reach $18.45 Billion in 2022, at a CAGR of 13.8%. BI estimates that 75 million IoT devices will be shipped for agricultural uses in 2020, at a CAGR of 20%. IoT devices can be of great help in enhancing crop productivity science it can be used to monitor soil pH, temperature and other variables. Internet of things, with its real-time, accurate and shared characteristics, will bring great changes to the agricultural supply chain and provide a critical technology for establishing a smooth flow of agricultural logistics (Wang and liu, 2014).

The key advantages of using IoT in enhancing farming are as follows:

1. Enhanced water management.
2. Continuous monitoring of land so that early stage precautions can be taken.
3. Save time, reduced labor cost, enhanced productivity make agriculture more productive.
4. Easy crop monitoring.
5. Farmer can easily identify soil moisture content and pH level and seeds can be shown accordingly.
6. Sensors and RFID chips aids to recognize the diseases occurred in plants and crops. RFID tags send the EPC (information) to the reader and are shared across the internet. The farmer or scientist can access this information from a remote place and take necessary actions; automatically crops can be protected from coming diseases (Jain *et al.*, 2012).
7. Crop sales will be increased in global market. Farmer can easily connect to the global market without restriction of any geographical area.

IoT Agriculture Apps

Agricultural IoT apps track the data of wireless sensors and ready it for predictive analytics. Here are a few IoT agriculture apps which are pioneering the second wave of green revolution:

- Phenonet Project (Openiot)

  Different wheat varieties are examined using Openiot by measuring air temperature, soil temperature and moisture. Farmers can forecast harvesting time, improve nutritional value of plants and adjust irrigation schedule.

- CLAAS Equipment

  One of the world leaders for agricultural engineering tools, CLAAS produces equipments which run on autopilot, which provide advice on increasing crop flow and on reducing the losses. Farmers can customize the program to fit their needs or allow the program to optimize the equipment automatically.

- Precision Hawk's UAV Sensor Platform

  Precision Hawk has introduced an Unmanned Aerial Vehicle (UAV) which performs a series of land-related tasks previously left to manual labor. This involves surveying, imaging and mapping of the land.

Community Radio

It a type of broadcasting service that is specific to a certain area, broadcasting matter common among local audience which is ignored by commercial or mass media broadcasters (UNESCO 2002).

Community Radio Concept

Radio confined to a local community or small geographical area is called community radio. It a low power transmission channel and covers about 20-30 km of area. It operates in a society that uses similar means of livelihood, comparatively localized, yet liked with national and regional upliftment goals. It is a social forecasting system mainly focused on enhancing heterogeneity. It broadcast thing on the demand of the listeners i.e. it is a truly peoples radio it receives requests from the listeners and provide the experts view on that. This type of radio is fully reliable on Milan declaration on mass media according to which it has the accountability to help maintain world's cultural and regional diversity and should support their lawmaking, administrative and economic measures. Three essential principles are behind community radio viz. no profit making, society possession and management and society contribution. Community radio is characterized by its restricted local spread, low range transmission and programming content that shows the educational development and socio-cultural needs of the local society it facilitates.

Community Radio for Agricultural Development

Farming from the ancient times has been a high knowledge intensive sector and needs uninterrupted flow of information. Farmers hunt for reliable, convincing and practical knowledge from both recognized systems and conventional practices is constantly increasing in this changing global environment, to function effectively and compete finically. The frequent changes taking place around with unrestricted development, globalization, rapid change in climate, decreasing area under cultivation highlights the need of information and knowledge in the agricultural development. For sustainable growth and to attain the food security extension to combat hunger and malnutrition, education and communication facilities are the key factors. The task of information sharing is made difficult by various challenges like diverse socio-cultural background, language problem, geographical location and partial initiatives. Agriculture extension is a tool or scheme for distributing valuable information among farmers and assist farmers to acquire necessary information, skills and techniques to make IT efficiently. Presently advancement in ICTs has revolutionized agriculture extension by providing different technological choices like TV, internet, mobile, telephone etc. Many agriculture extension programmes have been implemented in India, since independence. Agriculture is a state subject matter and in addition to multisector extension programmes there is a long list of single sector extension models that has been tried to be implemented simultaneously and often in overlapping manner in public sector. Besides well-structured and carefully designed these efforts were not very successful, hence underwent high level of criticism as they were not able to meet their goals (Khanal, 2011). Among the different measures of mass communication tolls for agriculture extension such as radio, TV and print media, the significance of radio in agriculture extension cannot be neglected. On radio of discussion and training on various topics related to farming management and supplementary information about short duration crops is frequently broadcasting. These supplementary extension programmes are most effective in the dry period during which framers can think about considering alternative approaches. In the planting period particular advice is taken on the full range crops being locally planted about land preparation, planting time, weed management, irrigation management, harvesting and marketing. Use of radio for delivering information is just one dimension, radio can serve multiple purposes like enhancing social and economic development in rural areas by spreading knowledge about health, nutrition, sanitation etc.

Mobile Agriculture or Connected Agriculture

The information cost in rural area has been significantly reduced by mobile phones. This technology has been of great importance to the rural farmers as it has provided them an opportunity to attain knowledge and information regarding agricultural concerns to increase the agricultural productivity. Information regarding market, weather, transport and latest agricultural techniques has speed up with the use ICTs services in agriculture extension especially mobile services (Aker, 2011). Tentative decision making by farmers has been made much easier by the use of mobile phone. Social relationship and social interaction has been increased to a great extent by use of mobile phones and SMS and voice recording technology are having a great role in social relation development. So, mobile phones are thought to be of great importance in agricultural development. Connectivity and other benefits like mobility and security to owners has been provided by this technology (Bayes *et al.*, 1999; Goodman, 2005; Kwaku & Kweku, 2006 and Dooner, 2006). Farmers without mobile phones have to face many difficulties to get market related information as compared to farmers with mobile phones. As in Malaysia farmers are facing many difficulties to make any contact with the agricultural expert due to lack of communication facilities. This society is still dependent on traditional methods like pictures and voice enhancers. The use of this might is not appropriate for spreading the information hence communication is one of the most important thing which is lacking in farming community and is the main cause of various problems faced by the agrarian community (Duncombe, 2011; Hing *et al.*, 2012). But at present mobile usage is increasing among farming community and are able to share their marketing, weather and business information among themselves. Now farmers are able to contact the market brokers and nearby cities to sell their harvest directly. The use of technology has made the farmers up to date about market from the socio-economic network (Ilahiane, 2007). The availability of mobiles has provided a new line of imagination to the farmers to get information and sell their harvest in the market, contact and bargain any broker of their choice. Before mobiles, broadcasting media such as radio and TV was the only source of information for the farmers. Mobile phone technology has provided a quick access to connect to different communities and share valuable information. Educated farmers are able to make use of messaging services to get recent updated information regarding agricultural market, making them able to take valuable decisions (Murthy, 2009). The impact of mobile phones could be evaluated in the form of rise and fall in the agricultural sale (Tripathi, 2008).

Android Apps for Farmers Launched by Narendra Modi Government (www.sarkariyojna.co.in)

Narendra Modi Government has launched several android apps for farmers and agriculture market. The applications aim to provide information about the latest agriculture trends, equipment, technologies and methods being used. The android apps for farmers can be downloaded from the official website mkisan.gov.in or from the Google Play store. Below is the list of all android apps launched by Narendra Modi's Government for farmers.

1. **Kisan Suvidha**

   This app has been developed to aid farmers to provide information related to current weather or the weather coming in the next 5 days, market value, traders, agro counselling, plant safety, IPM practices etc.

2. **Pusa Krishi**

   Pusa Krishi app helps farmers to know the cultural practices of various types of crops and information about those.

3. **M Kisan Application**

   This android app allows agriculturists and stake holders to gain information provided by experts and government officials. The portal can be utilized even without registering.

4. **Shetkari Masik Android App**

   The app can be used to download Shetkari Masik magazine and can be read without internet connectivity. "Shetkari Masik" magazine is published in the agricultural sector by department of agriculture Maharashtra since 1965.

5. **Farm-o-pedia App**

   This app is developed for rural Gujarat and helpful for agriculturists and any other involved in agricultural business. It provides valuable information about crops as per soil and season, crop wise information, weather information and domestic animals' management.

6. **Crop Insurance Android App**

   Insurance premium for notified crops can be calculated using this app based on area, coverage amount and loan amount in case of loanee farmers. It also provides information about sum insured, premium details and subsidy information about advised crop in advised area.

7. **Agri Market**

   This mobile app uses GPS technology to provide market price of the crops within 50 km area. Market price of any other market can be obtained by putting the GPS off.

e-Agriculture

e-Agriculture is a rising area focused on improving agriculture and rural development through better information and communication processes. Moreover, it comprises of formulation, pattern, evolution, rating and use of novel methods to apply ICTs in rural area, with key focus on farming. Comparatively e-agriculture is a new term and is expected to change and revolutionize our farmer's agricultural knowledge. In India agriculture contributes 18.6% to GDP and nearly 59% Indians are dependent on agriculture for their livelihood. Initiatives taken up by private sector like contract framing has marketed Indian agriculture. The main sectors of agricultural industry are farming, irrigation management, manure application, insect management, harvesting, food transportation, safety, quality management and commercialization. Any system working on information and knowledge collection for decision making in industry should provide precise, ample and concise information in and on time. The information offered by system must be in user-friendly form, easy to access, cheap and secure from illegal access (Chauhan *et al.*, 2015).

Precision Agriculture

Precision agriculture is a technique used to manage farming using information technology to make sure that the requirements of crop and soil are optimized for higher productivity and improved health, keeping in mind environment sustainability as well as to ensure ample profit to farmers. Satellites are used to monitor the crop so it is also known as satellite farming and site specific crop management system (SSCM). Precision farming is based on specific equipment, software and IT facilities. This method involves real time monitoring of crop data about crop situation, soil and ambient air, side by side it monitors related information such as hyper-local weather forecasting, labor cost and equipment availability. The data so obtained is used by predictive software to make farmers aware about crop rotation, ideal sowing time, harvesting time and soil management. Moisture content and temperature of soil and nearby air in the field is measured with the help of sensors. Farmers are able to do real-time monitoring of individual plant with the images provided by satellites and automated drones. The images so obtained can be treated and combined with sensors and other information to gain guidance for instant and future decisions like water requirement of the field and suitable location for planting a particular crop. This ensures that farmers do not waste resources and optimum fertilizers and pesticides are added which ultimately leads to reduced cost and control the effect of farming on environment hence lead to sustainable agriculture. Earlier precision farming was restricted to mass scale operations which can support IT infrastructure and other technology resources to make the best use of technology. Now with the availability to mobile phone apps, smart sensors, drones and cloud computing, precision farming can be done as practiced at small scale even at family farms (Banu, 2015). Precision agriculture is still under consideration in developing nations and planned back up from public and private is needed to indorse rapid technology adoption. Fruitful acceptance, though, includes no less than three phases i.e. searching, examination and implementation. Precision farming outlook can tackle both financial and environmental matters related to today's agricultural production. Still the question remains about cost effective and most efficient way to use technology tolls presently we are having, but the idea of "doing the right thing in the right place at the right time" has a solid spontaneous appeal. To make most out of green revolution all possible efforts should be made to make the technology adoption successful to make the world food secure. At last the success of precision farming depends mainly on how efficiently and how fast the knowledge required to use the latest technology can be discovered.

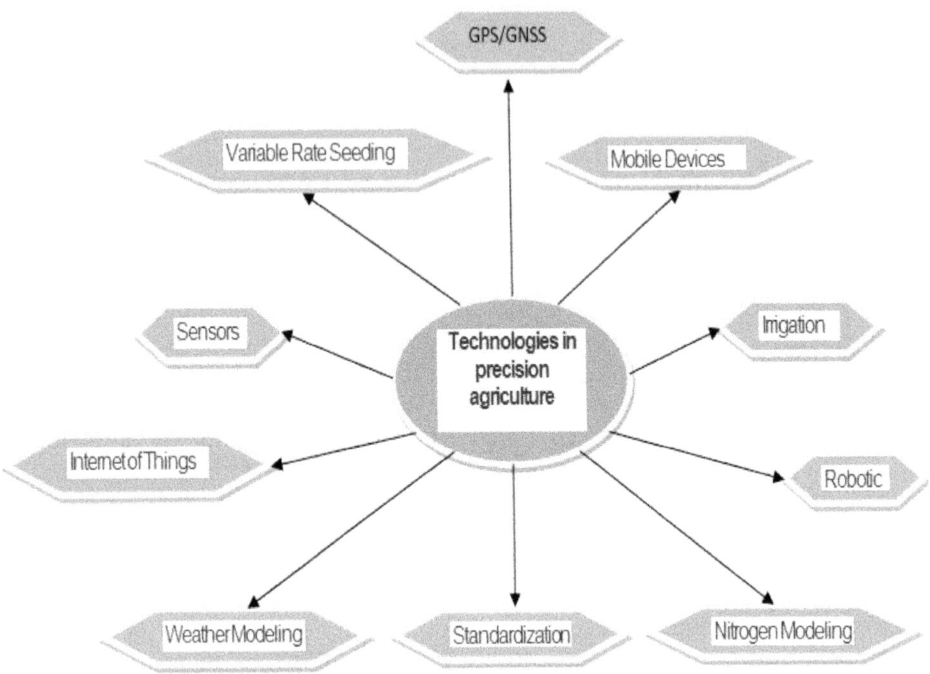

Indian Agriculture Extension System

For dissemination of information to the farming community by extension people the use of IT and electronic mass media is of high priority. In 1998, Innovations in Technology Dissemination (ITD) project was started under National Agriculture Technology Project (NATP) and the funding were made available by World Bank. The project is presently functional in twenty-eight districts of seven states viz. Andhra Pradesh, Bihar, Himachal Pradesh, Jharkhand, Maharashtra, Orissa and Punjab. The project pays emphasis on reforming public extension facilities and analyzing new measures for technology pass on. Under this project, Agriculture Technology Management Agency (ATMA) has been built up in each of the twenty-eight districts in seven states. ATMA is a listed association of key investors engaged in agricultural deeds for sustainable development in the district by clubbing research extension activities and dispersing day to day of Public Agricultural Technology Dissemination System. ATMA involves all the activities like research, training, development and extension activities run by community, personal and other organizations. State Agriculture Management and Extension Training Institute (SAMETI) is also reinforced to fulfil the exercise and capacity building requisite of the project. Training and capacity building of the agriculture officials is organized by National Institute of Agriculture Extension Management. ATMA model has been implemented in 252 districts of the nation, IT and media back has been intensively used for this.

Initiatives of ICT for Farming and Rural Development

Under National Agriculture Technology Project (NAPT), Information and Communications Technology (ICT) set-up is crafted in NARS by Academic Registration Information System (ARIS) in order to bring information management culture. Nearly 400 Academic Registration Information System (ARIS) cells have been established by NARC. These cells comprise of workstation for personal computer, server, UPS, switches, routers, LAN cable, internet connection, hubs and other major equipments. The basic foundation needed to link all Indian Council of Agricultural Research (ICAR) institutes and State Agricultural Universities (SAUs) has already been created. These cell are likely to enhance the use of IT in agricultural research, teaching and extension all over the nation. With the help of Library Improvement and Networking Programme of NATP the libraries of NARS has been improved with Information and Communications Technology (hardware, software, LAN. Internet, digitization, on-line/off-line resources etc.). Under ATMA, ICT has been implemented for farming extension activities. To strengthen the KVKs under NATP e-Extension 200 KVKs and 8 Zonal Coordinating Units were connected through internet and intranet for enabling these KVKs to deliver extension through internet. These KVKs will be established as data center.

Till date publically available Indian ICT services delivery models in farming sector are very less and most of them are private like knowledge centers of MSSRF (http://www.mssrf.org/specialprogrammes/mission_2007_NA/namain.htm), e-Choupal of ICT, ikisan of Nagarjuna Fertilisers and Chemicals Ltd. and Parrys corner. Besides a number of agribusiness corporates like TAFE, Mahindra and Mahindra and a many more are accepting ICT in their business. The initiatives of private sector are very significant and demanding representing the strong existence of corporate in agribusiness. Dairy Information Services Kiosk of NDDB and wired village WARANA are often quoted examples of cooperative sector.

The Major Initiatives of Government Based ICT Delivery Services are

i) Assam government has under taken a project ASHA (www.assamagribusiness.nic.in) which has spread all over the state with the help of large ICT network.

ii) Two different initiatives have been taken up by Kerala i.e. kissan kerala (www.kissankerala.net) and e-krishi (www.ekrishi.org/web/main/).

iii) Rajiv Internet Village services are started by Andhra Pradesh government in collaboration with ikishan to provide agribusiness services. IIT Hyderabad is working on e-sagu (http://agriculture.iit.net/esagu/esagu2004).

iv) Kisan Soochna Kendras are in under consideration of Uttrakhand government while first Agribusiness Information Center has been inaugurated by Haryana government.

v) Comprehensive Agrisnet has been pursued by Tamilnadu and Maharashtra. Telephonic help lines have been launched by several SAU's.

vi) Bhoomi project of Karnataka government is one of the most successful public sector projects and is under consideration in several other states.

ITC e-chaupal

Attempts made by grain trading company

- 4000 e-chaupals (kiosks) were started in villages to support grain procurement and agriculture.
- ICT pays for grain trading

n-Logue

Terrestrial wireless was used for connecting every village to district by local service providers

- Village kiosks
- No subsidy, possible loans from banks
- Key was video conferencing
- Main focus was on education, health and living standard
- Operation in 2000 villages and 40 districts

Other Initiatives

Grant/Assistance Driven

- MS Swaminathan Center (Centered on agri. and fishery use in Pondicherry)
- Tara-haat (Centered on rural business)
- Akshaya (In Kerala with government assistance)
- Gyaandoot* (In Madhya Pradesh focused on e-governance)
- Rural e-seva* (In Andhra Pradesh near east Godavari focused on e-governance)
- Warana* Wired village (By NIC in Maharasthra)

* These projects are presently managed by n-Logue

For Profit Initiatives

- **Drishtee** (works on already available infrastructure)

  ICTs have several models in Indian agriculture which have made noteworthy change in the supply of facilities in Indian agriculture like, setting up of Kissan Call Centers, Gyandoot Project, Bhoomi Project, Village Knowledge Centers, AGMARKNET etc.

Kissan Call Centers (KCCs)

Launched by Department of Agriculture and Co-operation on January 21, 2004. The principal technology taken in are:

- Computer cabinets with internet facility.
- Telephone line with high bandwidth (preferably 128 kbps ISDN line).
- Headphone aided telephones with teleconferencing facility (if needed).

The main objective of KCCs is to provide extension services in local language to agriculturists. The help line number (1551) can be dialed by the farmers to get the information from the agriculture graduates. If framer is not satisfied with the answer initially provided, call can be transferred to level II and level III executives making KCCs an important gateway for farmers. The services are free of cost and they are provided answers in the local language. If needed extension scientists or agriculture scientists visit the farmer's field to resolve the problem.

Village Knowledge Centers (VKCs)

M S Swaminathan research foundation launched village knowledge centers in Pondicherry in 1998. Making the rural population of Pondicherry food secure is the main aim behind the establishment of VKCs. For the fulfillment of this aim, technical information regarding agricultural inputs is provided the farmers. It provides information on daily basis about the market price from government as well as private bodies. It takes care about providing quality seeds and guides farmers about crop rotation system, fertilizer and pesticide use. VKCs have identified 13 districts with huge potential for agri-business in Pondicherry where government is going to invest Rs. 170 cr.

AGMARKETNET

Directorate of Marketing and Inspection (DMI) and National Information Center (NIC) jointly developed AGMARKETNET. AGMARKETNET is sponsored by DMI and NIC. With the establishment of nation-wide information network, the efficiency of marketing has been increased because timely information of about various market activities is obtained and distributed.

How Information is acquired by AGMARKETNET?

AGMARKETNET is linked with 670 farming practice market and 40 state agricultural marketing boards and directorates. The whole market AGMARK portal provides information on daily basis to its concerned states and the information is send to AGMARKETNET for the state's AGMARK. National Information System (NIS) maintains all the softwares. Agricultural and Processed Food Product Export Development Authority (APEDA), National Agricultural Co-operative Marketing Federation of India Ltd. (NAFED), Food Corporation of India (FCI), Central Warehousing, and Small Farmers Agri-Business Consortium (SFAC) are the principal operators of AGMARKETNET portal.

Boundaries in ICT Implementation and Feasible Solutions

Apart from huge capability of ICT in improving agriculture, there are drawbacks that make implementation and spread of ICT difficult in agricultural sector. Limitations in implementation and use of ICT in agriculture and rural area are explained by various scientists (Rao, 2003; Mittal, 2012). These limitations are as follows:

Lack of Knowledge about Advantages of ICT

People of rural area lack computer and internet facilities leading to lack of awareness about ICT benefits. On the other hand, providers of ICT and government officials are doubtful about the choice of rural population to say yes to ICT.

Clumsy and Messy Development of Systems

Specialized agency should be created for development of information system for improvement of agriculture and management system. Supporting the agricultural community is the main aim of the agency. Such agencies have consultative roles in the areas like user interface, overall system design, and content delivery mechanism and information kiosks setting standards.

Language Barriers and System Use Easiness

The easiness of system use to implement ICT in agriculture determines the success strategy of ICT implementation system. In most of the cases ISs supporting farming are not difficult to handle and proper internet facilities are not available in the rural area. A graphical presentation should be delivered to make the things understandable and easy to use. Language barrier is another problem associated with the use of ISs services by rural population. Native language based command line should be made in order to handle this problem.

Connectivity

In developing countries, expenditure on computers and internet is not in the affordable range of the poor or rural population of the country. Internet access availability is also low in rural area as compared to urban area because most of the internet service providers provide service in urban areas. Although in the last few years a great advancement has been made in the connectivity of rural areas. For the successful implementation of ICT in rural areas proper internet connection is mandatory. With the huge investment of private ISPs in town and cities now it has become possible to connect villages in large numbers. As satellite technology is very expensive the cellular network is the most powerful network to connect the remotest rural areas.

Network Bandwidth

Available bandwidth is another limitation even when telephone and other communication services exist. It is not possible to load graphics in low bandwidth which are required for simple representation things which are understandable to rural population and farmers. Transfer of dynamic information from locations and storage of static information in kiosks could be an answer to this problem.

Information Distribution Points

For effective use of internet services, massive use of information kiosks is vital. This type of kiosks should be planned as electronic supermarket. Other services to people living in rural area like distance learning, training, rural e-mail center, expert chat sessions and e-government are provided in addition to information services. Transformation of rural information kiosks to communication gateway for farmers and the other population is the main aim of these programmes.

Conclusion

Agriculturist and government officials working for agricultural improvement should to be capable of making efficient use of Information and Communications Technology in order to handle new situations which may arise by the whole or incomplete deregulation of agricultural market, decline of protective events of government, opening of agriproduct market, up and down in agricultural environment and use export opportunities. Better decision making enabled by quality information can be helpful in improving the quality of rural life. The difference or inequality among urban and rural life can be minimized by making the rural areas digitally efficient. Rapid advancement in ICT ensures development and spreading of digital services in agriculture. National strategies should be formulated for application and usage of ICT in agriculture. The strategy formulation process can be catalyzed by national coordination agencies with a consultative role. Single institution alone is not capable of successful implementation of ICT in rural areas and agriculture. Therefore, industries like fertilizer and food which are having a high influence on agriculture should mutually start and inspire application of ICT in agriculture.

References

Aker, J.C. 2011. Dial "A" for agriculture: a review of information and communication technologies for agricultural extension in developing countries. *Agri Eco*, 42 (6): 631-647.

Anderson, J.R. and Gershon F. 2007. Handbook of Agricultural Economics. *Agri Ext*, 3: 2343-2378.

Banu, S. 2015. Precision Agriculture: Tomorrow's Technology for Today's Farmer. *J Food Process Technol*, 6:1-6.

Bayes, A., Von, B.J. and Akhter, R. 1999. Village Pay Phones and Poverty Reduction: Insights from a Grameen Bank initiative in Bangladesh. ZEF discussion Papers on Development Policy No. 8 Centre for development Research, Bonn.

Byerlee, D., Janvry, Alain, d. **and Sadoulet, E. 2009.** Agriculture for Development: Toward a New Paradigm. *Annu Rev Resour Econ,* 1:15-31

Calvert, P. 1990. The Communicators Handbook Techniques and Technology Maupin House Gainesville, FL, USA

Carpenter, W.L. 1983. Communication Handbook. The Interstate Printers and Publishers, Inc, Danville.

Jain, D.P., Krishna, V. and Saritha, V. 2012. A Study on Internet of Things based Applications. http://www.businessinsider.com/internet-of-things-smart-agriculture-2016-10?IR=T

Donner, J. 2006. The social and economic implications of mobile telephony in Rwanda: An ownership/access typology. *Knowledge, Technology & Policy,* 19 (2):17-28.

Dunaway, D. 2002. Jankowski, Nicholas W.; Prehn, Ole. eds. "Community Radio at the Beginning of the 21st Century: Commercialism vs. Community Power" (pdf). Community Media in the Information Age: Perspectives and Prospects (Cresskill, NJ: Hampton Press). http://www.javnost-thepublic.org/media/datoteke/1998-2-dunaway.pdf. Retrieved 2009-02-15

Duncombe, R. 2011. Researching impact of mobile phones for development: concepts, methods and lessons for practice, *Info technol Develop,* 17(4):268-288.

Ekoja, I. 2003. Farmer's access to agricultural information in Nigeria. *Bull Am Soc Info Sci Technol,* 29(6): 21- 23.

Feder, Gershon & Richard E. Just & David Z. 1985. Adoption of Agricultural Innovations in Developing Countries: A Survey. *EDCC,* 33(2): 255-98.

Feder, G., Rinku, M. and Jamie, B.Q. 2004. Sending Farmers Back to School: The Impact of Farmer Field Schools in Indonesia. *Rev Agric Econ,* 26(1):45-62.

Foster, A. and Mark R. 2010. Microeconomics of Technology Adoption. *Annu. Rev. Econom,* 2:395-424.

Goodman, J. 2005. Linking Mobile Phone Ownership and Use to Social Capital in Rural South Africa and Tanzania, Vodafone Policy Paper Series, Number 2.

Ilahiane, H. 2007. "Impacts of Information and Communication Technologies in Agriculture: Farmers and Mobile Phones in Morocco." Paper presented at the Annual Meetings of the American Anthropological Association, December,1 Washington, DC.

Illinois. Jim Chase: The Evolution of the Internet of Things. White Paper, Texas Instruments, September, 2013.

Johnston M. A. 1986. The Value of World's Communication. London: Yaxien Press.

Khanal, S.R. 2011. Role of radio on agricultural development: A Review. *Bodhi: An Interdisciplinary Journal,* 5:201 -206

Kwaku Kyem, P. A., Kweku. & Le Maire, P. 2006. Transforming recent gains in the digital divide into digital opportunities: Africa and the boom in mobile Phone, Central Connecticut State University, USA. *EJISDC*, 28:35- 41.

Mittal, S.C. 2012. Role of Information Technology in Agriculture and its Scope in India, Available at: http://125.19.12.220/applications/Brihaspat.nsf/6dca49b7264f71ce65256a81003ad1cb/82f2c15ccd4dd9a065256b37001af3fe/$FILE/it_fai.pdf

Murthy, C.S.H.N. 2009. Use of convergent mobile technologies for sustainable economic transformation in the lives of small farmers in rural India. The Turk. *Online J. Dist. Educ*, 10(3): 32-4.

Oakley, P. and C. Garforth. 1985. Guide to Extension Training, FAO, Rome, Italy.

Olawoye, J.E. 1996. Agricultural Production in Nigeria, In: T. and Okiki, A. (eds), Utilising Research Findings to Increase Food Production-What the Mass Media Should Do in Tamming Hunger: The Role of Mass Media. Proceedings of the One-Day Seminar Organised by the Oyo State Chapter of the Media Forum for Agriculture, IITA, Ibadan.

Chauhan, R.M. 2015. Advantages and Challeging in E Agriculture. *Oriental J Comp Sci Technol*, 8(3): 228-233.

Rao, S.S. 2003. Information Systems in Indian Rural Communities. *J Comput Inform Syst*, 44:48-56.

Samah, B.A., Shaffril, H.A.M. Hassan, M.D.S. Hassan, M.A. and Ismail, N. 2009. Contribution of information and communication technology in increasing agro-based entreprenuers productivity in Malaysia. *J Agri Social Sci*, 5: 93-97.

Vinayak, N.M. and Pooja, K.A. 2016. Role of IoT in Agriculture. *IOSR J Comp Engineering*, 56-57

Xiaohui, W. and Nannan, L. 2014. The application of internet of things in agricultural means of production supply chain management. *J Chem Pharm Res*, 6(7):2304-2310.

www.ingramcontent.com/pod-product-compliance
Ingram Content Group UK Ltd.
Pitfield, Milton Keynes, MK11 3LW, UK
UKHW021436280726
14060UKWH00001BA/110